Large Deviations for Stochastic Processes

Mathematical
Surveys
and
Monographs

Volume 131

Large Deviations for Stochastic Processes

Jin Feng
Thomas G. Kurtz

American Mathematical Society

2010 *Mathematics Subject Classification.* Primary 60F10, 47H20; Secondary 60J05, 60J25, 60J35, 49L25.

For additional information and updates on this book, visit
www.ams.org/bookpages/surv-131

Library of Congress Cataloging-in-Publication Data

Large deviations for stochastic processes / Jin Feng, Thomas G. Kurtz.
p. cm. — (Mathematical surveys and monographs ; v. 131)
Includes bibliographical references.
ISBN-13: 978-0-8218-4145-7 (alk. paper)
ISBN-10: 0-8218-4145-9 (alk. paper)
1. Large deviations. 2. Semigroups of operators. 3. Markov processes. 4. Stochastic processes. 5. Viscosity solution. I. Feng, Jin, 1969– II. Kurtz, Thomas G. III. Series: Mathematical surveys and monographs ; no. 131.

QA273.67.L37 2006
519.2—dc22 2006045899

AMS softcover ISBN: 978-1-4704-1870-0

∞ The paper used in this book is acid-free and falls within the guidelines
established to ensure permanence and durability.
Visit the AMS home page at `http://www.ams.org/`

10 9 8 7 6 5 4 3 2 1 19 18 17 16 15 14

Contents

Preface

This work began as a research paper intended to show how the convergence of nonlinear semigroups associated with a sequence of Markov processes implied the large deviation principle for the sequence. We expected the result to be of little utility for specific applications, since classical convergence results for nonlinear semigroups involve hypotheses that are very difficult to verify, at least using classical methods. We should have recognized at the beginning that the modern theory of viscosity solutions provides the tools needed to overcome the classical difficulties. Once we did recognize that convergence of the nonlinear semigroups could be verified, the method evolved into a unified treatment of large deviation results for Markov processes, and the research "paper" steadily grew into the current volume.

There are many approaches to large deviations for Markov processes, but this book focuses on just one. Our general title reflects both the presentation in Part 1 of the theory of large deviations based on the large deviation analogue of the compactness theory for weak convergence, material that is the foundation of several of the approaches, and by the generality of the semigroup methods for Markov processes.

The goal of Part 2 is to develop an approach for proving large deviations, in the context of metric-space-valued Markov processes, using convergence of generators in much the same spirit as for weak convergence (e.g. Ethier and Kurtz [**36**]). This approach complements the usual method that relies on asymptotic estimates obtained through Girsanov transformations.

The usefulness of the method is best illustrated through examples, and Part 3 contains a range of concrete examples.

We would like to thank Alex de Acosta, Paul Dupuis, Richard Ellis, Wendell Fleming, Jorge Garcia, Markos Katsoulakis, Jim Kuelbs, Peter Ney, Anatolii Puhalskii and Takis Souganidis for a number of helpful conversations, and Peter Ney and Jim Kuelbs for organizing a long-term seminar at the University of Wisconsin - Madison on large deviations that provided much information and insight. In particular, the authors' first introduction to the close relationship between the theory of large deviations and that of weak convergence came through a series of lectures that Alex de Acosta presented in that seminar.

This work was supported in part by NSF grants DMS-9804816, DMS-9971571, DMS-0205034 and DMS-0503983.

Notation

(1) (E,r). A complete, separable metric space.
(2) $\mathcal{B}(E)$. The σ-algebra of all Borel subsets of E.
(3) $A \subset M(E) \times M(E)$. An operator identified with its graph as a subset in $M(E) \times M(E)$.
(4) $B(E)$. The space of bounded, Borel measurable functions. Endowed with the norm $\|f\| = \sup_{x\in E} |f(x)|$, $(B(E), \|\cdot\|)$ is a Banach space.
(5) $B_{loc}(E)$. The space of locally bounded, Borel measurable functions, that is, functions in $M(E)$ that are bounded on each compact.
(6) $B_\epsilon(x) = \{y \in E : r(x,y) < \epsilon\}$. The ball of radius $\epsilon > 0$ and center $x \in E$.
(7) *buc*-convergence, *buc*-approximable, *buc*-closure, closed and dense, and *buc*- lim, See Definition A.6.
(8) $C(E)$. The space of continuous functions on E.
(9) $C_b(E) = C(E) \cap B(E)$.
(10) $C(E, \overline{R})$. The collection of functions that are continuous as mappings from E into $\overline{R}$ with the natural topology on $\overline{R}$.
(11) $C_c(E)$. For E locally compact, the functions that are continuous and have compact support.
(12) $C^k(\mathcal{O})$, for $\mathcal{O} \subset R^d$ open and $k = 1, 2, \ldots, \infty$. The space of functions whose derivatives up to kth order are continuous in $\mathcal{O}$.
(13) $C_c^k(\mathcal{O}) = C^k(\mathcal{O}) \cap C_c(\mathcal{O})$.
(14) $C^{k,\alpha}(\mathcal{O})$, for $\mathcal{O} \subset R^d$ open, $k = 1, 2, \ldots$, and $\alpha \in (0, 1]$. The space of functions $f \in C^k(\mathcal{O})$ satisfying

$$\|f\|_{k,\alpha} = \sup_{0\leq\beta\leq k} \sup_{\mathcal{O}} \left|\frac{\partial^\beta f}{\partial x^\beta}\right| + \sup_{|\beta|=k} \left[\frac{\partial^\beta f}{\partial x^\beta}\right]_\alpha < \infty,$$

where

$$\left[f\right]_\alpha = \sup_{x\neq y} \frac{|f(x) - f(y)|}{|x-y|^\alpha}, \qquad \alpha \in (0,1].$$

(15) $C_{loc}^{k,\alpha}(\mathcal{O})$. The space of functions $f \in C(\mathcal{O})$ such that $f|_D \in C^{k,\alpha}(D)$ for every bounded open subset $D \subset \mathcal{O}$.
(16) $C_E[0,\infty)$. The space of E-valued, continuous functions on $[0,\infty)$.
(17) $\widehat{C}(U)$, for U locally compact. The space of continuous functions vanishing at infinity.
(18) $\mathcal{D}(A) = \{f : \exists (f,g) \in A\}$. The domain of an operator A.
(19) $\mathcal{D}^+(A) = \{f \in \mathcal{D}(A), f > 0\}$.
(20) $\mathcal{D}^{++}(A) = \{f \in \mathcal{D}(A), \inf_{y\in E} f(y) > 0\}$.
(21) $D_E[0,\infty)$. The space of E-valued, cadlag (right continuous with left limit) functions on $[0,\infty)$ with the Skorohod topology, unless another topology is specified. (See Ethier and Kurtz [**36**], Chapter 3).

(22) $\mathcal{D}(\mathcal{O})$, for $\mathcal{O} \subset R^d$ open. The space $C_c^\infty(\mathcal{O})$ with the topology giving the space of Schwartz test functions. (See D.1).
(23) $\mathcal{D}'(\mathcal{O})$. The space of continuous linear functionals on $\mathcal{D}(\mathcal{O})$, that is, the space of Schwartz distributions.
(24) g^* (respectively g_*). The upper semicontinuous (resp. lower semicontinuous) regularization of a function g on a metric space (E, r). The definition is given by (6.2) (resp. (6.3)).
(25) $\limsup_{n\to\infty} G_n$ and $\liminf_{n\to\infty} G_n$ for a sequence of sets G_n. Definition 2.4 in Section 2.3.
(26) $H_\rho^k(R^d)$, for $\rho \in \mathcal{P}(R^d)$. A weighted Sobolev space. See Appendix D.5.
(27) $K(E) \subset C_b(E)$. The collection of nonnegative, bounded, continuous functions.
(28) $K_0(E) \subset K(E)$. The collection of strictly positive, bounded, continuous functions.
(29) $K_1(E) \subset K_0(E)$. The collection of bounded continuous functions satisfying $\inf_{x\in E} f(x) > 0$.
(30) $M(E)$. The R-valued, Borel measurable functions on E.
(31) $M^u(E)$. The space of $f \in M(E)$ that are bounded above.
(32) $M^l(E)$. The space of $f \in M(E)$ that are bounded below.
(33) $M(E, \overline{R})$. The space of Borel measurable functions with values in $\overline{R}$ and $f(x) \in R$ for at least one $x \in E$.
(34) $M_E[0, \infty)$. The space of E-valued measurable functions on $[0, \infty)$.
(35) $M^{d\times d}$. The space of $d \times d$ matrices.
(36) $M^u(E, \overline{R}) \subset M(E, \overline{R})$ (respectively, $C^u(E, \overline{R}) \subset C(E, \overline{R})$). The collection of Borel measurable (respectively continuous) functions that are bounded above (that is, $f \in M^u(E, \overline{R})$ implies $\sup_{x\in E} f(x) < \infty$).
(37) $M^l(E, \overline{R}) \subset M(E, \overline{R})$ (respectively, $C^l(E, \overline{R}) \subset C(E, \overline{R})$). The collection of Borel measurable (respectively continuous) functions that are bounded below.
(38) $\mathcal{M}(E)$. The space of (positive) Borel measures on E.
(39) $\mathcal{M}_f(E)$. The space of finite (positive) Borel measures on E.
(40) $\mathcal{M}_m(U)$, U a metric space. The collection of $\mu \in \mathcal{M}(U \times [0, \infty))$ satisfying $\mu(U \times [0, t]) = t$ for all $t \geq 0$.
(41) $\mathcal{M}_m^T(U)$ $(T > 0)$. The collection of $\mu \in \mathcal{M}(U \times [0, T])$ satisfying $\mu(U \times [0, t]) = t$ for all $0 \leq t \leq T$.
(42) $\mathcal{P}(E) \subset \mathcal{M}_f(E)$. The space of probability measures on E.
(43) $\overline{R} = [-\infty, \infty]$.
(44) $\mathcal{R}(A) = \{g : \exists (f, g) \in A\}$. The range of an operator A.
(45) $T\#\rho = \gamma$. $\gamma \in \mathcal{P}(E)$ is the push-forward (Definition D.1) of $\rho \in \mathcal{P}(E)$ by the map T.

Introduction

CHAPTER 1

Introduction

The theory of large deviations is concerned with the asymptotic estimation of probabilities of rare events. In its basic form, the theory considers the limit of normalizations of $\log P(A_n)$ for a sequence of events with asymptotically vanishing probability. To be precise, for a sequence of random variables $\{X_n\}$ with values in a metric space (S, d), we are interested in the *large deviation principle* as formulated by Varadhan [**119**]

DEFINITION 1.1. [Large Deviation Principle] $\{X_n\}$ satisfies a *large deviation principle* (LDP) if there exists a lower semicontinuous function $I : S \to [0, \infty]$ such that for each open set A,

$$\liminf_{n\to\infty} \frac{1}{n} \log P\{X_n \in A\} \geq -\inf_{x\in A} I(x), \tag{1.1}$$

and for each closed set B,

$$\limsup_{n\to\infty} \frac{1}{n} \log P\{X_n \in B\} \leq -\inf_{x\in B} I(x). \tag{1.2}$$

I is called the *rate function* for the large deviation principle. A rate function is *good* if for each $a \in [0, \infty)$, $\{x : I(x) \leq a\}$ is compact.

Beginning with the work of Cramér [**16**] and including the fundamental work on large deviations for stochastic processes by Freidlin and Wentzell [**52**] and Donsker and Varadhan [**33**], much of the analysis has been based on change of measure techniques. In this approach, a reference measure is identified under which the events of interest have high probability, and the probability of the event under the original measure is estimated in terms of the Radon-Nikodym derivative relating the two measures.

More recently Puhalskii [**97**], O'Brien and Vervaat [**91**], de Acosta [**24**] and others have developed an approach to large deviations analogous to the Prohorov compactness approach to weak convergence of probability measures.

DEFINITION 1.2. $\{X_n\}$ *converges in distribution* to X (that is, the distributions $P\{X_n \in \cdot\}$ *converge weakly* to $P\{X \in \cdot\}$) if and only if $\lim_{n\to\infty} E[f(X_n)] = E[f(X)]$ for each $f \in C_b(S)$.

The analogy between the two theories becomes much clearer if we recall the following equivalent formulation of convergence in distribution.

PROPOSITION 1.3. *$\{X_n\}$ converges in distribution to X if and only if for each open set A,*

$$\liminf_{n\to\infty} P\{X_n \in A\} \geq P\{X \in A\},$$

or equivalently, for each closed set B,

$$\limsup_{n\to\infty} P\{X_n \in B\} \le P\{X \in B\}.$$

Our main theme is the development of this approach to large deviation theory as it applies to sequences of cadlag stochastic processes. The proof of weak convergence results typically involves verification of relative compactness or *tightness* for the sequence and the unique characterization of the possible limit distribution. The analogous approach to large deviations involves verification of *exponential tightness* (Definition 3.2) and unique characterization of the possible rate function.

In Part 1, we present results on exponential tightness and give Puhalskii's analogue of the Prohorov compactness theorem. We also give complete exponential tightness analogues of standard tightness criteria for cadlag stochastic processes and show that the rate function for a sequence of processes is determined by the rate functions for their finite dimensional distributions, just as the limit distribution for a weakly convergent sequence of processes is determined by the limits of the finite dimensional distributions.

In Part 2, we focus on Markov processes and give general large deviation results based on the convergence of corresponding nonlinear semigroups. Again we have an analogy with the use of the convergence of linear semigroups in proofs of weak convergence of Markov processes. The success of this approach depends heavily on the idea of a viscosity solution of a nonlinear equation. We also exploit control theoretic representations of the limiting semigroups to obtain useful representations of the rate function.

In Part 3, we demonstrate the effectiveness of these methods in a wide range of large deviation results for Markov processes, including classical Freidlin-Wentzell theory, random evolutions, and infinite dimensional diffusions.

1.1. Basic methodology

1.1.1. Analogy with the theory of weak convergence. In both the weak convergence and large deviation settings, proofs consist of first verifying a compactness condition and then showing that there is only one possible limit. For weak convergence, the first step is usually accomplished by verifying *tightness* which, by Prohorov's theorem, implies relative compactness. The corresponding condition for large deviations is *exponential tightness* (Definition 3.2). Puhalskii [**97**] (and in more general settings, O'Brien and Vervaat [**91**] and de Acosta [**24**]) has shown that exponential tightness implies the existence of a subsequence along which the large deviation principle holds. (See Theorem 3.7.) For stochastic processes, the second step of these arguments can be accomplished by verifying weak convergence (or the large deviation principle) for the finite dimensional distributions and showing that the limiting finite dimensional distributions (finite dimensional rate functions) determine a unique limiting distribution (rate function) for the process distributions.

We extend this analogous development in a number of ways. For the Skorohod topology on the space of cadlag sample paths in a complete, separable metric space, we give a complete exponential tightness analogue of the tightness conditions of Kurtz [**71**] and Aldous [**2**] (Theorem 4.1). We then extend the characterization of the rate function in terms of the finite-dimensional rate functions (Puhalskii [**97**], Theorem 4.5, and de Acosta [**2**], Lemma 3.2) to allow the rate function to be finite on paths with discontinuities (Section 4.7). Finally, we apply these results to Markov processes, using the asymptotic behavior of generators and semigroups associated with the Markov processes to verify exponential tightness for the processes (Section 4.5) and the large deviation principle for the finite dimensional distributions (Chapters 5, 6, 7). These arguments are again analogous to the use of convergence of generators and semigroups to verify tightness and convergence of finite dimensional distributions in the weak convergence setting.

1.1.2. Nonlinear semigroups and viscosity methods. Results of Varadhan and Bryc (Proposition 3.8) relate large deviations for sequences of random variables to the asymptotic behavior of functionals of the form $\frac{1}{n}\log E[e^{nf(X_n)}]$. For Markov processes,

$$V_n(t)f(x) = \frac{1}{n}\log E[e^{nf(X_n(t))}|X(0) = x] \tag{1.3}$$

defines a nonlinear semigroup, and large deviations for sequences of Markov processes can be studied using the asymptotic behavior of the corresponding sequence of nonlinear semigroups. Viscosity methods for nonlinear equations play a central role. Fleming and others (cf. [**44, 45, 40, 48, 46**]) have used this approach to prove large deviation results for X_n at single time points and exit times. We develop these ideas further, showing how convergence of the semigroups and their generators H_n can be used to obtain both exponential tightness and the large deviation principle for the finite dimensional distributions. These results then imply the pathwise large deviation principle.

1.1.3. Control theory. The limiting semigroup usually admits a variational form known as the *Nisio semigroup* in control theory. Connections between control problems and large deviation results were first made by Fleming [**44, 45**] and developed further by Sheu [**108, 109**]. Dupuis and Ellis [**35**] systematically develop these connections showing that, in many situations, one can represent a large class of functionals of the processes as the minimal cost functions of stochastic control problems and then verify convergence of the functionals to the minimal cost functions of limiting deterministic control problems. This convergence can then be used to obtain the desired large deviation result.

Variational representations for the sequence of functionals can be difficult to obtain, and in the present work, we first verify convergence of the semigroups by methods that only require convergence of the corresponding generators and conditions on the limiting generator. Working only with the sequence of generators frequently provides conditions that are easier to verify than conditions that give a convergent sequence of variational representations. Variational representations for the limit are still important, however, as they provide methods for obtaining simple representations of the large deviation rate function. In Chapter 8, we discuss methods for obtaining such representations. In particular, we formulate a set of conditions at the infinitesimal level, so that by verifying a variational representation for the limit generator, we obtain the variational structure of the semigroup that gives the large deviation rate function.

A generator convergence approach directly based on the sequence of control problems and Girsanov transformations was discussed in Feng [**41**] in a more restrictive setting. The present work avoids explicit use of a Girsanov transformation, and it greatly generalizes [**41**] making the approach much more applicable.

1.2. The basic setting for Markov processes

1.2.1. Notation. Throughout, (E, r) will be a complete, separable metric space, $M(E)$ will denote the space of real-valued, Borel measurable functions on E, $B(E) \subset M(E)$, the space of bounded, Borel measurable function, $C_b(E) \subset B(E)$, the space of bounded continuous functions, and $\mathcal{P}(E)$, the space of probability measures on E. We identify an operator A with its graph and, for example, write $A \subset C_b(E) \times C_b(E)$ if the domain $\mathcal{D}(A)$ and range $\mathcal{R}(A)$ are contained in $C_b(E)$. An operator can be multi-valued and nonlinear.

The space of E-valued, cadlag functions on $[0, \infty)$ with the Skorohod topology will be denoted by $D_E[0, \infty)$. Define $q(x, y) = 1 \wedge r(x, y)$, and note that q is a metric on E that is equivalent to r. The following metric gives the Skorohod topology . (See Ethier and Kurtz [**36**], Chapter 3). Let Λ' be the collection of strictly increasing functions mapping $[0, \infty)$ onto $[0, \infty)$ and satisfying

$$\gamma(\lambda) \equiv \sup_{0 \le s < t} |\log \frac{\lambda(t) - \lambda(s)}{t - s}| < \infty. \tag{1.4}$$

For $x, y \in D_E[0, \infty)$, define

$$d(x, y) = \inf_{\lambda \in \Lambda'} \{\gamma(\lambda) \vee \int_0^\infty e^{-u} \sup_{t \ge 0} q(x(\lambda(t) \wedge u), y(t \wedge u)) du\}. \tag{1.5}$$

1.2.2. Continuous-time Markov processes. An E-valued Markov process X is usually characterized in terms of its *generator*, a linear operator $A \subset B(E) \times B(E)$. One approach to the characterization of X is to require that all processes of the form

$$g(X(t)) - g(X(0)) - \int_0^t Ag(X(s)) ds \tag{1.6}$$

be martingales with respect to some filtration $\{\mathcal{F}_t\}$ independent of g. This requirement defines the (linear) martingale problem of Stroock and Varadhan [**116**], that is, a process X satisfying this condition is said to be a solution of the *martingale problem* for A. If $\nu \in \mathcal{P}(E)$ and $X(0)$ has distribution ν, X is a solution of the martingale problem for (A, ν). We say that *uniqueness* holds for the martingale problem for (A, ν) if any two solutions have the same finite dimensional distributions. The martingale problem for A is *well-posed* if for each $\nu \in \mathcal{P}(E)$, there exists a unique solution of the martingale problem for (A, ν).

If g is bounded away from zero, it is easy to see that (1.6) is a martingale if and only if

$$g(X(t)) \exp\{-\int_0^t \frac{Ag(X(s))}{g(X(s))} ds\}$$

is a martingale. Consequently, if we define $\mathcal{D}(\mathcal{H}) = \{f \in B(E) : e^f \in \mathcal{D}(A)\}$ and set

$$\mathcal{H}f = e^{-f} A e^f, \tag{1.7}$$

then we can define the *exponential martingale problem* by requiring

$$\exp\{f(X(t)) - f(X(0)) - \int_0^t \mathcal{H}f(X(s)) ds\} \tag{1.8}$$

to be a martingale with respect to a filtration independent of f. One can always assume that the domain $\mathcal{D}(A)$ contains constants, and hence if $g \in \mathcal{D}(A)$, then there exists a constant c such that $g+c \in \mathcal{D}(A)$ is positive and bounded away from zero. Setting $f = \log(g+c)$, (1.8) is a martingale if and only if (1.6) is a martingale. It follows that X is a solution of the linear martingale problem for A if and only if it is a solution of the exponential martingale problem for $\mathcal{H}$.

Weak convergence results for a sequence $\{X_n\}$ of Markov processes can be based on convergence of the corresponding operator semigroups $\{T_n(t)\}$, where

$$T_n(t)g(x) = E[g(X_n(t))|X_n(0) = x].$$

At least formally, T_n satisfies

$$\frac{d}{dt}T_n(t)g = A_nT_n(t)g, \quad T_n(0)g = g,$$

where A_n is the generator for X_n. An analogous approach to large deviation results is suggested by Fleming [**45**] using the nonlinear contraction (in the sup norm) semigroup V_n defined in (1.3). Again, at least formally, V_n should satisfy

$$\frac{d}{dt}V_n(t)f = \frac{1}{n}\mathcal{H}_n(nV_n(t)f),$$

which leads us to define

$$H_nf = \frac{1}{n}\mathcal{H}_n(nf) = \frac{1}{n}e^{-nf}A_ne^{nf}. \tag{1.9}$$

Note that

$$\exp\{nf(X_n(t)) - nf(X(0)) - \int_0^t nH_nf(X_n(s))ds\} \tag{1.10}$$

is a martingale.

1.2.3. Discrete-time Markov processes. Consideration of discrete-time Markov processes leads to slightly different formulations of the two martingale problems. Let $\{Y_k, k \geq 0\}$ be a time homogeneous Markov chain with state space E and transition operator

$$Tf(x) = E[f(Y_{k+1})|Y_k = x].$$

For $\epsilon > 0$, define $X(t) = Y_{[t/\epsilon]}$. Then, setting $\mathcal{F}_t^X = \sigma(X(s) : s \leq t)$, for $g \in B(E)$,

$$\begin{aligned} g(X(t)) - g(X(0)) - &\sum_{k=0}^{[t/\epsilon]-1} (Tg(Y_k) - g(Y_k)) \\ &= g(X(t)) - g(X(0) - \int_0^{[t/\epsilon]\epsilon} \epsilon^{-1}(T-I)g(X(s))ds \end{aligned}$$

is an $\{\mathcal{F}_t^X\}$-martingale, and for $f \in B(E)$,

$$\exp\{f(X(t)) - f(X(0)) - \int_0^{[t/\epsilon]\epsilon} \epsilon^{-1}\log e^{-f}Te^f(X(s))ds\}$$

is an $\{\mathcal{F}_t^X\}$-martingale. For a sequence, $X_n(t) = Y^n_{[t/\epsilon_n]}$, we define

$$A_ng = \epsilon_n^{-1}(T_n - I)g \tag{1.11}$$

and

$$H_nf = \frac{1}{n\epsilon_n}\log e^{-nf}T_ne^{nf} = \frac{1}{n\epsilon_n}\log(e^{-nf}(T_n - I)e^{nf} + 1), \tag{1.12}$$

so that

$$g(X_n(t)) - g(X_n(0)) - \int_0^{[t/\epsilon_n]\epsilon_n} A_n g(X_n(s))ds$$

and

$$\exp\{nf(X_n(t)) - nf(X(0)) - \int_0^{[t/\epsilon_n]\epsilon_n} nH_n f(X_n(s))ds\}$$

are martingales.

In Chapter 3, we give conditions for exponential tightness that are typically easy to check in the Markov process setting using the convergence (actually, only the boundedness) of H_n. In particular, $\sup_n \|H_n f\| < \infty$ for a sufficiently large class of f implies exponential tightness, at least if the state space is compact. In Section 4.7, we give conditions under which exponential tightness and the large deviation principle for finite dimensional distributions imply the large deviation principle in the Skorohod topology. In Chapters 5, 6 and 7, we give conditions under which convergence of the sequence of semigroups $\{V_n\}$ implies the large deviation principle for the finite dimensional distributions and conditions under which convergence of $\{H_n\}$ implies convergence of $\{V_n\}$. Consequently, convergence of $\{H_n\}$ is an essential ingredient in the results.

1.3. Related approaches

There are several approaches in the literature that are closely related to the methods we develop here. We have already mentioned the control theoretic methods developed in detail by Dupuis and Ellis [**35**] that play a key role in the rate function representation results of Chapter 8. Two other approaches share important characteristics with our methods.

1.3.1. Exponential martingale method of de Acosta.

de Acosta [**26, 27**] develops a general approach to large deviation results that, in the case of Markov processes, exploits the exponential martingale (1.10). Let λ be a measure on $[0,1]$, and assume that $e^{af} \in \mathcal{D}(A_n)$ for all $a \in R$. Then the fact that (1.10) is a martingale implies that

$$\exp\{\int_0^1 f(X_n(s))\lambda(ds) - \lambda[0,1]f(X_n(0)) - \int_0^1 \mathcal{H}_n e^{\lambda(s,1]f}(X_n(s))ds\}$$

has expectation 1. Setting

$$\Phi_n(x,\lambda,f) = \lambda[0,1]f(x(0)) + \int_0^1 \mathcal{H}_n e^{\lambda(s,1]f}(x(s))ds,$$

convergence of H_n implies (at least formally) that

$$\frac{1}{n}\Phi_n(x,n\lambda,f) \to \Phi(x,\lambda,f) = \lambda[0,1]f(x(0)) + \int_0^1 He^{\lambda(s,1]f}(x(s))ds.$$

Comparing this convergence to the conditions of Theorems 2.1 and 3.1 of [**26**] indicates a close relationship between the methods developed here and those developed by de Acosta, at least at the formal computational level.

1.3.2. The maxingale method of Puhalskii. Assume that the semigroups V_n converge to V. Then V is nonlinear in the sense that, in general,

$$V(t)(af + bg) \neq aV(t)f + bV(t)g.$$

However, defining

$$a \oplus b = \max\{a, b\}, \quad a \odot b = a + b,$$

and using $(\oplus, \odot)$ as operations on R in place of the usual $(+, \times)$, V becomes "linear"

$$V(t)((a \odot f) \oplus (b \odot g)) = (a \odot V(t)f) \oplus (b \odot V(t)g),$$

and defining

$$a \oplus_n b = \frac{1}{n}\log(e^{na} + e^{nb}), \quad a \odot_n b = a + b,$$

V_n is linear under $(\oplus_n, \odot_n)$:

$$V_n(t)((a \odot_n f) \oplus_n (b \odot_n g)) = (a \odot_n V_n(t)f) \oplus_n (b \odot_n V_n(t)g).$$

Furthermore,

$$\lim_{n\to\infty} a \oplus_n b = a \oplus b, \quad a \odot_n b = a \odot b.$$

The change of algebra (semi-ring, to be precise) on R produces a linear structure for V (or the V_n). Results analogous to those of linear analysis, such as the Riesz representation theorem, hold and can be used to study the properties of V. The counterpart of a measure is an *idempotent measure* . (See Puhalskii [**100**].) Taking this view point and mimicking the martingale problem approach in weak convergence theory for stochastic processes, Puhalskii [**99, 100**] defines a *maxingale problem* and develops large deviation theory for semimartingales.

The main result of [**100**], Theorem 5.4.1, can be stated roughly as follows: let $\{X_n\}$ be a sequence of Markov processes and H_n be the operators so that

$$M^f_{s,n}(t) = e^{nf(X_n(t)) - nf(X_n(s)) - \int_s^t H_n f(X_n(r))dr}$$

is a mean one martingale for each $f \in \mathcal{D}(H_n)$, each $s \geq 0$ and $t \geq s$, as in (1.10). Assuming that $H = \lim_{n\to\infty} H_n$ in an appropriate sense, under additional conditions on H and the convergence, $\{X_n\}$ is exponentially tight and along any subsequence of $\{X_n\}$, there exists a further subsequence such that the large deviation principle holds with some rate function $I(x)$. Then $\Pi(A) = \sup_{x\in A} e^{-I(x)}$ defines an idempotent probability measure on the trajectory space of the processes. (See Chapter 1 of [**100**]. Note that an idempotent measure is not a measure in the usual sense.) Π then solves a *maxingale problem* for H:

$$e^{f(x(t)) - f(x(0)) - \int_0^t Hf(x(s))ds}$$

is a maxingale under $\Pi(dx)$ for every $f \in \mathcal{D}(H)$ in the sense that

(1.13)
$$\frac{\sup_{\{x:x(r)=z(r),0\leq r\leq s\}} \exp\{f(x(t)) - f(x(s)) - \int_s^t Hf(x(r))dr - I(x(\cdot))\}}{\sup_{\{x:x(r)=z(r),0\leq r\leq s\}} \exp\{-I(x(\cdot))\}} = 1,$$

for every $0 \leq s \leq t$ and every trajectory z. If the maxingale problem for H has a unique idempotent probability measure Π_0 as solution, then $\Pi_0 = \Pi$ and the large deviation principle holds for $\{X_n\}$ with rate function I.

1.4. Examples

As explained in Section 1.2, an essential ingredient of our result is the convergence of $\{H_n\}$. As the following examples show, at least formally, the basic calculations for verifying this convergence may be quite simple. The calculations given here are frequently heuristic; however, we give rigorous results for these and other examples in Part 3.

EXAMPLE 1.4. [**Freidlin-Wentzell Theory - I**] The best-known example of large deviations for Markov processes comes from the work of Freidlin and Wentzell on diffusions with small diffusion coefficient. (See Chapters 3 and 4 in [**52**].) Consider a sequence of d-dimensional diffusion processes with X_n satisfying the Itô equation

$$X_n(t) = x + \frac{1}{\sqrt{n}} \int_0^t \sigma(X_n(s))dW(s) + \int_0^t b(X_n(s))ds.$$

Let $a(x) = \sigma(x) \cdot \sigma^T(x)$. Then the (linear) generator is

$$A_n g(x) = \frac{1}{2n} \sum_{ij} a_{ij}(x)\partial_i\partial_j g(x) + \sum_i b_i(x)\partial_i g(x),$$

where we can take $\mathcal{D}(A_n)$ to be the collection of functions of the form $c + f$ where $c \in R$ and $f \in C_c^2(R^d)$, the space of twice continuously differentiable functions with compact support in R^d.

$$\mathcal{H}_n f(x) = \frac{1}{2n} \sum_{ij} a_{ij}(x)\partial_i\partial_j f(x) + \frac{1}{2n} \sum_{ij} a_{ij}(x)\partial_i f(x)\partial_j f(x) + \sum_i b_i(x)\partial_i f(x)$$

and

$$H_n f(x) = \frac{1}{2n} \sum_{ij} a_{ij}(x)\partial_i\partial_j f(x) + \frac{1}{2} \sum_{ij} a_{ij}(x)\partial_i f(x)\partial_j f(x) + \sum_i b_i(x)\partial_i f(x).$$

Consequently, $Hf = \lim_{n\to\infty} H_n f$ is

$$Hf(x) = \frac{1}{2}(\nabla f(x))^T \cdot a(x) \cdot \nabla f(x) + b(x) \cdot \nabla f(x). \tag{1.14}$$

We can identify the rate function I in a simple form by finding a variational representation of H. First, we introduce a pair of functions on $R^d \times R^d$:

$$H(x,p) \equiv \frac{1}{2}p^T \cdot a(x) \cdot p + b(x) \cdot p = \frac{1}{2}|\sigma^T(x) \cdot p|^2 + b(x) \cdot p, \tag{1.15}$$

$$L(x,q) \equiv \sup_{p\in R^d} \{p \cdot q - H(x,p)\}.$$

$H(x,p)$ is convex in p, and H and L are dual Fenchel-Legendre transforms. In particular,

$$H(x,p) = \sup_{q\in R^d} \{p \cdot q - L(x,q)\}.$$

Therefore

$$Hf(x) = H(x, \nabla f(x)) = \mathbf{H}f(x) \equiv \sup_{u\in R^d} \{Af(x,u) - L(x,u)\},$$

where $Af(x,u) = u\nabla f(x)$, for $f \in C_c^2(R^d)$. This implies that H is a Nisio semigroup generator studied in control theory (e.g. Chapter 8). Applying Corollary 8.28

or Corollary 8.29, we arrive at the rate function for $\{X_n\}$ represented in a variational form:

$$I(x) = I_0(x_0) + \inf_{\{u:(x,u)\in\mathcal{J}\}} \int_0^\infty L(x(s), u(s))ds,$$

where I_0 is the rate function for $\{X_n(0)\}$ and $\mathcal{J}$ is the collection of solutions of the functional equation

$$f(x(t)) - f(x(0)) = \int_0^t Af(x(s), u(s))ds, \quad \forall f \in C_c^2(R^d).$$

In the present setting, this equation reduces to $\dot{x}(t) = u(t)$, so

$$(1.16)\quad I(x) = \begin{cases} I_0(x(0)) + \int_0^\infty L(x(s), \dot{x}(s))ds & \text{if } x \text{ is absolutely continuous} \\ \infty & \text{otherwise} \end{cases}$$

One can use other variational representations of the operator H and arrive at different expressions for the rate function I. For example, if we choose

$$Af(x, u) = u \cdot (\sigma^T(x)\nabla f(x)) + b(x) \cdot \nabla f(x), \quad f \in C_0^2(R^d),$$

$$L(x, u) = \frac{1}{2}|u|^2,$$

and define

$$\mathbf{H}f(x) = \sup_{u\in R^d} \{Af(x, u) - L(x, u)\},$$

then $H = \mathbf{H}$ and the rate function can be expressed as

$$I(x) = I_0(x(0)) + \frac{1}{2}\inf\{\int_0^\infty |u(s)|^2 ds : u \in L^2[0, \infty)$$

$$\text{and } x(t) = x(0) + \int_0^t b(x(s))ds + \int_0^t \sigma(x(s))u(s)ds\}.$$

EXAMPLE 1.5. [**Freidlin-Wentzell theory – II**]Wentzell [**128, 129**] has studied the jump process analogue of the small diffusion problem. Again, let $E = R^d$, and consider, for example,

$$(1.17)\qquad A_n g(x) = n\int_{R^d} (g(x + \frac{1}{n}z) - g(x))\eta(x, dz),$$

where for each $x \in R^d$, $\eta(x, \cdot)$ is a measure on $\mathcal{B}(R^d)$ and for each $B \in \mathcal{B}(R^d)$, $\eta(\cdot, B)$ is a Borel-measurable function. (For the moment, ignore the problem of defining an appropriate domain.) Then

$$\mathcal{H}_n f(x) = n\int_{R^d} (e^{f(x+\frac{1}{n}z)-f(x)} - 1)\eta(x, dz)$$

and

$$H_n f(x) = \int_{R^d} (e^{n(f(x+\frac{1}{n}z)-f(x))} - 1)\eta(x, dz).$$

Assuming $\int_E e^{\alpha\cdot z}\eta(x, dz) < \infty$, for each $\alpha \in R^d$, $Hf = \lim_{n\to\infty} H_n f$ is given by

$$Hf(x) = \int_{R^d} (e^{\nabla f(x)\cdot z} - 1)\eta(x, dz).$$

More generally, one can consider generators of the form

$$A_n g(x) = n\int_{R^d}(g(x+\frac{1}{n}z) - g(x) - \frac{1}{n}z\cdot\nabla g(x))\eta(x,dz) + \frac{1}{2n}\sum_{ij} a_{ij}(x)\partial_i\partial_j g(x) + b(x)\cdot\nabla g(x).$$

Then

$$\begin{aligned}\mathcal{H}_n f(x) &= n\int_{R^d}(e^{f(x+\frac{1}{n}z)-f(x)} - 1 - \frac{1}{n}z\cdot\nabla f(x))\eta(x,dz) + b(x)\cdot\nabla f(x)\\ &\quad +\frac{1}{2n}\sum_{ij} a_{ij}(x)\partial_i f(x)\partial_j f(x + \frac{1}{2n}\sum_{ij} a_{ij}(x)\partial_i\partial_j f(x))\end{aligned}$$

and

$$\begin{aligned}H_n f(x) &= \int_{R^d}(e^{n(f(x+\frac{1}{n}z)-f(x))} - 1 - z\cdot\nabla f(x))\eta(x,dz) + b(x)\cdot\nabla f(x)\\ &\quad +\frac{1}{2}\sum_{ij} a_{ij}(x)\partial_i f(x)\partial_j f(x) + \frac{1}{2n}\sum_{ij} a_{ij}(x)\partial_i\partial_j f(x).\end{aligned}$$

Now, assuming $\int_{R^d}(e^{\alpha\cdot z} - 1 - \alpha\cdot z)\eta(x,dz) < \infty$, for each $\alpha\in R^d$, $Hf = \lim_{n\to\infty} H_n f$ is given by

$$Hf(x) = \int_{R^d}(e^{\nabla f(x)\cdot z} - 1 - z\cdot\nabla f(x))\eta(x,dz) + \frac{1}{2}\sum_{ij} a_{ij}(x)\partial_i f(x)\partial_j f(x) + b(x)\cdot\nabla f(x).$$

A variational representation of H can be constructed as in Example 1.4. Define

$$H(x,p) = \int_{R^d}(e^{p\cdot z} - 1 - z\cdot p)\eta(x,dz) + \frac{1}{2}|\sigma^T(x)p|^2 + b(x)\cdot p$$

and

$$L(x,q) = \sup_{p\in R^d}\{p\cdot q - H(x,p)\}.$$

Then

$$Hf(x) = H(x,\nabla f(x)) = \mathbf{H}f(x) \equiv \sup_{u\in R^d}\{Af(x,u) - L(x,u)\},$$

where

$$Af(x,u) = u\cdot\nabla f(x), \quad f\in C_c^2(R^d),$$

and the rate function is given by (1.16).

EXAMPLE 1.6. [**Partial sum processes**] Functional versions of the classical large deviation results of Cramér [**16**] have been considered by a number of authors, including Borovkov [**11**] and Mogulskii [**87**]. Let $\xi_1, \xi_2, \ldots$ be independent and identically distributed R^d-valued random variables with distribution ν. Define

$$X_n(t) = \beta_n^{-1}\sum_{k=1}^{[t/\epsilon_n]}\xi_k.$$

Then, as in (1.11),

$$A_n f(x) = \epsilon_n^{-1}(\int_{R^d} f(x+\beta_n^{-1}z)\nu(dz) - f(x))$$

and

$$H_n f(x) = \frac{1}{n\epsilon_n} \log \int_{R^d} e^{n(f(x+\beta_n^{-1}z)-f(x))}\nu(dz).$$

If $\int_{R^d} e^{\alpha|z|}\nu(dz) < \infty$, for each $\alpha > 0$, $\epsilon_n = n^{-1}$, and $\beta_n = n$, then $H_n f \to Hf$ given by

$$Hf(x) = \log \int_{R^d} e^{\nabla f(x)\cdot z}\nu(dz).$$

If $\int_{R^d} z\nu(dz) = 0$, $\int_{R^d} e^{\alpha|z|}\nu(dz) < \infty$ for some $\alpha > 0$, $\epsilon_n\beta_n^2 = n$, and $\epsilon_n n \to 0$, then $H_n f \to Hf$ given by

$$Hf(x) = \frac{1}{2}\int_{R^d} (\nabla f(x)\cdot z)^2 \nu(dz).$$

Once again, a variational representation of H and the rate function I can be constructed using Fenchel-Legendre transforms. Define

$$H(p) = \frac{1}{2}\int_{R^d} (z\cdot p)^2\nu(dz)$$

and

$$L(q) = \sup_{p\in R^d}\{p\cdot q - H(p)\}.$$

Then

$$I(x) = \begin{cases} \int_0^\infty L(\dot{x}(s))ds & \text{if } x \text{ is absolutely continuous} \\ \infty & \text{otherwise.} \end{cases}$$

EXAMPLE 1.7. [**Levy processes**] If η in Example 1.5 does not depend on x, then the process is a Lévy process, that is, a process with stationary, independent increments. Lynch and Sethuraman [**82**] and Mogulskii [**88**] consider the real-valued case. de Acosta [**25**] considers Banach space-valued processes.

In de Acosta's setting, let E be a Banach space and E^* its dual. Recall that for $f \in C^1(E)$, the gradient $\nabla f(x) = \nabla_x f(x)$ is the unique element in E^* satisfying

$$f(x+z) - f(x) = \langle z, \nabla f(x)\rangle + o(|z|), \quad \forall z \in E.$$

Let $F \subset E^*$ separate points, and let $\mathcal{D}(A)$ be the collection of functions of the form

$$f(x) = g(\langle x, \xi_1\rangle, \ldots, \langle x, \xi_m\rangle),$$

where $\xi_1, \ldots, \xi_m \in F$ and $g \in C_c^2(R^m)$. Then,

$$\nabla f(x) = \sum_{k=1}^m \partial_k g(\langle x, \xi_1\rangle, \ldots, \langle x, \xi_m\rangle)\xi_k \in E^*.$$

Assuming $\int_E 1 \wedge \langle z, \xi\rangle^2 \eta(dz) < \infty$, for every $\xi \in F$, we have

$$\begin{aligned} A_n g(x) &= n\int_E \Big(g(\langle x + \frac{1}{n}z, \xi_1\rangle, \ldots, \langle x + \frac{1}{n}z, \xi_m\rangle) - g(\langle x, \xi_1\rangle, \ldots, \langle x, \xi_m\rangle) \\ &\qquad -\frac{1}{n} I_{\{\|z\|\le 1\}} \sum_{k=1}^m \langle z, \xi_k\rangle \partial_k g(\langle x, \xi_1\rangle, \ldots, \langle x, \xi_m\rangle)\Big)\eta(dz) \\ &\qquad + \sum_{k=1}^m \langle b, \xi_k\rangle \partial_k g(\langle x, \xi_1\rangle, \ldots, \langle x, \xi_m\rangle), \\ &= n\int_E \Big(g(\langle x + \frac{1}{n}z, \xi_1\rangle, \ldots, \langle x + \frac{1}{n}z, \xi_m\rangle) - g(\langle x, \xi_1\rangle, \ldots, \langle x, \xi_m\rangle) \\ &\qquad -\frac{1}{n} I_{\{\|z\|\le 1\}} \langle z, \nabla f(x)\rangle\Big)\eta(dz) + \langle b, \nabla f(x)\rangle, \end{aligned}$$

$$\mathcal{H}_n f(x) = n\int_E (e^{f(x+\frac{1}{n}z)-f(x)} - 1 - \frac{1}{n} I_{\{\|z\|\le 1\}}\langle z, \nabla f(x)\rangle)\eta(dz) + \langle b, \nabla f(x)\rangle,$$

and

$$H_n f(x) = \int_E (e^{n(f(x+\frac{1}{n}z)-f(x))} - 1 - \frac{1}{n} I_{\{\|z\|\le 1\}}\langle z, \nabla f(x)\rangle)\eta(dz) + \langle b, \nabla f(x)\rangle.$$

Now, assuming $\int_E (e^{\langle \alpha, z\rangle} - 1 - I_{\{|z|\le 1\}}\langle \alpha, z\rangle)\eta(dz) < \infty$, for each $\alpha \in E^*$, $Hf = \lim_{n\to\infty} H_n f$ is given by

$$Hf(x) = \int_E (e^{\langle z, \nabla f(x)\rangle} - 1 - I_{\{\|z\|\le 1\}}\langle z, \nabla f(x)\rangle)\eta(dz) + \langle b, \nabla f(x)\rangle.$$

Define

$$H(p) = \int_E (e^{\langle z, p\rangle} - 1 - I_{|z|\le 1}\langle z, p\rangle)\eta(dz) + \langle b, z\rangle, \quad p \in E^*,$$

and

$$L(q) = \sup_{p\in E^*} \{\langle q, p\rangle - H(p)\}, \quad q \in E.$$

H is a convex function and L is its Fenchel-Legendre transform, and hence,

$$H(p) = \sup_{q\in E}\{\langle q, p\rangle - L(q)\}.$$

Consequently

$$Hf(x) = \mathbf{H}f(x) \equiv \sup_{u\in E}\{Af(x,u) - L(u)\},$$

where

$$Af(x,u) = \langle u, \nabla f(x)\rangle,$$

and the rate function is again given by

$$I(x) = \begin{cases} I_0(x(0)) + \int_0^\infty L(\dot{x}(s))ds & \text{if } x \text{ is absolutely continuous} \\ \infty & \text{otherwise.} \end{cases}$$

EXAMPLE 1.8. [**Random evolutions - I**] Let B be the generator for a Markov process Y with state space E_0, and let $Y_n(t) = Y(nt)$. Let X_n satisfy

$$\dot{X}_n(t) = F(X_n(t), Y_n(t)),$$

where $F : R^d \times E_0 \to R^d$. If Y is ergodic with stationary distribution π, then

$$\lim_{n\to\infty} X_n = X,$$

where X satisfies $\dot{X} = \overline{F}(X)$, with $\overline{F}(x) = \int F(x,y)\pi(dy)$. Averaging results of this type date back at least to Hasminskii [**66**]. Corresponding large deviation theorems have been considered by Freidlin [**49, 50**]. (See Freidlin and Wentzell [**52**], Chapter 7.)

For simplicity, assume B is a bounded operator

$$Bg(y) = \lambda(y)\int (g(z) - g(y))\mu(y, dz). \tag{1.18}$$

Then

$$A_n g(x,y) = F(x,y)\cdot\nabla_x g(x,y) + n\lambda(y)\int (g(x,z) - g(x,y))\mu(y,dz),$$

$$\mathcal{H}_n f(x,y) = F(x,y)\cdot\nabla_x f(x,y) + n\lambda(y)\int (e^{f(x,z)-f(x,y)} - 1)\mu(y,dz)$$

and

$$H_n f(x,y) = F(x,y)\cdot\nabla_x f(x,y) + \lambda(y)\int (e^{n(f(x,z)-f(x,y))} - 1)\mu(y,dz).$$

Let $f_n(x,y) = f(x) + \frac{1}{n}h(x,y)$. Then

$$\lim_{n\to\infty} H_n f_n(x,y) = F(x,y)\cdot\nabla_x f(x) + \lambda(y)\int (e^{h(x,z)-h(x,y)} - 1)\mu(y,dz).$$

Consequently, H is multivalued.

One approach to dealing with this limit is to select h so that the limit is independent of y, that is, to find functions $h(x,y)$ and $g(x)$ such that

$$F(x,y)\cdot\nabla_x f(x) + \lambda(y)\int (e^{h(x,z)-h(x,y)} - 1)\mu(y,dz) = g(x). \tag{1.19}$$

Multiply both sides of the equation by $e^{h(x,y)}$, and fix x. One needs to solve an equation of the form

$$\alpha(y)\overline{h}(y) + \lambda(y)\int (\overline{h}(z) - \overline{h}(y))\mu(y,dz) = \gamma\overline{h}(y), \tag{1.20}$$

where $\overline{h}$ plays the role of $e^{h(x,y)}$ and hence must be positive. Of course, (1.20) is just an eigenvalue problem for the operator

$$Q = (\alpha I + B). \tag{1.21}$$

If E_0 is finite, that is, Y is a finite Markov chain, and B is irreducible, then the Perron-Froebenius theorem implies the existence of a unique (up to multiplication by a constant) strictly positive $\overline{h}$ which is the eigenfunction corresponding to the largest eigenvalue for Q. Note that in (1.19), it is only the ratio $\overline{h}(z)/\overline{h}(y)$ that is relevant, so the arbitrary constant cancels.

For linear semigroups, this approach to convergence of generators was introduced in Kurtz [**71**] and has been developed extensively by Papanicolaou and Varadhan [**93**] and Kushner (for example, [**79**]). For nonlinear semigroups, Kurtz [**72**] and Evans [**37**] have used similar arguments.

If we use $\gamma(\alpha(\cdot))$ to denote the largest eigenvalue for the Q in (1.21), then we arrive at a single valued limit operator

$$Hf(x) = \gamma(F(x,\cdot)\cdot\nabla f(x)).$$

Define

$$H(x,p) = \gamma(F(x,\cdot)\cdot p)$$

and

$$L(x,q) = \sup_{p\in R^d}\{p\cdot q - H(x,p)\},$$

for $x,p,q \in R^d$. There are various probabilistic representations for the principle eigenvalue γ. (See Appendix B.) It follows from these representations that H is convex in p, so

$$Hf(x) = H(x,\nabla f(x)) = \mathbf{H}f(x) \equiv \sup_{u\in R^d}\{Af(x,u) - L(x,u)\},$$

where $Af(x,u) = u\cdot\nabla f(x)$, for $f \in C_c^1(R^d)$.

By the results in Chapter 8, (1.16) gives a representation of the rate function. Another representation of the rate function can be of special interest. Donsker and Varadhan [**32**] generalize the classical Rayleigh-Ritz formula and obtain the following variational representation of the principal eigenvalue for Q when B is a general operator satisfying the maximum principle:

$$\gamma(\alpha) = \sup_{\mu\in\mathcal{P}(E_0)}\{\int_{E_0}\alpha d\mu - I_B(\mu)\},$$

where

$$I_B(\mu) = -\inf_{f\in\mathcal{D}(B),\inf_y f(y)>0}\int_{E_0}\frac{Bf}{f}d\mu. \tag{1.22}$$

Therefore, setting

$$Af(x,\mu) = \int_{E_0}F(x,y)d\mu(dy)\cdot\nabla f(x),$$

H satisfies

$$Hf(x) = \mathbf{H}f(x) \equiv \sup_{\mu\in\mathcal{P}(E_0)}\{Af(x,\mu) - I_B(\mu)\}.$$

By the methods in Chapter 8, the rate function can also be represented as

$$I(x) = I_0(x(0)) + \inf\{\int_0^\infty I_B(\mu(s))ds : \\ \dot{x}(t) = \int_{E_0}F(x(t),\cdot)d\mu(t)\},\mu(t)\in\mathcal{P}(E_0),t\geq 0\},$$

if x is absolutely continuous, and $I(x) = \infty$ otherwise.

EXAMPLE 1.9. [**Random evolutions - II**] Let B be as in (1.18), let Y be the corresponding Markov process, and let $Y_n(t) = Y(n^3t)$. Let X_n satisfy

$$\dot{X}_n(t) = nF(X_n(t),Y_n(t)).$$

Suppose that Y is ergodic with stationary distribution π and that

$$\int_{E_0}F(x,y)\pi(dy) = 0. \tag{1.23}$$

Then

$$A_ng(x,y) = nF(x,y)\cdot\nabla_x g(x,y) + n^3\lambda(y)\int(g(x,z)-g(x,y))\mu(y,dz),$$

$$\mathcal{H}_nf(x,y) = nF(x,y)\cdot\nabla_x f(x,y) + n^3\lambda(y)\int(e^{f(x,z)-f(x,y)}-1)\mu(y,dz),$$

and

$$H_n f(x,y) = nF(x,y)\cdot\nabla_x f(x,y) + n^2\lambda(y)\int (e^{n(f(x,z)-f(x,y))}-1)\mu(y,dz).$$

Let $f_n(x,y) = f(x) + \frac{1}{n^2}h_1(x,y) + \frac{1}{n^3}h_2(x,y)$, and assume that

$$Bh_1(x,\cdot)(y) = -F(x,y)\cdot\nabla_x f(x).$$

If E_0 is finite and B is irreducible, then (1.23) ensures the existence of h_1. Note that $h_1(x,y)$ will be of the form $\alpha(x,y)\cdot\nabla_x f(x)$, where α is a vector function satisfying

$$B\alpha(x,\cdot)(y) = -F(x,y).$$

It follows that

$$\begin{aligned}\lim_{n\to\infty} H_n f_n(x,y) &= \frac{\lambda(y)}{2}\int_{E_0}(h_1(x,z)-h_1(x,y))^2\mu(y,dz)\\ &\quad+\lambda(y)\int_{E_0}(h_2(x,z)-h_2(x,y))\mu(y,dz),\end{aligned}$$

and again we have a multivalued limit.

As before, one approach is to select h_2 so that the limit is independent of y. If E_0 is finite, (1.23) ensures that h_2 exists. In any case, if h_2 exists, the limit will be

$$Hf(x) = \int_{E_0}\frac{\lambda(y)}{2}\int_{E_0}((\alpha(x,z)-\alpha(x,y))\cdot\nabla_x f(x))^2\mu(y,dz)\pi(dy).$$

Note that H is of the same form as H in Example 1.4, indicating that this "random evolution" behaves like the "small diffusion" process. If $Y_n(t) = Y(n^3 t)$ is replaced by $Y_n(t) = Y(n^2 t)$, then X_n will converge in distribution to a diffusion. (See, for example, Ethier and Kurtz [**36**], Chapter 12.)

The results in Example 1.4 give variational representations of the rate function.

EXAMPLE 1.10. [**Periodic diffusions**] Baldi [**6**] has considered large deviations for models of the following type. Let σ be a periodic $d\times d$-matrix-valued function (for each $1\le i\le d$, there is a period $p_i>0$ such that $\sigma(y)=\sigma(y+p_ie_i)$ for all $y\in R^d$), and let X_n satisfy the Itô equation

$$dX_n(t) = \frac{1}{\sqrt{n}}\sigma(\alpha_n X_n(t))dW(t),$$

where $\alpha_n>0$ and $\lim_{n\to\infty} n^{-1}\alpha_n=\infty$. Let $a=\sigma\sigma^T$. Then

$$A_n f(x) = \frac{1}{2n}\sum_{ij} a_{ij}(\alpha_n x)\frac{\partial^2}{\partial x_i\partial x_j}f(x),$$

$$\mathcal{H}_n f(x) = \frac{1}{2n}\sum_{ij} a_{ij}(\alpha_n x)\partial_i\partial_j f(x) + \frac{1}{2n}\sum_{ij} a_{ij}(\alpha_n x)\partial_i f(x)\partial_j f(x),$$

and

$$H_n f(x) = \frac{1}{2n}\sum_{ij} a_{ij}(\alpha_n x)\partial_i\partial_j f(x) + \frac{1}{2}\sum_{ij} a_{ij}(\alpha_n x)\partial_i f(x)\partial_j f(x).$$

Let $f_n(x) = f(x)+\epsilon_n h(x,\alpha_n x)$, where $\epsilon_n = n\alpha_n^{-2}$. Noting that, by assumption, $\epsilon_n\alpha_n = n\alpha_n^{-1}\to 0$, if we select h with the same periods in y as the a_{ij} so that

$$\frac{1}{2}\sum_{ij} a_{ij}(y)\left(\frac{\partial^2}{\partial y_i\partial y_j}h(x,y)+\partial_i f(x)\partial_j f(x)\right) = g(x), \tag{1.24}$$

for some g independent of y, then

$$\lim_{n\to\infty} H_n f_n(x,y) = g(x).$$

For each x, the desired $h(x,\cdot)$ is the solution of the linear partial differential equation

$$\frac{1}{2}\sum_{ij} a_{ij}(y)\frac{\partial^2}{\partial y_i \partial y_j} h(x,y) = g(x) - \frac{1}{2}\sum_{i,j} a_{ij}(y)\partial_i f(x)\partial_j f(x) \tag{1.25}$$

on $[0,p_1]\times\cdots\times[0,p_d]$ with periodic boundary conditions extended periodically to all of R^d. For h to exist, we must have

$$g(x) = \frac{1}{2}\sum_{ij}\overline{a}_{ij}\partial_i f(x)\partial_j f(x), \tag{1.26}$$

where $\overline{a}_{ij}$ is the average of a_{ij} with respect to the stationary distribution for the diffusion on $[0,p_1]\times\cdots\times[0,p_d]$ whose generator is

$$A_0 f(y) = \frac{1}{2}\sum_{i,j} a_{ij}(y)\frac{\partial^2}{\partial y_i \partial y_j} f(y),$$

with periodic boundary conditions. To see that (1.26) must hold, simply integrate both sides of (1.25) by the stationary distribution. In particular,

$$h(x,y) = \frac{1}{2}\sum_{ij} h_{ij}(y)\partial_i f(x)\partial_j f(x),$$

where h_{ij} satisfies

$$A_0 h_{ij}(y) = \overline{a}_{ij} - a_{ij}(y).$$

If A_0 is uniformly elliptic and f is C^3, then, with sufficient regularity on the a_{ij}, $h \in C^2$ will exist.

By (1.26), the limit Hf is a special case of Hf in Example 1.4. Consequently, we can identify the rate function as in that example.

EXAMPLE 1.11. [**Donsker-Varadhan theory for occupation measures**] Large deviation theory for occupation measures (see Donsker and Varadhan [**33**]) is closely related to Example 1.8. Let Y be a Markov process with generator B and state space E_0, and for $n = 1,2,\ldots$, define

$$\Gamma_n(C,t) = \frac{1}{n}\int_0^{nt} I_C(Y(s))ds.$$

Then Z_n defined by $Z_n(t) = (Y(nt), \Gamma_n(\cdot,t))$ is a Markov process with state space $E = E_0 \times \mathcal{M}_F(E_0)$, where $\mathcal{M}_F(E_0)$ is the space of finite measures on E_0. Let $h \in C_b(E_0 \times R^m)$ be differentiable in the real variables and satisfy $h(\cdot,x) \in \mathcal{D}(B)$ for $x \in R^m$. For $\alpha_i \in C_b(E_0)$, $i = 1,\ldots,m$, let

$$g(y,z) = h(y,\langle\alpha_1,z\rangle,\ldots,\langle\alpha_m,z\rangle), \quad (y,z)\in E_0\times\mathcal{M}_F(E_0),$$

where $\langle\alpha,z\rangle = \int_{E_0}\alpha dz$. The generator for Z_n satisfies

$$A_n g(y,z) = nBh(y,\langle\alpha_1,z\rangle,\ldots,\langle\alpha_m,z\rangle) + \sum_{i=1}^m \alpha_i(y)\partial_i h(y,\langle\alpha_1,z\rangle,\ldots,\langle\alpha_m,z\rangle),$$

where $\partial_i h(y, x_1, \ldots, x_m) = \frac{\partial}{\partial x_i} h(y, x_1, \ldots, x_m)$. For definiteness, let B be a diffusion operator

$$Bg(y) = \frac{1}{2}\sum_{i,j} a_{ij}(y)\frac{\partial^2}{\partial y_i \partial y_j} g(y) + \sum_i b_i(y)\frac{\partial}{\partial y_i} g(y).$$

Then for $f(y,z) = h(y, \langle \alpha_1, z\rangle, \ldots, \langle \alpha_m, z\rangle)$,

$$\mathcal{H}_n f(y,z) = A_n f(y,z) + \frac{n}{2}\sum_{i,j} a_{ij}(y)\frac{\partial}{\partial y_i} f(y,z)\frac{\partial}{\partial y_j} f(y,z)$$

and

$$H_n f(y,z) = A_n f(y,z) + \frac{n^2}{2}\sum_{i,j} a_{ij}(y)\frac{\partial}{\partial y_i} f(y,z)\frac{\partial}{\partial y_j} f(y,z).$$

If we let $f(z) = h(\langle \alpha_1, z\rangle, \ldots, \langle \alpha_m, z\rangle)$ and

$$f_n(y,z) = h(\langle \alpha_1, z\rangle, \ldots, \langle \alpha_m, z\rangle) + \frac{1}{n} h_0(y, \langle \alpha_1, z\rangle, \ldots, \langle \alpha_m, z\rangle),$$

then

$$\begin{aligned} &\lim_{n\to\infty} H_n f_n(y,z) \\ (1.27) \quad &= \sum_{i=1}^m \alpha_i(y)\partial_i h(\langle \alpha_1, z\rangle, \ldots, \langle \alpha_m, z\rangle) + Bh_0(y, \langle \alpha_1, z\rangle, \ldots, \langle \alpha_m, z\rangle) \\ &\quad + \frac{1}{2}\sum_{i,j} a_{ij}(y)\frac{\partial}{\partial y_i} h_0(y, \langle \alpha_1, z\rangle, \ldots, \langle \alpha_m, z\rangle)\frac{\partial}{\partial y_j} h_0(y, \langle \alpha_1, z\rangle, \ldots, \langle \alpha_m, z\rangle). \end{aligned}$$

As in Example 1.8, we have a multivalued limit, and the rate function representation is also similar to the representation in Example 1.8.

EXAMPLE 1.12. [**Stochastic reaction-diffusion equations**] We consider a stochastic reaction-diffusion equation on a rescaled lattice, which is a Ginzburg-Landau model of non-conservative type. Let $F \in C^2(R)$, $\sup_r |F''(r)| < \infty$, and $\mathcal{O} = [0,1)^d$ with periodic boundary. We discretize $\mathcal{O}$ into

$$\Lambda_m \equiv \{0, \frac{1}{m}, \ldots, \frac{k}{m}, \ldots, 1 - \frac{1}{m}\}^d \subset \mathcal{O},$$

also with periodic boundary conditions (that is, $0 = 1$). Let $m = m(n)$ depend on n so that $\lim_{n\to\infty} m(n) = \infty$. When there is no possibility of confusion, we simply write m in place of $m(n)$. For $\rho \in R^{|\Lambda_m|}$, define

$$\begin{aligned} \nabla^i_m \rho(x) &= \frac{m}{2}(\rho(x + \frac{1}{m}e_k) - \rho(x - \frac{1}{m}e_k)); \\ \nabla_m \rho(x) &= (\nabla^1_m \rho(x), \ldots, \nabla^d_m \rho(x)) \\ (1.28) \quad \Delta_m \rho(x) &= \nabla_m \cdot \nabla_m \rho(x) \\ &= (\frac{m}{2})^2 \sum_{k=1}^d \Big(\rho(x + \frac{2}{m}e_k) - 2\rho(x) + \rho(x - \frac{2}{m}e_k)\Big), \end{aligned}$$

and for a vector-valued function,

$$\xi(x) = \Big(\xi_1(x), \ldots, \xi_d(x)\Big),$$

define

$$\mathrm{div}_m\xi(x) = \nabla_m \cdot \xi(x) = \sum_{k=1}^{d} \nabla_m^k \rho_k(x).$$

We consider a finite system of stochastic differential equations

$$dY_n(t,x) = (\Delta_m Y_n)(t,x)dt - F'(Y_n(t,x))dt + \frac{m^{d/2}}{\sqrt{n}} dB(t,x), \tag{1.29}$$

where $x \in \Lambda_m$ and $\{B(t,x) : x \in \Lambda_m\}$ are independent standard Brownian motions.

Let $E_n = R^{|\Lambda_m|}$. For $p \in C^\infty(\mathcal{O})$ and $\rho \in E_n$, define

$$\langle \rho, p \rangle_m \equiv \sum_{x \in \Lambda_m} p(x)\rho(x)m^{-d},$$

and

$$\langle \rho, p \rangle \equiv \int_{\mathcal{O}} p(x)\rho(x)dx, \qquad \rho \in L^2(\mathcal{O}).$$

For $p_1, \ldots, p_l \in C^\infty(\mathcal{O})$ and $\varphi \in C_c^\infty(R^l)$, let $\vec{\gamma} = (\gamma_1, \ldots, \gamma_l)$ and

$$\begin{aligned} f_n(\rho) &= \varphi(\langle \rho, \gamma_1 \rangle_m, \ldots, \langle \rho, \gamma_l \rangle_m) = \varphi(\langle \rho, \vec{\gamma} \rangle_m), \\ f(\rho) &= \varphi(\langle \rho, \vec{\gamma} \rangle), \end{aligned} \tag{1.30}$$

and

$$\begin{aligned} A_n f_n(\rho) &= \sum_{i=1}^{l} \partial_i \varphi(\langle \rho, \vec{\gamma} \rangle_m) \Big(\sum_{x \in \Lambda_m} \gamma_i(x)(\Delta_m \rho(x) - F'(\rho(x)))m^{-d} \Big) \\ &\quad + \frac{1}{2n} \sum_{i,j=1}^{l} \partial_{ij}^2 \varphi(\langle \rho, \vec{\gamma} \rangle_m) \Big(\sum_{x \in \Lambda_m} \gamma_i(x)\gamma_j(x)m^{-d} \Big). \end{aligned}$$

Then, by Itô's formula, Y_n is an E_n-valued solution to the martingale problem for A_n and

$$\begin{aligned} H_n f_n(\rho) &\equiv \frac{1}{n} e^{-nf} A_n e^{nf}(\rho) \\ &= \sum_{i=1}^{l} \partial_i \varphi(\langle \rho, \vec{\gamma} \rangle_m) \Big(\sum_{x \in \Lambda_m} \gamma_i(x)(\Delta_m \rho(x) - F'(\rho(x)))m^{-d} \Big) \\ &\quad + \frac{1}{2} \sum_{i,j=1}^{l} \partial_i \varphi(\langle \rho, \vec{\gamma} \rangle_m) \partial_j \varphi(\langle \rho, \vec{\gamma} \rangle_m) \Big(\sum_{x \in \Lambda_m} \gamma_i(x)\gamma_j(x)m^{-d} \Big) \\ &\quad + \frac{1}{2n} \sum_{i,j=1}^{l} \partial_{ij}^2 \varphi(\langle \rho, \vec{\gamma} \rangle_m) \Big(\sum_{x \in \Lambda_m} \gamma_i(x)\gamma_j(x)m^{-d} \Big). \end{aligned}$$

Define $\eta_n : E_n \to E = L^2(\mathcal{O})$ by

$$\eta_n(\rho_n)(x_1, \ldots, x_d) \equiv \sum_{i_1, \ldots, i_d = 0}^{m-1} \rho_n\Big(\frac{i_1}{m}, \ldots, \frac{i_d}{m}\Big) \prod_{j=1}^{d} \mathbf{1}_{[i_j/m, (i_j+1)/m)}(x_j). \tag{1.31}$$

For $\rho_n \in E_n$ and $\rho \in E$ satisfying $\|\eta_n(\rho_n) - \rho\|_{L^2(\mathcal{O})} \to 0$, we have

$$
\begin{aligned}
(1.32) \quad H_n f_n(\rho_n) &\to Hf(\rho) \\
&\equiv \sum_{i=1}^{l} \partial_i\varphi(\langle\rho,\vec{\gamma}\rangle)\Big(\langle\rho,\Delta\gamma_i\rangle - \langle F'(\rho),\gamma_i\rangle\Big) \\
&\quad + \frac{1}{2}\sum_{i,j=1}^{l} \partial_i\varphi(\langle\rho,\vec{\gamma}\rangle)\partial_j\varphi(\langle\rho,\vec{\gamma}\rangle)\int_{\mathcal{O}}\gamma_i(x)\gamma_j(x)dx \\
&= \langle\rho,\Delta\frac{\delta f}{\delta\rho}\rangle - \langle F'(\rho),\frac{\delta f}{\delta\rho}\rangle + \frac{1}{2}\|\frac{\delta f}{\delta\rho}\|^2_{L^2(\mathcal{O})}, \\
&= H(\rho,\frac{\delta f}{\delta\rho}),
\end{aligned}
$$

where

$$H(\rho,p) = \langle\Delta\rho + F'(\rho),p\rangle + \frac{1}{2}\|p\|^2_{L^2(\mathcal{O})}, \quad p \in C^\infty(\mathcal{O}),$$

and

$$\frac{\delta f}{\delta\rho} = \sum_{i=1}^{l} \partial_i\varphi(\langle\rho,\vec{\gamma}\rangle)\gamma_i. \tag{1.33}$$

By analogy with Example 1.4, we expect $\{\eta_n(Y_n)\}$ to satisfy a large deviation principle in $L^2(\mathcal{O})$. Let

$$L(\rho,u) = \sup_{p\in C^\infty(\mathcal{O})}\{\langle u,p\rangle - H(\rho,p)\}.$$

The rate function is analogous to that of (1.16) and should be given by

$$I(\rho) = I_0(\rho(0)) + \int_0^\infty L(\rho(s),\dot{\rho}(s))ds.$$

A rigorous proof of this statement is given in Chapter 13, Theorem 13.7.

The equation for Y_n is a discretized version of the stochastic partial differential equation given in weak form by

$$
\begin{aligned}
\int_{\mathcal{O}}\varphi(x)X_n(t,x)dx &= \int_{\mathcal{O}}\varphi(x)X_n(0,x)dx + \int_{\mathcal{O}\times[0,t]}\Delta\varphi(x)X_n(s,x)dxds \\
(1.34) \qquad &\quad - \int_{\mathcal{O}\times[0,t]}\varphi(x)F'(X_n(s,x))dxds \\
&\quad + \frac{1}{\sqrt{n}}\int_{\mathcal{O}\times[0,t]}\varphi(x)W(dx\times ds),
\end{aligned}
$$

where $W(dx \times ds)$ is a space-time Gaussian white noise. Unfortunately, (1.34) does not have a function-valued solution when $d \geq 2$. To study the large deviation problem in $L^2(\mathcal{O})$, as described above, even for the discretized equation (1.29), requires that m go to infinity slowly enough with n. (See Theorem 13.7.)

Equation (1.34) is also known as a stochastically perturbed Allen-Cahn equation. It has been used in material science to study time evolution of material density when the total mass is not conserved. More generally, in statistical physics, when F is non-convex, the equation is used to model phase transition (Spohn [**113**]). When $d = 1$, (1.34) is well-defined and Sowers [**112**] gives the large deviation principle.

EXAMPLE 1.13. [**Stochastic Cahn-Hilliard equations**] Spohn [**113**] introduces a type of Ginzburg-Landau model with a conserved quantity motivated by the study of phase transitions. Let $\mathcal{O}$ and F be as in Example 1.12. Formally, the equation can be written

$$\partial_t X_n(t,x) = \nabla \cdot \Big(\nabla(-\Delta X_n(t,x) + F'(X_n(t,x))) + \frac{1}{\sqrt{n}}\partial_t\partial_x \mathbf{W}(t,x)\Big) \tag{1.35}$$

or in a more easily interpretable, but still formal, weak form

$$\begin{aligned}\int_{\mathcal{O}} \varphi(x)X_n(t,x)dx &= \int_{\mathcal{O}} \varphi(x)X_n(0,x)dx - \int_{\mathcal{O}\times[0,t]} \Delta^2\varphi(x)X_n(s,x)dxds \\ &\quad + \int_{\mathcal{O}\times[0,t]} \Delta\varphi(x)F'(X_n(s,x))dxds \\ &\quad + \frac{1}{\sqrt{n}}\sum_{k=1}^{d}\int_{\mathcal{O}\times[0,t]} \partial_{x_k}\varphi(x)W_k(dx\times ds),\end{aligned} \tag{1.36}$$

where the W_k are independent Gaussian white noises and $\mathbf{W} = (W_1,\ldots,W_d)$. Without the stochastic term, (1.35) is known as the Cahn-Hilliard equation and originally arose in material science where the solution gives the evolution of a material density.

In these applications, (1.35) is derived as a phenomenological model using rough and heuristic arguments. Frequently, only asymptotic properties are of practical interest. Indeed, as in the previous example, (1.35) does not have a function-valued solution. Therefore, for the large deviation problem here, we consider a discretized version.

Let ∇_m and Δ_m be defined according to (1.28), and consider the system

$$dY_n(t;x) = \mathrm{div}_m\Big(\nabla_m(-\Delta_m Y_n(t,x) + F'(Y_n(t,x)))dt + \frac{m^{d/2}}{\sqrt{n}}d\mathbf{B}(t,x)\Big), \quad x\in\Lambda_m, \tag{1.37}$$

where

$$\mathbf{B}(t,x) = \Big(B_1(t,x),\cdots,B_d(t,x)\Big), \quad x\in\Lambda_m,$$

with the $B_j(\cdot,x)$ independent, R^d-valued standard Brownian motions. Note that the solution satisfies

$$\sum_{x\in\Lambda_m} Y_n(t,x) = \sum_{x\in\Lambda_m} Y_n(0,x), \quad t\geq 0,$$

so we may as well choose the state spaces

$$E_n \equiv \{\rho\in R^{|\Lambda_m|} : \sum_{x\in\Lambda_m}\rho(x) = 0\}, \quad E\equiv\{\rho\in L^2(\mathcal{O}) : \int_{\mathcal{O}}\rho(x)dx = 0\}. \tag{1.38}$$

The map η_n in (1.31) embeds E_n into E. For functions of the form (1.30), by Itô's formula,

$$\begin{aligned}A_n f(\rho) &= \sum_{i=1}^{l}\partial_i\varphi(\langle\rho,\vec{\gamma}\rangle_m)\Big(\langle\gamma_i,\Delta_m(-\Delta_m\rho + F'(\rho))\rangle_m\Big) \\ &\quad + \frac{1}{2n}\sum_{i,j=1}^{l}\partial^2_{ij}\varphi(\langle\rho,\vec{\gamma}\rangle_m)\sum_{z\in\Lambda_m}\nabla_m\gamma_i(z)\cdot\nabla_m\gamma_j(z)m^{-d}.\end{aligned} \tag{1.39}$$

Therefore

$$\begin{aligned} H_n f(\rho) &\equiv \frac{1}{n} e^{-nf} A_n e^{nf}(\rho) \\ &= \sum_{i=1}^{l} \partial_i \varphi(\langle \rho, \vec{\gamma} \rangle_m) \Big(\langle \gamma_i, \Delta_m(-\Delta_n \rho + F'(\rho)) \rangle_m \Big) \\ &\quad + \frac{1}{2} \sum_{i,j=1}^{l} \partial_i \varphi(\langle \rho, \vec{\gamma} \rangle_m) \partial_j \varphi(\langle \rho, \vec{\gamma} \rangle_m) \big(\sum_{x \in \Lambda_m} \nabla_m \gamma_i(x) \cdot \nabla_m \gamma_j(x) m^{-d} \big) \\ &\quad + \frac{1}{2n} \sum_{i,j=1}^{l} \partial_{ij}^2 \varphi(\langle \rho, \vec{\gamma} \rangle_m) \sum_{z \in \Lambda_m} \Big(\nabla_m \gamma_i(z) \cdot \nabla_m \gamma_j(z) m^{-d} \Big). \end{aligned}$$

As in (1.33), let

$$\frac{\delta f}{\delta \rho} = \sum_{i=1}^{l} \partial_i \varphi(\langle \rho, \vec{\gamma} \rangle) \gamma_i.$$

Then for every $\rho_n \in E_n$ satisfying $\lim_{n\to\infty} \|\eta_n(\rho_n) - \rho\|^2_{L^2(\mathcal{O})} = 0$,

$$\lim_{n\to\infty} H_n f(\rho_n) = Hf(\rho) = H(\rho, \frac{\delta f}{\delta \rho}), \tag{1.40}$$

where

$$H(\rho, p) = \langle \Delta(-\Delta\rho + F'(\rho)), p \rangle + \frac{1}{2} \|\nabla p\|^2_{L^2(\mathcal{O})}, \quad \rho \in L^2(\mathcal{O}), p \in C^\infty(\mathcal{O}). \tag{1.41}$$

Following the method of rate function identification in the finite dimensional Example 1.4, we expect that the rate function can be identified using the same arguments as in Example 1.12. The proof of this result is given in Chapter 13. See Theorem 13.13.

Bertini, Landam, and Olla [**10**] proved a large deviation principle for a variant of (1.37) in which the discrete version of $\nabla \mathbf{W}$ is replaced by a discrete version of ΔW, where W is a scalar Brownian sheet. The current form of (1.37) was treated using a slightly different technique in Feng [**42**].

EXAMPLE 1.14. [**Weakly interacting stochastic particles**] For $n = 1, 2, \ldots$, we consider the finite system of stochastic differential equations

$$dX_{n,i}(t) = -\nabla\Psi\Big(X_{n,i}(t)\Big) - \frac{1}{n}\sum_{j=1}^{n} \nabla\Phi\Big(X_{n,i}(t) - X_{n,j}(t)\Big)dt + dW_i(t), \tag{1.42}$$

where $W_i(t)$, $i = 1, 2, \ldots, n$, are independent, R^d-valued, standard Brownian motions. Define the empirical measure-valued processes

$$\rho_n(t) = \frac{1}{n} \sum_{k=1}^{n} \delta_{X_{n,k}(t)}. \tag{1.43}$$

Then under appropriate growth conditions on Ψ and Φ, a law of large numbers (the McKean-Vlasov limit) holds. Specifically, ρ_n converges to a probability measure-valued solution of

$$\frac{\partial}{\partial t}\rho = \frac{1}{2}\Delta\rho + \nabla \cdot (\rho \nabla \Psi) + \nabla \cdot (\rho \nabla(\rho * \Phi)), \tag{1.44}$$

where $\rho * \Phi(x) = \int_{R^d} \Phi(x-y)\rho(dy)$. We consider the corresponding large deviation principle.

When $\Phi(z) = \theta|z|^2/2$, $\theta > 0$, (1.42) is known as the ferromagnetic Curie-Weiss model

$$dX_{n,k}(t) = -\nabla\Psi(X_{n,k}(t)) - \frac{\theta}{n}\sum_{j=1}^{n}(X_{n,k}(t) - X_{n,j}(t))dt + dW_k(dt). \tag{1.45}$$

The large deviation problem for this model, as well as for a larger class of models, has been considered by Dawson and Gärtner in a series of publications. See [**23**] and [**22**] and the references therein. In connection with gas kinetics in statistical mechanics, the system (1.42) with a general semi-convex Φ gives a microscopic statistical foundation for certain deterministic models of the evolution of a spatially homogeneous gas in a granular media. In the infinite system limit, $n \to \infty$, the law of large numbers gives a nonlinear partial differential equation modeling the evolution. See Section 9.6 of Villani [**127**] for a discussion of these models.

Our methods applied to this problem seem more involved than those of Dawson and Gärtner [**23**]; however, our interest is not only in the large deviation problem, but also in the well-posedness of an associated nonlinear equation (1.49) it induces. This nonlinear equation is a special case of a Hamilton-Jacobi equation in the space of probability measures and is closely related to the Hamilton-Jacobi equations in Banach spaces studied by Crandall and Lions [**21, 20**] and Tataru [**117**]. The previous work, however, requires the state space to be a subset of a Banach space satisfying the Radon-Nikodym property. We note that even though the space of probability measures with Lebesgue density is a bounded subset of $L^1(R^d)$ (i.e. $\int_{R^d}\rho(x)dx = 1$), $L^1(R^d)$ does not satisfy the Radon-Nikodym property. (See, for example, [**31**], page 31.)

Assume that $\Phi, \Psi \in C^2(R^d)$ and that $|\nabla\Phi(z)|$ has sub-quadratic growth as $|z| \to \infty$. More conditions on Φ, Ψ will be imposed later. To simplify, we only consider the case where Φ is an even function (i.e. $\Phi(x) = \Phi(-x)$).

Let d be the (order) 2-Wasserstein metric on E (i.e. (D.14) with $p = 2$). Then (E, d) is a complete separable metric space and $\rho_n \to \rho_0$ in (E, d) if and only if $\rho_n \Rightarrow \rho_0$, in the sense of weak convergence of probability measures, and $\int_{R^d}|x|^2 d\rho_n \to \int_{R^d}|x|^2 d\rho_0$ (Lemma D.17). Define

$$E_n \equiv \{\rho(dx) \equiv \frac{1}{n}\sum_{k=1}^{n}\delta_{x_k}(dx), x_k \in R^d, k = 1, 2, \ldots\} \subset E.$$

Let r_n be the restriction of d to E_n. For fixed n, the corresponding topology is just the topology of weak convergence.

For each n, $\rho_n(\cdot)$ in (1.43) is a probability-measure-valued Markov process. We calculate its generator next. Define

$$B(\rho)p(x) = \frac{1}{2}\Delta p(x) - \nabla(\Psi + \Phi * \rho)(x)\cdot\nabla p(x), \quad \forall p \in C^2(R^d), \rho \in \mathcal{P}(R^d).$$

Let $p_1, \ldots, p_l \in C_c^2(R^d)$, and define

$$\mathbf{p} = (p_1, \ldots, p_l), \qquad \langle \mathbf{p}, \rho\rangle = (\int_{R^d} p_1 d\rho, \ldots, \int_{R^d} p_l d\rho).$$

Then for

$$f(\rho) = \varphi(\langle \mathbf{p}, \rho\rangle), \qquad \varphi \in C^2(R^l), \tag{1.46}$$

$$A_n f(\rho) = \sum_{i=1}^{l} \partial_i \varphi(\langle \mathbf{p}, \rho \rangle) \langle \rho, B(\rho) p_i \rangle + \frac{1}{2n} \sum_{i,j=1}^{l} \partial_{ij}^2 \varphi(\langle \mathbf{p}, \rho \rangle) \langle \rho, (\nabla p_i)^T \nabla p_j \rangle$$

Therefore

$$(1.47) \quad \begin{aligned} H_n f(\rho) &\equiv \frac{1}{n} e^{-nf} A_n e^{nf}(\rho) \\ &= \sum_{i=1}^{l} \partial_i \varphi(\langle \mathbf{p}, \rho \rangle) \langle \rho, B(\rho) p_i \rangle \\ &\quad + \frac{1}{2} \sum_{i,j=1}^{l} \partial_i \varphi(\langle \mathbf{p}, \rho \rangle) \partial_j \varphi(\langle \mathbf{p}, \rho \rangle) \langle \rho, (\nabla p_i)^T \nabla p_j \rangle \\ &\quad + \frac{1}{2n} \sum_{i,j=1}^{l} \partial_{ij}^2 \varphi(\langle \mathbf{p}, \rho \rangle) \langle \rho, (\nabla p_i)^T \nabla p_j \rangle. \end{aligned}$$

Setting

$$\frac{\delta f}{\delta \rho} = \sum_{i=1}^{l} \partial_i \varphi(\langle \mathbf{p}, \rho \rangle) p_i,$$

for $\rho_n \in E_n, \rho \in E$ satisfying $d(\rho_n, \rho) \to 0$,

$$\lim_{n \to \infty} H_n f(\rho_n) = H f(\rho) = H(\rho, \frac{\delta f}{\delta \rho}),$$

where

$$(1.48) \qquad H(\rho, p) = \langle \rho, B(\rho) p \rangle + \frac{1}{2} \int |\nabla p|^2 d\rho = \langle B^*(\rho)\rho, p \rangle + \frac{1}{2} \int |\nabla p|^2 d\rho,$$

and for each $\gamma \in E$, $B^*(\gamma) : \rho \in \mathcal{P}(R^d) \to \mathcal{D}'(R^d)$ is defined by

$$B^*(\gamma)\rho = \frac{1}{2}\Delta\rho + \nabla \cdot (\rho \nabla(\Psi + \gamma * \Phi)).$$

Large deviation results for this example are discussed in Chapter 13 (Theorem 13.37). The key here is to establish a uniqueness result for weak solutions of

$$(1.49) \qquad (I - \alpha H) f = h, \quad \alpha > 0,$$

for sufficiently many $h \in C_b(E)$. This is achieved in two steps in Example 9.35 and in Section 13.3.4. See Theorem 13.32.

1.5. An outline of the major obstacles

From the discussion in the previous sections, we see that one of our main assumptions should be the convergence of $\{H_n\}$ to a limit operator H in the sense that for each $f \in \mathcal{D}(H)$, there exist $f_n \in \mathcal{D}(H_n)$ such that

$$(1.50) \qquad f_n \to f, \quad H_n f_n \to H f,$$

where the type of convergence may depend on the particular problem. To rigorously formulate a large deviation program using generator convergence and nonlinear semigroup theory requires some work, but it is straightforward (Chapter 5) provided we assume that the limit operator H satisfies a *range condition*. Specifically, we need existence of solutions of

$$(I - \alpha H)f = h, \tag{1.51}$$

for all sufficiently small $\alpha > 0$ and a large enough collection of functions h. Unfortunately, there are very few examples for which this condition can actually be verified in the classical sense, that is, with solutions f satisfying the differentiability requirements assumed in the examples above. We overcome this difficulty by using the weak (viscosity) solution theory introduced by Crandall and Lions [**19**]. The range condition is replaced by the requirement that a *comparison principle* (Definitions 6.4 and 7.3) holds for (1.51). Assuming that

$$(I - \alpha H_n)f_n = h_n, \tag{1.52}$$

for a sequence h_n converging to h, we adapt a technique of Barles and Perthame [**7**] [**8**] to prove that the comparison principle implies the convergence of f_n. This technique does not require *a priori* estimates on the regularity of f.

There is a well developed, powerful theory for verifying the comparison principle, at least when the state space is a subset of R^d or a compact metric space (Chapter 6). Indeed, the approach has been applied in ad hoc ways to various large deviation examples (e.g. [**48**], [**40**]) for quite some time. When the state space is not compact, however, major obstacles arise. First, we need a good definition of viscosity solution in this context. The definition should be weak enough to allow extension of the Barles-Perthame arguments to noncompact metric spaces, yet strong enough so that the comparison principle can be proved for interesting examples. Our choice here is a definition (Definition 7.1) that preserves a nonlinear version of the maximum principle for H. This definition is different from existing definitions [**21**], [**117**], [**20**] in the literature on partial differential equations, where structural information about the equation is usually built into the definition. Second, the Barles and Perthame limiting procedure may not always work when E is not compact. This procedure is compatible with a variant of uniform convergence over compacts, but without other information, the usual estimates needed to derive convergence of solutions of (1.52) require the convergence in (1.50) to be *uniform* over the whole space. In general, uniformity will not hold for a sufficiently large collection of f. The uniformity requirement can be relaxed if we can verify the *exponential compact containment condition* (Condition 2.8). In applications, this condition can frequently be verified by a stochastic Lyapunov function technique (Section 4.6).

The general versions of our large deviation theorems are given in Chapter 7. A short version is summarized in Chapter 2. Further generalizations of these results are also given. For instance, we discuss situations where certain functionals of the Markov processes satisfy a large deviation principle, while the full processes may not. We also discuss the use of a general notion of convergence of test functions and operators, to handle processes with multiple scales.

In most of the examples, the most difficult technical argument comes in verifying the *comparison principle* (Definitions 6.4 and 7.3). Verification is an analytic issue and often gives the impression of being rather involved and disconnected from the probabilistic large deviation problems. This disconnect is not always the case. Specific probabilistic structures can give insight into the solution of the analytic problem. For instance, the large deviation theory for the interacting particle system in Example 1.14 and stochastic equations in Examples 1.12 and 1.13) leads to Hamilton-Jacobi operators, (1.48), (1.32), (1.41), that describe the evolution of optimally controlled partial differential equations. The corresponding comparison theories in the literature ([**21**], [**117**], [**20**]) are technical and limited. Moreover, these theories do not apply for Example 1.14. We solve the problem by devising simple, new comparison techniques exploiting the special structure in these problems. As a result, we not only arrive at large deviation principles, but also obtain simpler proofs of analytic viscosity solution results.

CHAPTER 2

An overview

The purpose of this chapter is to provide a rough "road map" for reading the general theory that follows. We collect versions of the main results regarding large deviations for Markov processes and describe some motivation for our approach. Proofs of these results will be given in later chapters. The results collected here may not be the sharpest or the fullest versions in the paper. Indeed, for most of the results, further generalizations are given later. For example, the majority of the material in Chapters 3 and 4 applies to general processes which may be non-Markovian; Chapter 5 contains a pure semigroup formulation (as opposed to the viscosity solution approach given here) of the large deviation results; Chapters 6 and 7 have a number of new results regarding viscosity solutions and their convergence, many of which may have application in areas besides large deviations.

A typical application requires the following steps:

(1) *Verify convergence of the sequence of operators H_n and derive the limit operator H.* Many examples of this convergence are given in the Introduction. In general, convergence will be in the extended limit or graph sense. (See Definition A.12.) In some examples, the limit is described in terms of a pair of operators, $(H_\dagger, H_\ddagger)$, where, roughly, $H_\dagger$ is the $\limsup$ of $\{H_n\}$ and $H_\ddagger$ is the $\liminf$.

(2) *Verify exponential tightness.* The convergence of H_n (in fact, boundedness of $\{H_n f_n\}$ for an appropriate collection of sequences $\{f_n\}$) typically gives exponential tightness, provided one can verify the exponential compact containment condition (4.10). Lyapunov function arguments can be used to verify this condition (Section 4.6). Alternatively, one can avoid verifying the compact containment condition by compactifying the state space and verifying the large deviation principle in the compactified space. If the rate function is infinite for every path that hits a point added in the compactification, then the large deviation principle holds for the original space (Theorem 4.11).

(3) *Verify the comparison principle for the limiting operator H* (or the pair $(H_\dagger, H_\ddagger)$). The comparison principle asserts a strong form of uniqueness for the equation $(I - \alpha H)f = h$. If the comparison principle holds for a sufficiently large class of functions h, then one can conclude that the nonlinear semigroups $\{V_n\}$ converge. Convergence of the semigroups implies the large deviation principle for the finite dimensional distributions of the sequence of processes, which by exponential tightness, then gives the large deviation principle in $D_E[0, \infty)$. The rate function is characterized by the limiting semigroup. Chapter 9 discusses concrete techniques for verifying the comparison principle in a number of situations. These techniques are general, yet practical enough to cover every example in the Introduction.

(4) *Construct a variational representation for H.* Typically we can identify the limiting semigroup as the Nisio semigroup for a control problem. The control problem then gives an alternative and more explicit representation of the rate function (Chapter 8).

In Chapters 10 to 13, we apply these methods to the examples in the Introduction.

2.1. Basic setup

Let (E_n, r_n), $n = 1, 2, \ldots$ and (E, r) be complete separable metric spaces.

a) (Continuous time Markov processes) Let $A_n \subset B(E_n) \times B(E_n)$, and assume the following: For each $n = 1, 2, \ldots$, existence and uniqueness hold for the $D_{E_n}[0,\infty)$-martingale problem for (A_n, μ) for each initial distribution $\mu \in \mathcal{P}(E_n)$. Letting $P^n_y \in \mathcal{P}(D_{E_n}[0,\infty))$ denote the distribution of the solution of the martingale problem for A_n starting from $y \in E_n$, the mapping $y \mapsto P^n_y$ is Borel measurable taking the weak topology on $\mathcal{P}(D_{E_n}[0,\infty))$. For each n, Y_n is a solution of the martingale problem for A_n.

b) (Discrete time processes) Let $\{\widetilde{Y}_n(k), k = 0, 1, \cdots\}$ be a time homogeneous Markov chain with state space E_n and transition operator T_n on $B(E_n)$:

$$T_n f(y) = E[f(\widetilde{Y}_n(k+1))|\widetilde{Y}_n(k) = y], \quad f \in B(E_n).$$

Let $\epsilon_n > 0$, and define

$$Y_n(t) \equiv \widetilde{Y}_n([\frac{t}{\epsilon_n}]).$$

Then

$$E[f(Y_n(t))|Y_n(0) = y] = T_n^{[t/\epsilon_n]} f(y), \quad f \in B(E_n).$$

In either case, we suppose $\eta_n : E_n \to E$ is Borel measurable and assume $X_n \equiv \eta_n(Y_n) \in D_E[0,\infty)$. We are interested in establishing the large deviation principle for X_n. For $f \in M(E)$, we set $\eta_n f = f \circ \eta_n$.

REMARK 2.1. In many applications, it suffices to take $E_n = E$ and $\eta_n(y) = y$. More generally, however, E_n may be a discrete set that is asymptotically dense in E or a higher dimensional space in which the large deviation principle is only verified for certain components.

2.2. Compact state spaces

We first assume that E is compact which simplifies a number of technical issues. This assumption is not as restrictive as it may first appear. Many results in R^d can be obtained by first verifying the large deviation principle in the one-point compactification $E = R^d \cup \{\infty\}$.

The notion of a viscosity solution for a nonlinear equation is central to the results that follow.

DEFINITION 2.2 (Viscosity Solution). Let E be a compact metric space, and $H \subset C(E) \times B(E)$. Fix $h \in C(E)$ and $\alpha > 0$. Let $f \in B(E)$ and define $g = \alpha^{-1}(f - h)$, that is, $f - \alpha g = h$. Then

a) f is a *viscosity subsolution* of $(I-\alpha H)f = h$ if and only if f is upper semicontinuous and for each $(f_0, g_0) \in H$ such that $\sup_x (f - f_0)(x) = \|f - f_0\|$, there exists $x_0 \in E$ satisfying

$$(f - f_0)(x_0) = \|f - f_0\| \tag{2.1}$$

and

$$\alpha^{-1}(f(x_0) - h(x_0)) = g(x_0) \le (g_0)^*(x_0). \tag{2.2}$$

b) f is a *viscosity supersolution* of $(I-\alpha H)f = h$ if and only if f is lower semicontinuous and for each $(f_0, g_0) \in H$ such that $\sup_x (f_0 - f)(x) = \|f_0 - f\|$, there exists $x_0 \in E$ satisfying

$$(f_0 - f)(x_0) = \|f - f_0\| \tag{2.3}$$

and

$$\alpha^{-1}(f(x_0) - h(x_0)) = g(x_0) \ge (g_0)_*(x_0). \tag{2.4}$$

The *comparison principle* holds for h if for each subsolution $\overline{f}$ and each supersolution $\underline{f}$, $\overline{f} \le \underline{f}$.

THEOREM 2.3. *Suppose E is compact.*

Step 1: *(Convergence of H_n) In the continuous time case, let*

$$H_n f \equiv \frac{1}{n} e^{-nf} A_n e^{nf}, \quad e^{nf} \in \mathcal{D}(A_n).$$

In the discrete time case, let

$$H_n f \equiv \frac{1}{n\epsilon_n} \log e^{-nf} T_n e^{nf} = \frac{1}{n\epsilon_n} \log(1 + e^{-nf}(T_n - I)e^{nf}).$$

Suppose $H \subset C(E) \times B(E)$ with $\mathcal{D}(H)$ dense in $C(E)$, and

$$H \subset ex\text{-}\lim_{n\to\infty} H_n$$

in the sense of Definition A.12. That is, for each $(f, g) \in H$, there exists $(f_n, g_n) \in H_n$ such that

$$\lim_{n\to\infty} \|\eta_n f - f_n\| + \|\eta_n g - g_n\| = 0.$$

Step 2: *(Exponential tightness) Under the convergence assumptions on $\{H_n\}$, exponential tightness of $\{X_n\}$ holds.*

Step 3: *(The comparison principle) Let $\alpha_0 > 0$, and assume that for each $0 < \alpha < \alpha_0$, there exists a subset $D_\alpha \subset C(E)$ such that D_α is dense in $C(E)$ and that, for each $h \in D_\alpha$, the comparison principle (Definition 6.4) holds for viscosity sub and super solutions of*

$$(I - \alpha H)f = h. \tag{2.5}$$

Then a viscosity solution $R_\alpha h$ exists for each $h \in D_\alpha$, and R_α extends continuously to all of $C(E)$.

In addition, suppose that $\{X_n(0)\}$ satisfies a large deviation principle with a good rate function I_0.

Then

a) *$\{X_n\}$ satisfies the large deviation principle with a good rate function I.*

b) *The limit*

$$V(t)h \equiv \lim_{m\to\infty} R^m_{t/m} h$$

exists and defines a strongly continuous operator semigroup on $C(E)$, *and*

$$(2.6)\quad I(x) = \sup_{\{t_1,\cdots,t_k\}\subset\Delta_x^c} \sup_{f_1,\cdots,f_k\in C_b(E)} \Big(I_0(x(0)) + \sum_{i=1}^k f_i(x(t_i)) - V(t_i - t_{i-1})f_i(x(t_{i-1}))\Big),$$

for $x \in D_E[0,\infty)$, *where* Δ_x *denotes the set of discontinuities of* x.

This theorem is proved in Chapter 6.

Consider the classical Freidlin and Wentzell diffusion problem (Example 1.4). Then

$$A_n g(x) = \frac{1}{2n}\sum_{ij} a_{ij}(x)\partial_i\partial_j g(x) + \sum_i b_i(x)\partial_i g(x),$$

and

$$H_n f(x) = \frac{1}{2n}\sum_{ij} a_{ij}(x)\partial_i\partial_j f(x) + \frac{1}{2}\sum_{ij} a_{ij}(x)\partial_i f(x)\partial_j f(x) + \sum_i b_i(x)\partial_i f(x).$$

Assuming a and b are bounded on bounded subsets, the convergence

$$Hf = \lim_{n\to\infty} H_n f$$

for

$$Hf(x) = \frac{1}{2}(\nabla f(x))^T \cdot a(x)\cdot\nabla f(x) + b(x)\cdot\nabla f(x)$$

is immediate for all $f \in D_d = \{f \in C(E) : f|_{R^d} - f(\infty) \in C_c^2(R^d)\}$. (Taking $E = R^d \cup \{\infty\}$, $H_n f(\infty) = Hf(\infty) = 0$.) Exponential tightness then holds in $D_E[0,\infty)$. Assuming the large deviation principle holds for $\{X_n(0)\}$, the conditions of Theorem 2.3 are then satisfied for this example provided we can verify the comparison principle. Note that H can be written $Hf(x) = H(x,\nabla f(x))$. Chapter 9 gives a detailed discussion of conditions under which the comparison principle holds for H of this form. In particular, if a and b are bounded and continuous and $a(x)$ is positive definite for all x, then Lemmas 9.15 and 9.16 give the desired result.

The conclusion of Theorem 2.3 is less than satisfactory, since $\{V(t)\}$ is only determined implicitly and the rate function is expressed in terms of $\{V(t)\}$. For our example, however, we can write

$$Hf(x) = \sup_u (u\cdot\sigma^T(x)\nabla f(x) + b(x)\nabla f(x) - |u|^2/2),$$

where $\sigma\sigma^T = a$. This representation suggests that V is the Nisio semigroup for a control problem. In particular, we should have

$$V(t)f(x_0) = \sup(f(x(t)) - \int_0^t \frac{1}{2}|u(s)|^2 ds),$$

where the supremum is over x and u satisfying

$$(2.7)\qquad x(t) = x(0) + \int_0^t \sigma(x(s))u(s)ds + \int_0^t b(x(s))ds, \quad x(0) = x_0.$$

It then follows, at least under boundedness and continuity assumptions on σ and b, that

$$I(x) = \inf\{I_0(x(0)) + \int_0^t \frac{1}{2}|u(s)|^2 ds\},$$

where the infimum is over all u such that (x, u) satisfies (2.7).

More generally, we look for a linear operator $A \subset C(E) \times C(E \times U)$, where U is another complete, separable metric space, and a lower semicontinuous $L : E \times U \to [0, \infty)$ such that

$$Hf(x) = \sup_{u \in U}(Af(x, u) - L(x, u)).$$

The corresponding control problem may use relaxed controls, that is measures λ on $U \times [0, \infty)$ where $\lambda(U \times [0, t]) = t$. The control problem requires

$$f(x(t)) = f(x(0)) + \int_{U \times [0,t]} Af(x(s), u)\lambda(du \times ds), \tag{2.8}$$

for all $f \in \mathcal{D}(A)$, and V is given by

$$V(t)f(x_0) = \sup_{(x,\lambda)} (f(x(t)) - \int_{U \times [0,t]} L(x(s), u)\lambda(du \times ds),$$

where the supremum is over (x, λ) satisfying (2.8) with $x(0) = x_0$. Some additional constraint on λ is also possible. The rate function then satisfies

$$I(x) = \inf\{I_0(x(0)) + \int_{U \times [0,t]} L(x(s), u)\lambda(du \times ds)\}.$$

See Theorem 8.14.

2.3. General state spaces

We now allow E to be an arbitrary complete, separable metric space. In this more general setting, it is often difficult to verify uniform convergence of $H_n f_n$ and, consequently, weaker notions of convergence are useful. One such notion is convergence of bounded sequences of functions, uniformly on compact sets, which we refer to as *buc*-convergence (see Definition A.6) and denote *buc*-lim.

With reference to Examples 1.8 - 1.11, we also see that the natural limit of H_n may have a range in functions defined on a larger space than E (call it E'). Consequently, we have mappings $\eta_n : E_n \mapsto E$, $\eta'_n : E_n \mapsto E'$, and $\gamma : E' \mapsto E$ which are consistent in the sense that $\eta_n = \gamma \circ \eta'_n$. We require that η_n, η'_n are Borel measurable and γ is continuous and onto (i.e. $E = \gamma(E')$). We also work with a notion of convergence which is essentially *buc*-convergence for our sequence of spaces $\{E_n\}$.

Recall the definitions of set convergence.

DEFINITION 2.4. Define

$$\liminf_{n \to \infty} G_n = \{x \in E : \exists x_n \in G_n, \ni \lim_{n \to \infty} x_n = x\},$$

and

$$\limsup_{n \to \infty} G_n = \{x \in E : \exists n_1 < n_2 < \cdots \text{ and } x_{n_k} \in G_{n_k} \ni \lim_{k \to \infty} x_{n_k} = x\}.$$

If $G \equiv \limsup_n G_n = \liminf_n G_n$, then we say G_n converges to G and write $G = \lim_{n \to \infty} G_n$.

The following notion of convergence will be central to our main results.

DEFINITION 2.5. [LIM convergence induced by an index set $\mathcal{Q}$] Let $\mathcal{Q}$ be an index set. For $n = 1, 2, \ldots$ and $q \in \mathcal{Q}$, let $K_n^q \subset E_n$ satisfy the following:

(1) For $q_1, q_2 \in \mathcal{Q}$, there exists $q_3 \in \mathcal{Q}$ such that $K_n^{q_1} \cup K_n^{q_2} \subset K_n^{q_3}$.
(2) For each $x \in E$, there exists $q \in \mathcal{Q}$ such that $x \in \liminf_{n\to\infty} \eta_n(K_n^q)$.
(3) For each $q \in \mathcal{Q}$, there exist a compact $\widehat{K}^q \subset E$

$$\lim_{n\to\infty} \sup_{y\in K_n^q} \inf_{x\in\widehat{K}^q} r(\eta_n(y), x) = 0,$$

and a compact $\widehat{K}'^q \subset E'$ such that

$$\lim_{n\to\infty} \sup_{y\in K_n^q} \inf_{x\in\widehat{K}'^q} r'(\eta_n'(y), x) = 0. \tag{2.9}$$

(4) For each compact $K \subset E$, there exists $q \in \mathcal{Q}$, such that $K \subset K^q \equiv \liminf_{n\to\infty} \eta_n(K_n^q)$.

For $f_n \in B(E_n)$ and $f \in B(E)$, define $f = \mathrm{LIM} f_n$ if and only if $\sup_n \|f_n\| < \infty$ and

$$\lim_{n\to\infty} \sup_{y\in K_n^q} |f_n(y) - \eta_n f(y)| = 0,$$

where $\eta_n f = f \circ \eta_n$.

If the requirements in the above definition hold, the compact sets $\widehat{K}^q$ and $\widehat{K}'^q$ can always be chosen to be $\widehat{K}^q = \limsup_{n\to\infty} \eta_n(K_n^q)$ and $\widehat{K}'^q = \limsup_{n\to\infty} \eta_n'(K_n^q)$. Note that the definition of LIM given here is a special case of the abstract definition given in Section A.2. (See Condition A.13.) We give two useful examples of LIM.

EXAMPLE 2.6. If $E_n = E$, $\mathcal{Q}$ is the class of compact sets in E, and $K_n^K = K$ for every $n = 1, 2, \ldots$ and $K \in \mathcal{Q}$, then LIM is just *buc*-convergence.

EXAMPLE 2.7. With reference to Examples 1.12 and 1.13, let η_n be given by (1.31). For $m = m(n)$, let $E = E' = L^2(\mathcal{O})$, and let $\mathcal{Q}$ be the class of compact sets in E. For each $\rho \in E$, define its projection into $E_n = R^{|\Lambda_m|}$ by

(2.10)

$$(\pi_n\rho)(x) = m^d \int_{y_1=\frac{i_1}{m}}^{\frac{i_1+1}{m}} \cdots \int_{y_d=\frac{i_d}{m}}^{\frac{i_d+1}{m}} \rho(y_1, \ldots, y_d) dy_d \ldots dy_1, \quad x = (\frac{i_1}{m}, \ldots, \frac{i_d}{m}) \in \Lambda_m.$$

Then $\eta_n(\pi_n(\rho)) \in E$, and assuming $m(n) \to \infty$, for every compact set $K \subset L^2(\mathcal{O})$,

$$\lim_{n\to\infty} \sup_{\rho\in K} \|\eta_n(\pi_n(\rho)) - \rho\|_{L^2(\mathcal{O})} = 0. \tag{2.11}$$

If, for each $K \in \mathcal{Q}$, we define

$$K_n^K = \pi_n(K) \equiv \{\pi_n(\rho) : \rho \in K \subset E\} \subset E_n = R^{|\Lambda_{m(n)}|},$$

then all the requirements for $\{K_n^K : K \in \mathcal{Q}\}$ in Definition 2.5 are satisfied. In particular, for each $K \in \mathcal{Q}$, by (2.11) and the above definition of K_n^K,

$$\lim_{n\to\infty} \sup_{\gamma_n\in K_n^K} \inf_{\gamma\in K} \|\eta_n(\gamma_n) - \gamma\|_{L^2(\mathcal{O})} \le \lim_{n\to\infty} \sup_{\rho\in K} \|\eta_n(\pi_n(\rho)) - \rho\|_{L^2(\mathcal{O})} = 0.$$

Therefore, we can choose $\widehat{K}^K = K$ to satisfy the third requirement in Definition 2.5. Finally, since

$$\sup_{\gamma\in K_n^K} |f_n(\gamma) - \eta_n f(\gamma)| = \sup_{\rho\in K} |f_n(\pi_n(\rho)) - f(\eta_n(\pi_n(\rho)))|,$$

if $f \in C(E)$, $f = \text{LIM} f_n$ is implied by the following two conditions:

(1) $\sup_n \|f_n\| < \infty$,

(2) for $\gamma_n \in R^{|\Lambda_{m(n)}|}$ satisfying $\eta_n(\gamma_n) \to \rho$ in $\|\cdot\|_{L^2(\mathcal{O})}$, $\lim_{n\to\infty} f_n(\gamma_n) = f(\rho)$.

Dropping compactness and generalizing the notion of convergence requires extra work and part of the extra work comes in verifying the following condition.

CONDITION 2.8. *For each $q \in \mathcal{Q}$, $T > 0$, and $a > 0$, there exists $\widehat{q}(q,a,T) \in \mathcal{Q}$ such that*

(2.12)

$$\limsup_{n\to\infty} \sup_{y\in K_n^q} \frac{1}{n} \log P\{Y_n(t) \notin K_n^{\widehat{q}(q,a,T)}, \text{ for some } 0 \le t \le T | Y_n(0) = y\} \le -a.$$

The Lyapunov function techniques discussed in Section 4.6 are frequently useful for verifying this condition.

EXAMPLE 2.9. We illustrate the Lyapunov function approach for Example 1.12. Definitions and properties of the projection $\pi_m : E \to R^{|\Lambda_m|} \equiv E_n$ and of LIM convergence are discussed in Example 2.7. In this context, Condition 2.8 becomes: for each compact $K_0 \subset L^2(\mathcal{O})$, $T > 0$, and $a > 0$, there exists another compact $K_1 \subset L^2(\mathcal{O})$ such that

(2.13)

$$\limsup_{n\to\infty} \sup_{\rho_0\in\pi_m(K_0)} \frac{1}{n} \log P(\exists t, 0 \le t \le T, \ni Y_n(t) \notin \pi_m(K_1) | Y_n(0) = \rho_0) \le -a.$$

Example 1.12 is motivated by a physical context. We can associate with the model (1.29) a *free energy functional* of the form

$$\mathcal{E}_m(\rho) \equiv \frac{1}{2} \sum_{x\in\Lambda_m} |\nabla_m \rho(x)|^2 m^{-d} + \sum_{x\in\Lambda_m} F(\rho(x)) m^{-d}, \qquad \rho \in E_n \equiv R^{|\Lambda_m|}, \tag{2.14}$$

where $\nabla_m g(x)$ is the vector with components

$$\nabla_m^k \rho(x) = \frac{m}{2}(\rho(x + m^{-1}e_k) - \rho(x - m^{-1}e_k)).$$

$\mathcal{E}_m$ is the desired Lyapunov function.

Define

$$\begin{aligned} H_n\mathcal{E}_m(\rho) &\equiv \frac{1}{n} e^{-n\mathcal{E}_m} A_n e^{n\mathcal{E}_m}(\rho) \\ &= \langle \Delta_m\rho - F'(\rho), -\Delta_m\rho + F'(\rho)\rangle_m + \frac{1}{2}\|\Delta_m\rho - F'(\rho)\|^2_{L^2(\Lambda_m)} \\ &\quad + \frac{m^d}{2n} \sum_{x\in\Lambda_m} (\frac{d}{2} m^{2-d} + F''(\rho(x)) m^{-d}) \\ &= -\frac{1}{2}\|\Delta_m\rho - F'(\rho)\|^2_{L^2(\Lambda_m)} + \frac{1}{2n}\Big(\frac{1}{2} d m^{2+d} + m^d \sum_{x\in\Lambda_m} F''(\rho(x)) m^{-d}\Big). \end{aligned} \tag{2.15}$$

Then by Itô's formula,

$$\exp\{n\mathcal{E}_m(Y_n(t)) - n\mathcal{E}_m(Y_n(0)) - \int_0^t nH_n\mathcal{E}_m(Y_n(s))ds\}$$

is a positive local martingale (hence a supermartingale), and we can apply Lemma 4.20 to obtain (2.13).

If $m^{2+d} = O(n)$, then

$$\sup_n \sup_{\rho \in R^{|\Lambda_m|}} H\mathcal{E}_m(\rho) < \infty. \tag{2.16}$$

Let $C > C_0 > 0$ and $\rho_0 \in R^{|\Lambda_m|}$ be such that $\mathcal{E}_m(\rho_0) \leq C_0 < \infty$. By the finiteness of Λ_m, $\{\rho \in R^{|\Lambda_m|} : \mathcal{E}_m(\rho) < C\}$ is open in E_n. By Lemma 4.20,

$$P(\exists t \in [0,T], \mathcal{E}_m(Y_n(t)) \geq C | Y_n(0) = \rho_0) \leq e^{-n(C-C_0)+nT \sup_n \sup_{\rho \in L^2(\Lambda_m)} H_n \mathcal{E}_m(\rho)}.$$

If

$$F(r) \geq c_1 + c_2 r^2 \text{ for some } c_1 \in R, c_2 > 0, \tag{2.17}$$

then

$$K_C \equiv \text{cl}\,(\cup_n \{\eta_n(\rho) : \rho \in E_n, \mathcal{E}_m(\rho) \leq C < \infty\})$$

is a compact subset of $E = L^2(\mathcal{O})$. Therefore, for every $a, T, C_0 > 0$, by selecting C large enough, we can find a compact set $K_{a,T,C_0} \subset L^2(\mathcal{O})$ such that

$$\sup_{\{\rho \in R^{|\Lambda_m|} : \mathcal{E}_m(\rho) \leq C_0\}} P(\exists t, 0 \leq t \leq T, \eta_n(Y_n(t)) \notin K_{a,T,C_0}, | Y_n(0) = \rho) \leq e^{-na}. \tag{2.18}$$

Next, we extend (2.18) to (2.13) by relaxing the requirements on the initial data. For this purpose, we need to make use of a stability result regarding equation (1.29) which is proved in Lemma 13.4: For $T > 0$, there exists $C_T > 0$ such that if Y_n and Z_n are solutions of (1.29),

$$\sup_{0 \leq t \leq T} \|Y_n(t) - Z_n(t)\|_{L^2(\Lambda_m)} \leq C_T \|Y_n(0) - Z_n(0)\|_{L^2(\Lambda_m)} \quad a.s. \tag{2.19}$$

Noting that the collection of $\rho \in L^2(\mathcal{O})$ such that

$$\sup_n \mathcal{E}_m(\pi_m(\rho)) < \infty \tag{2.20}$$

is dense in $L^2(\mathcal{O})$, for compact $K_0 \subset L^2(\mathcal{O})$ and $\epsilon > 0$, there exist $\rho_0^{(1)}, \ldots, \rho_0^{(N)}$ such that $K_0 \subset \cup_{k=1}^N B(\rho_0^{(k)}, C_1^{-1}\epsilon)$ and

$$C_0^\epsilon \equiv \max_{1 \leq k \leq N} \sup_m \mathcal{E}_m(\pi_m(\rho_0^{(k)})) < \infty. \tag{2.21}$$

Therefore, by (2.18) and (2.19), there exists a compact K_{a,T,C_0^ϵ} such that

$$\begin{aligned}
&\sup_{\{\rho \in \pi_m(K_0)\}} P(\exists t, 0 \leq t \leq T, \eta_n(Y_n(t)) \notin K^\epsilon_{a,T,C_0^\epsilon}, | Y_n(0) = \rho) \\
&\leq \sup_{\{\rho \in R^{|\Lambda_m|} : \mathcal{E}_m(\rho) \leq C_0^\epsilon\}} P(\exists t, 0 \leq t \leq T, \eta_n(Y_n(t)) \notin K_{a,T,C_0}, | Y_n(0) = \rho) \\
&\leq e^{-na},
\end{aligned}$$

where

$$K^\epsilon_{a,T,C_0^\epsilon} = \{\rho \in E : \inf_{\gamma \in K_{a,T,C_0^\epsilon}} \|\rho - \gamma\|_{L^2(\mathcal{O})} < \epsilon\}.$$

For a sequence $\epsilon_l \to 0$, let

$$K_1 = \text{cl} \cap_l K^{\epsilon_l}_{a+l,T,C_0^{\epsilon_l}}.$$

Then K_1 is compact, and since $\rho = \pi_m(\eta_n(\rho))$ for every $\rho \in E_n$, $\{\eta_n(Y_n(t)) \in K_1\} \subset \{Y_n(t) \in \pi_m(K_1)\}$ and (2.13) holds.

For most φ, (1.32) will be unbounded and the convergence of $H_n f_n$ will not be uniform over E_n. Examples of this type lead us to introduce a pair of limiting operators: $H_\dagger$, corresponding (roughly) to the limit when $\{H_n f_n\}$ is bounded above and $H_\ddagger$ corresponding to $\{H_n f_n\}$ being bounded below. (See the Convergence Condition 7.11.)

We have the following counterpart of Theorem 2.3. Note that since we are no longer assured that the extrema in Definition 2.2 are achieved, more general definitions of viscosity solution and the comparison principle are also required. (See Definitions 7.1 and 7.3.)

THEOREM 2.10. *Let (E_n, r_n), (E', r'), (E, r), η_n, η'_n, and γ be as above, and let* LIM *be given by Definition 2.5.*

Step 1: *(Convergence of H_n) In the continuous time case, let*

$$H_n f \equiv \frac{1}{n} e^{-nf} A_n e^{nf}, \quad e^{nf} \in \mathcal{D}(A_n).$$

In the discrete time case, let

$$H_n f \equiv \frac{1}{n\epsilon_n} \log e^{-nf} T_n e^{nf} = \frac{1}{n\epsilon_n} \log(1 + e^{-nf}(T_n - I)e^{nf}).$$

Let $H_\dagger \subset C^l(E, \overline{R}) \times M^u(E', \overline{R})$ and $H_\ddagger \subset C^u(E, \overline{R}) \times M^l(E', \overline{R})$, and assume that $\{H_n\}$ satisfies the Convergence Condition 7.11.

Step 2: *(Exponential tightness) Suppose $\mathcal{D}(H_\dagger) \cap \mathcal{D}(H_\ddagger) \cap C_b(E)$ is* buc-*dense in $C_b(E)$ and Condition 2.8 is satisfied. Then $\{X_n\}$ is exponentially tight.*

Step 3: *(The comparison principle) Let $\alpha_0 > 0$, and assume that for each $0 < \alpha < \alpha_0$, there exists a subset $D_\alpha \subset C_b(E)$ such that $C_b(E)$ is the buc-closure (Definition A.6) of D_α and that for each $h \in D_\alpha$, the comparison principle holds for viscosity subsolutions of $(I - \alpha H_\dagger)f = h$ and supersolutions of $(I - \alpha H_\ddagger)f = h$. Then a viscosity solution $R_\alpha h$ exists for each $h \in D_\alpha$, and R_α extends continuously (in terms of* buc-convergence*) to all of $C_b(E)$.*

In addition, we suppose that $\{X_n(0)\}$ satisfies a large deviation principle with a good rate function I_0.

Then

a) *$\{X_n\}$ satisfies the large deviation principle with a good rate function I.*

b) *The limit*

$$V(t)h \equiv \lim_{m \to \infty} R^m_{t/m} h$$

exists and defines an operator semigroup on $C_b(E)$, and the rate function is given by (2.6).

The above result is a special case of Theorem 7.18. The assumption that $\limsup_{n\to\infty} \eta'_n(K^q_n)$ is compact in E' can be replaced by weaker conditions. See Theorem 7.24. Theorem A.8 gives conditions under which $C_b(E)$ is the buc-closure of a subset D.

Finally, the control representation method for simplifying the rate function (Theorem 8.14 and Corollary 8.28) continues to apply for non-compact E.

Part 1

The general theory of large deviations

CHAPTER 3

Large deviations and exponential tightness

We discuss basic properties of the large deviation principle for metric space-valued random variables. Section 3.1 collects well-known definitions and general properties; Section 3.2 develops rate function identification using a class of functions that *isolates points*; Section 3.3 derives a form of the rate function in product space that will be very useful later in the stochastic processes setting.

3.1. Basic definitions and results

Let (S,d) be a metric space. For $n = 1,2,\ldots$, let X_n be a S-valued random variable and let P_n denote its distribution on the Borel subsets of S, that is, $P_n(A) = P\{X_n \in A\}$. Recall that $\{X_n\}$ satisfies the large deviation principle (Definition 1.1) with rate function I if for each open set A,

$$-\inf_{x\in A} I(x) \le \liminf_{n\to\infty} \frac{1}{n}\log P\{X_n \in A\}, \tag{3.1}$$

and for each closed set B,

$$\limsup_{n\to\infty} \frac{1}{n}\log P\{X_n \in B\} \le -\inf_{x\in B} I(x). \tag{3.2}$$

REMARK 3.1. a) If I satisfies (3.1) and (3.2), then the lower semicontinuous regularization of I,

$$I_*(x) = \lim_{\epsilon\to 0}\inf_{y\in B_\epsilon(x)} I(y),$$

also satisfies these inequalities. We will require I to be lower semicontinuous, so $I = I_*$.

b) Requiring I to be lower semicontinuous, the large deviation principle with rate function I implies

$$-I(x) = \lim_{\epsilon\to 0}\liminf_{n\to\infty}\frac{1}{n}\log P\{X_n\in B_\epsilon(x)\} = \lim_{\epsilon\to 0}\limsup_{n\to\infty}\frac{1}{n}\log P\{X_n \in \overline{B}_\epsilon(x)\}, \tag{3.3}$$

and it follows that I is uniquely determined.

c) If (3.3) holds for all $x \in S$, then (3.1) holds for all open sets A and (3.2) holds for all compact sets B. This statement (that is the large deviation principle with "closed" replaced by "compact") is called the *weak large deviation principle.* (O'Brien and Vervaat [**91**] say *vague* rather than weak.)

d) The large deviation principle is equivalent to the assertion that for each $A \in \mathcal{B}(S)$,

$$-\inf_{x\in A^o} I(x) \le \liminf_{n\to\infty}\frac{1}{n}\log P\{X_n\in A\} \le \limsup_{n\to\infty}\frac{1}{n}\log P\{X_n \in A\} \le -\inf_{x\in\overline{A}} I(x),$$

where A^o is the interior of A and $\overline{A}$ is the closure.

e) Lower semicontinuity of I is equivalent to $\{x : I(x) \leq c\}$ being closed for each $c \in R$. If $\{x : I(x) \leq c\}$ is compact for each $c \in R$, we say that I is *good.*

In weak convergence theory, a sequence $\{P_n\}$ is tight if for each $\epsilon > 0$ there exists a compact subset $K \subset S$ such that $\inf_n P_n(K) \geq 1 - \epsilon$. The analogous concept in large deviation theory is exponential tightness.

DEFINITION 3.2. (Exponential Tightness) A sequence of probability measures $\{P_n\}$ on S is said to be *exponentially tight* if for each $a > 0$, there exists a compact set $K_a \subset S$ such that

$$\limsup_{n\to\infty} \frac{1}{n} \log P_n(K_a^c) \leq -a.$$

A sequence $\{X_n\}$ of S-valued random variables is exponentially tight if the corresponding sequence of distributions is exponentially tight.

de Acosta [**25**] (see the proof of Lemma 4.1) and O'Brien [**89**] (Theorem 3.3) have observed the following equivalence. For $\eta > 0$ and $K \subset S$, let $K^\eta = \{x : \inf_{y\in K} d(x,y) < \eta\}$.

LEMMA 3.3. *For each $n = 1, 2, \ldots$, let P_n be a tight probability measure on S. (Recall that if S is complete and separable, then every probability measure is tight.) Then $\{P_n\}$ is exponentially tight if and only if for each $a > 0$ and $\eta > 0$ there exists a compact $K_{a,\eta} \subset S$ such that*

$$\limsup_{n\to\infty} \frac{1}{n} \log P_n((K_{a,\eta}^\eta)^c) \leq -a.$$

PROOF. Since for each n, P_n is tight, without loss of generality, we can assume

$$\sup_n \frac{1}{n} \log P_n((K_{a,\eta}^\eta)^c) \leq -a,$$

and hence

$$P_n((K_{a,\eta}^\eta)^c) \leq e^{-na}.$$

Let $\widehat{K}_a$ be the closure of $\cap_k K_{a+k,k^{-1}}^{k^{-1}}$. Then $\widehat{K}_a$ is compact (it is complete and totally bounded) and

$$P_n(\widehat{K}_a^c) \leq \sum_{k=1}^\infty e^{-na-nk} \leq \frac{e}{e-1} e^{-na},$$

so

$$\limsup_{n\to\infty} \frac{1}{n} \log P_n(\widehat{K}_a^c) \leq -a.$$

□

EXAMPLE 3.4. We illustrate the use of Lemma 3.3 with the following lemma on sums of Banach space-valued random variables. (See de Acosta [**25**].) Let E be a Banach space, and let $\xi_1, \xi_2, \ldots$ be independent and identically distributed E-valued random variables satisfying

$$E[e^{\alpha\|\xi_k\|}] < \infty, \quad \text{for all } \alpha > 0.$$

Define

$$X_n = \frac{1}{n} \sum_{k=1}^n \xi_k.$$

We claim that $\{X_n\}$ is exponentially tight.

To check the claim, let $\{x_i\}$ be dense in E, and for each m, let

$$K_m = \{\sum_{i=1}^m q_i x_i : q_i \geq 0, \sum_{i=1}^m q_i \leq 1\}.$$

Clearly, K_m is compact. For $a, \epsilon > 0$, let $\alpha > a/\epsilon$ and select m so that

$$\alpha\epsilon - \log E[e^{\alpha\|\xi_k\| I_{\{\xi_k \notin \cup_{i=1}^m B_\epsilon(x_i)\}}}] > a. \tag{3.4}$$

Define $\gamma_k = \|\xi_k\| I_{\{\xi_k \notin \cup_{i=1}^m B_\epsilon(x_i)\}}$. For $i = 1, \ldots, m$, let $D_i = B_\epsilon(x_i) \cap \cap_{j=1}^{i-1} B_\epsilon(x_j)^c$. (Note that the D_i are disjoint and $\cup_{i=1}^m D_i = \cup_{i=1}^m B_\epsilon(x_i)$.) Finally, note that

$$\widehat{X}_n = \sum_{i=1}^m \frac{1}{n} \sum_{k=1}^n I_{\{\xi_k \in D_i\}} x_i \in K_m,$$

and

$$X_n - \widehat{X}_n = \sum_{i=1}^m \frac{1}{n} \sum_{k=1}^n I_{\{\xi_k \in D_i\}} (\xi_k - x_i) + \frac{1}{n} \sum_{k=1}^n I_{\{\xi_k \notin \cup_{i=1}^m B_\epsilon(x_i)\}} \xi_k. \tag{3.5}$$

By (3.5),

$$\|X_n - \widehat{X}_n\| \leq \epsilon + \frac{1}{n} \sum_{k=1}^n \gamma_k,$$

and by (3.4),

$$P\{\frac{1}{n} \sum_{k=1}^n \gamma_k > \epsilon\} \leq \frac{E[e^{\alpha \sum_{k=1}^n \gamma_k}]}{e^{\alpha n \epsilon}} \leq e^{-na}.$$

Consequently,

$$\limsup_{n\to\infty} \frac{1}{n} \log P\{X_n \notin K_m^{2\epsilon}\} \leq -a,$$

and since ϵ and a are arbitrary, Lemma 3.3 gives the exponential tightness of $\{X_n\}$.

One consequence of exponential tightness is that if a sequence is exponentially tight, then the large deviation principle holds if and only if the weak large deviation principle holds. In addition, if $\{P_n\}$ is exponentially tight and the large deviation lower bound (3.1) holds, then the rate function I is good. (See, for example, Dembo and Zeitouni [**29**], Lemma 1.2.18.) Lemma 3.5 gives the converse.

LEMMA 3.5. *For each $n = 1, 2, \ldots$, let P_n be a tight probability measure on S. If $\{P_n\}$ satisfies the large deviation principle with good rate function I, then $\{P_n\}$ is exponentially tight.*

PROOF. Let $a > 0$ and $K_a = \{x : I(x) \leq a\}$. Then for $\eta > 0$,

$$\limsup_{n\to\infty} \frac{1}{n} \log P_n((K_a^\eta)^c) \leq - \inf_{x \in (K_a^\eta)^c} I(x) \leq -a,$$

and exponential tightness follows by Lemma 3.3. □

Another useful observation is that if S is a product space, then a sequence of tight probability measures $\{P_n\} \subset \mathcal{P}(S)$ is exponentially tight if and only if the marginal distributions are exponentially tight.

LEMMA 3.6. *Let $(S_1, d_1), (S_2, d_2), \ldots$ be metric spaces, and let $S = S_1 \times S_2 \times \cdots$ with (for example) $d(x, y) = \sum_k 2^{-k} 1 \wedge d_k(x_k, y_k)$. For $n = 1, 2, \ldots$, let P_n be a tight probability measure on S. Then $\{P_n\}$ is exponentially tight if and only if each marginal sequence $\{P_n^k\} \subset \mathcal{P}(S_k)$ is exponentially tight.*

PROOF. Suppose $\{P_n\}$ is exponentially tight. Then for $a > 0$ there exists compact K_a such that

$$\limsup_{n\to\infty} \frac{1}{n} \log P_n(K_a^c) \leq -a.$$

Without loss of generality, we can assume that K_a is of the form $K_a = K_a^1 \times K_a^2 \times \cdots$, where each K_a^k is compact in S_k. It follows immediately that

$$\limsup_{n\to\infty} \frac{1}{n} \log P_n^k((K_a^k)^c) \leq -a.$$

Conversely, suppose that for each k, $\{P_n^k\}$ is exponentially tight. Then for $a > 0$ there exist compact sets K_a^k such that

$$P_n^k((K_a^k)^c) \leq e^{-nak}.$$

Setting $K_a = K_a^1 \times K_a^2 \times \cdots$, we have

$$P_n(K_a^c) \leq \sum_{k=1}^{\infty} e^{-nak} = \frac{e^{-na}}{1 - e - na},$$

and hence

$$\limsup_{n\to\infty} \frac{1}{n} \log P_n(K_a^c) \leq -a.$$

□

Exponential tightness plays the same role in large deviation theory as tightness does in weak convergence theory (Chapter 3, Ethier and Kurtz [**36**]). The following analogue of the Prohorov compactness theorem is essentially Theorem P of Puhalskii [**97**]. More general results have been given by O'Brien and Vervaat [**91**] and de Acosta (see Theorem 2.1 of [**24**]).

THEOREM 3.7. *Let (S, d) be a metric space and $\{P_n\}$ a sequence of tight probability measures on the Borel σ-algebra of S. Suppose that $\{P_n\}$ is exponentially tight. Then there exists a subsequence $\{n_k\}$ along which the large deviation principle holds with a good rate function.*

PROOF. Select compact K_k, $k = 1, 2, \ldots$ so that

$$\limsup_{n\to\infty} \frac{1}{n} \log P_n(K_k^c) \leq -k.$$

By the compactness of the K_k, there is a countable set $\{x_i\} \subset \cup_k K_k$ that is dense in $\cup_k K_k$. Let $1 \geq \epsilon_1 > \epsilon_2 > \cdots > 0$ satisfy $\lim_{j\to\infty} \epsilon_j = 0$. Then we can select a subsequence $\{n_l\}$ such that

$$-I_{\epsilon_j}(x_i) = \lim_{l\to\infty} \frac{1}{n_l} \log P_{n_l}(B_{\epsilon_j}(x_i))$$

exists for all i and j. Define $I_0(x_i) = \lim_{j\to\infty} I_{\epsilon_j}(x_i)$. For $x \in S_0 \equiv \mathrm{cl} \cup_k K_k$, define

$$I(x) = \lim_{\epsilon\to 0} \inf_{x_i \in B_\epsilon(x)} I_0(x_i),$$

and for $x \in S - S_0$, define $I(x) = \infty$. We claim that the large deviation principle holds for the subsequence $\{P_{n_l}\}$ with rate function I.

If $x \in S - S_0$, then the definition of K_k implies

$$\lim_{\epsilon \to 0} \limsup_{l \to \infty} \frac{1}{n_l} \log P\{X_{n_l} \in B_\epsilon(x)\} = -\infty$$

giving (3.3).

Consider $x \in S_0$. Let $\epsilon_j + d(x, x_i) < \epsilon$. Then

$$\liminf_{l \to \infty} \frac{1}{n_l} \log P\{X_n \in B_\epsilon(x)\} \geq -I_{\epsilon_j}(x_i),$$

and hence

$$\lim_{\epsilon \to 0} \liminf_{l \to \infty} \frac{1}{n_l} \log P\{X_n \in B_\epsilon(x)\} \geq -I(x).$$

Similarly, if $\epsilon > 0$ and $d(x, x_i) + \epsilon < \epsilon_j$, then

$$\limsup_{l \to \infty} \frac{1}{n_l} \log P\{X_n \in B_\epsilon(x)\} \leq -I_{\epsilon_j}(x_i),$$

and hence, since we can select $(x_{i_k}, \epsilon_{j_k})$ so that $x_{i_k} \to x$, $j_k \to \infty$, and $I_{\epsilon_{j_k}}(x_{i_k}) \to I(x)$,

$$\lim_{\epsilon \to 0} \limsup_{l \to \infty} \frac{1}{n_l} \log P\{X_n \in B_\epsilon(x)\} \leq -I(x).$$

It follows that (3.3) holds for all $x \in S$. But (3.3) implies the weak large deviation principle and exponential tightness then gives the full large deviation principle. □

The following moment characterization of the large deviation principle due to Varadhan and Bryc is central to our consideration of large deviation theorems for Markov processes.

PROPOSITION 3.8. *Let $\{X_n\}$ be a sequence of S-valued random variables.*

a) *(Varadhan Lemma) Suppose that $\{X_n\}$ satisfies the large deviation principle with a good rate function I. Then for each $f \in C_b(S)$,*

$$\lim_{n \to \infty} \frac{1}{n} \log E[e^{nf(X_n)}] = \sup_{x \in S}\{f(x) - I(x)\}. \tag{3.6}$$

b) *(Bryc formula) Suppose that the sequence $\{X_n\}$ is exponentially tight and that the limit*

$$\Lambda(f) = \lim_{n \to \infty} \frac{1}{n} \log E[e^{nf(X_n)}] \tag{3.7}$$

exists for each $f \in C_b(S)$. Then $\{X_n\}$ satisfies the large deviation principle with good rate function

$$I(x) = \sup_{f \in C_b(S)}\{f(x) - \Lambda(f)\}. \tag{3.8}$$

REMARK 3.9. a) We will refer to Λ as the *rate transform* for $\{X_n\}$.
b) See Theorems 4.3.1 and 4.4.2 in Dembo and Zeitouni [**29**].

PROOF. Let $x \in E$, and for $\epsilon > 0$, let $\delta_\epsilon > 0$ satisfy

$$\overline{B}_\epsilon(x) \subset \{y : f(y) > f(x) - \delta_\epsilon\}.$$

By the continuity of f, we can assume that $\lim_{\epsilon\to 0}\delta_\epsilon = 0$. Then

$$P\{X_n \in \overline{B}_\epsilon(x)\} \le e^{n(\delta_\epsilon - f(x))}E[e^{nf(X_n)}],$$

so

$$\lim_{\epsilon\to 0}\liminf_{n\to\infty}\frac{1}{n}\log P\{X_n \in \overline{B}_\epsilon(x)\} \le -f(x) + \liminf_{n\to\infty}\frac{1}{n}\log E[e^{nf(X_n)}] \tag{3.9}$$

and

$$\lim_{\epsilon\to 0}\limsup_{n\to\infty}\frac{1}{n}\log P\{X_n \in \overline{B}_\epsilon(x)\} \le -f(x) + \limsup_{n\to\infty}\frac{1}{n}\log E[e^{nf(X_n)}]. \tag{3.10}$$

For $x \in S$ and $\epsilon > 0$, let $f_{\epsilon,x} \in C_b(S)$ satisfy $f_{\epsilon,x}(y) \le f_{\epsilon,x}(x) - r\mathbf{1}_{B_\epsilon^c(x)}(y)$ for all $y \in S$. Then

$$e^{-nr} + P\{X_n \in B_\epsilon(x)\} \ge e^{-nf_{\epsilon,x}(x)}E[e^{nf_{\epsilon,x}(X_n)}],$$

so

$$\limsup_{n\to\infty}\frac{1}{n}\log P\{X_n \in B_\epsilon(x)\} \ge -f_{\epsilon,x}(x) + \limsup_{n\to\infty}\frac{1}{n}\log E[e^{nf_{\epsilon,x}(X_n)}]$$

and

$$\liminf_{n\to\infty}\frac{1}{n}\log P\{X_n \in B_\epsilon(x)\} \ge -f_{\epsilon,x}(x) + \liminf_{n\to\infty}\frac{1}{n}\log E[e^{nf_{\epsilon,x}(X_n)}]. \tag{3.11}$$

If (3.7) holds for each $f \in C_b(S)$, then by (3.10) and (3.11), (3.3) holds with

$$I(x) = \sup_{f\in C_b(S)}\{f(x) - \Lambda(f)\}$$

implying the weak large deviation principle. Assuming exponential tightness, we have Part (b).

If (3.3) holds, then by (3.9)

$$\sup_{x\in S}(f(x) - I(x)) \le \liminf_{n\to\infty} E[e^{nf(X_n)}].$$

If I is good, $\{X_n\}$ is exponentially tight and there exists a compact K such that

$$\limsup_{n\to\infty}\frac{1}{n}\log E[e^{nf(X_n)}] = \limsup_{n\to\infty}\frac{1}{n}\log E[e^{nf(X_n)}\mathbf{1}_K(X_n)].$$

For each $\epsilon > 0$, there exists and a finite collection $x_1, \ldots, x_m \in K$ such that $K \subset \cup_{i=1}^m B_\epsilon(x_i)$ and $f(y) \le \max\{\epsilon + f(x_i) : y \in B_\epsilon(x_i)\}$. Consequently,

$$\begin{aligned}
&\limsup_{n\to\infty}\frac{1}{n}\log E[e^{nf(X_n)}\mathbf{1}_K(X_n)] \\
&\quad\le \limsup_{n\to\infty}\frac{1}{n}\log E[\sum_{i=1}^m e^{n(\epsilon+f(x_i))}\mathbf{1}_{B_\epsilon(x_i)}(X_n)] \\
&\quad\le \max_{1\le i\le m}(f(x_i) + \epsilon + \limsup_{n\to\infty}\frac{1}{n}\log P\{X_n \in B_\epsilon(x_i)\}),
\end{aligned}$$

and hence for each $\epsilon > 0$, there exists $x_\epsilon \in K$ such that

$$\limsup_{n\to\infty}\frac{1}{n}\log E[e^{nf(X_n)}\mathbf{1}_K(X_n)] \le f(x_\epsilon) + \epsilon + \limsup_{n\to\infty}\frac{1}{n}\log P\{X_n \in B_\epsilon(x_\epsilon)\}.$$

Since K is compact, we can select a subsequence along which $x_\epsilon \to x_0 \in K$ as $\epsilon \to 0$. If $r(x_\epsilon, x_0) + \epsilon < \delta$, then

$$\begin{aligned} f(x_\epsilon) + \epsilon + \limsup_{n\to\infty} \frac{1}{n} \log P\{X_n \in B_\epsilon(x_\epsilon)\} \\ \leq f(x_\epsilon) - f(x_0) + \epsilon + f(x_0) + \limsup_{n\to\infty} \frac{1}{n} \log P\{X_n \in B_\delta(x_0)\}, \end{aligned}$$

and it follows that

$$\limsup_{n\to\infty} \frac{1}{n} \log E[e^{nf(X_n)} \mathbf{1}_K(X_n)] \leq f(x_0) - I(x_0)$$

giving Part (a). □

COROLLARY 3.10. *Suppose that the sequence $\{X_n\}$ is exponentially tight and that the limit*

$$\Lambda(f) = \lim_{n\to\infty} \frac{1}{n} \log E[e^{nf(X_n)}] \tag{3.12}$$

exists for each $f \in C_b(S)$. Then

$$I(x) = - \inf_{f\in C_b(S),\, f(x)=0} \Lambda(f)$$

PROOF. The result follows from (3.8) and the fact that $\Lambda(f + c) = \Lambda(f) + c$ for any constant c. □

In weak convergence theory, the continuous mapping theorem plays a key role. The analogous theorem for large deviation theory is the *contraction principle*, Dembo and Zeitouni [**29**], Theorem 4.2.1. The following result of Puhalskii ([**97**] Theorem 2.2) extends the usual statement of the contraction principle to discontinuous mappings in exactly the same way that the continuous mapping theorem extends to discontinuous functions. Theorem 4.2.23 of [**29**] and Garcia [**53**] give different approaches to the extension of the contraction principle to discontinuous functions.

LEMMA 3.11. *For $n = 1, 2, \ldots$, let X_n be an S-valued random variable with a tight distribution. Suppose that $\{X_n\}$ satisfies a large deviation principle with good rate function I. Let (S', d') be a metric space and suppose $F : S \mapsto S'$ is measurable and that F is continuous at $x \in S$ for each x with $I(x) < \infty$. Define $Y_n = F \circ X_n$. Then $\{Y_n\}$ satisfies a large deviation principle with good rate function*

$$I'(y) = \inf\{I(x) : F(x) = y\}.$$

In particular, if y is not in the range of F, then $I'(y) = \infty$.

PROOF. Let $K_m = \{x : I(x) \leq m\}$. Then F is continuous at each point in K_m and hence $F(K_m)$ is compact in S'. Exponential tightness for $\{Y_n\}$ follows since for each $\eta > 0$, there exists $\epsilon > 0$ such that $x \in K_m^\epsilon$ implies $F(x) \in F(K_m)^\eta$ and hence $\{Y_n \notin F(K_m)^\eta\} \subset \{X_n \notin K_m^\epsilon\}$ and

$$\begin{aligned} \limsup \frac{1}{n} \log P\{Y_n \notin F(K_m)^\eta\} &\leq \limsup \frac{1}{n} \log P\{X_n \notin K_m^\epsilon\} \\ &\leq - \inf_{x\in(K_m^\epsilon)^c} I(x) \leq -m. \end{aligned}$$

Let I' denote the rate function for $\{Y_n\}$ (or a subsequence). Suppose $F(x) = y$. If $I(x) < \infty$, then for each $\epsilon > 0$, there exists $\delta > 0$ such that $F(B_\delta(x)) \subset B_\epsilon(y)$. Consequently,

$$\begin{aligned} I'(y) &= -\lim_{\epsilon\to 0}\liminf_{n\to\infty}\frac{1}{n}\log P\{Y_n \in B_\epsilon(y)\} \\ &\leq -\lim_{\delta\to 0}\liminf_{n\to\infty}\frac{1}{n}\log P\{X_n \in B_\delta(x)\} = I(x) \end{aligned}$$

and hence

$$I'(y) \leq \inf\{I(x) : F(x) = y\}.$$

If $I'(y) \leq m < \infty$, then

$$I'(y) = -\lim_{\epsilon\to 0}\liminf_{n\to\infty}\frac{1}{n}\log P\{Y_n \in \overline{B}_\epsilon(y)\} \geq \lim_{\epsilon\to 0}\inf_{x\in \mathrm{cl}F^{-1}(\overline{B}_\epsilon(y))} I(x)$$

and there exist $x_\epsilon \in \mathrm{cl}F^{-1}(\overline{B}_\epsilon(x))$ such that $\lim_{\epsilon\to 0} I(x_\epsilon) \leq I'(y) \leq m$. Since $\{x : I(x) \leq m\}$ is compact, we may assume that $x_\epsilon \to x_0$. But

$$I(x_0) \leq \lim_{\epsilon\to\infty} I(x_\epsilon) \leq I'(y).$$

Since $I(x_0) < \infty$, F is continuous at x_0, and since $x_\epsilon \in \mathrm{cl}F^{-1}(\overline{B}_\epsilon(y))$, there exist $x'_\epsilon \in F^{-1}(\overline{B}_\epsilon(y))$ such that $x'_\epsilon \to x_0$ and hence $F(x_0) = \lim_{\epsilon\to 0} F(x'_\epsilon) = y$. Consequently,

$$I'(y) \geq \inf\{I(x) : F(x) = y\}.$$

Note that $K'_m = \{y : I'(y) \leq m\} = \{F(x) : I(x) \leq m\}$. Since F is continuous on $\{x : I(x) \leq m\}$, K'_m is compact. □

LEMMA 3.12. *Let (S, d) and $(\widehat{S}, \widehat{d})$ be metric spaces. Suppose that $S \subset \widehat{S}$, $\mathcal{B}(S) \subset \mathcal{B}(\widehat{S})$, and if $\{x_n\} \subset S$ converges to $x \in S$ under d, then it also converges under $\widehat{d}$. Suppose $\{X_n\}$ satisfies a large deviation principle in $(\widehat{S}, \widehat{d})$ with rate function $\widehat{I}$, $P\{X_n \in S\} = 1$, and $\{X_n\}$ is exponentially tight in (S, d). Then $\{X_n\}$ satisfies a large deviation principle in S with rate function given by the restriction of $\widehat{I}$ to S.*

PROOF. Let I be the rate function for the large deviation principle for $\{X_n\}$ (or a subsequence) in (S, d). Since the injection map $F : S \mapsto \widehat{S}$ is continuous, we must have $\widehat{I}(x) = I(x)$ for $x \in S$ and $\widehat{I}(x) = \infty$ for $x \in \widehat{S} - S$. □

The following lemma, which is essentially Theorem 4.2.13 of Dembo and Zeitouni [**29**], is the exponential analogue of a weak convergence result known in the statistical literature as Slutsky's theorem. (See, for example, Ethier and Kurtz [**36**], Corollary 3.3.3.)

LEMMA 3.13. *For $n = 1, 2, \ldots$, let (X_n, Y_n) be an $S\times S$-valued random variable. Suppose that for each $\eta > 0$,*

$$\limsup_{n\to\infty}\frac{1}{n}\log P\{d(X_n, Y_n) > \eta\} = -\infty. \tag{3.13}$$

Then the following hold:

a) *$\{X_n\}$ is exponentially tight if and only if $\{Y_n\}$ is exponentially tight.*

b) *If* $\{X_n\}$ *(or, equivalently,* $\{Y_n\}$*) is exponentially tight, then for each* $f \in C_b(E)$

$$\limsup_{n\to\infty} \frac{1}{n} \log E[e^{nf(X_n)}] = \limsup_{n\to\infty} \frac{1}{n} \log E[e^{nf(Y_n)}] \tag{3.14}$$

and

$$\liminf_{n\to\infty} \frac{1}{n} \log E[e^{nf(X_n)}] = \liminf_{n\to\infty} \frac{1}{n} \log E[e^{nf(Y_n)}], \tag{3.15}$$

and hence, the large deviation principle holds for $\{X_n\}$ *if and only if it holds for* $\{Y_n\}$*. If the large deviation principle holds,* $\{X_n\}$ *and* $\{Y_n\}$ *have the same rate function.*

PROOF. Part (a) is an immediate consequence of Lemma 3.3. As in the proof of Proposition 3.17, the collection of f for which (3.14) (and, similarly, (3.15)) holds is *buc*-closed. (See Appendix A.2.) Consequently, it suffices to consider $f \in C_u(S)$, the space of bounded, uniformly continuous functions. Let $\omega_f(\epsilon) = \sup\{|f(x) - f(y)| : d(x,y) \le \epsilon\}$ be the modulus of continuity for f (which, by uniform continuity, satisfies $\lim_{\epsilon\to 0} \omega_f(\epsilon) = 0$), and let K be a compact subset of S satisfying $\limsup_{n\to\infty} \frac{1}{n} \log P\{X_n \in K^c\} < -2\|f\|$. Then, by (3.13), $\limsup_{n\to\infty} \frac{1}{n} \log P\{d(X_n, Y_n) > \eta\} < -2\|f\|$, so for $\eta > 0$,

$$\begin{aligned}
\limsup_{n\to\infty} \frac{1}{n} \log E[e^{nf(X_n)}] &= \limsup_{n\to\infty} \frac{1}{n} \log E[I_{\{d(X_n,Y_n)\le\eta\}} e^{nf(X_n)}] \\
&\le \limsup_{n\to\infty} \frac{1}{n} \log E[I_{\{d(X_n,Y_n)\le\eta\}} e^{n(f(Y_n)+\omega_f(\eta))}] \\
&= \omega_f(\eta) + \limsup \frac{1}{n} \log E[e^{nf(Y_n)}]
\end{aligned}$$

and similarly

$$\limsup_{n\to\infty} \frac{1}{n} \log E[e^{nf(X_n)}] \ge -\omega_f(\eta) + \limsup_{n\to\infty} \frac{1}{n} \log E[e^{nf(Y_n)}].$$

Letting $\eta \to 0$, (3.14)follows. The proof of (3.15) is similar. □

The following variation of Lemma 3.13 is slightly more complicated to state, but may be more natural to apply in certain situations.

LEMMA 3.14. *For each* n*, let* X_n *be an* S*-valued random variable with a tight distribution. Suppose that for each* $\eta > 0$ *and* $a > 0$*, there exists a sequence* $\{X_n^{\eta,a}\}$ *of* S*-valued random variables such that for each* n*,* $X_n^{\eta,a}$ *has a tight distribution,* $\{X_n^{\eta,a}\}$ *satisfies the large deviation principle with good rate function* $I^{\eta,a}$*, and*

$$\limsup_{n\to\infty} \frac{1}{n} \log P\{d(X_n, X_n^{\eta,a}) > \eta\} \le -a.$$

Then $\{X_n\}$ *satisfies the large deviation principle with good rate function* I *satisfying*

$$I(x) = \lim_{\delta\to 0} \liminf_{\eta\to 0} \liminf_{a\to\infty} \inf_{y\in B_\delta(x)} I^{\eta,a}(y) = \lim_{\delta\to 0} \limsup_{\eta\to 0} \limsup_{a\to\infty} \inf_{y\in B_\delta(x)} I^{\eta,a}(y) \tag{3.16}$$

PROOF. By Lemma 3.5, $\{X_n^{\eta,a}\}$ is exponentially tight, and Lemma 3.3 then gives exponential tightness for $\{X_n\}$. By Theorem 3.7, we may as well assume that $\{X_n\}$ satisfies the large deviation principle, and show that the rate function is uniquely determined by (3.16). Observe that for $\epsilon > 0$,

$$\frac{1}{n}\log P\{X_n \in B_\epsilon(x)\} \leq \frac{1}{n}\log(P\{X_n^{\eta,a} \in B_{\epsilon+\eta}(x)\} + P\{d(X_n, X_n^{\eta,a}) > \eta\}).$$

Consequently, by (3.3),

$$-I(x) \leq (-a) \vee (-\inf_{y\in B_\delta(x)} I^{\eta,a}(y))$$

for any $\delta > \eta$, and hence

$$I(x) \geq \lim_{\delta\to 0}\limsup_{\eta\to 0}\limsup_{a\to\infty}\inf_{y\in B_\delta(x)} I^{\eta,a}(y). \tag{3.17}$$

Similarly,

$$\frac{1}{n}\log P\{X_n^{\eta,a} \in B_\delta(x)\} \leq \frac{1}{n}\log(P\{X_n \in B_{\delta+\eta}(x)\} + P\{d(X_n, X_n^{\eta,a}) > \eta\}),$$

and hence

$$-\inf_{y\in B_\delta(x)} I^{\eta,a}(y) \leq (-a) \vee (-\inf_{y\in \overline{B}_{\delta+\eta}(x)} I(y))$$

and

$$I(x) = \lim_{\delta\to 0}\lim_{\eta\to 0}\inf_{y\in\overline{B}_{\delta+\eta}(x)} I(y) \leq \lim_{\delta\to 0}\liminf_{\eta\to 0}\liminf_{a\to\infty}\inf_{y\in B_\delta(x)} I^{\eta,a}(y). \tag{3.18}$$

Together (3.17) and (3.18) give (3.16). □

3.2. Identifying a rate function

To prove a large deviation principle, it is, in general, enough to verify (3.7) for a class of functions that is much smaller than $C_b(S)$. Let

$$\Gamma = \{f : (3.7) \text{ holds}\}.$$

DEFINITION 3.15. A collection of functions $D \subset C_b(S)$ will be called *rate function determining* if whenever $D \subset \Gamma$ for an exponentially tight sequence, the sequence satisfies the large deviation principle with good rate function

$$I(x) = \sup_{f\in D}\{f(x) - \Lambda(f)\}. \tag{3.19}$$

The following observations are useful in identifying rate function determining classes.

LEMMA 3.16. *a) For positive sequences $\{a_n\}$ and $\{b_n\}$,*

$$\lim_{n\to\infty} |\frac{1}{n}\log(a_n + b_n) - \frac{1}{n}\log(\max\{a_n, b_n\})| = 0.$$

b) Let $f \in C_b(S)$, and let $\{F_n\}$ be a sequence of events such that

$$\limsup_{n\to\infty}\frac{1}{n}\log P(F_n^c) < -2\|f\|.$$

Then

$$\lim_{n\to\infty} |\frac{1}{n}\log E[e^{nf(X_n)}] - \frac{1}{n}\log E[I_{F_n}e^{nf(X_n)}]| = 0.$$

PROOF. Part (a) is immediate. For part (b), note that

$$\liminf_{n\to\infty}\frac{1}{n}\log E[e^{nf(X_n)}]\geq\inf_x f(x)\geq-\|f\|$$

and

$$\limsup_{n\to\infty}\frac{1}{n}\log E[I_{F_n^c}e^{nf(X_n)}]<-\|f\|,$$

so the conclusion follows from part (a). □

See Section A.2 for definition of *buc*-topology used in the next proposition.

PROPOSITION 3.17. *Let $\{X_n\}$ be exponentially tight. Then*

a) $\Gamma=\{f:(3.7)\text{ holds}\}$ *is buc-closed.*

b) *If $C_b(S)$ is the buc-closure of D, then D is rate function determining.*

PROOF. Suppose f is *buc*-approximable by functions in Γ. Let $C_{f,\Gamma}$ be as in Definition A.6. Let K be compact and satisfy $\limsup_{n\to\infty}\frac{1}{n}\log P\{X_n\in K^c\}\leq-3C_{f,\Gamma}$. Then, for $\epsilon>0$, there exists $g\in\Gamma$ such that $\|g\|\leq C_{f,\Gamma}$ and $\sup_{x\in K}|f(x)-g(x)|\leq\epsilon$. Observe that

$$\limsup_{n\to\infty}\frac{1}{n}\log E[I_{K^c}(X_n)e^{ng(X_n)}]\leq\sup_{x\in E}g(x)-3C_{f,\Gamma}\leq-2C_{f,\Gamma},$$

and hence that

$$\begin{aligned}\lim_{n\to\infty}\frac{1}{n}\log E[e^{ng(X_n)}] &= \lim_{n\to\infty}\frac{1}{n}\log\left(E[I_Ke^{ng(X_n)}]+E[I_{K^c}e^{ng(X_n)}]\right)\\ &= \lim_{n\to\infty}\frac{1}{n}\log E[I_Ke^{ng(X_n)}].\end{aligned}$$

Since $f(x)\leq(g(x)+\epsilon)I_K(x)+I_{K^c}(x)\|f\|$,

$$\begin{aligned}&\limsup_{n\to\infty}\frac{1}{n}\log E[e^{nf(X_n)}]\\ &\leq\limsup_{n\to\infty}\frac{1}{n}\log\left(E[I_K(X_n)e^{n(g(X_n)+\epsilon)}+I_{K^c}(X_n)e^{n\|f\|}]\right)\\ &\leq\max\left\{\limsup_{n\to\infty}\frac{1}{n}\log E[I_K(X_n)e^{n(g(X_n)+\epsilon)}],\|f\|+\limsup_{n\to\infty}\log P\{X_n\in K^c\}\right\}\\ &\leq\max\{\Lambda(g)+\epsilon,-2C_{f,\Gamma}\}=\Lambda(g)+\epsilon.\end{aligned}$$

Similarly,

$$\begin{aligned}&\liminf_{n\to\infty}\frac{1}{n}\log E[e^{nf(X_n)}]\\ &\geq\liminf_{n\to\infty}\frac{1}{n}\log\left(E[I_K(X_n)e^{n(g(X_n)-\epsilon)}+I_{K^c}(X_n)e^{-n\|f\|}]\right)\\ &\geq\max\left\{\liminf_{n\to\infty}\frac{1}{n}\log E[I_K(X_n)e^{n(g(X_n)-\epsilon)}],-\|f\|+\liminf_{n\to\infty}\log P\{X_n\in K^c\}\right\}\\ &\geq\max\{\Lambda(g)-\epsilon,-3C_{f,\Gamma}\}=\Lambda(g)-\epsilon.\end{aligned}$$

The existence of $\Lambda(f)$ and part (a) follow.

For part (b), let $\Gamma_{x,c}=\{f:f(x)-\Lambda(f)\leq c\}$. We claim that $\Gamma_{x,c}$ is *buc*-closed. Suppose that f is *buc*-approximable by $\Gamma_{x,c}$. Let $C_{f,\Gamma_{x,c}}$ be as in Definition A.6, and let $K\subset S$ be compact such that $x\in K$ and $\limsup_{n\to\infty}P\{X_n\in K^c\}\leq-3C_{f,\Gamma_{x,c}}$.

As in the proof of part (a), for $\epsilon > 0$, there exists $g \in \Gamma_{x,c}$ such that $|f(x)-g(x)| \le \epsilon$ and $\Lambda(g) - \epsilon \le \Lambda(f) \le \Lambda(g) + \epsilon$. Consequently,

$$f(x) - \Lambda(f) \le c + 2\epsilon,$$

and since ϵ is arbitrary, $f \in \Gamma_{x,c}$.

Let $D \subset C_b(S)$ and suppose that $\Lambda(f)$ exists for $f \in D$. Let $\overline{D}$ denote the *buc*-closure of D. Then by part (a), $\Lambda(f)$ exists for $f \in \overline{D}$. Set $c = \sup_{f\in D}(f(x)-\Lambda(f))$. Since $\Gamma_{x,c}$ is *buc*-closed, we have $c = \sup_{f\in\overline{D}}(f(x) - \Lambda(f))$, and since in part (b), $\overline{D} = C_b(S)$, the result follows. □

DEFINITION 3.18. A collection of functions $D \subset C_b(S)$ *isolates points* in S, if for each $x \in S$, each $\epsilon > 0$, and each compact $K \subset S$, there exists $f \in D$ satisfying $|f(x)| < \epsilon$, $\sup_{y\in K} f(y) \le 0$, and

$$\sup_{y\in K\cap B_\epsilon^c(x)} f(y) < -\frac{1}{\epsilon}.$$

If f satisfies these conditions, we will say that f *isolates x relative to ϵ and K*.

A collection of functions $D \subset C_b(S)$ is *bounded above* if $\sup_{f\in D}\sup_y f(y) < \infty$.

The following lemma is immediate.

LEMMA 3.19. *Let $C(c,S) = \{f \in C_b(S) : \sup_y f(y) \le c\}$. If $D \subset C(c,S)$ is dense in $C(c,S)$ in the compact uniform topology, then D is bounded above and isolates points.*

PROPOSITION 3.20. *If D is bounded above and isolates points, then D is rate function determining.*

PROOF. Suppose $\{X_n\}$ is exponentially tight, and let $\Gamma = \{f \in C_b(S) :$ (3.7) holds$\}$. It is sufficient to show that for any subsequence along which the large deviation principle holds, the corresponding rate function is given by (3.19). Consequently, we may as well assume $\Gamma = C_b(S)$. Note that $\Lambda(f)$ is a monotone function of f.

Let $c > 0$ satisfy $\sup_{f\in D}\sup_y f(y) \le c$. Fix $f_0 \in C_b(S)$ satisfying $f_0(x) = 0$, and let $\delta > 0$. Let $a > c + |\Lambda(f_0)|$, and let K_a be a compact set satisfying

$$\limsup_{n\to\infty} \frac{1}{n} \log P\{X_n \notin K_a\} \le -a.$$

Let $0 < \epsilon < \delta$ be such that $\sup_{y\in B_\epsilon(x)} |f_0(y)| \le \delta$ and $\inf_{y\in K} f_0(y) \ge -\epsilon^{-1}$. By the assumptions on D, there exists $f \in D$ satisfying $f \le cI_{K_a^c}$, $fI_{K_a} \le f_0 I_{K_a} + \delta$, and $|f(x)| < \epsilon$. It follows that

$$e^{nf} \le e^{nc} I_{K_a^c} + e^{n(f_0+\delta)},$$

and hence

$$\Lambda(f) \le \max(-(a-c), \Lambda(f_0) + \delta) = \Lambda(f_0) + \delta.$$

Consequently,

$$f(x) - \Lambda(f) \ge -\Lambda(f_0) - 2\delta,$$

and (3.19) follows. □

COROLLARY 3.21. *Fix $x \in S$, and suppose $\{f_m\} \subset D$ satisfies the following:*

a) $\sup_m \sup_y f_m(y) \le c < \infty$.

b) *For each compact K and each $\epsilon > 0$*

$$\lim_{m\to\infty} \sup_{y\in K} f_m(y) \le 0$$

and

$$\lim_{m\to\infty} \sup_{y\in K\cap B_\epsilon^c(x)} f_m(y) = -\infty.$$

c) $\lim_{m\to\infty} f_m(x) = 0$.

Then $I(x) = -\lim_{m\to\infty} \Lambda(f_m)$.

PROOF. Let $f_0 \in C_b(S)$ and $f_0(x) = 0$. The assumptions on the sequence $\{f_m\}$ imply that for each compact $K \subset S$ and $\delta > 0$ there exists m_0 such that for $m > m_0$

$$e^{nf_m} \le e^{nc} I_{K^c} + e^{n(f_0+\delta)},$$

and hence

$$\liminf_{m\to\infty}(f_m(x) - \Lambda(f_m)) \ge -\Lambda(f_0) - \delta.$$

Since $f_m(x) \to 0$, it follows that $I(x) = -\lim_{m\to\infty} \Lambda(f_m)$. □

3.3. Rate functions in product spaces

We now consider the construction of rate function determining classes for product spaces. Recall (Lemma 3.6) that a sequence $\{X_n\}$ with values in a product space is exponentially tight if and only if each of its marginal distributions is exponentially tight.

LEMMA 3.22. *Let $(S_1, d_1), \dots, (S_m, d_m)$ be complete separable metric spaces, and let $S = S_1 \times \dots \times S_m$ and $d(x, y) = \sum_{k=1}^m d_k(x_k, y_k)$. Suppose that for each k, $D_k \subset C_b(S_k)$ is bounded above and isolates points in S_k. Then $D = \{f_1 + \cdots + f_m) : f_k \in D_k\}$ is bounded above and isolates points in S.*

The proof of Lemma 3.22 is similar to the proof of the following lemma.

LEMMA 3.23. *Let $(S_1, d_1), (S_2, d_2), \dots$ be complete separable metric spaces, and let $S = S_1 \times S_2 \times \cdots$ and $d(x, y) = \sum_k 2^{-k}(d_k(x_k, y_k) \wedge 1)$. Suppose that for each k, $D_k \subset C_b(S_k)$ is bounded above by 1 and isolates points in S_k. Then $D = \{m^{-1}(f_1 + \cdots + f_m) : m \ge 1, f_k \in D_k\}$ is bounded above by 1 and isolates points in S.*

PROOF. Let $x \in S$, $\epsilon > 0$, and compact $K \subset S$. Select $m > 1$ such that $2^{-m} < \epsilon/2$. For $k = 1, \dots, m$, there exist compact $K_k \subset S_k$ such that $K \subset K_1 \times \cdots \times K_m \times S_{m+1} \times \cdots$. For each $k = 1, \dots, m$, let $f_k \in D_k$ isolate x_k relative to ϵ/m and K_k. Then

$$\frac{1}{m}\sum_{k=1}^m f_k \le \frac{1}{m}\sum_{k=1}^m I_{K_k^c} \le I_{K^c},$$

and

$$|\frac{1}{m}\sum_{k=1}^m f_k(x_k)| \le \frac{\epsilon}{m}.$$

If $y \in K \cap B_\epsilon^c(x)$, then there exists $k \leq m$ such that $y_k \in K_k \cap B_{\epsilon/2}^c(x_k)$. Consequently, we have

$$\sup_{y \in K \cap B_\epsilon^c(x)} \frac{1}{m} \sum_{k=1}^m f_k(y_k) \leq \sup_{y \in K \cap B_\epsilon^c(x)} \frac{1}{m} \min_{1 \leq k \leq m} f_k(y_k) \leq -\frac{1}{\epsilon},$$

and $f = m^{-1}(f_1 + \cdots + f_m)$ isolates x relative to ϵ and K. □

The following variation of Lemma 3.23 gives a rate function determining class with somewhat simpler structure.

LEMMA 3.24. *Let $(S_1, d_1), (S_2, d_2), \ldots$ be complete separable metric spaces, and let $S = S_1 \times S_2 \times \cdots$ and $d(x,y) = \sum_k 2^{-k}(d_k(x_k, y_k) \wedge 1)$. Suppose that for each k, $D_k \subset C_b(S_k)$ is bounded above and isolates points in S_k. Then $D = \{f_1 + \cdots + f_m : m \geq 1, f_k \in D_k\}$ is a rate function determining class for S.*

PROOF. Suppose $\{X_n\}$ is exponentially tight in S and the large deviation principle holds with rate function I. Letting X_n^k denote the kth component of X_n, the large deviation principle holds for $\{(X_n^1, \ldots, X_n^m)\}$. Let I_m denote the corresponding rate function. By Lemma 3.22 and Proposition 3.20,

$$I_m(x_1, \ldots, x_m) = \sup_{f_1 \in D_1, \ldots, f_m \in D_m} (f_1(x_1) + \cdots + f_m(x_m) - \Lambda(f_1 + \cdots + f_m)),$$

where

$$\Lambda(f_1 + \cdots + f_m) = \lim_{n \to \infty} \frac{1}{n} E[e^{n(f_1(X_n^1) + \cdots + f_m(X_n^m))}].$$

It follows easily that $I(x) \geq I_m(x_1, \ldots, x_m)$, and in fact, by the contraction principle, Lemma 3.11,

$$I_m(x_1, \ldots, x_m) = \inf\{I(y) : y_1 = x_1, \ldots, y_m = x_m\}.$$

Consequently, by the lower semicontinuity of I, it follows that

$$\lim_{m \to \infty} I_m(x_1, \ldots, x_m) = I(x)$$

and hence

$$I(x) = \sup_{f \in D}\{f(x) - \Lambda(f)\}.$$

□

The following result is particularly useful in considering Markov processes.

PROPOSITION 3.25. *Suppose $\{(X_n, Y_n)\}$ is exponentially tight in the product space $(S_1 \times S_2, d_1 + d_2)$. Let $\mu_n \in \mathcal{P}(S_1 \times S_2)$ be the distribution of (X_n, Y_n) and let $\mu_n(dx \times dy) = \eta_n(dy|x)\mu_n^1(dx)$, that is, μ_n^1 is the S_1-marginal of μ_n and η_n gives the conditional distribution of Y_n given X_n. Suppose that for each $f \in C_b(S_2)$*

$$\Lambda_2(f|x) = \lim_{n \to \infty} \frac{1}{n} \log \int_{S_2} e^{nf(y)} \eta_n(dy|x) \tag{3.20}$$

exists, that the convergence is uniform for x in compact subsets of S_1, and that $\Lambda_2(f|x)$ is a continuous function of x. For $x \in S_1$ and $y \in S_2$, define

$$I_2(y|x) = \sup_{f \in C_b(S_2)} (f(y) - \Lambda_2(f|x)).$$

If $\{X_n\}$ *satisfies the large deviation principle with good rate function* I_1, *then* $\{(X_n, Y_n)\}$ *satisfies the large deviation principle with good rate function*

$$I(x,y) = I_1(x) + I_2(y|x). \tag{3.21}$$

REMARK 3.26. The uniformity of convergence in (3.20) can be replaced by the assumption that there exists a sequence $K_n \subset S_1$ such that

$$\lim_{n\to\infty} \sup_{x\in K_n} |\Lambda_2(f|x) - \frac{1}{n}\log \int_{S_2} e^{nf(y)}\eta_n(dy|x)| = 0$$

and

$$\lim_{n\to\infty} \frac{1}{n} \log \mu_n^1(K_n^c) = -\infty.$$

PROOF. By the assumption of exponential tightness, it is enough to show that for any subsequence along which the large deviation principle holds, the rate function is given by (3.21). Consequently, we may as well assume that the large deviation principle holds for the original sequence.

Observe that by the assumptions on the convergence in (3.20), for $f(x,y) = f_1(x) + f_2(y)$, we have

$$\Lambda(f) = \Lambda_1(f_1 + \Lambda_2(f_2|\cdot))$$

We apply Corollary 3.21 taking, for definiteness

$$f_{m_1,m_2}(x,y) = f^1_{m_1}(x) + f^2_{m_2}(y) = -(m_1^2 d_1(x_0,x)) \wedge m_1 - (m_2^2 d_2(y_0,y)) \wedge m_2.$$

Recall that $\Lambda(f+c) = \Lambda(f) + c$, and observe that by the continuity of $\Lambda_2(f^2_{m_2}|x)$ as a function of x,

$$\widehat{f}^1_{m_1} = f^1_{m_1} + \Lambda_2(f^2_{m_2}|\cdot) - \Lambda_2(f^1_{m_1}|x_0)$$

satisfies the conditions of Corollary 3.21 with $S = S_1$ and $x = x_0$. Then

$$\begin{aligned} I(x_0,y_0) &= \lim_{m_1,m_2\to\infty} \Lambda(f_{m_1,m_2}) \\ &= \lim_{m_2\to\infty}\lim_{m_1\to\infty} \left(\Lambda_1(f^1_{m_1} + \Lambda_2(f^2_{m_2}|\cdot) - \Lambda_2(f^1_{m_1}|x_0)) + \Lambda_2(f^2_{m_2}|x_0)\right) \\ &= I_1(x_0) + I_2(y_0|x_0) \end{aligned}$$

giving the desired result. □

CHAPTER 4

Large deviations for stochastic processes

We focus on stochastic processes with cadlag trajectories. Throughout, (E, r) will be a complete, separable metric space, $D_E[0, \infty)$ will be the space of cadlag, E-valued functions with the Skorohod (J_1) topology, and $C_E[0, \infty)$ will be the space of continuous, E-valued functions. Theorems 4.1 and 4.4 in Section 4.1 characterize exponential tightness for these processes. These results are direct analogs of standard tightness theorems in the weak convergence theory. The effects of changes of time-scale on large deviation results are discussed in Section 4.2.

Exponential tightness is usually easier to verify if the state space is compact. Frequently, results can be obtained by first compactifying the state space, verifying the large deviation principle in the compact space, and then inferring the large deviations principle in the original space from the result in the compact space. Section 4.3 gives results supporting this approach.

Restricted to $C_E[0, \infty)$, the Skorohod topology is just the compact uniform topology. If the sequence of processes is asymptotically continuous in a strong enough sense, then we see in Section 4.4 that the large deviation principle in the Skorohod topology implies the large deviation principle in the compact uniform topology.

The main exponential tightness results are developed further in Sections 4.5 and 4.6 in the context of martingale problems. Section 4.7 gives the large deviation analog of the fact that tightness in the Skorohod topology plus convergence of the finite dimensional distributions implies weak convergence of the processes. Here, exponential tightness plus the large deviation principle for the finite dimensional distributions implies the large deviation principle for the processes. The rate function is identified as the appropriate supremum of the finite dimensional rate functions. The results can be viewed as a variation of the projective limit method (Dawson and Gärtner [**22**], de Acosta [**24**]) adapted to the Skorohod topology.

4.1. Exponential tightness for processes

We now consider exponential tightness for a sequence of processes $\{X_n\}$ in $D_E[0, \infty)$. Let X_n be adapted to a right-continuous, complete filtration $\{\mathcal{F}_t^n\}$. Let $S^n(T)$ be the collection of all $\{\mathcal{F}_t^n\}$-stopping times bounded by T, and let $S_0^n(T) \subset S^n(T)$ be the sub-collection of discrete stopping times. Recall that each $\tau \in S^n(T)$ can be approximated by a decreasing sequence in $S_0^n(T)$. The following theorem is the analogue for exponential tightness of Ethier and Kurtz [**36**], Theorem 3.8.6. The weak convergence version of condition (b) is due to Kurtz [**71**] and the weak convergence version of condition (c) is due to Aldous [**2**]. For $\delta > 0$ and $T > 0$, define the modulus of continuity in $D_E[0, \infty)$ by

$$w'(x, \delta, T) = \inf_{\{t_i\}} \max_i \sup_{s,t \in [t_{i-1}, t_i)} r(x(s), x(t)),$$

where the infimum is over $\{t_i\}$ satisfying

$$0 = t_0 < t_1 < \cdots < t_{m-1} < T \leq t_m$$

and $\min_{1\leq i\leq n}(t_i - t_{i-1}) > \delta$. See Ethier and Kurtz [**36**], Section 3.6, for a discussion of the properties of w'. We define $q(x,y) = 1 \wedge r(x,y)$. Note that q is a metric equivalent to r.

THEOREM 4.1. *Let $\mathcal{T}_0$ be a dense subset of $[0,\infty)$. Suppose that for each $t \in \mathcal{T}_0$, $\{X_n(t)\}$ is exponentially tight. Then the following are equivalent.*

a) *$\{X_n\}$ is exponentially tight in $D_E[0,\infty)$.*

b) *For each $T > 0$, there exist $\beta > 0$ and random variables $\gamma_n(\delta,\lambda,T)$, $\delta,\lambda > 0$, satisfying*

$$E[e^{n\lambda q^\beta(X_n(t+u),X_n(t))\wedge q^\beta(X_n(t),X_n(t-v))}|\mathcal{F}_t^n] \leq E[e^{\gamma_n(\delta,\lambda,T)}|\mathcal{F}_t^n], \tag{4.1}$$

for $0 \leq t \leq T$, $0 \leq u \leq \delta$, and $0 \leq v \leq t \wedge \delta$, such that for each $\lambda > 0$,

$$\lim_{\delta\to 0}\limsup_{n\to\infty}\frac{1}{n}\log E[e^{\gamma_n(\delta,\lambda,T)}] = 0 \tag{4.2}$$

and

$$\lim_{\delta\to 0}\limsup_{n\to\infty}\frac{1}{n}\log E[e^{n\lambda q^\beta(X_n(\delta),X_n(0))}] = 0. \tag{4.3}$$

c) *Condition (4.3) holds, and for each $T > 0$, there exists $\beta > 0$ such that for each $\lambda > 0$*

$$C_n(\delta,\lambda,T) \equiv \sup_{\tau\in S_0^n(T)}\sup_{u\leq\delta} E[\sup_{v\leq\delta\wedge\tau} e^{n\lambda q^\beta(X_n(\tau+u),X_n(\tau))\wedge q^\beta(X_n(\tau),X_n(\tau-v))}]$$

satisfies

$$\lim_{\delta\to 0}\limsup_{n\to\infty}\frac{1}{n}\log C_n(\delta,\lambda,T) = 0. \tag{4.4}$$

d) *For each $\epsilon > 0$ and $T > 0$*

$$\lim_{\delta\to 0}\limsup_{n\to\infty}\frac{1}{n}\log P\{w'(X_n,\delta,T) > \epsilon\} = -\infty. \tag{4.5}$$

REMARK 4.2. Note that (4.1) is implied by the simpler inequality

$$E[e^{n\lambda q^\beta(X_n(t+u),X_n(t))}|\mathcal{F}_t^n] \leq E[e^{\gamma_n(\delta,\lambda,T)}|\mathcal{F}_t^n]. \tag{4.6}$$

Puhalskii [**97**], Theorem 4.4, (see also Puhalskii [**100**], Theorem 3.2.3) gives a condition similar to (c) that would correspond to taking the supremum over u inside the expectation in the definition of $C_n(\delta,\lambda,T)$ (a condition that is substantially more difficult to verify). Under Puhalskii's condition, it is, in fact, unnecessary to consider stopping times, that is, the supremum over $\tau \in S_0^n(T)$ can be replaced by the supremum over $0 \leq t \leq T$, and C-exponential tightness (Definition 4.12) follows without additional assumptions. Theorem 4.2 of Puhalskii [**97**] (Theorem 3.2.1 of [**100**]) gives the equivalence of (a) and (d).

Note also that the minimum in the exponent can be replaced by the product. In particular, if $0 \leq a, b \leq 1$ and $\beta > 0$, then $a^\beta b^\beta \leq a^\beta \wedge b^\beta \leq a^{\beta/2}b^{\beta/2}$.

Before proving Theorem 4.1, we need the following analogue of Ethier and Kurtz [**36**], Lemma 3.8.4. We suppress the n to simplify the notation and let $a_\beta = 2^{(\beta-1)\vee 0}$ so that $q^\beta(x,y) \leq a_\beta(q^\beta(x,z) + q^\beta(z,y))$.

LEMMA 4.3. *Fix $\beta > 0$. For $\delta, \lambda, T > 0$, define*

$$C(\delta, \lambda, T) \equiv \sup_{\tau \in S_0(T)} \sup_{u \leq \delta} E[\sup_{v \leq \delta \wedge \tau} e^{\lambda q^\beta(X(\tau+u), X(\tau)) q^\beta(X(\tau), X(\tau-v))}].$$

Let $\tau \in S(T)$, and let $\widehat{\tau}$ be a stopping time with $\widehat{\tau} \geq \tau$ a.s. Then

$$E[\sup_{v \leq \delta \wedge \tau} e^{\lambda q^\beta(X(\widehat{\tau} \wedge (\tau+\delta)), X(\tau)) q^\beta(X(\tau), X(\tau-v))}] \leq C(3\delta, 24 a_\beta^4 \lambda, T + 2\delta).$$

PROOF. Let $\tau \in S(T+\delta)$, and let $M_\tau(\delta)$ denote the collection of $\mathcal{F}_\tau$-measurable random variables U satisfying $0 \leq U \leq \delta$. Note that $\tau + U \in S(T + 2\delta)$. It follows from Ethier and Kurtz [**36**] (3.8.23) and the Schwartz and Jensen inequalities, that for $\tau \in S(T + \delta)$ and $U \in M_\tau(\delta)$

$$\begin{aligned} E[\sup_{v \leq \delta \wedge \tau} & e^{\lambda q^\beta(X(\tau+U)), X(\tau)) q^\beta(X(\tau), X(\tau-v))}] \\ & \leq \sqrt{C(2\delta, 2a_\beta \lambda, T + \delta) C(3\delta, 8a_\beta^2 \lambda, T + 2\delta)} \\ & \leq C(3\delta, 8a_\beta^2 \lambda, T + 2\delta)\ . \end{aligned} \tag{4.7}$$

(Note that $C(\delta, \lambda, T)$ is nondecreasing in all three variables.) For $\tau \in S(T)$, observe that

$$\begin{aligned} q^\beta(X(\widehat{\tau} \wedge (\tau + \delta)), X(\tau)) & q^\beta(X(\tau), X(\tau - v)) \\ \leq a_\beta q^\beta & (X(\tau + \delta), X(\tau)) q^\beta(X(\tau), X(\tau - v)) \\ & + a_\beta^2 q^\beta(X(\tau + \delta), X(\widehat{\tau} \wedge (\tau + \delta))) q^\beta(X(\widehat{\tau} \wedge (\tau + \delta)), X(\tau)) \\ & + a_\beta^2 q^\beta(X(\tau + \delta), X(\widehat{\tau} \wedge (\tau + \delta))) q^\beta(X(\widehat{\tau} \wedge (\tau + \delta)), X(\tau - v)). \end{aligned}$$

Observing that $U = \tau + \delta - \widehat{\tau} \wedge (\tau + \delta)$ is $\mathcal{F}_{\widehat{\tau} \wedge (\tau+\delta)}$-measurable so that (4.7) can be applied, Hölder's inequality and (4.7 give

$$\begin{aligned} E[\sup_{v \leq \delta \wedge \tau} & e^{\lambda q^\beta(X(\widehat{\tau} \wedge (\tau+\delta)), X(\tau)) q^\beta(X(\tau), X(\tau-v))}] \\ & \leq \left(C(3\delta, 24a_\beta^3 \lambda, T + \delta) C(3\delta, 24 a_\beta^4 \lambda, T + 2\delta)^2\right)^{1/3} \\ & \leq C(3\delta, 24a_\beta^4 \lambda, T + 2\delta). \end{aligned}$$

□

PROOF. (of Theorem 4.1). (*a* implies *b*) Fix $\beta > 0$. For compact $K \subset D_E[0, \infty)$, define

$$\gamma_K(\delta, T) = \sup_{x \in K} \sup_{0 \leq t \leq T} \sup_{0 \leq u \leq \delta, 0 \leq v \leq \delta \wedge t} q^\beta(x(t + u), x(t)) \wedge q^\beta(x(t), x(t - v))$$

and note that $\lim_{\delta \to 0} \gamma_K(\delta, T) = 0$. By the assumption of exponential tightness, for each $\lambda > 0$, there is a compact K_λ satisfying

$$\limsup_{n \to \infty} \frac{1}{n} \log P\{X_n \notin K_\lambda\} \leq -2\lambda.$$

Then, setting $\gamma_n(\delta, \lambda, T) = n\lambda(\gamma_{K_\lambda}(\delta, T) + I_{\{X_n \notin K_\lambda\}})$, (4.1) is trivially satisfied and noting that

$$E[e^{n\lambda I_{\{X_n \notin K_\lambda\}}}] \leq 1 + e^{n\lambda} P\{X_n \notin K_\lambda\} \to 1,$$

we have

$$\lim_{n\to\infty}\frac{1}{n}\log E[e^{\gamma_n(\delta,\lambda,T)}]=\lambda\gamma_{K_\lambda}(\delta,T)$$

and (4.2) follows. A similar argument gives (4.3).

(b implies c) Note that if (4.1) holds for $0\le t\le T$, then the inequality also holds with t replaced by $\tau\in S_0^n(T)$, and hence, (4.2) implies (4.4).

(c implies d) We follow the development for weak convergence in Section 3.8 of Ethier and Kurtz [**36**]; however, the details are actually simpler in the present setting. Let $0<\epsilon\le 1$. Suppressing the index n for the moment, let $\tau_0=\sigma_0=0$ and for $k=1,2,\ldots$, define

$$\tau_k=\inf\{t>\tau_{k-1}:r(X(t),X(\tau_{k-1}))\ge\epsilon\},$$

if $\tau_{k-1}<\infty$, and

$$\sigma_k=\sup\{t\le\tau_k:r(X(t),X(\tau_k))\vee r(X(t-),X(\tau_k))\ge\epsilon\},$$

if $\tau_k<\infty$. Set $\sigma_k=\infty$ if $\tau_k=\infty$. It follows (see Section 3.8 of Ethier and Kurtz [**36**]) that $\min\{\tau_{k+1}-\sigma_k:\tau_k<T+\delta/2\}>\delta$ implies $w'(X,\delta/2,T)\le 2\epsilon$, so

$$P\{w'(X,\delta/2,T)>2\epsilon\}\le\sum_{k=1}^{[T/\delta]+2}P\{\tau_{k+1}-\sigma_k\le\delta,\tau_k<T+\delta/2\}.$$

Consequently, it is sufficient to show that

$$\lim_{\delta\to 0}\limsup_{n\to\infty}\max_k\frac{1}{n}\log P\{\tau_{k+1}-\sigma_k\le\delta,\tau_k<T+\delta/2\}=-\infty. \tag{4.8}$$

Let τ be any stopping time with $\tau\le T+\delta/2$, and define

$$\tau^+=\inf\{t>\tau:r(X(t),X(\tau))\ge\epsilon\}$$

and

$$\tau^-=\sup\{t\le\tau:r(X(t),X(\tau))\vee r(X(t-),X(\tau))\ge\epsilon\}.$$

Then for $0<\epsilon\le 1$ and any $\lambda>0$,

(4.9)

$$P\{\tau^+-\tau^-\le\delta\}\le e^{-n\lambda\epsilon^{2\beta}}E[\sup_{v\le\delta\wedge\tau}e^{n\lambda q^\beta(X_n(\tau^+\wedge(\tau+\delta)),X_n(\tau))q^\beta(X_n(\tau),X_n(\tau-v))}].$$

By Lemma 4.3, the right side of (4.9) is bounded by

$$e^{-n\lambda\epsilon^{2\beta}}C_n(3\delta,24a_\beta^4\lambda,T+3\delta).$$

By (4.4), it follows that

$$\lim_{\delta\to 0}\limsup_{n\to\infty}\sup_{\tau\in S(T+\delta/2)}\frac{1}{n}\log P\{\tau^+-\tau^-\le\delta\}\le-\lambda\epsilon^{2\beta}.$$

Since λ is arbitrary and for $\tau=\tau_k\wedge(T+\delta/2)$,

$$P\{\tau_{k+1}-\sigma_k\le\delta,\tau_k<T+\delta/2\}\le P\{\tau^+-\tau^-\le\delta\},$$

(4.8) follows.

(d implies a) Let $t_1,t_2,\ldots$ be some ordering of $\mathcal{T}_0$. For each $\lambda>0$ and $k=1,2,\ldots$, there exists $\delta_k>0$ and compact $\Gamma_k\subset E$ such that

$$P\{w'(X_n,\delta_k,k)>k^{-1}\}\le e^{-n\lambda k}$$

and

$$P\{X_n(t_k)\notin\Gamma_k\}\le e^{-n\lambda k},$$

for $n = 1, 2, \ldots$. Let $H_k^1 = \{x : w'(x, \delta_k, k) \le k^{-1}\}$ and $H_k^2 = \{x : x(t_k) \in \Gamma_k\}$. Let K_λ be the closure of $\cap_{k=1}^\infty H_k^1 \cap H_k^2$. Then K_λ is compact and

$$P\{X_n \notin K_\lambda\} \le 2\sum_{k=1}^{\infty} e^{-n\lambda k} \le \frac{2e^{-n\lambda}}{1-e^{-n\lambda}},$$

so $\limsup_{n\to\infty} \frac{1}{n} \log P\{X_n \notin K_\lambda\} \le -\lambda$. □

The next theorem from Schied [**106**] (see Theorem A.1 in [**30**]) reduces the verification of the exponential tightness of $\{X_n\}$ to that of $\{f(X_n)\}$ for real-valued functions f. The weak convergence analogue is in Kurtz [**73**]. (See Ethier and Kurtz [**36**], Theorem 3.9.1. Jakubowski [**62**] gives simpler conditions on the collection of functions and a simpler proof.).

THEOREM 4.4. *A sequence $\{X_n\}$ is exponentially tight in $D_E[0,\infty)$ if and only if*

a) *for each $T > 0$ and $a > 0$, there exists a compact $K_{a,T} \subset E$ such that*

$$\limsup_{n\to\infty} \frac{1}{n} \log P(\exists t \le T \ni X_n(t) \notin K_{a,T}) \le -a; \tag{4.10}$$

b) *there exists a family of functions $F \subset C(E)$ that is closed under addition and separates points in E such that for each $f \in F$, $\{f(X_n)\}$ is exponentially tight in $D_R[0,\infty)$.*

REMARK 4.5. We will refer to Condition (a) as the *exponential compact containment condition.*

PROOF. Necessity of the two conditions follows from the definition of exponential tightness and continuity of the mapping $x \in D_E[0,\infty) \to f \circ x \in D_R[0,\infty)$ for $f \in C(E)$.

Exponential tightness of $\{f \circ X_n\}$, for all $f \in F$ implies, by Lemma 3.6, that for $f_1, f_2, \ldots \in F$ $\{Z_n\} \equiv \{(f_1 \circ X_n, f_2 \circ X_n, \ldots)\}$ is exponentially tight in $D_R[0,\infty) \times D_R[0,\infty) \times \cdots$. Since F is closed under addition, exponential tightness in $D_{R^\infty}[0,\infty)$ then follows. (See Ethier and Kurtz [**36**], Problem 3.11.22.) For compact $K \subset E$, let $\{f_i^K\}$ separate points in K. Then

$$q_K(x,y) = \sum_{i=1}^{\infty} 2^{-i} 1 \wedge |f_i^K(x) - f_i^K(y)|$$

is a metric equivalent to r on K. In particular, there exists a nondecreasing function $\rho_K : [0,\infty) \to [0,\infty)$ with $\lim_{u\to 0} \rho_K(u) = 0$ such that $r(x,y) \le \rho_K(q_K(x,y))$, $x, y \in K$. Let $Z_n^K = (f_1^K \circ X_n, f_2^K \circ X_n, \ldots)$, and let the metric on R^∞ be given by $r_{R^\infty}(u,v) = \sum_{i=1}^\infty 2^{-i} 1 \wedge |u_i - v_i|$.

It follows that

$$\begin{aligned}&\{w'(X_n,\delta,T) > \epsilon\}\\ &\quad\subset \{X_n(t) \notin K,\ \text{some } t \le T\} \cup \{X_n(t) \in K, t \le T, \rho_K(w'(Z_n^K,\delta,T)) > \epsilon\},\end{aligned}$$

and the exponential tightness of $\{Z_n^K\}$ then implies

$$\lim_{\delta\to 0} \limsup_{n\to\infty} \frac{1}{n} \log P\{w'(X_n,\delta,T) > \epsilon\} \le \limsup_{n\to\infty} \frac{1}{n} \log P\{X_n(t) \notin K,\ \text{some } t \le T\}.$$

Since K is an arbitrary compact set, (4.10) then implies (4.5). □

The following localization result may simplify certain arguments, particularly for Markov processes, where if generators A and A_K satisfy $Af(x) = A_K f(x)$, $x \in K$, $f \in \mathcal{D}(A) = \mathcal{D}(A_K)$, then solutions of the corresponding martingale problems should agree in distribution up to the first time they leave K. (See Lemma 4.5.16 of Ethier and Kurtz [**36**].)

LEMMA 4.6. *Suppose $\{X_n\}$ satisfies the exponential compact containment condition and for $a > 0$, $T > 0$, let $K_{a,T}$ satisfy (4.10). For $x \in D_E[0,\infty)$, let $\tau_{a,T}(x) = \inf\{t : x(t) \notin K_{a,T}\}$. Suppose that for each $a > 0$, $T > 0$, and $n = 1, 2, \ldots$, $X_n^{a,T}(\cdot \wedge \tau_{a,T}(X_n^{a,T}))$ has the same distribution as $X_n(\cdot \wedge \tau_{a,T}(X_n))$ and that for each $a > 0$ and $T > 0$, $\{X_n^{a,T}\}$ satisfies the large deviation principle with good rate function $I^{a,T}$. Then $\{X_n\}$ satisfies the large deviation principle with good rate function given by*

$$I(x) = \lim_{\delta\to 0} \liminf_{a,T\to\infty} \inf_{y\in B_\delta(x)} I^{a,T}(y) = \lim_{\delta\to 0} \limsup_{a,T\to\infty} \sup_{y\in B_\delta(x)} I^{a,T}(y)$$

PROOF. Without loss of generality, we can assume that $X_n(\cdot \wedge \tau_{a,T}(X_n)) = X_n^{a,T}(\cdot \wedge \tau_{a,T}(X_n^{a,T}))$. (See Lemma 5.15 of Ethier and Kurtz [**36**].) Setting $X_n^{\eta,a} = X_n^{a,T}$ for $\eta = e^{-T}$, the lemma follows by Lemma 3.14. □

4.2. Large deviations under changes of time-scale

Continuous time large deviation problems are frequently approached by discrete time approximations. The following lemma demonstrates the generality of this approach (cf. Dupuis and Ellis [**35**] Lemma 10.3.4).

LEMMA 4.7. *For X_n in $D_E[0,\infty)$ and $\epsilon_n > 0$, define $Y_n(t) = X_n([t/\epsilon_n]\epsilon_n)$. Then the following hold.*

a) *If $\{X_n\}$ is exponentially tight, then $\{Y_n\}$ is exponentially tight.*

b) *If $\epsilon_n \to 0$ and $\{X_n\}$ is exponentially tight, then for each $f \in C_b(D_E[0,\infty))$*

$$\limsup_{n\to\infty} \frac{1}{n} \log E[e^{nf(X_n)}] = \limsup_{n\to\infty} \frac{1}{n} \log E[e^{nf(Y_n)}] \tag{4.11}$$

and

$$\liminf_{n\to\infty} \frac{1}{n} \log E[e^{nf(X_n)}] = \liminf_{n\to\infty} \frac{1}{n} \log E[e^{nf(Y_n)}], \tag{4.12}$$

and hence, the large deviation principle holds for $\{X_n\}$ if and only if it holds for $\{Y_n\}$ and if it holds, $\{X_n\}$ and $\{Y_n\}$ have the same rate function.

PROOF. Clearly, if $\{X_n\}$ satisfies the exponential compact containment condition, then so does $\{Y_n\}$. By the necessity of part (b) of Theorem 4.1, there exist $\gamma_n(\delta, \lambda, T)$ satisfying (4.1). Consequently, $\{Y_n\}$ will satisfy (4.1) with $\gamma_n(\delta, \lambda, T)$ replaced by

$$\widehat{\gamma}_n(\delta, \lambda, T) = \begin{cases} 0, & \delta < \epsilon_n \\ \gamma_n(2\delta, \lambda, T), & \delta \geq \epsilon_n. \end{cases}$$

Since $\widehat{\gamma}_n(\delta, \lambda, T)$ clearly satisfies (4.2), exponential tightness for $\{Y_n\}$ follows.

Part (b) will follow from Lemma 3.13 provided we verify (3.13). But, defining x^ϵ by $x^\epsilon(t) = x([t/\epsilon]\epsilon)$, if $K \subset D_E[0,\infty)$ is compact, then for the metric given by (1.5),

$$\lim_{\epsilon\to 0} \sup_{x\in K} d(x, x^\epsilon) = 0,$$

and (3.13) follows. (See Lemma 4.8 below.) □

Lemma 4.7 can be generalized to other time changes. The following property of the Skorohod topology provides the basis for these extensions.

LEMMA 4.8. *Let ζ be a cadlag, nonnegative, and nondecreasing function on $[0,\infty)$, and let $\eta > 0$ and $T > 0$ satisfy $\eta \geq \sup_{t\leq T} |\zeta(t) - t|$. Then for $\delta > 2\eta$ and the metric d given by (1.5),*

$$d(x, x\circ\zeta) \leq \log\frac{\delta}{\delta - 2\eta} \vee (w'(x,\delta,T+\eta) + \frac{\eta(T+\eta)}{\delta} + e^{-T}). \tag{4.13}$$

PROOF. Fix $\epsilon > 0$. Let $0 = t_0 < t_1 < \cdots < t_{m-1} < T+\eta \leq t_m$ satisfy

$$\max_i \sup_{s,t\in[t_{i-1},t_i)} r(x(s), x(t)) \leq w'(x,\delta,T+\eta) + \epsilon.$$

For each i, define $\tau_i = \inf\{t : \zeta(t) \geq t_i\}$, and for $0 \leq t \leq t_m$, let $\lambda(t)$ be the linear interpolation of the points (t_i, τ_i). For $t > t_m$, let $\lambda'(t) = 1$. Since $|t_i - \tau_i| \leq \eta$, $\frac{\delta}{\delta+2\eta} \leq \lambda'(t) \leq \frac{\delta}{\delta-2\eta}$, and $\gamma(\lambda) \leq \log\frac{\delta}{\delta-2\eta}$. For $t \vee \lambda(t) \leq T+\eta$,

$$r(x\circ\zeta\circ\lambda(t), x(t)) \leq w'(x,\delta,T+\eta) + \epsilon,$$

and it follows that

$$\int_0^T e^{-u} \sup_{t\geq 0} r(x\circ\zeta(\lambda(t)\wedge u), x(t\wedge u))du \leq w'(x,\delta,T+\eta) + \epsilon + 2m\eta.$$

Note that $2m\eta$ bounds the Lebesgue measure of the set of u for which there exists a t and a t_i satisfying

$$\lambda(t)\wedge u < t_i \leq t\wedge u \quad \text{or} \quad t\wedge u < t_i \leq \lambda(t)\wedge u \tag{4.14}$$

since either of the inequalities in (4.14) implies $|u - t_i| \leq \eta$. Since $m \leq (T+\eta)/\delta$ and $\epsilon > 0$ is arbitrary, (4.13) follows. □

LEMMA 4.9. *For each n, let X_n be a process in $D_E[0,\infty)$, let Λ_n be a nonnegative, nondecreasing process independent of X_n, and define*

$$Y_n(t) = X_n(\Lambda_n(t)).$$

Suppose that for each $t > 0$ and $\eta > 0$,

$$\lim_{n\to\infty}\frac{1}{n}\log P\{\sup_{s\leq t}|s - \Lambda_n(s)| > \eta\} = -\infty. \tag{4.15}$$

a) *If $\{X_n\}$ is exponentially tight, then $\{Y_n\}$ is exponentially tight.*
b) *If $\{X_n\}$ is exponentially tight, then for each $f \in C_b(D_E[0,\infty))$*

$$\limsup_{n\to\infty}\frac{1}{n}\log E[e^{nf(X_n)}] = \limsup_{n\to\infty}\frac{1}{n}\log E[e^{nf(Y_n)}] \tag{4.16}$$

and

$$\liminf_{n\to\infty}\frac{1}{n}\log E[e^{nf(X_n)}] = \liminf_{n\to\infty}\frac{1}{n}\log E[e^{nf(Y_n)}], \tag{4.17}$$

and hence, the large deviation principle holds for $\{X_n\}$ if and only if it holds for $\{Y_n\}$ and if it holds, $\{X_n\}$ and $\{Y_n\}$ have the same rate function.

REMARK 4.10. An obvious question is what happens to $\{Y_n\}$ if Λ_n does not converge as in 4.15? Russell [**103**] gives a systematic discussion of this question.

PROOF. Lemma 4.8 implies (3.13), and the result follows by Lemma 3.13. □

4.3. Compactification

The requirement in Theorem 4.1 that $\{X_n(t)\}$ be exponentially tight for each $t \in \mathcal{T}_0$ and the stronger requirement in Condition (a) of Theorem 4.4 are, of course, trivially satisfied if E is compact. As in the weak convergence setting, these conditions can sometimes be finessed in the noncompact case by working with some compactification $\widehat{E}$ of E. The following result gives conditions under which the large deviation principle in $D_{\widehat{E}}[0,\infty)$ can be used to obtain the large deviation principle in $D_E[0,\infty)$. The result also covers situations where the topology on $\widehat{E}$ is weaker than the topology on E. For example, $\widehat{E}$ could be the space of probability measures on a metric space with the weak topology while E is the space of probability measures on the same metric space having a density with respect to some fixed reference measure ν with the metric on E given by the $L_1(\nu)$-norm.

THEOREM 4.11. *Suppose $(\widehat{E},\widehat{r})$ and (E,r) are metric spaces, $E \subset \widehat{E}$, $\mathcal{B}(E) \subset \mathcal{B}(\widehat{E})$, and if $\{z_n\} \subset E$ converges to $z \in E$ under r, then it also converges under $\widehat{r}$ (that is, r generates a stronger topology on E than $\widehat{r}$). Let $\{P_n\}$ be a sequence of probability measures on $D_{(\widehat{E},\widehat{r})}[0,\infty)$ such that for each n, $P_n(D_{(E,r)}[0,\infty)) = 1$. (We include the metric in the notation for clarity. Note that $D_{(E,r)}[0,\infty) \subset D_{(E,\widehat{r})}[0,\infty)$, but the two spaces need not be equal.) Let*

$$\widehat{\Gamma} = D_{(\widehat{E},\widehat{r})}[0,\infty) - D_{(E,r)}[0,\infty).$$

Suppose that the large deviation principle for $\{P_n\}$ holds in $D_{\widehat{E}}[0,\infty)$ with good rate function $\widehat{I}$.

a) *If the exponential compact containment condition holds in (E,r), then $\widehat{I}(x) = \infty$, for all $x \in \widehat{\Gamma}$, and the large deviation principle holds for $\{P_n\}$ on $D_{(E,r)}[0,\infty)$.*
b) *Suppose that the topology on E generated by $\widehat{r}$ is the same as the topology generated by r so*

$$\widehat{\Gamma} = \{x \in D_{\widehat{E}}[0,\infty) : \exists t \geq 0 \ni x(t) \text{ or } x(t-) \in \widehat{E} - E\}.$$

If $\widehat{I}(x) = \infty$ for each $x \in \widehat{\Gamma}$, then the large deviation principle holds for $\{P_n\}$ on $D_E[0,\infty)$ with rate function I given by $I(x) = \widehat{I}(x)$ for $x \in D_E[0,\infty)$. In particular, the exponential compact containment condition holds in E.

PROOF. Note that if K is compact in (E,r), then it is also compact in $(\widehat{E},\widehat{r})$ (but not conversely) and the topology on K given by r is the same as the topology on K given by $\widehat{r}$. In particular, $D_{(K,r)}[0,\infty) = D_{(K,\widehat{r})}[0,\infty)$. If $\mathcal{K} \subset D_{(\widehat{E},\widehat{r})}[0,\infty)$ is compact, then $\mathcal{K}_K = \mathcal{K} \cap D_{(K,r)}[0,\infty)$ is a compact subset of $D_{(E,r)}[0,\infty)$. (Apply the characterization of Skorohod convergence given in [**36**], Proposition 3.6.5, and the equivalence of r and $\widehat{r}$ on K.)

Under the conditions of Part (a), for $a,\eta > 0$ and $e^{-T} < \eta$, let $K = K_{a,T}$ be a compact subset of E (under r) satisfying (4.10), and let $\mathcal{K}$ be a compact subset of $D_{(\widehat{E},\widehat{r})}[0,\infty)$ satisfying

$$\limsup_{n\to\infty} \frac{1}{n} \log P\{X_n \notin \mathcal{K}\} \leq -a.$$

Let $\mathcal{K}_K = \mathcal{K} \cap D_{(K,r)}[0,\infty)$. Since $\mathcal{K}_K^\epsilon \supset \mathcal{K} \cap (D_{(K,r)}[0,\infty))^\epsilon$,

$$\begin{aligned}
&\limsup_{n\to\infty} \frac{1}{n} \log P\{X_n \notin \mathcal{K}_K^\epsilon\} \\
&\quad \leq \max\{\limsup_{n\to\infty} \frac{1}{n} P\{X_n \notin \mathcal{K}\}, \limsup_{n\to\infty} \frac{1}{n} \log P\{X_n \notin (D_{(K,r)}[0,\infty))^\epsilon\} \\
&\quad \leq -a,
\end{aligned}$$

where the ϵ-"fattening" is with respect to the metric (1.5) given by r. It follows that $\{X_n\}$ is exponentially tight, and Part (a) follows by Lemma 3.12.

For Part (b), suppose that $\widehat{I}(x) = \infty$ for each $x \in \widehat{\Gamma}$. If A is an open subset of $D_E[0,\infty)$, then by the equivalence of the topologies, there exists an open subset $\widehat{A}$ of $D_{\widehat{E}}[0,\infty)$ such that $A = \widehat{A} \cap D_E[0,\infty)$. Consequently,

$$\begin{aligned}
\liminf_{n\to\infty} \frac{1}{n} \log P_n(A) &= \liminf_{n\to\infty} \frac{1}{n} \log P_n(\widehat{A} \cap D_E[0,\infty)) \\
&= \liminf_{n\to\infty} \frac{1}{n} \log P_n(\widehat{A}) \\
&\geq -\inf_{x\in\widehat{A}} \widehat{I}(x) \\
&= -\inf_{x\in A} \widehat{I}(x),
\end{aligned}$$

where the last equality follows from the fact that $x \in \widehat{A} - A$ implies $\widehat{I}(x) = \infty$. Similarly, if B is a closed subset of $D_E[0,\infty)$, then there exists a closed subset $\widehat{B}$ of $D_{\widehat{E}}[0,\infty)$ such that $B = \widehat{B} \cap D_E[0,\infty)$. Consequently,

$$\begin{aligned}
\limsup_{n\to\infty} \frac{1}{n} \log P_n(B) &= \limsup_{n\to\infty} \frac{1}{n} \log P_n(\widehat{B} \cap D_E[0,\infty)) \\
&= \limsup_{n\to\infty} \frac{1}{n} \log P_n(\widehat{B}) \\
&\leq -\inf_{x\in\widehat{B}} \widehat{I}(x) \\
&= -\inf_{x\in B} \widehat{I}(x),
\end{aligned}$$

and Part (b) follows. □

4.4. Large deviations in the compact uniform topology

DEFINITION 4.12. [C-exponential tightness] A sequence of stochastic processes $\{X_n\}$ that is exponentially tight in $D_E[0,\infty)$ is *C-exponentially tight* if for each $\eta > 0$ and $T > 0$,

$$\limsup_{n\to\infty} \frac{1}{n} \log P(\sup_{s\leq T} r(X_n(s), X_n(s-)) \geq \eta) = -\infty. \tag{4.18}$$

The weak convergence analogue of this condition is simply that

$$\sup_{s\leq T} r(X_n(s), X_n(s-)) \Rightarrow 0$$

for each $T > 0$ which implies that any weak limit point of $\{X_n\}$ must be continuous. The definition of C-exponential tightness given here is not the same as that given in Puhalskii **[98]** (Definition 3.2.2 of **[100]**); however, the following theorem shows that the two definitions are equivalent.

THEOREM 4.13. *An exponentially tight sequence $\{X_n\}$ in $D_E[0,\infty)$ is C-exponentially tight if and only if each rate function I that gives the large deviation principle for a subsequence $\{X_{n(k)}\}$, satisfies $I(x) = \infty$ for each $x \in D_E[0,\infty)$ such that $x \notin C_E[0,\infty)$.*

PROOF. Let $A_{\eta,T} = \{x \in D_E[0,\infty) : \sup_{s\le T} r(x(s),x(s-)) \geq \eta\}$ and $B_{\eta,T} = \{x \in D_E[0,\infty) : \sup_{s<T} r(x(s),x(s-)) > \eta\}$. Then $A_{\eta,T}$ is closed and $B_{\eta,T}$ is open. If $\{X_n\}$ is exponentially tight and (4.18) holds, then for any subsequence satisfying the large deviation principle with a rate function I (see Theorem 3.7), we have

$$\begin{aligned} -\inf_{x\in B_{\eta,T}} I(x) &\leq \liminf_{k\to\infty} \frac{1}{n(k)} \log P(X_{n(k)} \in B_{\eta,T}) \\ &\leq \limsup_{k\to\infty} \frac{1}{n(k)} \log P(X_{n(k)} \in A_{\eta,T}) = -\infty. \end{aligned}$$

Therefore, $I(x) = \infty$ for each $x \in B_{\eta,T}$. By the arbitrariness of η and T, $I(x) = \infty$ for every $x \in D_E[0,\infty) - C_E[0,\infty)$.

Now, assume that $\{X_n\}$ is exponentially tight but that (4.18) does not hold. Then there exists a subsequence $n(k)$ such that

$$\liminf_{k\to\infty} \frac{1}{n(k)} \log P(X_{n(k)} \in A_{\eta,T}) > -\infty, \tag{4.19}$$

and by exponential tightness a subsubsequence along which the large deviation principle holds with a rate function I. It follows that

$$-\inf_{x\in A_{\eta,T}} I(x) \geq \liminf_{k\to\infty} \frac{1}{n(k)} \log P(X_{n(k)} \in A_{\eta,T}) > -\infty,$$

that is, there exists $x \in A_{\eta,T} \subset D_E[0,\infty) - C_E[0,\infty)$ such that $I(x) < \infty$. □

With reference to (1.5), for $x, y \in D_E[0,\infty)$, let

$$d_u(x,y) = \int_0^\infty e^{-u} \sup_{t\le u} q(x(t),y(t))du.$$

Then d_u is a metric corresponding to the compact uniform topology. Let $\mathcal{S}_E$ be the Borel σ-algebra of $(D_E[0,\infty),d)$. Note that $f_x(y) = d_u(x,y)$ defines a $\mathcal{S}_E$-measurable function and hence $B^u_\epsilon(x) = \{y \in D_E[0,\infty) : d_u(x,y) < \epsilon\}$ and $\overline{B}^u_\epsilon(x) = \{y \in D_E[0,\infty) : d_u(x,y) \leq \epsilon\}$ are in $\mathcal{S}_E$. On $C_E[0,\infty)$, d_u and d are equivalent metrics, and in the theory of weak convergence, if $X_n \Rightarrow X$ in the Skorohod topology and X is continuous, then $X_n \Rightarrow X$ in the compact uniform topology in the sense that

$$\lim_{n\to\infty} E[f(X_n)] = E[f(X)]$$

for every bounded function that is d_u-continuous and $\mathcal{S}_E$-measurable. C-exponential tightness is the large deviation analogue of the limit being continuous in weak convergence.

THEOREM 4.14. *Suppose that $\{P_n\}$ is a sequence of probability measures on $D_E[0,\infty)$ that is C-exponentially tight and that the large deviation principle holds with rate function I (in the Skorohod topology). Then the large deviation principle holds in the compact uniform topology with the same rate function.*

PROOF. The result is an immediate consequence of the contraction principle, Lemma 3.11. Take $S = (D_E[0,\infty), d)$, $S' = (D_E[0,\infty), d_u)$, and $F(x) = x$. Then F is continuous at each point in $C_E[0,\infty)$ and by Theorem 4.13, $I(x) = \infty$ if $x \notin C_E[0,\infty)$. □

4.5. Exponential tightness for solutions of martingale problems

DEFINITION 4.15. Let A be the generator of a Markov process, and let $\mathcal{H}_\dagger$ be the collection of $(f,g) \in M(E) \times M(E)$ such that for every solution X of the martingale problem for A,

$$Z_{f,g}(t) = e^{f(X(t)) - f(Y(0)) - \int_0^t g(X(s))ds} \tag{4.20}$$

is a cadlag supermartingale, or, in the discrete time case with $X(t) = Y_{[t/\epsilon]}$,

$$Z_{f,g}(t) = e^{f(X(t)) - f(X(0)) - \int_0^{[t/\epsilon]\epsilon} g(X(s))ds} \tag{4.21}$$

is a supermartingale.

Of course, $\mathcal{H}$ defined in (1.7) is a subset of $\mathcal{H}_\dagger$, and, for example, if A gives a diffusion and L is the second order differential operator such that $Af = Lf$ for $f \in \mathcal{D}(A)$, then for $f \in C^2$ and $h = e^{-f}Le^f$, (4.20) is a local martingale by Itô's formula. Since every nonnegative local martingale is a supermartingale, $(f, e^{-f}Le^f) \in \mathcal{H}_\dagger$. More generally, if $f_k \in \mathcal{D}(\mathcal{H})$, $\inf_k \inf_x f_k(x) > -\infty$, $\sup_k \sup_x \mathcal{H}f_k(x) < \infty$, and for each $x \in E$, $(f_k(x), \mathcal{H}f_k(x)) \to (f(x), h(x))$, then (4.20) is a supermartingale.

For $n = 1, 2, \ldots$, let E_n be a metric space and let $\eta_n : E_n \to E$ be a Borel measurable mapping into a complete, separable metric space E. Let $A_n \subset B(E_n) \times B(E_n)$ and $H_n = \{(f, \frac{1}{n}e^{-nf}g) : f \in B(E_n), (e^{nf}, g) \in A_n\}$. Let $H_{\dagger,n}$ be the collection of $(f,g) \in M(E_n) \times M(E_n)$ such that

$$e^{nf(Y_n(t)) - nf(Y_n(0)) - \int_0^t ng(Y_n(s))ds}$$

is a supermartingale for every solution Y_n of the martingale problem for A_n. Of course, $H_n \subset H_{\dagger,n}$.

Similarly, in discrete time, let T_n be a transition operator on $B(E_n)$, $\epsilon_n > 0$, and

$$H_n = \frac{1}{n\epsilon_n} \log e^{-nf} T_n e^{nf}, \quad f \in B(E_n).$$

Then $H_{\dagger,n}$ is the collection of $(f,g) \in B(E_n) \times B(E_n)$ such that

$$e^{nf(Y_n(t)) - nf(Y_n(0)) - \int_0^{[t/\epsilon_n]\epsilon_n} ng(Y_n(s))ds}$$

is a supermartingale for every Markov chain with time step ϵ_n satisfying

$$E[f(Y_n(t))|Y_n(0) = y] = T_n^{[t/\epsilon_n]} f(y).$$

In other words,

$$e^{-g(x)} T_n e^{nf}(x) \le e^{nf(x)}, \quad x \in E_n.$$

As above, $q = r \wedge 1$.

DEFINITION 4.16. $D \subset C_b(E)$ *approximates the metric* q if for each compact $K \subset E$ and $z \in K$, there exists $f_n \in D$ such that

$$\lim_{n\to\infty} \sup_{x\in K} |f_n(x) - q(x,z)| = 0.$$

We want to give conditions implying exponential tightness for the sequence $X_n = \eta_n(Y_n)$. For $f \in B(E)$, define

$$\eta_n f = f \circ \eta_n.$$

We assume that X_n is cadlag.

COROLLARY 4.17. *Let Y_n, X_n, H_n, and $H_{\dagger,n}$ be as above in either the continuous or discrete time case, and in the discrete time case, assume that $\epsilon_n \to 0$. Let $F \subset C_b(E)$ and $S \subset R$. Suppose that*

a) *$\{X_n\}$ satisfies the exponential compact containment condition.*
b) *Either F is closed under addition and separates points in E and $S = R$, or F approximates the metric q and $S = (0, \infty)$.*
c) *For each $\lambda \in S$ and $f \in F$, there exists $\{(f_n, g_n)\}$ such that $(\lambda f_n, g_n) \in H_{\dagger,n}$, $\sup_n \|f_n\| < \infty$, for each compact $K \subset E$,*

$$\lim_{n\to\infty} \sup_{x\in\eta_n^{-1}(K)} |f_n(x) - \eta_n f(x)| = 0 \tag{4.22}$$

and

$$\sup_n \sup_{x\in\eta_n^{-1}(K)} g_n(x) = C_\lambda(f, K) < \infty. \tag{4.23}$$

Then $\{X_n\}$ is exponentially tight.

REMARK 4.18. In many of the examples mentioned in the Introduction, $E_n = E$, η_n is the identity map, and it suffices to let $f_n = f$. In others including Examples 1.9 and 1.12, the more general condition is needed.

PROOF. First, assume that F is closed under addition and separates points in E and $S = R$. We consider the continuous-time case. The proof for the discrete-time case is essentially the same.

Since for $\lambda > 0$,

$$\begin{aligned} &E[e^{n\lambda|f(X_n(t+s))-f(X_n(t))|}|\mathcal{F}_t^n] \\ &\quad\le E[e^{n\lambda(f(X_n(t+s))-f(X_n(t)))}|\mathcal{F}_t^n] + E[e^{-n\lambda(f(X_n(t+s))-f(X_n(t)))}|\mathcal{F}_t^n], \end{aligned}$$

by Theorems 4.1 and 4.4, it is enough to show that for each $\lambda \in R$, $T > 0$ and n, there exists $\gamma_n(s, \lambda, T)$, increasing in s and satisfying

$$E[\exp\{n\lambda f(X_n(t+s)) - n\lambda f(X_n(t))\}|\mathcal{F}_t^n] \le E[\exp\{\gamma_n(s, \lambda, T)\}|\mathcal{F}_t^n]$$

for $0 \le t \le T$, such that

$$\lim_{s\to 0} \limsup_{n\to\infty} \frac{1}{n} \log E[e^{\gamma_n(s,\lambda,T)}] = 0.$$

For compact $K \subset E$, let $\Gamma_n(K,T) = \{X_n(t) \in K, t \le T\} = \{Y_n(t) \in \eta_n^{-1}(K), t \le T\}$. If $f \in F$ and $(\lambda f_n, g_n) \in H_n$ satisfies Condition (c), then

$$\begin{aligned} &E[I_{\Gamma_n(K,T)} e^{n\lambda f_n(Y_n(t+s))-n\lambda f_n(Y_n(t))}|\mathcal{F}_t^n] \\ &\quad\le e^{nC_\lambda(f,K)s} E[e^{n\lambda f_n(Y_n(t+s))-n\lambda f_n(Y_n(t))-\int_0^s ng_n(Y_n(t+u))du}|\mathcal{F}_t^n] \\ &\quad\le e^{nC_\lambda(f,K)s}. \end{aligned} \tag{4.24}$$

By assumption, there exists a compact set $K = K(\lambda, f, T) \subset E$ such that

$$\limsup_{n\to\infty} \frac{1}{n} \log P(\exists t \le T, X_n(t) \notin K) \le -(1 + 2|\lambda|\|f\|).$$

Therefore, for $t, s \geq 0$, $t+s \leq T$,

$$\begin{aligned}
&E[e^{n\lambda f(X_n(t+s))-n\lambda f(X_n(t))}|\mathcal{F}_t^n] \\
&\quad \leq E[e^{n\lambda f(X_n(t+s))-n\lambda f(X_n(t))}I_{\Gamma_n(K,T)}|\mathcal{F}_t^n] \\
&\qquad +E[e^{n\lambda f(X_n(t+s))-n\lambda f(X_n(t))}I_{\Gamma_n(K,T)^c}|\mathcal{F}_t^n] \\
&\quad \leq e^{2n|\lambda|\sup_{y\in\eta_n^{-1}(K)}|\eta_n f(y)-f_n(y)|}E[I_{\Gamma_n(K,T)}e^{n\lambda f_n(Y_n(t+s))-n\lambda f_n(Y_n(t))}|\mathcal{F}_t^n] \\
&\qquad +e^{2n|\lambda|\|f\|}P\{\Gamma_n(K,T)^c|\mathcal{F}_t^n\} \\
&\quad \leq e^{2n|\lambda|\sup_{y\in\eta_n^{-1}(K)}|f_n(y)-\eta_n f(y)|+nC_\lambda(f,K)s}+e^{2n\lambda\|f\|}P\{\Gamma_n(K,T)^c|\mathcal{F}_t^n\}.
\end{aligned}$$

Taking

$$\gamma_n(s,\lambda,T)=\log\{e^{2n|\lambda|\sup_{y\in\eta_n^{-1}(K)}|f_n(y)-\eta_n f(y)|+nC_\lambda(f,K)s}+e^{2n|\lambda|\|f\|}I_{\Gamma_n(K,T)^c}\},$$

then

$$\begin{aligned}
&\limsup_{n\to\infty}\frac{1}{n}\log E[e^{\gamma_n(s,\lambda,T)}] \\
&\quad =\max\{C_\lambda(f,K)s,2|\lambda|\|f\|+\limsup_{n\to\infty}\frac{1}{n}\log P\{\Gamma_n(K,T)^c\}\} \\
&\quad =C_\lambda(f,K)s.
\end{aligned}$$

Hence, by Theorems 4.1 and 4.4, the conclusion follows.

Now assume that F approximates the metric q and $S=(0,\infty)$. For $\lambda\in S$, select compact $K\subset E$ so that

$$\limsup_{n\to\infty}\frac{1}{n}\log P(\exists t\leq T, X_n(t)\notin K)\leq -(1+\lambda).$$

For each $\epsilon>0$, there exist $N_\epsilon>0$, $\{\xi_1,\ldots,\xi_{N_\epsilon}\}\subset E$, and $\{f_1^\epsilon,\ldots,f_{N_\epsilon}^\epsilon)\in F$ such that $\min_{k=1,\ldots,N_\epsilon}q(x,\xi_k)<\epsilon$, for all $x\in K$, and $\max_{1\leq k\leq N_\epsilon}\sup_{z\in K}|q(z,\xi_k)-f_k(z)|\leq\epsilon$.

For $y\in K$, there exists $\xi_k\in\{\xi_1,\ldots,\xi_N\}$ such that $q(y,\xi_k)<\epsilon$. Then, for $x\in K$,

$$q(x,y)\leq q(x,\xi_k)+q(y,\xi_k)\leq q(x,\xi_k)+\epsilon\leq q(x,\xi_k)-q(y,\xi_k)+2\epsilon,$$

and

$$q(x,y)\leq f_k^\epsilon(x)-f_k^\epsilon(y)+4\epsilon.$$

It follows that

$$e^{n\lambda q(x,y)}\leq e^{4n\lambda\epsilon}\sum_{k=1}^{N(\epsilon)}e^{n\lambda f_k(x)-n\lambda f_k(y)}.$$

By (4.22) and (4.23), there exist $(\lambda f_{n,k}^\epsilon, g_{n,k}^\epsilon)\in H_{\dagger,n}$ such that for n sufficiently large

$$\max_k\sup_{y\in\eta_n^{-1}(K)}|f_{n,k}^\epsilon(y)-\eta_n f_k^\epsilon(y)|\leq\epsilon$$

and $C_\lambda(f_k^\epsilon, K) \equiv \max_k \sup_{y\in\eta_n^{-1}(K)} |g_{n,k}^\epsilon(y)| < \infty$, and hence,

$$\begin{aligned}
&I_{\Gamma_n(K,T)} e^{n\lambda q(X_n(t+s),X_n(t))} \\
&\leq I_{\Gamma_n(K,T)} e^{6\epsilon n\lambda} \sum_{k=1}^{N_\epsilon} e^{n\lambda f_{n,k}^\epsilon(Y_n(t+s)) - n\lambda f_{n,k}^\epsilon(Y_n(t))} \\
&\leq I_{\Gamma_n(K,T)} e^{6\epsilon n\lambda + sn\vee_k C_\lambda(f_k^\epsilon,K)} \\
&\qquad \sum_{k=1}^{N_\epsilon} e^{n\lambda f_{n,k}^\epsilon(Y_n(t+s)) - n\lambda f_{n,k}^\epsilon(Y_n(t)) - \int_t^{t+s} ng_{n,k}^\epsilon(Y_n(r))dr} \\
&\leq e^{6\epsilon n\lambda + sn\vee_k C_\lambda(f_k^\epsilon,K)} \sum_{k=1}^{N_\epsilon} e^{n\lambda f_{n,k}^\epsilon(Y_n(t+s)) - n\lambda f_{n,k}^\epsilon(Y_n(t)) - \int_t^{t+s} ng_{n,k}^\epsilon(Y_n(r))dr}.
\end{aligned}$$

Therefore

$$\begin{aligned}
E[e^{n\lambda q(X_n(t+s),X_n(t))}|\mathcal{F}_t^n] &\leq E[I_{\Gamma_n(K,T)} e^{n\lambda q(X_n(t+s),X_n(t))}|\mathcal{F}_t^n] \\
&\qquad + E[I_{\Gamma_n(K,T)^c} e^{n\lambda q(X_n(t+s),X_n(t))}|\mathcal{F}_t^n] \\
&\leq e^{6\epsilon n\lambda + sn\vee_{k=1}^{N_\epsilon} C_\lambda(f_k^\epsilon,K)} + e^{n\lambda} P\{\Gamma_n(K,T)^c|\mathcal{F}_t^n\}.
\end{aligned}$$

For $\delta > 0$, let $\epsilon_0(\delta) = \inf\{\epsilon > 0 : \vee_{k=1}^{N_\epsilon} C_\lambda(f_k^\epsilon, K) \leq \delta^{-1/2}\}$. Note that $\epsilon_0(\delta) \to 0$ as $\delta \to 0$, and select $\epsilon(\delta)$ such that $\epsilon(\delta) < \epsilon_0(\delta) + \delta$ and $\vee_{k=1}^{N_{\epsilon(\delta)}} C_\lambda(f_k^{\epsilon(\delta)}, K) \leq \delta^{-1/2}$. Finally, define

$$\gamma_n(\delta, \lambda, T) = \log\{e^{6\epsilon(\delta)n\lambda + \delta n\vee_{k=1}^{N_{\epsilon(\delta)}} C_\lambda(f_k^{\epsilon(\delta)},K)} + e^{n\lambda} I_{\Gamma_n(K,T)^c}\},$$

and observe that

$$\begin{aligned}
&\limsup_{n\to\infty} \frac{1}{n} \log E[e^{\gamma_n(s,\lambda,T)}] \\
&= \max\{6\epsilon(\delta)\lambda + \delta \vee_{k=1}^{N_{\epsilon(\delta)}} C_\lambda(f_k^{\epsilon(\delta)}, K), \lambda + \limsup_{n\to\infty} \frac{1}{n} \log P(\Gamma_n(K,T)^c)\} \\
&= 6\epsilon(\delta)\lambda + \delta \vee_{k=1}^{N_{\epsilon(\delta)}} C_\lambda(f_k^{\epsilon(\delta)}, K).
\end{aligned}$$

Since the right side goes to zero as $\delta \to 0$, the conclusion follows by Theorem 4.1. □

The following technical modification of Corollary 4.17 is occasionally useful. The proof is essentially the same.

COROLLARY 4.19. *Let Y_n, X_n, H_n, and $H_{\dagger,n}$ be as above in either the continuous or discrete time case, and in the discrete time case, assume that $\epsilon_n \to 0$. Let $F \subset C_b(E)$ and $S \subset R$. Let $\mathcal{Q}$ be an index set and for $n = 1, 2, \ldots$ and $q \in \mathcal{Q}$, let K_n^q be a measurable subset of E_n. Suppose that*

a) *For each $q \in \mathcal{Q}$, $\cup_n \eta_n(K_n^q)$ is relatively compact in E.*
b) *For each $a > 0$ and $T > 0$, there exists $q(a,T) \in \mathcal{Q}$ such that*

$$\limsup_{n\to\infty} \frac{1}{n} \log P\{\exists t \leq T \ni Y_n(t) \notin K_n^{q(a,T)}\} \leq -a. \tag{4.25}$$

(In particular, $\{X_n\}$ satisfies the exponential compact containment condition.)

c) *Either F is closed under addition and separates points in E and $S = R$, or F approximates the metric q and $S = (0,\infty)$.*

d) *For each $\lambda \in S$ and $f \in F$, there exists $\{(f_n, g_n)\}$ such that $(\lambda f_n, g_n) \in H_{\dagger,n}$, $\sup_n \|f_n\| < \infty$, for each $q \in \mathcal{Q}$,*

$$\lim_{n\to\infty} \sup_{x\in K_n^q} |f_n(x) - \eta_n f(x)| = 0 \tag{4.26}$$

and

$$\sup_n \sup_{x\in K_n^q} g_n(x) = C_\lambda(f, K) < \infty. \tag{4.27}$$

Then $\{X_n\}$ is exponentially tight.

4.6. Verifying compact containment

The usual approach to verifying a property like the exponential compact containment condition (Condition (a) of Theorem 4.4 and Corollary 4.17) is with a Lyapunov function technique.

LEMMA 4.20. *Let K be compact and $G \supset K$ be open. Let A be the generator for a Markov process, and let $\mathcal{H}_\dagger$ be given by Definition 4.15. Let $(f, g) \in \mathcal{H}_\dagger$, and define*

$$\beta \equiv \inf_{x\in G^c} f(x) - \sup_{x\in K} f(x), \quad \gamma \equiv \sup_{x\in G} g(x) \vee 0.$$

Let X be a solution of the martingale problem for A, and suppose that $Z_{f,g}$ defined in (4.20) is right continuous. Then for $T > 0$,

$$P\{X(0) \in K, X(t) \notin G \text{ some } t \le T\} \le P\{X(0) \in K\}e^{-\beta+T\gamma}. \tag{4.28}$$

REMARK 4.21. Suppose, as in Corollary 4.17, Y takes valued in E_0, $\eta : E_0 \to E$, $X = \eta(Y)$, and $(f, g) \in \mathcal{H}_\dagger \subset M(E_0) \times M(E_0)$. If $K \subset E$ is compact, $G \supset K$ is open and

$$\beta \equiv \inf_{y\in\eta^{-1}(G)^c} f(y) - \sup_{y\in\eta^{-1}(K)} f(y), \quad \gamma \equiv \sup_{y\in\eta^{-1}(G)} g(y),$$

then for $T > 0$,

$$P\{X(0) \in K, X(t) \notin G \text{ some } t \le T\} \le P\{X(0) \in K\}e^{-\beta+T\gamma}. \tag{4.29}$$

PROOF. Let

$$\tau = I_{\{X(0)\in K\}} \inf\{t : X(t) \notin G\}.$$

Then, since $Z_{f,g}$, given in (4.20), is a right continuous supermartingale, by the optional sampling theorem

$$\begin{aligned} P\{X(0) \in K, X(t) \notin G \text{ some } t \le T\}e^{\beta-T\gamma} + P\{X(0) \notin K\} &\le E[Z_{f,g}(\tau \wedge T)] \\ &\le 1, \end{aligned}$$

and (4.28) follows. □

The following lemma is an immediate consequence of the previous lemma.

LEMMA 4.22. *Let X_n be a solution of the martingale problem for A_n. Let K be compact, and let $G \supset K$ be open. Let $(f_n, g_n) \in H_{\dagger,n} = \{(f, g) : (nf, ng) \in \mathcal{H}_{\dagger,n}\}$, and assume that the corresponding supermartingale is right continuous. Define*

$$\beta(K, G) \equiv \liminf_{n\to\infty}(\inf_{x\in G^c} f_n(x) - \sup_{x\in K} f_n(x)), \quad \gamma(G) \equiv \limsup_{n\to\infty} \sup_{x\in G} g_n(x).$$

Then

$$\limsup_{n\to\infty}\frac{1}{n}\log P\{X_n(t)\notin G \text{ some } t\le T\}$$
$$\le \max\{-\beta(K,G)+T\gamma(G),\limsup_{n\to\infty}\frac{1}{n}\log P\{X_n(0)\notin K\}\}.$$

If E is locally compact, Lemma 4.22 can be applied to open G with compact closure to verify the exponential compact containment condition.

EXAMPLE 4.23. Suppose in Example 1.5 there exists $C>0$ such that

$$\sum_i x_i b_i(x)\le C(1+|x|^2),\quad |a_{ij}(x)|\le C(1+|x|^2), \tag{4.30}$$

and there exists $\alpha>0$ such that

$$C(\eta,\alpha)\equiv\sup_{x\in R^d}\int_{R^d}\left(\frac{|z|}{1+|x|}\right)^2 e^{\alpha\frac{|z|}{1+|x|}}\eta(x,dz)<\infty. \tag{4.31}$$

Then for $\delta>0$ satisfying $6\delta<\alpha$, let $f_n(x)=f(x)=\delta\log(1+|x|^2)$ and

$$g_n(x)=\frac{1}{2n}\sum_{ij}a_{ij}(x)\partial_i\partial_j f(x)+\frac{1}{2}\sum_{ij}a_{ij}(x)\partial_i f(x)\partial_j f(x)+\sum_i b_i(x)\partial_i f(x)$$
$$+\int_{R^d}(e^{n\delta\log\left(1+\frac{n^{-1}2x\cdot z+n^{-2}|z|^2}{1+|x|^2}\right)}-1-\frac{2\delta x\cdot z}{1+|x|^2})\eta(x,dz).$$

The derivative terms are uniformly bounded in x and n by a constant depending on C in (4.30) and δ. The integral term satisfies

$$\int_{R^d}(e^{n\delta\log\left(1+\frac{n^{-1}2x\cdot z+n^{-2}|z|^2}{1+|x|^2}\right)}-1-\frac{2\delta x\cdot z}{1+|x|^2})\eta(x,dz)$$
$$\le\int_{R^d}(e^{n\delta\log\left(1+\frac{n^{-1}2x\cdot z+n^{-2}|z|^2}{1+|x|^2}\right)}-e^{\frac{2\delta x\cdot z}{1+|x|^2}})\eta(x,dz)$$
$$+\int_{R^d}(e^{\frac{2\delta x\cdot z}{1+|x|^2}}-1-\frac{2\delta x\cdot z}{1+|x|^2})\eta(x,dz).$$

The second term on the right is bounded by $2\delta^2C(\eta,4\delta)<\infty$. Setting $\Gamma_{n,x}=\{z:|z|>n(1+|x|)\}$, the first term on the right is bounded by

$$\int_{\{z:|z|\le n(1+|x|)\}}e^{\delta\frac{|2x\cdot z|+n^{-1}|z|^2}{1+|x|^2}}\frac{\delta|z|^2}{n(1+|x|^2)}\eta(x,dz)$$
$$+\int_{\{z:|z|>n(1+|x|)\}}\left(1+\frac{n^{-1}2x\cdot z+n^{-2}|z|^2}{1+|x|^2}\right)^{n\delta}\eta(x,dz)$$
$$\le\frac{2\delta}{n}C(\eta,6\delta)+\int_{\Gamma_{n,x}}\left(1+\frac{4}{n}\frac{|z|}{1+|x|}+\frac{2}{n^2}\left(\frac{|z|}{1+|x|}\right)^2\right)^{n\delta}\eta(x,dz)$$
$$\le\frac{2\delta}{n}C(\eta,6\delta)+\int_{\Gamma_{n,x}}e^{4\delta\frac{|z|}{1+|x|}}\eta(x,dz)$$
$$\le\frac{2\delta}{n}C(\eta,6\delta)+\frac{1}{n^2}C(\eta,4\delta),$$

where $C(\eta,4\delta)\le C(\eta,6\delta)<\infty$.

It follows that $(f_n, g_n) \in H_{\dagger,n}$ and for any bounded G,

$$\gamma(G) \leq \limsup_{n\to\infty} \sup_{x\in R^d} g_n(x) < \infty.$$

Suppose $K = \{x : |x| \leq c\}$. For $\beta > 0$, let d satisfy $\delta \log(1+d^2) = \beta + \delta \log(1+c^2)$. Then for

$$G = \{x : |x| < d\},$$

the conditions of Lemma 4.22 are satisfied. Consequently, if $\{X_n(0)\}$ is exponentially tight, the exponential compact containment condition holds for $\{X_n\}$.

Furthermore, if (4.31) holds for all $\alpha > 0$, Conditions (b) and (c) of Corollary 4.17 are easy to check, and hence $\{X_n\}$ is exponentially tight.

If E is not locally compact, then, with Lemma 3.3 in mind, it may be useful to work with a sequence of open sets $\{G_k\}$.

LEMMA 4.24. *Let X_n be a solution of the martingale problem for A_n. Let $K \subset E$ be compact, and for $k = 1, 2, \ldots$, let G_k be open and contained in a finite union of open balls of radius k^{-1}. Let $\widehat{K}$ be the closure of $\cap_k G_k$. (Note that $\widehat{K}$ will be complete and totally bounded, hence compact.) Assume that $K \subset \widehat{K}$ Let $(f_{k,n}, g_{k,n}) \in H_{\dagger,n}$, and assume that the corresponding supermartingale is right continuous. Define*

$$\beta_{k,n} \equiv \inf_{x\in G_k^c} f_{k,n}(x) - \sup_{x\in K} f_{k,n}(x), \quad \gamma_{k,n} \equiv \sup_{x\in G_k} g_{k,n}(x).$$

If $\gamma = \sup_{k,n} \gamma_{k,n} < \infty$ and $\beta_{k,n} \geq \beta + \frac{2}{n} \log k$, then

$$\begin{aligned}&\limsup_{n\to\infty} \frac{1}{n} \log P\{X_n(t) \notin \widehat{K} \text{ some } t \leq T\}\\ &\qquad \leq \max\{-\beta + \gamma T, \limsup_{n\to\infty} \frac{1}{n} \log P\{X_n(0) \notin K\}\}.\end{aligned}$$

PROOF. By Lemma 4.20,

$$P\{X_n(0) \in K, X_n(t) \notin G_k \text{ some } t \leq T\} \leq P\{X_n(0) \in K\} e^{-n\beta_{k,n}+nT\gamma_{k,n}},$$

and hence

$$\begin{aligned}(4.32)\quad &\limsup_{n\to\infty} \frac{1}{n} \log P\{X_n(t) \notin \widehat{K} \text{ some } t \leq T\}\\ &\leq \max\{\limsup_{n\to\infty} \frac{1}{n} \log \sum_k e^{-n\beta_{k,n}+n\gamma_{k,n}T}, \limsup_{n\to\infty} \frac{1}{n} \log P\{X_n(0) \notin K\}\}.\end{aligned}$$

Under the assumptions on γ and β, the right side of (4.32) is bounded by

$$\max\{-\beta + \gamma T, \limsup_{n\to\infty} \frac{1}{n} \log P\{X_n(0) \notin K\}\}.$$

□

4.7. Finite dimensional determination of the process rate function

The following lemma shows that finite dimensional rate functions give lower bounds for process rate functions.

LEMMA 4.25. *Let $\{X_n\}$ satisfy the large deviation principle with rate function I. Let $x \in D_E[0,\infty)$, and let Δ_x denote the set of discontinuities of x. If $t_1,\ldots,t_m \in \Delta_x^c$ and the large deviation principle holds for $\{(X_n(t_1),\ldots,X_n(t_m))\}$ with rate function $I_{t_1,\ldots,t_m}$, then*

$$I(x) \geq I_{t_1,\ldots,t_m}(x(t_1),\ldots,x(t_m)). \tag{4.33}$$

Consequently, if for each finite subset $\{t_1,\ldots,t_m\} \subset \mathcal{T}_0$, the large deviation principle holds for $\{(X_n(t_1),\ldots,X_n(t_m))\}$ with rate function $I_{t_1,\ldots,t_m}$, then

$$I(x) \geq \sup_{\{t_i\}\subset\mathcal{T}_0\cap\Delta_x^c} I_{t_1,\ldots,t_m}(x(t_1),\ldots,x(t_m)). \tag{4.34}$$

PROOF. For $\epsilon > 0$ and $t_1,\ldots,t_m \notin \Delta_x$, there exists $\epsilon' > 0$ such that

$$B_{\epsilon'}(x) \subset \{y : (y(t_1),\ldots,y(t_m)) \in B_\epsilon^m((x(t_1),\ldots,x(t_m)))\}$$

where B_ϵ^m denotes a ball in E^m. Consequently, (4.33) follows from (3.3). □

The assumption that $t_1,\ldots,t_m \in \Delta_x^c$ can be dropped in (4.33) if the right side is replaced by the minimum of $I_{t_1,\ldots,t_m}(y_1,\ldots,y_m)$ over the 2^m choices where y_i is $x(t_i)$ or $x(t_i-)$. This minimum is in fact needed unless other conditions are introduced. For example, let $P\{X_n = I_{[1+1/n,\infty)}\} = 1$. Then for $x = I_{[1,\infty)}$, $I(x) = 0$, but $I_1(1) = \infty$.

Our next goal is to estimate I from above in terms of finite dimensional rate functions. Let $y \in D_E[0,\infty)$. Suppose that $0 \leq t_1 < t_2 < \cdots$ satisfies $\lim_{i\to\infty} t_i = \infty$, and define Py by

$$Py(t) = y(t_1), \quad 0 \leq t < t_2,$$

and for $t \geq t_2$,

$$Py(t) = y(t_i), \quad t_i \leq t < t_{i+1}.$$

Note that if we set $\zeta(t) = t_1$ for $0 \leq t < t_2$ and $\zeta(t) = t_i$ for $t_i \leq t < t_{i+1}$, $i = 2,3,\ldots$, then $Py(t) = y \circ \zeta(t)$.

LEMMA 4.26. *Let $\mathcal{T}_0$ be dense in $[0,\infty)$. Let $x \in D_E[0,\infty)$. For $\epsilon > 0$ and compact $K \subset D_E[0,\infty)$, there exist $t_1,\ldots,t_m \in \mathcal{T}_0$ and $\epsilon' > 0$ such that*

$$B_\epsilon(x) \supset \{y : (y(t_1),\ldots,y(t_m)) \in B_{\epsilon'}^m((x(t_1),\ldots,x(t_m)))\} \cap K. \tag{4.35}$$

PROOF. Without loss of generality, we can assume that there exist compact $\Gamma_T \subset E$ and a function $h(\delta,T) > 0$ defined for $T,\delta > 0$ satisfying $\lim_{\delta\to 0} h(\delta,T) = 0$ such that

$$K = \{y : y(t) \in \Gamma_T, t \leq T, w'(y,\delta,T+\delta) \leq h(\delta,T)\}.$$

Let $2\eta < \delta$, and let $0 \leq t_1 < t_2 < \cdots$ satisfy $t_i \in \mathcal{T}_0$, $t_i \to \infty$, $t_2 \leq \eta$, and $t_{i+1} - t_i \leq \eta$. Select m so that $t_m \geq T + \delta$. If y is in the set on the right of (4.35), then by Lemma 4.8,

$$\begin{aligned}
d(x,y) &\leq d(x,Px) + d(Px,Py) + d(Py,y) \\
&\leq d(x,Px) + \epsilon' + (w'(y,\delta,T+\eta) + \frac{\eta(T+\eta)}{\delta} + e^{-T}) \vee \log\frac{\delta}{\delta - 2\eta} \\
&\leq d(x,Px) + \epsilon' + (h(\delta,T) + \frac{\eta(T+\eta)}{\delta} + e^{-T}) \vee \log\frac{\delta}{\delta - 2\eta}.
\end{aligned}$$

Select $\epsilon' < \epsilon/5$, T so that $e^{-T} \leq \epsilon'$, and δ so that $h(\delta,T) \leq \epsilon'$. Then select η so that $d(x,Px) \leq \epsilon'$, $\eta(T+\eta)/\delta \leq \epsilon'$, and $\log(\delta/(\delta - 2\eta)) \leq 3\epsilon'$. Then $d(x,y) < \epsilon$ for all y in the set on the right of (4.35). □

LEMMA 4.27. *Let $\{X_n\}$ be exponentially tight in $D_E[0,\infty)$, and assume that the large deviation principle holds with rate function I (otherwise select a subsequence). Let $\mathcal{T}_0$ be dense in $[0,\infty)$, and suppose that for each finite subset $\{t_1,\ldots,t_m\}\subset\mathcal{T}_0$, the large deviation principle holds for $\{(X_n(t_1),\ldots,X_n(t_m))\}$ with rate function $I_{t_1,\ldots,t_m}$. Then either $\liminf_{n\to\infty}\frac{1}{n}\log P\{X_n\in B_\epsilon(x)\}=-\infty$ and (by exponential tightness)*

$$\sup_{\{t_i\}\subset\mathcal{T}_0} I_{t_1,\ldots,t_m}(x(t_1),\ldots,x(t_m))=\infty=I(x) \tag{4.36}$$

or

$$\liminf_{n\to\infty}\frac{1}{n}\log P\{X_n\in B_\epsilon(x)\}\geq -I_{t_1,\ldots,t_m}(x(t_1),\ldots,x(t_m)) \tag{4.37}$$

and

$$I(x)\leq \sup_{\{t_i\}\subset\mathcal{T}_0} I_{t_1,\ldots,t_m}(x(t_1),\ldots,x(t_m)) \tag{4.38}$$

PROOF. Lemma 4.26 implies that for each $\epsilon>0$ and compact set $K\subset D_E[0,\infty)$, there exists $t_1,\ldots,t_m\in\mathcal{T}_0$ and $\epsilon'>0$ such that

$$\begin{aligned}
&\liminf_{n\to\infty}\frac{1}{n}\log P\{X_n\in B_\epsilon(x)\}\vee\liminf_{n\to\infty}\frac{1}{n}\log P\{X_n\in K^c\}\\
&\quad=\liminf_{n\to\infty}\frac{1}{n}\log P(\{X_n\in B_\epsilon(x)\cup K^c\})\}\\
&\quad\geq\liminf_{n\to\infty}\frac{1}{n}\log P\{(X(t_1),\ldots,X(t_m))\in B^m_{\epsilon'}(x(t_1),\ldots,x(t_m))\}\\
&\quad\geq -I_{t_1,\ldots,t_m}(x(t_1),\ldots,x(t_m)).
\end{aligned}$$

Since for each $a>0$ there exists a compact K_a such that

$$\limsup_{n\to\infty}\frac{1}{n}\log P\{X_n\in K_a^c\}\leq -a,$$

the lemma follows. □

Lemmas 4.25 and 4.27 give the following theorem.

THEOREM 4.28. *Assume that $\{X_n\}$ is exponentially tight in $D_E[0,\infty)$ and that for each $0\leq t_1<t_2<\cdots<t_m$, $\{(X_n(t_1),\ldots,X_n(t_m))\}$ satisfies the large deviation principle in E^m with rate function $I_{t_1,\ldots,t_m}$. Then $\{X_n\}$ satisfies the large deviation principle in $D_E[0,\infty)$ with good rate function*

$$I(x)=\sup_{\{t_i\}\subset\Delta_x^c} I_{t_1,\ldots,t_m}(x(t_1),\ldots,x(t_m)). \tag{4.39}$$

PROOF. Let I be the rate function corresponding to a subsequence of $\{X_n\}$ for which the large deviation principle holds. Exponential tightness implies I will be good. Taking $\mathcal{T}_0=\Delta_x^c$, Lemmas 4.25 and 4.27 imply that I satisfies (4.39), which in turn implies that there is only one possible rate function and hence the large deviation principle holds for the full sequence $\{X_n\}$ with rate function given by (4.39). □

The following corollary is a consequence of Lemma 3.23.

COROLLARY 4.29. *Suppose that $D \subset C_b(E)$ is bounded above and isolates points. (See Definition 3.18.) Assume that $\{X_n\}$ is exponentially tight in $D_E[0,\infty)$ and that for each $0 \le t_1 \le \ldots \le t_m$ and $f_1, \ldots, f_m \in D$*

$$\Lambda(t_1,\ldots,t_m,f_1,\ldots,f_m) = \lim_{n\to\infty} \frac{1}{n} \log E[e^{n(f_1(X_n(t_1))+\cdots+f_m(X_n(t_m)))}]$$

exists. Then $\{X_n\}$ satisfies the large deviation principle in $D_E[0,\infty)$ with good rate function

$$I(x) = \sup_m \sup_{\{t_1,\ldots,t_m\}\subset\Delta_x^c} \sup_{f_1,\ldots,f_m\in D} \{f_1(x(t_1)) + \cdots + f_m(x(t_m)) - \Lambda(t_1,\ldots,t_m,f_1,\ldots,f_m)\}.$$

If C-exponential tightness holds, then Theorem 4.28 can be simplified. The next theorem is essentially Puhalskii [**97**], Theorem 4.5 (Theorem 3.2.8 of [**100**]). He only considers R^d-valued processes, but his proof extends easily to the metric-valued case. See also de Acosta [**24**] for generalizations to projective systems.

THEOREM 4.30. *Assume that $\{X_n\}$ is C-exponentially tight in $D_E[0,\infty)$. Let $\mathcal{T}_0$ be a dense subset of $[0,\infty)$, and suppose that for each $0 \le t_1 \le \ldots \le t_m \in \mathcal{T}_0$, $\{(X_n(t_1),\ldots,X_n(t_m))\}$ satisfies the large deviation principle in E^m with rate function $I_{t_1,\ldots,t_m}$. Then $\{X_n\}$ satisfies the large deviation principle in $D_E[0,\infty)$ with good rate function*

$$I(x) = \sup_{\{t_i\}\subset\mathcal{T}_0} I_{t_1,\ldots,t_m}(x(t_1),\ldots,x(t_m)). \tag{4.40}$$

PROOF. By C-exponential tightness, if $x \notin C_E[0,\infty)$, then $I(x) = \infty$ and (4.40) follows by (4.36). Otherwise, (4.40) follows by (4.34) and (4.38). □

Part 2

Large deviations for Markov processes and semigroup convergence

CHAPTER 5

Large deviations for Markov processes and nonlinear semigroup convergence

The main result of this chapter, Theorem 5.15, shows that the convergence of Fleming's log-exponential nonlinear semigroup implies the large deviation principle for Markov processes in a metric space. The convergence of the semigroup immediately implies the large deviation principle for the one dimensional distributions by Bryc's lemma, Proposition 3.8. The semigroup property then gives the large deviation principle for the finite dimensional distributions, which, after verifying exponential tightness, implies the pathwise large deviation principle by Theorem 4.28.

In most applications, we need infinitesimal criteria for the semigroup convergence, and Corollaries 5.19, 5.20, 5.21, 5.22 are derived for this purpose. Their proofs depend on a resolvent estimate (Lemma 5.9) and a semigroup estimate (Lemma 5.12). These estimates play an important role in Chapter 7 as well.

The semigroup convergence results in this chapter are based on the Crandall-Liggett theory of nonlinear semigroups. Application of these theorems depends on verification of a range condition (Condition 5.1) that is typically difficult or impossible. This difficulty is overcome by the introduction of viscosity solution methods in Chapters 6 and 7.

Theorem 5.15 gives the rate function in terms of the limiting nonlinear semigroup. Since the semigroup is usually not known explicitly, this representation may not be either intuitive or computationally useful. This difficulty is overcome by the introduction of the control representation of the semigroup (and hence, of the rate function) in Chapter 8.

5.1. Convergence of sequences of operator semigroups

Let B be a Banach space. A (possibly nonlinear) operator $H \subset B \times B$ is *dissipative* if for each $\alpha > 0$ and $(f_1, g_1), (f_2, g_2) \in H$, $\|f_1 - f_2 - \alpha(g_1 - g_2)\| \geq \|f_1 - f_2\|$. $\overline{H}$ will denote the closure of H as a graph in $B \times B$, $\mathcal{D}(H) = \{f : \exists (f, g) \in H\}$, and $\mathcal{R}(H) = \{g : \exists (f, g) \in H\}$. If H is dissipative, then $\overline{H}$ is dissipative, and for $\alpha > 0$,

$$\overline{\mathcal{R}(I - \alpha H)} = \mathcal{R}(I - \alpha \overline{H})$$

and $J_\alpha \equiv (I - \alpha \overline{H})^{-1}$ is a contraction operator, that is for $f_1, f_2 \in \overline{\mathcal{R}(I - \alpha H)}$

$$\|J_\alpha f_1 - J_\alpha f_2\| \leq \|f_1 - f_2\|.$$

See Miyadera [**86**] for general background information on dissipative operators.

The following condition will be referred to as the *range condition* on H:

CONDITION 5.1. *There exists* $\alpha_0 > 0$ *such that*

$$\mathcal{D}(H) \subset \overline{\mathcal{R}(I - \alpha H)},$$

for all $0 < \alpha < \alpha_0$.

A dissipative operator H is called m-*dissipative* if $\mathcal{R}(I - \alpha H) = B$ for all $\alpha > 0$. The following is Lemma 2.13 of Miyadera [**86**].

LEMMA 5.2. *Let* H *be dissipative. If* $\mathcal{R}(I - \alpha_0 H) = B$ *for some* $\alpha_0 > 0$, *then* $\mathcal{R}(I - \alpha H) = B$ *for all* $\alpha > 0$, *that is,* H *is m-dissipative.*

The primary consequence of the range condition for a dissipative operator is the generation theorem of Crandall and Liggett [**18**].

PROPOSITION 5.3. *(Crandall-Liggett Theorem). Let* H *be dissipative and satisfy the range condition 5.1. Then for each* $f \in \overline{\mathcal{D}(H)}$,

$$S(t)f = \lim_{m\to\infty}(I - \frac{t}{m}\overline{H})^{-m}f = \lim_{m\to\infty}(I - \frac{1}{m}\overline{H})^{-[mt]}f$$

exists and defines a contraction semigroup on $\overline{\mathcal{D}(H)}$.

LEMMA 5.4. *Let* V *be a contraction operator on* B, *that is* $V : B \mapsto B$ *and* $\|Vf - Vg\| \leq \|f - g\|$, $f, g \in B$. *For* $\epsilon > 0$, *define*

$$Hf = \epsilon^{-1}(Vf - f).$$

Then H *is dissipative and satisfies the range condition 5.1. If* $\{S(t)\}$ *is the semigroup corresponding to* H *given by Proposition 5.3, then for* $f \in B$,

$$\|S(t)f - V^{[t/\epsilon]}f\| \leq (\epsilon + \sqrt{t\epsilon})\|Hf\|$$

PROOF. See Miyadera [**86**] Lemma 3.16. □

With the examples in the Introduction in mind, we state a convergence theorem for nonlinear semigroups. Note that the estimate in Lemma 5.4 implies the corresponding result for a sequence of discrete-time semigroups for which the time-step $\epsilon_n \to 0$.

PROPOSITION 5.5. *Let* E_n, E *be arbitrary metric spaces, and let* $\eta_n : E_n \mapsto E$ *be Borel measurable mappings. Define the bounded linear operator* $\eta_n : B(E) \mapsto B(E_n)$ *by*

$$\eta_n f = f \circ \eta_n.$$

Let $H_n \subset B(E_n) \times B(E_n)$ *and* $H \subset B(E) \times B(E)$ *be dissipative, and suppose that each satisfies the range condition 5.1 with the same* $\alpha_0 > 0$. *Let* $\{S_n(t)\}$ *denote the semigroup generated by* H_n *and* $\{S(t)\}$ *denote the semigroup generated by* H.

a) *Suppose that*

$$H \subset ex\text{-}\lim_{n\to\infty} H_n$$

in the sense that for each $(f, g) \in H$, *there exists* $(f_n, g_n) \in H_n$ *satisfying*

$$\|\eta_n f - f_n\| + \|\eta_n g - g_n\| \to 0.$$

Then, for $f \in \overline{\mathcal{D}(H)}$ *and* $f_n \in \overline{\mathcal{D}(H_n)}$ *satisfying* $\|f_n - \eta_n f\| \to 0$ *and* $T > 0$,

$$\lim_{n\to\infty}\sup_{t\leq T}\|\eta_n S(t)f - S_n(t)f_n\| = 0.$$

b) *Let* LIM *satisfy Condition A.13 for* $J_n \subset B(E_n)$ *and* $J \subset B(E)$, *and let* $H_n \subset J_n \times J_n$. *Suppose that*

$$H \subset ex\text{-LIM}H_n$$

and that for $f_n, g_n \in J_n$ *satisfying*

$$\text{LIM}f_n = \text{LIM}g_n = f_0,$$

we have

$$\text{LIM}(S_n(t)f_n - S_n(t)g_n) = 0, \qquad t > 0,$$

and

$$\text{LIM}((I - \alpha H_n)^{-1}f_n - (I - \alpha H_n)^{-1}g_n) = 0, \qquad 0 < \alpha < \alpha_0.$$

Then, for $f \in \overline{\mathcal{D}(H)}$ *and* $f_n \in \overline{\mathcal{D}(H_n)}$ *satisfying* $f = \text{LIM}f_n$, *and* $t_n \to t$,

$$S(t)f = \text{LIM}S_n(t_n)f_n.$$

REMARK 5.6. This result extends the linear semigroup convergence theory of Trotter [**118**] and Kurtz [**69, 70**] to the setting of the Crandall-Liggett theorem. As noted earlier, convergence of linear semigroups can be used to prove weak convergence results in much the same way as we will use convergence of nonlinear semigroups to prove large deviation theorems. Applications to diffusion approximations is one of the motivations for Trotter's original work.

PROOF. These results follow by directly applying Theorem 3.2 in Kurtz [**72**]. For Part (a), take

$$\mathcal{L} = \{\langle f, \{f_n\}\rangle : f \in B(E), f_n \in B(E_n), \sup_n \|f_n\| < \infty\}$$

with norm $\|\langle f, \{f_n\}\rangle\| = \|f\| \vee \sup_n \|f_n\|$. Define a linear mapping $P \subset \mathcal{L} \times B(E)$ by

$$P = \{(\langle f, \{f_n\}\rangle, f) \in \mathcal{L} \times B(E) : \lim_{n\to\infty} \|\eta_n f - f_n\| = 0\}.$$

It is easy to see that $\mathcal{D}(P)$ is closed and P is a continuous linear mapping. Define $\mathcal{H} \subset \mathcal{L} \times \mathcal{L}$ by

$$\mathcal{H} = \{(\langle f, \{f_n\}\rangle, \langle g, \{g_n\}\rangle) \in \mathcal{L} \times \mathcal{L} : (f, g) \in \overline{H}, (f_n, g_n) \in \overline{H_n}\}.$$

Then $\mathcal{H}$ is dissipative on $\mathcal{L}$ and satisfies the range condition, hence generates a semigroup $\{\mathcal{T}(t)\}$. For $\langle f, \{f_n\}\rangle \in \overline{\mathcal{D}(\mathcal{H})}$,

$$\mathcal{T}(t)\langle f, \{f_n\}\rangle = \langle S(t)f, \{S_n(t)f_n\}\rangle.$$

For $0 < \alpha < \alpha_0$, let $\mathcal{J}_\alpha = (I - \alpha\mathcal{H})^{-1}$ on

$$\mathcal{C} = \{\langle f, \{f_n\}\rangle : f \in \overline{\mathcal{D}(H)}, f_n \in \overline{\mathcal{D}(H_n)}\}.$$

Note that $\mathcal{C}$ is closed and that for each $\langle f, \{f_n\}\rangle \in \mathcal{C}$,

$$\mathcal{J}_\alpha\langle f, \{f_n\}\rangle = (I - \alpha\mathcal{H})^{-1}\langle f, \{f_n\}\rangle = \langle (I - \alpha H)^{-1}f, \{(I - \alpha H_n)^{-1}f_n\}\rangle.$$

For any $\langle f, \{f_n\}\rangle, \langle f, \{g_n\}\rangle \in \mathcal{C} \cap \mathcal{D}(P)$,

$$\|(I - \alpha H_n)^{-1}f_n - (I - \alpha H_n)^{-1}g_n\| \le \|f_n - g_n\| \le \|f_n - \eta_n f\| + \|g_n - \eta_n f\| \to 0.$$

Therefore $P(\mathcal{J}_\alpha\langle f, \{f_n\}\rangle - \mathcal{J}_\alpha\langle f, \{g_n\}\rangle) = 0$.

The conditions of Lemma 3.1 in Kurtz [**72**], with $\mathcal{B} = \mathcal{J}_\alpha$, follow by straightforward calculations. It is also easy to verify that the conditions hold for $\mathcal{B} = \mathcal{T}(t)$. Therefore by (3.4) of Theorem 3.2 of Kurtz [**72**],

$$\lim_{n\to\infty} \|\eta_n S(t)f - S_n(t)f_n\| = 0,$$

for each $f \in \overline{\mathcal{D}(H)}$ and $f_n \in \overline{\mathcal{D}(H_n)}$ satisfying $\lim_{n\to\infty} \|\eta_n f - f_n\| = 0$. The fact that the convergence is uniform on bounded time intervals follows from the strong continuity of the semigroup $\{\mathcal{T}(t)\}$. (Note that the definition of A in Theorem 3.2 of Kurtz [**72**], which corresponds to H here, can be replaced by the assumption that

$$A \subset \{(Px, Py) : (x, y) \in \mathcal{H} \cap \mathcal{D}(P) \times \mathcal{D}(P)\}.)$$

Part (b) follows through the same arguments by taking

$$\mathcal{L} = \{\langle f, \{f_n\}\rangle : f \in J, f_n \in J_n, \sup_n \|f_n\| < \infty\},$$

and

$$P = \{(\langle f, \{f_n\}\rangle, f) \in \mathcal{L} \times J : f = \text{LIM} f_n\}.$$

□

5.2. Applications to large deviations

LEMMA 5.7. *Let E be a complete separable metric space and $B(E)$ be the space of bounded measurable functions. Define a bounded linear operator $A : B(E) \mapsto B(E)$ by*

$$Af(x) = \lambda(x) \int_E (f(y) - f(x)) q(x, dy), \tag{5.1}$$

where $0 \le \lambda(x) \le \overline{\lambda} \equiv \sup_x \lambda(x) < \infty$ and q is a transition function on E. (For each $x \in E$, $q(x, \cdot) \in \mathcal{P}(E)$, and for each $B \in \mathcal{B}(E)$, $q(\cdot, B) \in B(E)$.)

Define

$$\begin{aligned} Hf &= e^{-f} A e^f; \\ A^g f &= e^{-g} A(f e^g) - (e^{-g} f) A e^g; \\ Lg &= A^g g - Hg; \\ H_K f(x) &= \sup_{\|g\| \le K} \{A^g f(x) - Lg(x)\}. \end{aligned}$$

Then

$$Hf(x) = \sup_{g \in B(E)} \{A^g f(x) - Lg(x)\}, \tag{5.2}$$

and $H_K f = Hf$ for $\|f\| \le K$. Furthermore, H_K is Lipschitz on $B(E)$ with Lipschitz constant $M_K = 2\overline{\lambda} e^{2K}$ and is m-dissipative on $B(E)$. Finally,

$$\|(I - \alpha H_K)^{-1} h\| \le \|h\|,$$

H is m-dissipative on $B(E)$, and the sup *in (5.2) is attained when $g = f$.*

PROOF. For $f \in B(E)$, let $x_n \in E$ satisfy $\lim_{n\to\infty} f(x_n) = \sup_y f(y)$. Then

$$\begin{aligned} \limsup_{n\to\infty} Af(x_n) &= \limsup_{n\to\infty} \lambda(x_n) \int_E (f(y) - f(x_n)) q(x_n, dy) \\ &\le \limsup_{n\to\infty} \lambda(x_n)(\sup_y f(y) - f(x_n)) = 0, \end{aligned} \tag{5.3}$$

so A is strongly dissipative. (See Definition A.16 and Lemma A.18.)

For $f, g \in B(E)$ and (fixed) $x \in E$, let

$$F(y) = e^{g(y)-g(x)}\{(f(y) - f(x)) - (g(y) - g(x))\} + e^{g(y)-g(x)} - e^{f(y)-f(x)}.$$

By the convexity of the exponential function, $F(y) \le 0$. Note also that

$$A^g f(x) = e^{-g(x)}\lambda(x) \int_E e^{g(y)}(f(y) - f(x))q(x, dy).$$

Since $F(x) = 0 = \sup_y F(y)$, by (5.3),

$$\begin{aligned} 0 \ge AF(x) &= \lambda(x) \int_E e^{g(y)-g(x)}(f(y) - f(x))q(x, dy) \\ &\quad -\lambda(x) \int_E e^{g(y)-g(x)}(g(y) - g(x))q(x, dy) \\ &\quad +\lambda(x) \int_E (e^{g(y)-g(x)} - e^{f(y)-f(x)})q(x, dy) \\ &= A^g f(x) - A^g g(x) + Hg(x) - Hf(x) \\ &= A^g f(x) - Lg(x) - Hf(x), \end{aligned}$$

and hence,

$$Hf(x) \ge A^g f(x) - Lg(x).$$

But by the definition of L,

$$Hf(x) = A^f f(x) - Lf(x),$$

and (5.2) follows. By this representation of H, it is also easy to see that $H_K f = Hf$ whenever $\|f\| \le K$.

To prove H_K is Lipschitz, let $f_1, f_2 \in B(E)$. For each $x \in E$, there exists $\|g_n\| \le K$ such that

$$H_K f_1(x) = \sup_{\|g\| \le K} \{A^g f_1(x) - Lg(x)\} \le A^{g_n} f_1(x) - Lg_n(x) + \frac{1}{n},$$

and hence

$$\begin{aligned} &(H_K f_1 - H_K f_2)(x) \\ \le\ & A^{g_n} f_1(x) - Lg_n(x) + \frac{1}{n} - A^{g_n} f_2(x) + Lg_n(x) \\ \le\ & A^{g_n}(f_1 - f_2)(x) + \frac{1}{n} \\ =\ & \lambda(x) \int_E e^{g_n(y)-g_n(x)}\{(f_1 - f_2)(y) - (f_1 - f_2)(x)\}q(x, dy) + \frac{1}{n} \\ \le\ & 2e^{2K} \sup_x \lambda(x)\|f_1 - f_2\| + \frac{1}{n}. \end{aligned}$$

It follows that

$$\|H_K f_1 - H_K f_2\| \le 2e^{2K} \sup_x \lambda(x)\|f_1 - f_2\|,$$

so H_K is Lipschitz with constant $M_K = 2e^{2K} \sup_x \lambda(x)$.

To see that H_K is strongly dissipative, we apply Lemma A.18. Let $x_n \in E$ be such that $(f_1 - f_2)(x_n) \geq \sup_y (f_1 - f_2)(y) - 1/n$. Then $(f_1 - f_2)(y) - (f_1 - f_2)(x_n) \leq 1/n$. It follows that there exist g_n with $\|g_n\| \leq K$ such that

$$\begin{aligned}&(H_K f_1 - H_K f_2)(x_n)\\ \leq\ & \lambda(x_n) \int_E e^{g_n(y) - g_n(x_n)} \{(f_1 - f_2)(y) - (f_1 - f_2)(x_n)\} q(x, dy) + \frac{1}{n}\\ \leq\ & 2e^{2K} \sup_x \lambda(x) \frac{1}{n} + \frac{1}{n}.\end{aligned}$$

Therefore $\limsup_{n\to\infty} (H_K f_1 - H_K f_2)(x_n) \leq 0$, and H_K is strongly dissipative.

Let $h \in B(E)$. For $0 < \alpha < 1/M_K = \alpha_K$, the mapping

$$P_K f = h + \alpha H_K f$$

is a strict contraction on $B(E)$. Hence, there exists a unique solution $f \in B(E)$ of $f = P_K f$, which implies

$$(I - \alpha H_K) f = h,$$

that is, $\mathcal{R}(I - \alpha H_K) = B(E)$ for $0 < \alpha < \alpha_K$. Since H_K is dissipative, by Lemma 5.2, H_K is m-dissipative.

By the m-dissipativity of H_K, for each $\alpha > 0$ and $h, h_0 \in B(E)$,

$$\|(I - \alpha H_K)^{-1} h - (I - \alpha H_K)^{-1} h_0\| \leq \|h - h_0\|.$$

Since $H_K 0 = 0$, $\|(I - \alpha H_K)^{-1} h\| \leq \|h\|$. Consequently, if $\|h\| \leq K$ and $f = (I - \alpha H_K)^{-1} h$, then $Hf = H_K f$ and

$$f - \alpha H f = h.$$

It follows that H is m-dissipative on $B(E)$.

The fact that the sup in (5.2) is attained when $g = f$ can be verified directly through the definition of Hf and Lf. □

Let $\{T(t)\}$ be the semigroup given by

$$T(t) f(x) = \int_E f(y) P(t, x, dy),$$

where $P(t, x, \Gamma)$ is the transition function for a Markov process that has cadlag paths. The *full generator* for $\{T(t)\}$ is the collection A of $(f, g) \in B(E) \times B(E)$ satisfying

$$T(t) f = f + \int_0^t T(s) g ds, \quad t \geq 0.$$

(See Ethier and Kurtz [**36**], Section 1.5.) The full generator satisfies $\mathcal{R}(I - \epsilon A) = B(E)$ for $\epsilon > 0$ with

$$(I - \epsilon A)^{-1} f = \int_0^\infty \epsilon^{-1} e^{-\epsilon^{-1} t} T(t) f dt.$$

Note that $(I - \epsilon A)^{-1}$ is given by a transition function

$$q_\epsilon(x, \Gamma) = \int_0^\infty \epsilon^{-1} e^{-\epsilon^{-1} t} P(t, x, \Gamma) dt.$$

Define the Yosida approximation of A by

$$A^\epsilon = \frac{1}{\epsilon}((I - \epsilon A)^{-1} - I) = A(I - \epsilon A)^{-1}.$$

Then A^ϵ is a bounded, linear dissipative operator on $\mathcal{R}(I-\epsilon A) = B(E)$ of the form (5.1).

We apply the semigroup convergence result in Part (b) of Proposition 5.5 using the definition of LIM given in Definition 2.5 and assuming Condition 2.8. Note that Condition 2.8 can be restated as follows: For each $q \in \mathcal{Q}$, $T > 0$, and $a > 0$, there exist $\widehat{q}(q, a, T)$ and $C(q, a, T)$ such that

$$\sup_{y \in K_n^q} P\{Y_n(t) \notin K_n^{\widehat{q}(q,a,T)} \text{ for some } 0 \le t \le T | Y_n(0) = y\} \le C(q, a, T)e^{-na}. \tag{5.4}$$

A natural setting for this notion of limit is one in which $E_n = E'$ is independent of n and $\eta_n \equiv \eta : E' \to E$, $\mathcal{Q}$ is the collection of compact subsets of E' and $K_n^K = K$ for $K \in \mathcal{Q}$. Another example would be to take E_n to be the d-dimensional lattice with mesh $\frac{1}{n}$ and $\eta_n : E_n \to R^d$ to be the natural embedding. Then one could take $K_n^K = E_n \cap K$ for compact $K \subset R^d$.

LEMMA 5.8. *Let E_n, E, η_n be defined as in Proposition 5.5. For each n, let $A_n \subset B(E_n) \times B(E_n)$ be of the form (5.1), that is*

$$A_n f(x) = \lambda_n(x) \int_E (f(y) - f(x)) q_n(x, dy).$$

Let Y_n be a solution of the martingale problem for A_n with distribution $P_n \in \mathcal{P}(D_{E_n}[0,\infty))$. Define $X_n(t) = \eta_n(Y_n(t))$, and let $H_n f = n^{-1}e^{-nf}A_n e^{nf}$. Suppose that Condition 2.8 is satisfied for $\{Y_n\}$.

For each $T > 0$, define $Q_{n,T} \in \mathcal{P}(D_{E_n}[0,T])$ by

$$\frac{dQ_{n,T}}{dP_n} = \exp\{n(f_n(Y_n(T)) - f_n(Y_n(0)) - \int_0^T H_n f_n(Y_n(s))ds)\}. \tag{5.5}$$

Then, under $Q_{n,T}$, $\{Y_n(t), 0 \le t \le T\}$ is a solution of the martingale problem for $A_n^{f_n}$ given by

$$\begin{aligned} A_n^{f_n} h(x) &= e^{-nf_n(x)} A_n(he^{nf_n})(x) - e^{-nf_n(x)} h(x) A_n e^{nf_n}(x) \\ &= \lambda_n(x) \int_{E_n} e^{n(f_n(y) - f_n(x))}(h(y) - h(x)) q_n(x, dy), \end{aligned} \tag{5.6}$$

and if $\sup_n(\|f_n\| + \|H_n f_n\|) < \infty$, Y_n with distribution $Q_{n,T}$ on $D_{E_n}[0,T]$ satisfies Condition 2.8.

PROOF. $Q_{n,T} \in \mathcal{P}(D_{E_n}[0,T])$ since

$$M_n(t) \equiv \exp\{n(f_n(Y_n(t)) - f_n(Y_n(0)) - \int_0^t H_n f_n(Y_n(s))ds)\}$$

is a mean one martingale under P_n. The fact that under $Q_{n,T}$, Y_n is a solution of the martingale problem for $A_n^{f_n}$ can be checked by applying Theorem III.20 of Protter [**96**].

For $q, \widehat{q} \in \mathcal{Q}$,

$$\begin{aligned}
&\frac{1}{n} \log \sup_{y \in K_n^q} Q_{n,T}(Y_n(t) \notin K_n^{\widehat{q}}, \quad \text{for some } 0 \le t \le T | Y_n(0) = y) \\
&= \frac{1}{n} \log \sup_{y \in K_n^q} E^{P_n}[\frac{dQ_{n,T}}{dP_n} I(Y_n(t) \notin K_n^{\widehat{q}}, \quad \text{for some } 0 \le t \le T) | Y_n(0) = y] \\
&\le \frac{1}{n} \log \Big(e^{2n\|f_n\| - nT \inf_{y \in E_n} H_n f_n(y)} \\
&\qquad \times \sup_{y \in K_n^q} P_n(Y_n(t) \notin K_n^{\widehat{q}}, \quad \text{for some } 0 \le t \le T) | Y_n(0) = y) \Big) \\
&\le 2\|f_n\| - T \inf_{y \in E_n} H_n f_n(y) \\
&\qquad + \frac{1}{n} \sup_{y \in K_n^q} \log P_n(Y_n(t) \notin K_n^{\widehat{q}}, \quad \text{for some } 0 \le t \le T) | Y_n(0) = y).
\end{aligned}$$

By (2.12), the conclusion follows. □

LEMMA 5.9. *Under the Conditions of Lemma 5.8, let $h_n^i \in B(E_n)$ and*

$$f_n^i = (I - \alpha H_n)^{-1} h_n^i \in B(E_n), \quad i = 1, 2.$$

Then for $q \in \mathcal{Q}$, $T > 0$, and $a > 0$, there exist $\widehat{q} = \widehat{q}(q, a, T, \sup_n \|h_n^1\|, \sup_n \|h_n^2\|) \in \mathcal{Q}$ and $C = C(q, a, T, \sup_n \|h_n^1\|, \sup_n \|h_n^2\|)$ such that

$$\sup_{y \in K_n^q} |f_n^1(y) - f_n^2(y)| \le (e^{-\alpha^{-1}T} + Ce^{-na}) \|h_n^1 - h_n^2\| + \sup_{y \in K_n^{\widehat{q}(q,a,T)}} |h_n^1(y) - h_n^2(y)|, \tag{5.7}$$

and hence

$$\text{LIM} h_n^1 = \text{LIM} h_n^2 = h \tag{5.8}$$

implies

$$\text{LIM}(f_n^1 - f_n^2) = 0. \tag{5.9}$$

REMARK 5.10. The f_n^is exist by Lemma 5.7.

The notion of LIM here follows that in Definition 2.5 by assuming Condition 2.8 with index set $\mathcal{Q}$.

PROOF. Define $A_n^g f = e^{-ng} A_n f e^{ng} - e^{-ng} f A_n e^{ng}$ (see (5.6)) and $L_n f = A_n^f f - H_n f$. By Lemma 5.7 (appropriately modified to take into account the factor of n),

$$H_n f(y) = \sup_{g \in B(E_n)} \{A_n^g f(y) - L_n g(y)\} = A_n^f f(y) - L_n f(y)$$

holds for each $f \in B(E_n)$. Therefore

$$h_n^i(y) = f_n^i - \alpha H_n f_n^i \le f_n^i - \alpha(A_n^g f_n^i(y) - L_n g(y)), \quad y \in E_n, g \in B(E_n),$$

and the equality is attained when $g = f_n^i$, $i = 1, 2$. Consequently,

$$f_n^i(y) = \sup_{g \in B(E_n)} (I - \alpha A_n^g)^{-1} (h_n^i - \alpha L_n g)(y) = (I - \alpha A_n^{f_n^i})^{-1} (h_n^i - \alpha L_n f_n^i)(y),$$

and

$$
\begin{aligned}
(5.10)\qquad & (f_n^1 - f_n^2)(y) \\
&= \sup_{g\in B(E_n)} (I-\alpha A_n^g)^{-1}(h_n^1 - \alpha L_n g)(y) \\
&\qquad - \sup_{g\in B(E_n)} (I-\alpha A_n^g)^{-1}(h_n^2 - \alpha L_n g)(y) \\
&\le (I-\alpha A_n^{f_n^1})^{-1}(h_n^1 - h_n^2)(y) \\
&= E^{Q_n^1}[\int_0^\infty \alpha^{-1}e^{-\alpha^{-1}t}\Big(h_n^1(Y_n(t)) - h_n^2(Y_n(t))\Big)dt|Y_n(0)=y],
\end{aligned}
$$

where Q_n^1 is the distribution under which Y_n is a solution of the martingale problem for $A_n^{f_n^1}$.

By the dissipativity of H_n and the fact that $(0,0)\in H_n$, $\|f_n^1\|\le\|h_n^1\|$ and

$$
-\inf_y H_n f_n^1(y) = \frac{1}{\alpha}\sup_y(h_n^1(y) - f_n^1(y)) \le \frac{1}{\alpha}\sup_n 2\|h_n^1\| < \infty.
$$

Consequently, by Lemma 5.8, there exist $\widehat{q}_1 = \widehat{q}_1(q,a,T,\sup_n\|h_n^1\|)$ and $C_1 = C_1(q,a,T,\sup_n\|h_n^1\|)$ satisfying (5.4) for Y_n with distribution Q_n^1. Then, setting

$$
G_n(\widehat{q},T) = \{Y_n(t)\notin K_n^{\widehat{q}} \text{ for some } 0\le t\le T\}
$$

$$
\begin{aligned}
(5.11)\quad & \sup_{y\in K_n^q}(f_n^1 - f_n^2)(y) \\
&\le \sup_{y\in K_n^q} E^{Q_n^1}[\int_0^\infty \alpha^{-1}e^{-\alpha^{-1}t}\Big(h_n^1(Y_n(t)) - h_n^2(Y_n(t))\Big)dt|Y_n(0)=y] \\
&\le \left(e^{-\alpha^{-1}T} + \sup_{y\in K_n^q} Q_n^1(G_n(\widehat{q},T)|Y_n(0)=y)\right)(0\vee \sup_{y\in E_n}(h_n^1(y) - h_n^2(y)) \\
&\qquad + \sup_{y\in K_n^{\widehat{q}_1}}(h_n^1(y) - h_n^2(y)) \\
&\le (e^{-\alpha^{-1}T} + C_1e^{-na})0\vee \sup_{y\in E_n}(h_n^1(y) - h_n^2(y)) \\
&\qquad + \sup_{y\in K_n^{\widehat{q}_1}}(h_n^1(y) - h_n^2(y)).
\end{aligned}
$$

There exist $C_2(q,a,T,\sup_n\|h_n^2\|)$ and $\widehat{q}_2(q,a,T,\sup_n\|h_n^2\|)$, so that a similar inequality holds with the roles of f_1 and f_2 reversed. Consequently (5.7) holds with $C = C_1(q,a,T,\sup_n\|h_n^1\|)\vee C_2(q,a,T,\sup_n\|h_n^2\|)$ and $\widehat{q}$ satisfying

$$
K_n^{\widehat{q}_1(q,a,T,\sup_n\|h_n^1\|)} \cup K_n^{\widehat{q}_2(q,a,T,\sup_n\|h_n^2\|)} \subset K_n^{\widehat{q}}.
$$

□

Let E_n, E, η_n be defined as in Proposition 5.5. For each n, let $A_n\subset B(E_n)\times B(E_n)$ be a full generator corresponding to a Markov transition semigroup $\{T_n(t)\}$. (In particular, A_n is a closed, linear dissipative operator satisfying $\mathcal{R}(I-\alpha A_n) = B(E_n)$ for some (hence all) $\alpha>0$.) Define $V_n(t)f = \frac{1}{n}\log T_n(t)e^{nf}$ and $H_nf = \frac{1}{n}e^{-nf}A_ne^{nf}$ for $e^{nf}\in\mathcal{D}(A_n)$, or more precisely, since A_n can be multivalued,

$$
H_n = \{(f, \frac{1}{n}e^{-nf}g) : (e^{nf}, g)\in A_n\}.
$$

Let $\epsilon_n > 0$, and define

$$A_n^{\epsilon_n} = \frac{1}{\epsilon_n}((I - \epsilon_n A_n)^{-1} - I) = A_n(I - \epsilon_n A_n)^{-1}.$$

Then $A_n^{\epsilon_n}$ is a bounded, linear dissipative operator on $\mathcal{R}(I - \epsilon_n A_n) = B(E_n)$ of the form (5.1), $T_n^{\epsilon_n}(t) = e^{tA_n^{\epsilon_n}}$ defines a strongly continuous, linear contraction semigroup on $B(E_n)$, and $V_n^{\epsilon_n}(t)f = \frac{1}{n}\log T_n^{\epsilon_n}(t)e^{nf}$ defines a strongly continuous, nonlinear contraction semigroup on $B(E_n)$. Define

$$H_n^{\epsilon_n} f = \frac{1}{n} e^{-nf} A_n^{\epsilon_n} e^{nf},$$

for $f \in B(E_n)$, and note that

$$H_n^{\epsilon_n} f = \lim_{t\to 0+} \frac{V_n^{\epsilon_n}(t) - I}{t} f$$

for $f \in B(E_n)$.

LEMMA 5.11. *Let Y_n be a solution of the martingale problem for A_n, let N be a unit Poisson process, and let $\{\Delta_1, \Delta_2, \ldots\}$ be independent, unit exponential random variables. Assume that the Y_n, N, and the Δ_k are independent. Define*

$$\lambda_{\epsilon_n}(t) = \epsilon_n \sum_{k=1}^{N(t/\epsilon_n)} \Delta_k$$

and

$$Y_n^{\epsilon_n}(t) = Y_n(\lambda_{\epsilon_n}(t)).$$

Then $Y_n^{\epsilon_n}$ is a solution of the martingale problem for $A_n^{\epsilon_n}$, and if $(f_n, g_n) \in H_n$,

$$\begin{aligned}\sup_{y\in E_n} &|V_n(t)f_n(y) - V_n^{\epsilon_n}(t)f_n(y)| \\ &\le e^{2n\|f_n\|}\|g_n\|E[|\lambda_{\epsilon_n}(t) - t|] \qquad (5.12)\\ &\le \sqrt{2\epsilon_n t}\,e^{2n\|f_n\|}\|g_n\|.\end{aligned}$$

PROOF. Since $(e^{nf_n}, e^{nf_n}g_n) \in A_n$,

$$E[e^{nf_n(Y_n(t))}|Y_n(0) = y] = e^{nf_n(y)} + nE[\int_0^t e^{nf_n(Y_n(r))} g_n(Y_n(r))dr|Y_n(0) = y]$$

and

$$E[e^{nf_n(Y_n^{\epsilon_n}(t))}|Y_n(0) = y] = e^{nf_n(y)} + nE[\int_0^{\lambda_{\epsilon_n}(t)} e^{nf_n(Y_n(r))} g_n(Y_n(r))dr|Y_n(0) = y].$$

Consequently,

$$\begin{aligned}\sup_{y\in E_n} &|T_n(t)e^{nf_n}(y) - T_n^{\epsilon_n}(t)e^{nf_n}(y)| \\ &\le \sup_{y\in E_n} nE[\int_{t\wedge\lambda_{\epsilon_n}(t)}^{t\vee\lambda_{\epsilon_n}(t)} e^{nf_n(Y_n(r))}|g_n(Y_n(r))|dr] \\ &\le ne^{n\|f_n\|}\|g_n\|E[|\lambda_{\epsilon_n}(t) - t|],\end{aligned}$$

and since $|\log a - \log b| \le |b - a|/a \wedge b$ and $T_n(t)e^{nf_n}$ and $T_n^{\epsilon_n}(t)e^{nf_n}$ are both bounded below by $e^{-n\|f_n\|}$,

$$\sup_{y\in E_n} |V_n(t)f_n(y) - V_n^{\epsilon_n}(t)f_n(y)| \le e^{2n\|f_n\|}\|g_n\|E[|\lambda_{\epsilon_n}(t) - t|].$$

The second inequality follows from the fact that

$$E[(\lambda_{\epsilon_n}(t)-t)^2]=2\epsilon_n t$$

□

LEMMA 5.12. *Suppose Y_n satisfies Condition 2.8, and define*

$$V_n(t)h_n^i(y)=\frac{1}{n}\log E^P[\exp\{nh_n^i(Y_n(t))\}|Y_n(0)=y], i=1,2.$$

Then for $q\in\mathcal{Q}$, $t>0$, and $a>0$, there exist $\widehat{q}(q,a,t)\in\mathcal{Q}$ and $C(q,a,t)$ such that

$$\begin{aligned}(5.13)\quad \sup_{y\in K_n^q}|V_n(t)h_n^1(y)-V_n(t)h_n^2(y)| &\le \frac{1}{n}\frac{C(q,a,t)e^{-n(a-\|h_n^1\|-\|h_n^2\|)}}{1-C(q,a,t)e^{-na}}\\ &\quad+\sup_{y\in K_n^{\widehat{q}(q,a,T)}}|h_n^1(y)-h_n^2(y)|,\end{aligned}$$

and hence

$$\mathrm{LIM}h_n^1=\mathrm{LIM}h_n^2=h.$$

implies

$$\mathrm{LIM}(V_n(t)h_n^1-V_n(t)h_n^2)=0. \tag{5.14}$$

PROOF. Since $V_n(t)(h+c)=V_n(t)h+c$, without loss of generality, we can assume that h_n^1 and h_n^2 are nonnegative. Note that if a, b, c, and d are positive and $a+b\ge c+d$, then

$$\log\frac{a+b}{c+d}\le\int_c^a\frac{1}{x}dx+\int_a^{a+b}\frac{1}{x}dx\le\log\frac{a}{c}+\frac{b}{a}.$$

Let $q\in\mathcal{Q}$. By (5.4), there exists $C(q,a,t)>0$ and $\widehat{q}(q,a,t)\in\mathcal{Q}$ such that, for n satisfying $e^{-na}C(q,a,t)<1$,

$$\begin{aligned}&\sup_{y\in K_n^q}|V_n(t)h_n^1(y)-V_n(t)h_n^2(y)|\\ &\le\sup_{y\in K_n^q}|\frac{1}{n}\log E^P[I_{\{Y_n(t))\in K_n^{\widehat{q}(q,a,t)}\}}e^{nh_n^1(Y_n(t))}|Y_n(0)=y]\\ &\qquad-\frac{1}{n}\log E^P[I_{\{Y_n(t)\in K_n^{\widehat{q}(q,a,t)}\}}e^{nh_n^2(Y_n(t))}|Y_n(0)=y]|\\ &\qquad+\frac{1}{n}\sup_{y\in K_n^q}\frac{e^{n(\|h_n^1\|+\|h_n^2\|)}P\{Y_n(t)\notin K_n^{\widehat{q}(q,a,t)}|Y_n(0)=y\}}{P\{Y_n(t)\in K_n^{\widehat{q}(q,a,t)}|Y_n(0)=y\}}\\ &\le\sup_{y\in K_n^{\widehat{q}}}|h_n^1(y)-h_n^2(y)|+\frac{1}{n}\frac{C(q,a,t)e^{-n(a-\|h_n^1\|-\|h_n^2\|)}}{1-C(q,a,t)e^{-na}},\end{aligned}$$

giving (5.13). Taking $a>\|h_n^1\|+\|h_n^2\|$, (5.14) follows. □

LEMMA 5.13. *Let H_n and $H_n^{\epsilon_n}$ be as above with $n\epsilon_n\to 0$. Let $H\subset B(E)\times B(E)$ be dissipative and satisfy the range condition 5.1. Then H generates a strongly continuous contraction semigroup on $\overline{\mathcal{D}(H)}$ given by*

$$V(t)f=\lim_{m\to\infty}(I-\frac{t}{m}\overline{H})^{-m}f.$$

a) *Suppose that*

$$H \subset ex\text{-}\lim_{n\to\infty} H_n$$

in the sense that for each $(f,g) \in H$, *there exist* $(f_n, g_n) \in H_n$ *satisfying*

$$\|f_n - \eta_n f\| + \|g_n - \eta_n g\| \to 0. \tag{5.15}$$

Then

$$H \subset ex\text{-}\lim_{n\to\infty} H_n^{\epsilon_n}$$

and for every $f \in \overline{\mathcal{D}(H)}$ *and* $f_n \in B(E_n)$ *satisfying* $\|\eta_n f - f_n\| \to 0$,

$$\lim_{n\to\infty} \|\eta_n V(t)f - V_n^{\epsilon_n}(t)f_n\| = 0 \tag{5.16}$$

and

$$\lim_{n\to\infty} \|\eta_n V(t)f - V_n(t)f_n\| = 0. \tag{5.17}$$

b) *Suppose* $\{Y_n\}$ *satisfies Condition 2.8. Let* LIM *be given by Definition 2.5, and suppose that*

$$H \subset ex\text{-}\mathrm{LIM} H_n$$

in the sense that for each $(f,g) \in H$, *there exists* $(f_n, g_n) \in H_n$ *satisfying*

$$f = \mathrm{LIM} f_n, \quad g = \mathrm{LIM} g_n.$$

Then

$$H \subset ex\text{-}\mathrm{LIM} H_n^{\epsilon_n}$$

and for every $f \in \overline{\mathcal{D}(H)}$ *and* $f_n \in B(E_n)$ *satisfying* $f = \mathrm{LIM} f_n$,

$$V(t)f = \mathrm{LIM} V_n^{\epsilon_n}(t)f_n, \tag{5.18}$$

and

$$V(t)f = \mathrm{LIM} V_n(t)f_n. \tag{5.19}$$

REMARK 5.14. Ordinarily, the dissipativity of H will follow from the fact that it is the limit (norm convergence) of dissipative operators, that is, if for $(f_i, g_i) \in H$, $i = 1, 2$ and $\epsilon \geq 0$, $\lim_{n\to\infty} \|\eta_n(f_1 - f_2 - \epsilon(g_1 - g_2))\| = \|f_1 - f_2 - \epsilon(g_1 - g_2)\|$ then the dissipativity of H will follow from the dissipativity of H_n and the convergence in (5.15).

PROOF. The operator $A_n^{\epsilon_n}$ is just the Yosida approximation of A_n, and by Lemma 1.2.4, Ethier and Kurtz [**36**], $A_n^{\epsilon_n}$ is a bounded, linear, dissipative operator and $\{T_n^{\epsilon_n}(t)\}$ forms a strongly continuous semigroup with

$$A_n^{\epsilon_n} f = \lim_{t\to 0+} \frac{T_n^{\epsilon_n}(t)f - f}{t}, \tag{5.20}$$

for $f \in \mathcal{D}(A_n^{\epsilon_n}) = B(E_n)$. The convergence in (5.20) implies that

$$H_n^{\epsilon_n} f = \lim_{t\to 0+} \frac{V_n^{\epsilon_n}(t)f - f}{t},$$

for $f \in B(E_n)$.

By Lemma 5.7, $H_n^{\epsilon_n}$ is m-dissipative, and hence

$$V_n^{\epsilon_n}(t)f = \lim_{m\to\infty} (I - \frac{t}{m} H_n^{\epsilon_n})^{-m} f. \tag{5.21}$$

(See Miyadera [**86**], Corollary 3.21.)

For Part (a), we show that (5.15) implies the existence of $(\widehat{f}_n, \widehat{g}_n) \in H_n^{\epsilon_n}$ satisfying

$$\|\widehat{f}_n - \eta_n f\| + \|\widehat{g}_n - \eta_n g\| \to 0.$$

For $(f, g) \in H$ and $(f_n, g_n) \in H_n$ satisfying (5.15), for n sufficiently large, $e^{nf_n}(1 - n\epsilon_n g_n) > 0$ and we can define $\widehat{f}_n$ by

$$e^{n\widehat{f}_n} = e^{nf_n}(1 - n\epsilon_n g_n).$$

If A_n is single valued, then

$$e^{nf_n}(1 - n\epsilon_n g_n) = (I - \epsilon_n A_n)e^{nf_n}$$

and

$$e^{n\widehat{f}_n} = (1 - \epsilon_n A_n)e^{nf_n}.$$

Recall that $(f_n, g_n) \in H_n$ if and only if $(e^{nf_n}, e^{nf_n}g_n) \in A_n$, so

$$\widehat{g}_n = H_n^{\epsilon_n}\widehat{f}_n = e^{n(f_n - \widehat{f}_n)}g_n,$$

and since

$$e^{n\widehat{f}_n - nf_n} = 1 - n\epsilon_n g_n \to 1,$$

we have $\|f_n - \widehat{f}_n\| \to 0$ and $\|\widehat{g}_n - \eta_n g\| \to 0$. Hence $H \subset ex\text{-}\lim_{n\to\infty} H_n^{\epsilon_n}$. Therefore, by Proposition 5.5, (5.16) holds.

Since ϵ_n can be any sequence going to zero sufficiently fast, we can assume that $\lim_{n\to\infty} \epsilon_n e^{nc} = 0$ for all $c > 0$. Then the limit in (5.17) follows by (5.12) and the fact that V_n is a contraction.

Similar arguments give Part (b). If $(f_n, g_n) \in H_n$ satisfy

$$f = \text{LIM} f_n, \qquad g = \text{LIM} g_n, \tag{5.22}$$

then there exist $(\widehat{f}_n, \widehat{g}_n) \in H_n^{\epsilon_n}$ satisfying

$$f = \text{LIM}\widehat{f}_n, \qquad g = \text{LIM}\widehat{g}_n.$$

Hence $H \subset ex\text{-LIM}H_n^{\epsilon_n}$. By Proposition 5.5 and Lemmas 5.9 and 5.12, (5.18) holds.

Let $f' \in \overline{\mathcal{D}(H)}$ and $(f, g) \in H$. There exist $(f_n, g_n) \in H_n$ satisfying (5.22). Then

$$\begin{aligned}\sup_{y\in K_n^q} |\eta_n V(t)f'(y) - V_n(t)\eta_n f'(y)| \le 2\|f' - f\| &+ \sup_{y\in K_n^q} |\eta_n V(t)f(y) - V_n^{\epsilon_n}(t)f_n(y)| \\ &+ \sup_{y\in K_n^q} |V_n^{\epsilon_n}(t)f_n(y) - V_n(t)f_n(y)| \\ &+ \sup_{y\in K_n^q} |V_n(t)f_n(y) - V_n(t)\eta_n f(y)|.\end{aligned}$$

Again, assuming that $\lim_{n\to\infty} \epsilon_n e^{nc} = 0$ for all $c > 0$, the second term on the right converges to zero by (5.18), the third term by (5.12), and the last term by Lemma 5.12. Consequently,

$$\limsup_{n\to\infty} \sup_{y\in K_n^q} |\eta_n V(t)f'(y) - V_n(t)\eta_n f'(y)| \le 2\|f' - f\|,$$

and since $f \in \mathcal{D}(H)$ is arbitrary, the left side converges to zero. Finally, (5.19) follows by Lemma 5.12. □

THEOREM 5.15. *Let (E_n, r_n), $n = 1, 2, \ldots$ and (E, r) be complete and separable metric spaces. Let $\eta_n : E_n \mapsto E$ be Borel measurable, and define $\eta_n : B(E) \mapsto B(E_n)$ by $\eta_n f = f \circ \eta_n$.*

Assume that one of the following holds:

a) *(Continuous-time case.) For each $n = 1, 2, \ldots$, let $A_n \subset \overline{C}(E_n) \times B(E_n)$, and assume that existence and uniqueness holds for the $D_E[0, \infty)$-martingale problem for (A_n, μ) for each initial distribution $\mu \in \mathcal{P}(E)$. For $y \in E_n$, let $P_y^n \in \mathcal{P}(D_{E_n}[0, \infty))$ be the distribution of the solution of the martingale problem for A_n starting from y, and assume that the mapping $y \mapsto P_y^n$ is Borel measurable taking the weak topology on $\mathcal{P}(D_{E_n}[0, \infty))$ (cf. Theorem 4.4.6, Ethier and Kurtz* [**36**]*). Let Y_n be a solution of the martingale problem for A_n.*

b) *(Discrete-time case.) For each $n = 1, 2, \ldots$, let T_n be a transition operator on $B(E_n)$ for a Markov chain and let $\epsilon_n > 0$. Let Y_n be a discrete-time Markov chain with time-step ϵ_n and transition operator T_n, that is,*

$$E[f(Y_n(t))|Y_n(0) = y] = T_n^{[t/\epsilon_n]} f(y),$$

and suppose $\epsilon_n \to 0$.

Define $X_n = \eta_n \circ Y_n$, and define $\{V_n(t)\}$ on $B(E_n)$ by

$$V_n(t)f(y) = \frac{1}{n} \log E[e^{nf(Y_n(t))}|Y_n(0) = y]. \tag{5.23}$$

Let $D \subset C_b(E)$ be closed under addition, and suppose that there exists an operator semigroup $\{V(t)\}$ on D such that

$$\|\eta_n V(t)f - V_n(t)f_n\| \to 0, \tag{5.24}$$

whenever $f \in D$, $f_n \in B(E_n)$ and $\|\eta_n f - f_n\| \to 0$. In the discrete-time case, assume also that

$$\|\eta_n V(t)f - V_n(t + \epsilon_n)f_n\| \to 0. \tag{5.25}$$

Let $\mu_n = P\{X_n(0) \in \cdot\}$. Suppose that $\{\mu_n\}$ satisfies a large deviation principle with good rate function I_0 on E, which is equivalent (see Proposition 3.8) to the existence of the rate transform

$$\Lambda_0(f) = \lim_{n\to\infty} \frac{1}{n} \log \int_E e^{nf} d\mu_n,$$

for each $f \in C_b(E)$, and

$$I_0(x_0) = \sup_{f \in C_b(E)} \{f(x_0) - \Lambda_0(f)\}, \quad x_0 \in E.$$

Then

a) *For each $0 \le t_1 < \cdots < t_k$ and $f_1, \ldots, f_k \in D$,*

$$\begin{aligned}&\lim_{n\to\infty} \frac{1}{n} \log E^{P^n}[e^{nf_1(X_n(t_1))+\ldots+nf_k(X_n(t_k))}] \\ &= \Lambda_0(V(t_1)(f_1 + V(t_2 - t_1)(f_2 + \ldots + V(t_k - t_{k-1})f_k)\ldots)).\end{aligned}$$

b) *Let $0 \le t_1 < \ldots < t_k$, and assume that*

$$\{(X_n(t_1), \ldots, X_n(t_k))\} \tag{5.26}$$

is exponentially tight in E^k. If D contains a set that is bounded above and isolates points, then (5.26) satisfies the large deviation principle with rate function

$$(5.27)\quad \begin{aligned} &I_{t_1,\ldots,t_k}(x_1,\ldots,x_k) \\ &\quad = \sup_{f_1,\ldots,f_k\in D}\{f_1(x_1)+\ldots+f_k(x_k) \\ &\qquad -\Lambda_0(V(t_1)(f_1+V(t_2-t_1)(f_2+\ldots V(t_k-t_{k-1})f_k)\ldots))\}. \end{aligned}$$

c) *If $\{X_n\}$ is exponentially tight in $D_E[0,\infty)$ and D contains a set that is bounded above and isolates points, then $\{X_n\}$ satisfies the large deviation principle in $D_E[0,\infty)$ with rate function*

$$I(x) = \sup_{\{t_i\}\subset\Delta_x^c} I_{t_1,\ldots,t_k}(x(t_1),\ldots,x(t_k)). \tag{5.28}$$

d) *If $\{X_n\}$ is C-exponentially tight and D contains a set that is bounded above and isolates points, then $\{X_n\}$ satisfies the large deviation principle with rate function*

$$I(x(\cdot)) = \sup_{t_1,\ldots,t_k} I_{t_1,\ldots,t_k}(x(t_1),\ldots,x(t_k)). \tag{5.29}$$

REMARK 5.16. Let $\{\{K_n^q\} : q\in\mathcal{Q}\}$ be as in Definition 2.5. If for each $t\geq 0$,

$$\inf_{q\in\mathcal{Q}}\limsup_{n\to\infty}\frac{1}{n}\log P\{Y_n(t)\notin K_n^q\} = -\infty,$$

then (5.24) can be replaced by the assumption: If $f\in D$ and $f_n\in B(E_n)$ satisfies $f=\mathrm{LIM}f_n$, then

$$V(t)f = \mathrm{LIM}V_n(t)f_n.$$

In the discrete-time case, also replace (5.25) by

$$V(t)f = \mathrm{LIM}V_n(t+\epsilon_n)f_n.$$

PROOF. We consider the continuous-time case. The proof for the discrete time case is similar.

Note that since D is closed under addition and $V(t): D\mapsto D$, if $f_1, f_2\in D$, then $f_1+V(t)f_2\in D$. By Ethier and Kurtz [**36**], Theorem 4.4.2, Y_n is a Markov process. It follows that for $f_1,\ldots,f_k\in D$,

$$\begin{aligned} &E^{P^n}[e^{nf_1(X_n(t_1))+\ldots+nf_k(X_n(t_k))}] \\ &\quad = E[E[e^{n\eta_n f_1(Y_n(t_1))+\ldots+n\eta_n f_k(Y_n(t_k))}|\mathcal{F}_{t_{k-1}}^{Y_n}]] \\ &\quad = E[e^{n\eta_n f_1(Y_n(t_1))+\ldots+nV_n(t_k-t_{k-1})\eta_n f_k(Y_n(t_{k-1}))}] \\ &\quad = \int e^{ng_n(y)}P_nY_n^{-1}(0)(dy) \\ &\quad \sim \int e^{n\eta_n V(t_1)(f_1+V(t_2-t_1)(f_2+\ldots+V(t_k-t_{k-1})f_k)\ldots)}(y)P_nY_n^{-1}(0)(dy), \end{aligned}$$

where

$$\begin{aligned} g_n(y) &= V_n(t_1)\Big(\eta_n f_1+V_n(t_2-t_1)(\eta_n f_2+\ldots \\ &\qquad +V_n(t_{k-1}-t_{k-2})(\eta_n f_{k-1}+V_n(t_k-t_{k-1})\eta_n f_k)\ldots\Big)(y) \end{aligned}$$

and hence

$$\lim_{n\to\infty}\frac{1}{n}\log E^{P^n}[e^{nf_1(X_n(t_1))+\ldots+nf_k(X_n(t_k))}]$$
$$=\Lambda_0(V(t_1)(f_1+V(t_2-t_1)(f_2+\ldots+V(t_k-t_{k-1})f_k)\ldots)).$$

Part (b) follows by Lemma 3.22 and Proposition 3.20; Parts (c) and (d) follow by Theorems 4.28 and 4.30. □

COROLLARY 5.17. *Let E_n, E, η_n, Y_n, X_n, V_n, D, V, and I_0 be as in Theorem 5.15. Suppose that D contains a set that is bounded above and isolates points and that $\lim_{n\to\infty}\eta_n(E_n)=E$. Define*

$$I_t(y|x)=\sup_{f\in D}(f(y)-V(t)f(x)). \tag{5.30}$$

Then, setting $t_0=0\le t_1<\cdots<t_k$,

$$I_{t_1,\ldots,t_k}(x_1,\ldots,x_k)=\inf_{x_0\in E}(I_0(x_0)+\sum_{i=1}^k I_{t_i-t_{i-1}}(x_i|x_{i-1}))$$

and

$$I(x)=\sup_{\{t_i\}\subset\Delta_x^c}(I_0(x(0))+\sum_{i=1}^k I_{t_i-t_{i-1}}(x(t_i)|x(t_{i-1}))). \tag{5.31}$$

If in addition $\{X_n\}$ is C-exponentially tight (cf. Theorem 4.13), then I satisfies

$$I(x)=\sup_{t_1<\cdots<t_k}(I_0(x(0))+\sum_{i=1}^k I_{t_i-t_{i-1}}(x(t_i)|x(t_{i-1}))). \tag{5.32}$$

REMARK 5.18. Remark 5.16 applies here as well.

PROOF. Note that

$$\frac{1}{n}\log E^{P_n}[e^{nf_k(X_n(t_k))}|X_n(t_1),\ldots,X_n(t_{k-1})]$$
$$=\frac{1}{n}E^{P_n}[V_n(t_k-t_{k-1})\eta_n f_k(Y_n(t_{k-1}))|X_n(t_1),\ldots,X_n(t_{k-1})]$$
$$=V(t_k-t_{k-1})f_k(X_n(t_{k-1}))+O(\|\eta_n V(t_k-t_{k-1})f_k-V_n(t_k-t_{k-1})\eta_n f_k\|),$$

and it follows by Proposition 3.25 and Remark 3.26 that

$$I_{t_1,\ldots,t_k}(x_1,\ldots,x_k)=I_{t_1,\ldots,t_{k-1}}(x_1,\ldots,x_{k-1})+I_{t_k-t_{k-1}}(x_k|x_{k-1}).$$

Similarly, the rate function for $(X_n(0),X_n(t_1))$ is

$$I_{0,t_1}(x_0,x_1)=I_0(x_0)+I_{t_1}(x_1|x_0),$$

and the corollary follows by induction and the contraction principle. □

Lemma 5.13 gives the semigroup convergence needed in Theorem 5.15.

COROLLARY 5.19. *Let E_n, E, η_n, A_n, Y_n, V_n, and X_n be as in the continuous-time case in Theorem 5.15. Define*

$$H_nf=\frac{1}{n}e^{-nf}A_ne^{nf}$$

on $\mathcal{D}(H_n)=\{f:e^{nf}\in\mathcal{D}(A_n)\}$, and let

$$H\subset ex\text{-}\lim_{n\to\infty}H_n. \tag{5.33}$$

Suppose that H is dissipative, $\mathcal{D}(H) \subset C_b(E)$, H satisfies the range condition 5.1, and that the (norm) closure of $\mathcal{D}(H)$ is closed under addition and contains a set that is bounded above and isolates points. Suppose that $\{X_n\}$ satisfies the exponential compact containment condition (Remark 4.5) and that $\{X_n(0)\}$ satisfies the large deviation principle with good rate function I_0.

Then

a) *H generates a semigroup $V(t)$ on $\overline{\mathcal{D}(H)}$, and for $f \in \overline{\mathcal{D}(H)}$ and $f_n \in B(E_n)$ satisfying $\|\eta_n f - f_n\| \to 0$,*

$$\lim_{n\to\infty} \|\eta_n V(t)f - V_n(t)f_n\| = 0.$$

b) *$\{X_n\}$ is exponentially tight and satisfies the large deviation principle with good rate function I defined by (5.30) and (5.31) with $D = \overline{\mathcal{D}(H)}$.*

PROOF. The exponential tightness of $\{X_n\}$ follows by Corollary 4.17, and the convergence of V_n follows by Lemma 5.13. Theorem 5.15 and Lemma 5.17 give the large deviation principle for $\{X_n = \eta_n \circ Y_n\}$ with rate function given by (5.31). □

COROLLARY 5.20. *Let E_n, E, η_n, ϵ_n, Y_n, and X_n be as in the discrete-time case in Theorem 5.15. Let T_n be the transition operator for Y_n, that is, $T_n f(y) = E[f(Y_n(\epsilon_n))|Y_n(0) = y]$, and define*

$$H_n f = \frac{1}{n\epsilon_n} \log e^{-nf} T_n e^{nf} = \frac{1}{n\epsilon_n} \log(e^{-nf}(T_n - I)e^{nf} + 1).$$

Assume

$$H \subset ex\text{-}\lim_{n\to\infty} H_n. \tag{5.34}$$

Suppose that H is dissipative, $\mathcal{D}(H) \subset C_b(E)$, H satisfies the range condition 5.1, and that the (norm) closure of $\mathcal{D}(H)$ is closed under addition and contains a set that is bounded above and isolates points. Suppose that $\{X_n\}$ satisfies the exponential compact containment condition (Remark 4.5) and that $\{X_n(0)\}$ satisfies the large deviation principle with good rate function I_0.

Define $\{V_n(t)\}$ on $B(E_n)$ by

$$V_n(t)f(y) = \frac{1}{n} \log E[e^{nf(Y_n(t))}|Y_n(0) = y].$$

Then

a) *H generates a semigroup $V(t)$ on $\overline{\mathcal{D}(H)}$, and*

$$\lim_{n\to\infty} \|\eta_n V(t)f - V_n(t)f_n\| + \|\eta_n V(t)f - V_n(t+\epsilon_n)f_n\| = 0,$$

whenever $f_n \in B(E_n)$, $f \in \overline{\mathcal{D}(H)}$ and $\|\eta_n f - f_n\| \to 0$.

b) *$\{X_n\}$ is exponentially tight and satisfies the large deviation principle with rate function I defined in (5.30) and (5.31) with $D = \overline{\mathcal{D}(H)}$.*

PROOF. The exponential tightness of $\{X_n\}$ follows by Corollary 4.17. Define

$$V_n(t)f = \frac{1}{n} \log T_n^{[t/\epsilon_n]} e^{nf}. \tag{5.35}$$

Then Proposition 5.5 and Lemma 5.4 imply

$$\|\eta_n V(t)f - V_n(t)f_n\| \to 0, \tag{5.36}$$

whenever $f_n \in B(E_n)$, $f \in \overline{\mathcal{D}(H)}$, and $\|\eta_n f - f_n\| \to 0$. The convergence of H_n ensures that $f \in \overline{\mathcal{D}(H)}$ and $\|\eta_n f - f_n\| \to 0$ imply $V_n(\epsilon_n)f_n = \frac{1}{n}\log T_n e^{nf_n}$ satisfies $\|V_n(\epsilon_n)f_n - \eta_n f\| \to 0$, and hence (5.25) holds. The Corollary then follows by Theorem 5.15. □

The conclusions of Corollaries 5.19, 5.20 also hold if we use LIM instead of norm convergence. Corollary 4.19 gives the exponential tightness.

COROLLARY 5.21. *Let* LIM *be given by Definition 2.5. Suppose that the conditions of Corollary 5.19 hold with (5.33) replaced by*

$$H \subset ex\text{-}\mathrm{LIM} H_n.$$

(See (A.5) in Section A.2). Assume that the Y_n satisfy Condition 2.8 and

$$\inf_{q\in\mathcal{Q}} \limsup_{n\to\infty} \frac{1}{n} \log P\{Y_n(0) \notin K_n^q\} = -\infty.$$

Then

a) *H generates a semigroup $V(t)$ on $\overline{\mathcal{D}(H)}$, and*

$$V(t)f = \mathrm{LIM} V_n(t) f_n$$

whenever $f \in \overline{\mathcal{D}(H)}$, $f_n \in B(E_n)$, and $f = \mathrm{LIM} f_n$.

b) *$\{X_n\}$ is exponentially tight and satisfies the large deviation principle with good rate function I given by (5.30) and (5.31) with $D = \overline{\mathcal{D}(H)}$.*

COROLLARY 5.22. *Let* LIM *be given by Definition 2.5. Suppose that the conditions of Corollary 5.20 hold with (5.34) replaced by*

$$H \subset ex\text{-}\mathrm{LIM} H_n.$$

(See (A.5) in Section A.2). Assume that the Y_n satisfy Condition 2.8 and

$$\inf_{q\in\mathcal{Q}} \limsup_{n\to\infty} \frac{1}{n} \log P\{Y_n(0) \notin K_n^q\} = -\infty.$$

Then

a) *H generates a semigroup $V(t)$ on $\overline{\mathcal{D}(H)}$, and*

$$V(t)f = \mathrm{LIM} V_n(t) f_n = \mathrm{LIM} V_n(t+\epsilon_n) f_n,$$

whenever $f_n \in B(E_n)$, $f \in \overline{\mathcal{D}(H)}$, and $f = \mathrm{LIM} f_n$.

b) *$\{X_n\}$ is exponentially tight and satisfies the large deviation principle with rate function I given by (5.30) and (5.31) with $D = \overline{\mathcal{D}(H)}$.*

CHAPTER 6

Large deviations and nonlinear semigroup convergence using viscosity solutions

In the preceding chapter, we proved a large deviation theorem using semigroup convergence by imposing a range condition on the limit operator. This condition is usually difficult to verify. It essentially requires that

$$(I - \alpha H)f = h \tag{6.1}$$

have a solution $f \in \mathcal{D}(H)$, for each $h \in \mathcal{D}(H)$ and small $\alpha > 0$. (Because of the dissipativity of H, if such a solution exists, it is unique.) To verify the solvability of (6.1) in the ordinary sense typically requires proving a regularity theorem for the solutions of a nonlinear equation. (See the examples in Chapter 1.) If the Markov processes are diffusions in R^d, for example, verifying that a solution is in $\mathcal{D}(H)$ involves verifying differentiability and growth conditions. Regularity results for these equations are not readily available.

In this chapter, we show that uniqueness of f in a weak sense (the viscosity sense) is sufficient for the semigroup convergence (Theorem 6.13), and hence for large deviation results (Theorem 6.14). In the setting we discuss, existence of these weak solutions follows by construction. This approach is quite effective in applications. In Chapter 9, we collect and develop techniques for proving uniqueness for viscosity solutions, and in the final chapters, we apply these techniques to obtain large deviation results for the examples introduced in Chapter 1.

Instead of the resolvent condition in (6.1), we could also formulate conditions in terms of the Cauchy problem (Feng [**41**])

$$-\frac{\partial}{\partial t}u(t,x) + (Hu(t,\cdot))(x) = 0 \quad \text{for } t > 0, \quad u(0,x) = f_0(x).$$

To set the ideas, we assume (E, r) is compact throughout this section. (These results also cover many locally compact situations. See the discussion on compactifying the state space in Section 4.3 and Theorem 4.11.) In Chapter 7, we remove the compactness assumption (at the cost of greater technicality) and generalize the results of this section to processes in complete, separable metric spaces.

For $g \in B(E)$, g^* will denote the *upper semicontinuous regularization* of g, that is,

$$g^*(x) = \lim_{\epsilon \to 0} \sup_{y \in B_\epsilon(x)} g(y), \tag{6.2}$$

and g_* will denote the *lower semicontinuous regularization*,

$$g_*(x) = \lim_{\epsilon \to 0} \inf_{y \in B_\epsilon(x)} g(y). \tag{6.3}$$

6.1. Viscosity solutions, definition and convergence

DEFINITION 6.1 (Viscosity Solution). Let E be a compact metric space, and $H \subset C(E) \times B(E)$. Fix $h \in C(E)$ and $\alpha > 0$. Let $f \in B(E)$ and define $g = \alpha^{-1}(f - h)$, that is, $f - \alpha g = h$. Then

a) f is a *viscosity subsolution* of (6.1) if and only if f is upper semicontinuous and for each $(f_0, g_0) \in H$ such that $\sup_x (f - f_0)(x) = \|f - f_0\|$, there exists $x_0 \in E$ satisfying

$$(f - f_0)(x_0) = \|f - f_0\| \tag{6.4}$$

and

$$\alpha^{-1}(f(x_0) - h(x_0)) = g(x_0) \le (g_0)^*(x_0). \tag{6.5}$$

b) f is a *viscosity supersolution* of (6.1) if and only if f is lower semicontinuous and for each $(f_0, g_0) \in H$ such that $\sup_x (f_0 - f)(x) = \|f_0 - f\|$, there exists $x_0 \in E$ satisfying

$$(f_0 - f)(x_0) = \|f - f_0\| \tag{6.6}$$

and

$$\alpha^{-1}(f(x_0) - h(x_0)) = g(x_0) \ge (g_0)_*(x_0). \tag{6.7}$$

A function $f \in C(E)$ is said to be a *viscosity solution* of (6.1) if it is both a subsolution and a supersolution.

If in the definition of subsolution (6.5) holds for every x_0 satisfying (6.4), f will be called a *strong* subsolution, and similarly, if in the definition of supersolution (6.7) holds for every x_0 satisfying (6.6), f will be called a *strong* supersolution.

REMARK 6.2. The basic goal of the definition of a viscosity solution is to extend H so that the desired solution is included in an extended domain while at the same time preserving the dissipativity of the operator.

Continuous viscosity solutions were introduced in Crandall and Lions [**19**] to study certain nonlinear partial differential equations. The concept was later extended to include discontinuous sub- super- solutions. See Barles and Perthame [**8**], Barles and Souganidis [**9**], Ishii [**60**], Ishii and Lions [**61**] and Fleming and Soner [**47**]. The last reference focuses on control theory. In addition to the existence issues, these articles offer a rich theory of comparison results (see the next definition). Definition 6.1 is an adaptation of this concept to general spaces.

When $E \subset R^d$ is compact, H is local, and $\mathcal{D}(H)$ is dense in $C(E)$, the regular and strong definitions are equivalent and also equivalent to that, say, in Barles and Souganidis [**9**] or Ishii and Lions [**61**] (take $F(x, u, Hu) = u(x) - \alpha(Hu)(x) - h(x)$). The relationship between strong and ordinary viscosity solutions is similar to the relationship between strong and ordinary dissipative operators (see Appendix A.3).

The operators H that arise as limits of the type discussed in the Introduction have the property that $(f, g) \in H$ and $c \in R$ implies $(f + c, g) \in H$. The following lemma simplifies the definition of viscosity solution.

LEMMA 6.3. *Suppose $(f, g) \in H$ and $c \in R$ implies $(f + c, g) \in H$. Then an upper semicontinuous function f is a viscosity subsolution of (6.1) if and only if for each $(f_0, g_0) \in H$, there exists $x_0 \in E$ such that*

$$f(x_0) - f_0(x_0) = \sup_x (f(x) - f_0(x_0)) \tag{6.8}$$

and

$$\alpha^{-1}(f(x_0) - h(x_0)) \equiv g(x_0) \le (g_0)^*(x_0). \tag{6.9}$$

Similarly, a lower semicontinuous function f is a viscosity supersolution of (6.1) if and only if for each $(f_0, g_0) \in H$, there exists $x_0 \in E$ such that

$$f_0(x_0) - f(x_0) = \sup_x (f_0(x) - f(x_0)) \tag{6.10}$$

and

$$\alpha^{-1}(f(x_0) - h(x_0)) \equiv g(x_0) \ge (g_0)_*(x_0). \tag{6.11}$$

PROOF. Let f be a subsolution and $(f_0, g_0) \in H$. Then there exists a constant c such that

$$\sup_x (f(x) - f_0(x) + c) = \|f - f_0 + c\|,$$

and hence, by the definition of subsolution, there exists $x_0 \in E$ such that

$$f(x_0) - f_0(x_0) + c = \sup_x (f(x) - f_0(x) + c) = \|f - f_0 + c\| \tag{6.12}$$

and (6.9) holds. Of course, (6.12) implies (6.8). The proof of the second part is similar. □

DEFINITION 6.4 (Comparison Principle). We say that

$$(I - \alpha H) f = h$$

satisfies a *comparison principle*, if $\overline{f}$ a viscosity subsolution and $\underline{f}$ a viscosity supersolution implies $\overline{f} \le \underline{f}$ on E.

We will see that the comparison principle can be used to replace the range condition in results analogous to Proposition 5.5 and Theorem 5.15. The following lemma shows that the comparison principle generalizes the range condition.

LEMMA 6.5. *Let E be a compact metric space and $H \subset C(E) \times B(E)$, and suppose that $(f, g) \in H$ and $c \in R$ implies $(f + c, g) \in H$. Then for each $h \in C(E) \cap \overline{\mathcal{R}(I - \alpha H)}$, the comparison principle holds for $f - \alpha H f = h$.*

PROOF. Suppose $\overline{f}$ is a viscosity subsolution, $(f_n, g_n) \in H$, and $\|f_n - \alpha g_n - h\| \to 0$. Set $h_n = f_n - \alpha g_n$. By Lemma 6.3, there exist x_n satisfying $\overline{f}(x_n) - f_n(x_n) = \sup_x(\overline{f}(x) - f_n(x))$ and

$$\begin{aligned}\sup_x(\overline{f}(x) - f_n(x)) &= \overline{f}(x_n) - f_n(x_n)\\ &\le h(x_n) + \alpha g_n^*(x_n) - f_n(x_n)\\ &= h(x_n) - (h_n)_*(x_n) \to 0.\end{aligned}$$

Similarly, if $\underline{f}$ is a supersolution of $f - \alpha H f = h$,

$$\liminf_{n\to\infty} \inf_x (\underline{f}(x) - f_n(x)) \ge 0,$$

and it follows that $\overline{f} \le \underline{f}$. □

Let $H \subset B(E) \times B(E)$. The *closure* of H is

$$\overline{H} = \{(f, g) \in B(E) \times B(E) : \exists \{(f_n, g_n)\} \subset H, \|f_n - f\| + \|g_n - g\| \to 0\}. \tag{6.13}$$

DEFINITION 6.6. $\widehat{H} \subset B(E) \times B(E)$ is a *viscosity extension* of H if $\widehat{H} \supset H$ and for each $h \in C(E)$, f is a viscosity subsolution (supersolution) of $(I - \alpha H)f = h$ if and only if f is a viscosity subsolution (supersolution) for $(I - \alpha\widehat{H})f = h$. In particular, the comparison principle holds for $(I - \alpha H)f = h$ if and only if the comparison principle holds for $(I - \alpha\widehat{H})f = h$

It is easy to check that $\overline{H}$ is a viscosity extension of H.

EXAMPLE 6.7. Let E be the one point compactification of R^d (we will write $E = R^d \cup \{\infty\}$). Let the operator H be given by (1.14) for $f \in D_d = \{f \in C(E) : f|_{R^d} - f(\infty) \in C_c^2(R^d)\}$, and let $H(x,p)$ be given by (1.15). Suppose $|a(x)| \le C(1+|x|^2)$ and $|b(x)| \le C(1+|x|)$, and let

$$\widehat{H} = \{(f, H(\cdot, \nabla f(\cdot))) : f \in C^1, f(\infty) \equiv \lim_{|x|\to\infty} f(x) \text{ exists }, \lim_{|x|\to\infty} |x||\nabla f(x)| = 0\}.$$

Then $\widehat{H}$ is a viscosity extension of H.

Let $f \in \mathcal{D}(\widehat{H})$. For simplicity, assume that $f(\infty) = 0$. Let $\{\zeta_n\} \subset C_c^2$ satisfy $\zeta_n(x) \equiv 1$ for $|x| \le n$, $0 \le \zeta_n \le 1$, and $\lim_{n\to\infty} \sup_x |x||\nabla\zeta_n(x)| = 0$, and let $\rho \in C_c^2$ satisfy $\rho \ge 0$ and $\int_{R^d} \rho(x)dx = 1$. Then f_n defined by

$$f_n(x) = \zeta_n(x) \int_{R^d} f(x-y)n^d\rho(ny)dy$$

is in $\mathcal{D}(H)$, $\|f - f_n\| \to 0$ (since f is uniformly continuous and vanishes at ∞), and

$$\nabla f_n(x) = \zeta_n(x) \int_{R^d} \nabla f(x-y)n^d\rho(ny)dy + \int_{R^d} f(x-y)n^d\rho(ny)dy\nabla\zeta_n(x)$$

satisfies

$$\lim_{n\to\infty} \sup_x |x||\nabla f(x) - \nabla f_n(x)| = 0.$$

By the growth conditions on a and b,

$$\lim_{n\to\infty} \|Hf_n - \widehat{H}f\| = 0.$$

In particular, $\widehat{H} \subset \overline{H}$. Suppose $x_n \to x \neq \infty$. Then

$$\limsup_{n\to\infty} Hf_n(x_n) = \limsup_{n\to\infty} H(x_n, \nabla f_n(x_n)) = \limsup_{n\to\infty} H(x_n, \nabla f(x_n)) \le H^*f(x),$$

and similarly for lim inf. If $|x_n| \to \infty$, then $Hf_n(x_n) \to 0 = H^*f(\infty) = H_*f(\infty)$.

LEMMA 6.8. *Suppose E is compact. Let $H \subset \widehat{H} \subset C(E) \times B(E)$. Suppose that for each $(f,g) \in \widehat{H}$ there exists $\{(f_n, g_n)\} \subset H$ such that $\lim_{n\to\infty} \|f_n - f\| = 0$ and for $x_n, x \in E$ such that $x_n \to x$,*

$$g_*(x) \le \liminf_{n\to\infty} g_n(x_n) \le \limsup_{n\to\infty} g_n(x_n) \le g^*(x). \tag{6.14}$$

Suppose that $(f,g) \in H$ implies $(f+c, g) \in H$ for all $c \in R$. Then $\widehat{H}$ is a viscosity extension of H.

PROOF. Clearly, if f is a viscosity subsolution for $(I - \alpha\widehat{H})f = h$, then f is a viscosity subsolution for $(I - \alpha H)f = h$. Suppose f is a viscosity subsolution for $(I - \alpha H)f = h$, and let $(f_0, g_0) \in \widehat{H}$. There exist $(f_n, g_n) \in H$ such that

$\|f_0 - f_n\| \to 0$ and (6.14) holds with g replaced by g_0. Since f_n is continuous and f is upper semicontinuous, there exist x_n such that

$$f(x_n) - f_n(x_n) = \sup_x (f(x) - f_n(x))$$

and $\alpha^{-1}(f(x_n) - h(x_n)) \leq g_n^*(x_n)$. By the compactness of E, there exists $x_0 \in E$ and a subsequence of $\{x_n\}$ along which $x_n \to x_0$. By the upper semicontinuity of f,

$$f(x_0)-f_0(x_0) \geq \lim_{n\to\infty} f(x_n)-f_n(x_n) = \lim_{n\to\infty} \sup_x (f(x)-f_n(x)) = \sup_x (f(x)-f_0(x)).$$

Hence the first inequality must be equality and, in particular, $f(x_n) \to f(x_0)$. Consequently,

$$\alpha^{-1}(f(x_0) - h(x_0)) = \lim_{n\to\infty} \alpha^{-1}(f(x_n) - h(x_n)) \leq \limsup_{n\to\infty} g_n^*(x_n) \leq g_0^*(x_0),$$

verifying that f is a subsolution for $(I - \alpha\widehat{H})f = h$. The proof is similar for supersolutions. □

LEMMA 6.9. *Let (E_n, r_n) be general metric spaces and (E, r) be a compact metric space. Let $\eta_n : E_n \mapsto E$ be Borel measurable, and let $E = \lim_{n\to\infty} \eta_n(E_n)$ in the sense that for each $x \in E$, there exist $z_n \in E_n$ such that $x = \lim_{n\to\infty} \eta_n(z_n)$. Define $\eta_n : B(E) \mapsto B(E_n)$ by $\eta_n f = f \circ \eta_n$.*

For each n, we assume $H_n \subset B(E_n) \times B(E_n)$ is a dissipative operator such that $(f, g) \in H_n$ and $c \in R$ implies $(f + c, g) \in H_n$. Let $H \subset C(E) \times B(E)$. Assume that for each $(f_0, g_0) \in H$, there exist $(f_{n,0}, g_{n,0}) \in H_n$ such that

$$\|f_{n,0} - \eta_n f_0\| \to 0 \tag{6.15}$$

and for each $x \in E$ and each sequence $z_n \in E_n$ satisfying $\eta_n(z_n) \to x$,

$$(g_0)_*(x) \leq \liminf_{n\to\infty} g_{n,0}(z_n) \leq \limsup_{n\to\infty} g_{n,0}(z_n) \leq (g_0)^*(x). \tag{6.16}$$

Let $\alpha > 0$ and $h \in C(E)$, and assume that the comparison principle holds for

$$(I - \alpha H)f = h. \tag{6.17}$$

Let $(f_n, g_n) \in H_n$, and define $f_n - \alpha g_n = h_n$. If $\|h_n - \eta_n h\| \to 0$, then there exists a unique viscosity solution f of (6.17), $f \in C(E)$, and

$$\lim_{n\to\infty} \|f_n - \eta_n f\| = 0.$$

If $(f_0, g_0) \in H$, then

$$\|f - f_0\| \leq \|h - (f_0 - \alpha g_0)\|. \tag{6.18}$$

REMARK 6.10. Note that $\|\eta_n g_0 - g_{n,0}\| \to 0$ implies (6.16).

PROOF. First, we claim $\sup_n \|f_n\| < \infty$. Let $\{(f_{n,0}, g_{n,0})\}$ be any sequence satisfying (6.15) and (6.16). By the dissipativity of H_n,

$$\|f_n - f_{n,0}\| \leq \|h_n - (f_{n,0} - \alpha g_{n,0})\|. \tag{6.19}$$

Therefore

$$\begin{aligned}
\limsup_{n\to\infty} \|f_n\| &\leq 2\limsup_{n\to\infty} \|f_{n,0}\| + \limsup_{n\to\infty} \|h_n\| + \alpha \limsup_{n\to\infty} \|g_{n,0}\| \\
&\leq 2\|f_0\| + \|h\| + \alpha \max\{\|(g_0)^*\|, \|(g_0)_*\|\} \\
&\leq 2\|f_0\| + \|h\| + \alpha\|g_0\|.
\end{aligned}$$

Now define $\overline{f}, \underline{f} \in B(E)$ by

$$\overline{f}(x) = \lim_{k\to\infty} \sup\{f_n(z) : n \geq k, r(x, \eta_n(z)) \leq \frac{1}{k}\},$$

and

$$\underline{f}(x) = \lim_{k\to\infty} \inf\{f_n(z) : n \geq k, r(x, \eta_n(z)) \leq \frac{1}{k}\}.$$

Then $\overline{f}$ is bounded and upper-semicontinuous, and $\underline{f}$ is bounded and lower-semicontinuous. We will prove that they are respectively sub- and supersolutions of (6.17). We only consider $\overline{f}$. The argument for $\underline{f}$ is similar.

Let $(f_0, g_0) \in H$. By assumption, there exists $(f_{n,0}, g_{n,0}) \in H_n$ such that (6.15) and (6.16) hold. By the dissipativity of H_n and Lemma 6.3, for each n, there exist $z_n^m \in E_n$ such that $\lim_{m\to\infty}(f_n - f_{n,0})(z_n^m) = \sup_z(f_n - f_{n,0})(z)$, and $\lim_{m\to\infty}(g_n - g_{n,0})(z_n^m) \leq 0$. (See Lemma A.19.) We can therefore select $m(n)$ such that $(f_n - f_{n,0})(z_n^{m(n)}) \geq \sup_z(f_n - f_{n,0})(z) - 1/n$, and $(g_n - g_{n,0})(z_n^{m(n)}) \leq 1/n$.

Define $x_n = \eta_n(z_n^{m(n)})$. Note that by the definition of $\overline{f}$, the compactness of E, and (6.15),

$$\sup_x(\overline{f} - f_0)(x) = \limsup_{n\to\infty} \sup_z (f_n - f_{n,0})(z) = \limsup_{n\to\infty}(f_n - f_{n,0})(z_n^{m(n)}).$$

Selecting a subsequence along which the last lim sup becomes a limit and x_n converges to a point x_0, by the definition of $\overline{f}$, (6.15), and the continuity of f_0,

$$\begin{aligned} (6.20) \qquad (\overline{f} - f_0)(x_0) &\geq \limsup_{n\to\infty}(f_n - f_{n,0})(z_n^{m(n)}) \\ &= \sup_x(\overline{f} - f_0)(x). \end{aligned}$$

From the above, along the selected subsequence, we also have $f_n(z_n^{m(n)}) \to \overline{f}(x_0)$. Define $\overline{g}$ by $\overline{f} - \alpha\overline{g} = h$. Then

$$g_n(z_n^{m(n)}) \to \overline{g}(x_0).$$

Hence

$$\overline{g}(x_0) = \lim_{n\to\infty} g_n(z_n^{m(n)}) \leq (g_0)^*(x_0),$$

and it follows that $\overline{f}$ is an upper semicontinuous subsolution of (6.17). Finally, by the comparison principle, $f \equiv \overline{f} = \underline{f} \in C(E)$, and f is a viscosity solution of (6.17).

To see that

$$\|\eta_n f - f_n\| \to 0,$$

we show that for each sequence $z_n \in E_n$,

$$(6.21) \qquad |f_n(z_n) - f(\eta_n(z_n))| \to 0.$$

Since E is compact, we may assume $\eta_n(z_n) \to \widetilde{x}$ for some $\widetilde{x} \in E$ (otherwise take a subsequence). For each $k > 0$, for n sufficiently large,

$$f_n(z_n) \leq \sup\{f_n(z) : n \geq k, r(\widetilde{x}, \eta_n(z)) < \frac{1}{k}\}.$$

Therefore

$$\limsup_{n\to\infty} f_n(z_n) \leq \overline{f}(\widetilde{x}) = f(\widetilde{x}) = \lim_{n\to\infty} f(\eta_n(z_n)).$$

Similarly one can prove

$$\liminf_{n\to\infty} f_n(z_n) \geq \underline{f}(\widetilde{x}) = f(\widetilde{x}) = \lim_{n\to\infty} f(\eta_n(z_n)),$$

and (6.21) follows.

To verify (6.18), let $(f_{n,0}, g_{n,0})$ be as in (6.19). Note that the left side of (6.19) converges to $\|f - f_0\|$. Consequently, either there exist z_n such that

$$\limsup_{n\to\infty}(h_n(z_n) - f_{n,0}(z_n) + \alpha g_{n,0}(z_n) \geq \|f - f_0\|$$

or

$$\liminf_{n\to\infty}(h_n(z_n) - f_{n,0}(z_n) + \alpha g_{n,0}(z_n) \leq -\|f - f_0\|.$$

Select a subsequence along which $\eta_n(z_n) \to x$. Then by (6.15) and (6.16), the assumption on h_n, and the continuity of h, either

$$h(x) - f_0(x) + \alpha(g_0)^*(x) \geq \|f - f_0\|$$

or

$$h(x) - f_0(x) + \alpha(g_0)_*(x) \leq -\|f - f_0\|.$$

In either case, (6.18) follows. □

The original ideas in the above proof were due to Barles and Perthame [**7**] and Barles and Souganidis [**9**]. See also Fleming and Soner [**47**]. The condition $E = \lim_{n\to\infty}\eta_n(E_n)$ can be removed if we assume the existence of a strong viscosity solution of equation (6.17).

EXAMPLE 6.11. Verification of the comparison principle is frequently very technical. To at least show that such a condition is plausible, we verify it in the simplest possible case, $Hf = \frac{1}{2}|\nabla f|^2$, the limit operator corresponding to large deviation for rescaled Brownian motion. We continue with $E = R^d \cup \{\infty\}$.

By the kind of closure argument illustrated in Example 6.7, we can take

$$\mathcal{D}(H) = \{f \in C^1 : f(\infty) \equiv \lim_{|x|\to\infty} f(x) \text{ exists and } \lim_{|x|\to\infty} |\nabla f(x)| = 0\}.$$

For $h \in C(E)$, let $\overline{f}$ be a subsolution and $\underline{f}$ a supersolution of $(I - \alpha H)f = h$. It is easy to check that every subsolution is a strong subsolution and every supersolution is a strong supersolution. (See Lemma 9.9.)

First, suppose that $\sup_x(\overline{f}(x) - \underline{f}(x)) > (\overline{f}(\infty) - \underline{f}(\infty))$. Consider

$$\Phi_n(x,y) = \overline{f}(x) - \underline{f}(y) - n\frac{|x-y|^2}{1+|x-y|^2},$$

extended so that Φ_n is upper semicontinuous on $E \times E$. In particular, $\Phi_n(\infty,\infty) = \overline{f}(\infty) - \underline{f}(\infty)$, $\Phi_n(\infty, y) = \overline{f}(\infty) - \underline{f}(y) - n$, $y \in R^d$, and $\Phi_n(x,\infty) = \overline{f}(x) - \underline{f}(\infty) - n$, $x \in R^d$. Since Φ_n is upper semicontinuous, there exist x_n, y_n such that

$$\Phi_n(x_n, y_n) = \sup_{x,y} \Phi_n(x,y) \tag{6.22}$$

and $x_n, y_n \in R^d$. Setting $f_0^n(x) = \underline{f}(y_n) + n\frac{|x-y_n|^2}{1+|x-y_n|^2}$, $f_0^n \in \mathcal{D}(H)$ and x_n satisfies $\overline{f}(x_n) - f_0^n(x_n) = \sup_x(\overline{f}(x) - f_0^n(x))$. Consequently, by (6.9),

$$\alpha^{-1}(\overline{f}(x_n) - h) \leq Hf_0^n(x_n) = n^2\frac{4|x_n - y_n|^2}{(1+|x_n - y_n|^2)^4},$$

and setting $f_1^n(y) = \overline{f}(x_n) - n\frac{|x_n-y|^2}{1+|x_n-y|^2}$, a similar argument gives

$$\alpha^{-1}(\underline{f}(y_n) - h) \geq Hf_1^n(y_n) = n^2\frac{4|x_n - y_n|^2}{(1+|x_n - y_n|^2)^4}.$$

It follows that

$$\alpha^{-1}(\overline{f}(x_n) - \underline{f}(y_n)) \leq 0.$$

A small amount of analysis (Lemma 9.2) gives $\lim_{n\to\infty} n\frac{|x_n-y_n|^2}{1+|x_n-y_n|^2} = 0$ and

$$0 \geq \lim_{n\to\infty}(\overline{f}(x_n) - \underline{f}(y_n)) = \lim_{n\to\infty}\Phi_n(x_n, y_n) \geq \sup_x(\overline{f}(x) - \underline{f}(x)). \tag{6.23}$$

We still need to consider the case

$$\overline{f}(\infty) - \underline{f}(\infty) = \sup_x(\overline{f}(x) - \underline{f}(x)).$$

Now we take

$$\Phi_n(x,y) = \overline{f}(x) - \underline{f}(y) - n(\frac{1}{1+|x|^2} + \frac{1}{1+|y|^2}),$$

and again take x_n and y_n satisfying (6.22). Now either or both x_n and y_n may be ∞. As before

$$\alpha^{-1}(\overline{f}(x_n) - h) \leq Hf_0^n(x_n) = n^2\frac{4|x_n|^2}{(1+|x_n|^2)^4}, \tag{6.24}$$

where the right side is zero if $x_n = \infty$. Similarly,

$$\alpha^{-1}(\underline{f}(y_n) - h) \geq Hf_1^n(y_n) = n^2\frac{4|y_n|^2}{(1+|y_n|^2)^4}. \tag{6.25}$$

Lemma 9.2 implies

$$\lim_{n\to\infty} n(\frac{1}{1+|x_n|^2} + \frac{1}{1+|y_n|^2}) = 0,$$

which in turn implies the right sides of (6.24) and (6.25) converge to zero. Consequently, we have

$$\limsup_{n\to\infty}(\overline{f}(x_n) - \underline{f}(y_n) \leq 0,$$

which as in (6.23) implies $\overline{f}(\infty) - \underline{f}(\infty) \leq 0.$

LEMMA 6.12. *Let E_n, E, η_n, H_n and H satisfy the assumptions of Lemma 6.9. Suppose that, for $i = 1, 2$, there exist $(f_n^i, g_n^i) \in H_n$ and $h_i \in C(E)$ such that $h_n^i \equiv f_n^i - \alpha g_n^i$ satisfies $\|h_n^i - \eta_n h_i\| \to 0$ and the comparison principle holds for*

$$(I - \alpha H)f_i = h_i. \tag{6.26}$$

Then for $i = 1, 2$ there exists a unique viscosity solution $f_i \in C(E)$ of (6.26) and $\|f_1 - f_2\| \leq \|h_1 - h_2\|$.

PROOF. The existence and uniqueness of viscosity solutions $f_i \in C(E)$ of (6.26) follows by Lemma 6.9 and the assumption that $\|\eta_n h_i - h_n^i\| \to 0$. It also follows that $\|\eta_n f_i - f_n^i\| \to 0$. Since H_n is dissipative, $\|f_n^1 - f_n^2\| \leq \|h_n^1 - h_n^2\|$. Noting $h_i, f_i \in C(E)$,

$$\|f_1 - f_2\| = \lim_{n\to\infty}\|\eta_n(f_1 - f_2)\| \leq \lim_{n\to\infty}\|f_n^1 - f_n^2\| \leq \lim_{n\to\infty}\|h_n^1 - h_n^2\| \leq \|h_1 - h_2\|.$$

□

THEOREM 6.13. *Let (E_n, r_n) be general metric spaces, (E, r) be a compact metric space, $\eta_n : E_n \mapsto E$ be Borel measurable, and $E = \lim_{n\to\infty} \eta_n(E_n)$ (cf. Lemma 6.9). Define $\eta_n : B(E) \mapsto B(E_n)$ by $\eta_n f = f \circ \eta_n$. For each $n = 1, 2, \ldots$, assume $H_n \subset B(E_n) \times B(E_n)$ is dissipative and satisfies the range condition 5.1 for $0 < \alpha < \alpha_0$ with α_0 independent of n. Assume that $(f, g) \in H_n$ and $c \in R$ implies $(f + c, g) \in H_n$.*

Let $H \subset C(E) \times B(E)$. Suppose that for each $(f_0, g_0) \in H$, there exist $(f_{n,0}, g_{n,0}) \in H_n$ such that $\|f_{n,0} - \eta_n f_0\| \to 0$ and for each $x \in E$ and each sequence $z_n \in E_n$ satisfying $\eta_n(z_n) \to x$,

$$(g_0)_*(x) \leq \liminf_{n\to\infty}(g_{n,0})(z_n) \leq \limsup_{n\to\infty}(g_{n,0})(z_n) \leq (g_0)^*(x). \tag{6.27}$$

Assume that for each $0 < \alpha < \alpha_0$,

$$C(E) \subset \lim_{n\to\infty} \mathcal{R}(I - \alpha H_n),$$

in the sense that for each $h \in C(E)$, there exist $h_n \in \mathcal{R}(I - \alpha H_n)$ such that $\|\eta_n h - h_n\| \to 0$, and that there exists a dense subset $D_\alpha \subset C(E)$ such that for each $h \in D_\alpha$, the comparison principle holds for

$$(I - \alpha H)f = h. \tag{6.28}$$

Then

a) *For each $h \in D_\alpha$, there exists a unique viscosity solution of (6.28) which we will denote by $f = R_\alpha h$.*
b) *R_α is a contraction and hence extends to all of $C(E)$.*
c) *The operator*

$$\widehat{H} = \cup_\alpha\{(R_\alpha h, \frac{R_\alpha h - h}{\alpha}) : h \in C(E)\} \tag{6.29}$$

is dissipative, $\overline{\mathcal{D}(\widehat{H})} \supset \mathcal{D}(H)$, and $\widehat{H}$ satisfies the range condition 5.1.
d) *$\widehat{H}$ defined in (6.29) generates a strongly continuous contraction semigroup $\{V(t)\}$ on $\overline{\mathcal{D}(\widehat{H})}$ given by*

$$V(t)f = \lim_{n\to\infty} R^n_{t/n} f. \tag{6.30}$$

e) *Letting $\{V_n(t)\}$ denote the semigroup on $\overline{\mathcal{D}(H_n)}$ generated by H_n, for each $f \in \overline{\mathcal{D}(\widehat{H})}$ and $f_n \in \overline{\mathcal{D}(H_n)}$ satisfying $\|f_n - \eta_n f\| \to 0$,*

$$\lim_{n\to\infty} \|V_n(t) f_n - \eta_n V(t) f\| = 0.$$

PROOF. For $0 < \alpha < \alpha_0$ and $h \in D_\alpha$, since $h \in \lim_{n\to\infty} \mathcal{R}(I - \alpha H_n)$, by Lemma 6.9 there exists a unique viscosity solution $f \in C(E)$ of

$$(I - \alpha H)f = h \tag{6.31}$$

(giving Part (a)), and there exist $(f_n, g_n) \in H_n$ such that

$$\|f_n - \eta_n f\| + \|(f_n - \alpha g_n) - \eta_n h\| \to 0. \tag{6.32}$$

By Lemma 6.12

$$\|R_\alpha h_1 - R_\alpha h_2\| \leq \|h_1 - h_2\|, \tag{6.33}$$

for $h_1, h_2 \in D_\alpha$, and R_α extends to all of $C(E)$, giving Part (b).

If (f_n^i, g_n^i), f^i and h^i, $i = 1, 2$, are as in (6.32) (perhaps, for different values of α), then $\|f_n^1 - f_n^2\| \to \|f^1 - f^2\|$ and

$$\begin{aligned}
&\lim_{n\to\infty} \|(f_n^1 - \alpha g_n^1) - (f_n^2 - \alpha g_n^2)\| \\
= &\lim_{n\to\infty} \|(f_n^1 - \alpha \frac{f_n^1 - \eta_n h^1}{\alpha^1}) - (f_n^2 - \alpha \frac{f_n^2 - \eta_n h^2}{\alpha^2})\| \\
= &\|(f^1 - \alpha \frac{f^1 - h^1}{\alpha^1}) - (f^2 - \alpha \frac{f^2 - h^2}{\alpha^2}\|,
\end{aligned}$$

so the dissipativity of $\widehat{H}$ follows from the dissipativity of H_n. For $(f_0, g_0) \in H$, (6.18) implies

$$\|R_\alpha h - f_0\| \le \|h - f_0 + \alpha g_0\|.$$

Taking $h = f_0$, we see that $\lim_{\alpha\to 0} R_\alpha f_0 = f_0$, and hence $\overline{\mathcal{D}(\widehat{H})} \supset \mathcal{D}(H)$.

By Lemma 6.9, $\mathcal{D}(\widehat{H}) \subset C(E)$. For $0 < \alpha < \alpha_0$, $\mathcal{R}(I - \alpha\widehat{H}) \supset D_\alpha$ and hence $\overline{\mathcal{R}(I - \alpha\widehat{H})} = C(E) \supset \mathcal{D}(\widehat{H})$. Part (d) follows by Proposition 5.3. Part (e) then follows by Proposition 5.5. □

6.2. Large deviations using viscosity semigroup convergence

THEOREM 6.14. *Let (E_n, r_n) be complete, separable metric spaces and (E, r) be a compact metric space. Let $\eta_n : E_n \mapsto E$ be Borel measurable, and define $\eta_n : B(E) \mapsto B(E_n)$ by $\eta_n f = f \circ \eta_n$. Assume that $E = \lim_{n\to\infty} \eta_n(E_n)$.*

Assume that one of the following holds:

a) *(Continuous-time case.) For each $n = 1, 2, \ldots$, $A_n \subset B(E_n) \times B(E_n)$ and existence and uniqueness holds for the $D_{E_n}[0,\infty)$-martingale problem for (A_n, μ) for each initial distribution $\mu \in \mathcal{P}(E_n)$. Letting $P_y^n \in \mathcal{P}(D_{E_n}[0,\infty))$ denote the distribution of the solution of the martingale problem for A_n starting from $y \in E_n$, the mapping $y \mapsto P_y^n$ is Borel measurable taking the weak topology on $\mathcal{P}(D_{E_n}[0,\infty))$ (cf. Theorem 4.4.6, Ethier and Kurtz* [**36**]*). Y_n is a solution of the martingale problem for A_n, and*
$$H_n f = \frac{1}{n} e^{-nf} A_n e^{nf}, \quad e^{nf} \in \mathcal{D}(A_n).$$
b) *(Discrete-time case.) For each $n = 1, 2, \ldots$, T_n is a transition operator on $B(E_n)$ for a Markov chain, $\epsilon_n > 0$, Y_n is a discrete-time Markov chain with time-step ϵ_n and transition operator T_n, that is,*
$$E[f(Y_n(t))|Y_n(0) = y] = T_n^{[t/\epsilon_n]} f(y),$$
and
$$H_n f = \frac{1}{n\epsilon_n} \log e^{-nf} T_n e^{nf}, \quad f \in B(E_n).$$
The sequence $\epsilon_n \to 0$.

Define $\{V_n(t)\}$ on $B(E_n)$ by

$$V_n(t) f(y) = \frac{1}{n} \log E[e^{nf(Y_n(t))}|Y_n(0) = y].$$

Let $H \subset C(E) \times B(E)$ with $\mathcal{D}(H)$ dense in $C(E)$. Suppose that for each $(f, g) \in H$, there exist $(f_n, g_n) \in H_n$ such that

$$\|\eta_n f - f_n\| \to 0,$$

$\sup_n \|g_n\| < \infty$, *and for each* $x \in E$ *and sequence* $z_n \in E_n$ *satisfying* $\eta_n(z_n) \to x$,

$$(g)_*(x) \le \liminf_{n\to\infty} g_n(z_n) \le \limsup_{n\to\infty} g_n(z_n) \le (g)^*(x). \tag{6.34}$$

Fix $\alpha_0 > 0$. *Suppose that for each* $0 < \alpha < \alpha_0$, *there exists a dense subset* $D_\alpha \subset C(E)$ *such that for each* $h \in D_\alpha$, *the comparison principle holds for*

$$(I - \alpha H)f = h. \tag{6.35}$$

Define $X_n = \eta_n(Y_n)$. *Suppose that* $\{X_n(0)\}$ *satisfies a large deviation principle with a good rate function. Then*

a) $\widehat{H}$ *defined in (6.29) generates a semigroup on* $C(E)$ *as in (6.30).*

b) *In the continuous time case,*

$$\lim_{n\to\infty} \|\eta_n V(t)f - V_n(t)f_n\| = 0,$$

whenever $f \in C(E)$, $f_n \in B(E_n)$, *and* $\|\eta_n f - f_n\| \to 0$.

c) *In the discrete time case,*

$$\lim_{n\to\infty} \|\eta_n V(t)f - V_n(t)f_n\| + \|\eta_n V(t)f - V_n(\epsilon_n + t)f_n\| = 0,$$

whenever $f \in C(E)$, $f_n \in B(E_n)$ *and* $\|\eta_n f - f_n\| \to 0$.

d) $\{X_n\}$ *is exponentially tight and satisfies the large deviation principle with rate function* I *given by (5.30) and (5.31) with* $D = C(E)$ *and* $\{V(t)\}$ *as in Theorem 6.13.*

PROOF. For the continuous-time case, follow the proof of Corollary 5.19. Define $\widehat{A}_n$ $\widehat{A}_n^{\epsilon_n}$ and $\widehat{H}_n^{\epsilon_n}$. Then $\widehat{H}_n^{\epsilon_n}$ satisfies the range condition 5.1 and generates a semigroup $\{V_n^{\epsilon_n}(t)\}$. Also, recall that by Lemma 5.7, $\mathcal{R}(I - \alpha\widehat{H}_n^{\epsilon_n}) = B(E_n)$ for all $\alpha > 0$. Therefore $D_\alpha \subset \mathcal{R}(I - \alpha\widehat{H}_n^{\epsilon_n})$.

By Theorem 6.13,

$$\|V_n^{\epsilon_n}(t)f_n - \eta_n V(t)f\| \to 0,$$

whenever $f_n \in B(E_n)$, $f \in C(E)$, and $\|\eta_n f - f_n\| \to 0$. The remainder of the proof is the same as for Corollary 5.19.

The discrete-time case follows by direct application of Theorem 6.13, Lemma 5.4, and Theorem 5.15. $\square$

CHAPTER 7

Extensions of viscosity solution methods

We extend the results of Chapter 6 in several directions. First, we allow the state space to be a non-compact metric space. Since results in locally compact spaces can frequently be obtained by considering the one-point compactification of the space, the real goal of this extension is to be able to treat problems in Banach spaces or other non-locally compact infinite dimensional settings.

Second, we relax the boundedness assumptions on the domain and range of H. At first glance, this extension may appear odd, but it allows for greater flexibility in the choice of test functions needed in the verification of the comparison principle.

Third, we consider large deviations for processes obtained as projections or other transformations of Markov processes where the state space of the transformed process is lower dimensional than the original. Examples 1.8, 1.9, and 1.11, describe settings where these results are useful. In each of these examples, the natural limiting operator H satisfies $H \subset C(E) \times M(E')$, where E' is another metric space, and there is a natural mapping $\gamma : E' \mapsto E$. In the cited examples, $E' = E \times E_0$ for some metric space E_0, and γ is just the obvious projection. Theorem 7.24 gives the main result for this setting.

We highlight two technical points in this chapter. First, as mentioned at the beginning of Section 2.3, when E is non-compact, we need to work with notions of convergence of functions in $B(E)$ that are weaker than uniform convergence. As in Corollary 5.22, the corresponding semigroup convergence, in particular, the computation that gives us the large deviation principle for the finite dimensional distributions, requires the *a priori* estimates on the resolvent and semigroup given in Lemmas 5.9 and 5.12. Second, we need to extend the concept of viscosity solution (Definition 7.1), since we are no longer assured that suprema are actually achieved.

Throughout this chapter, we assume (E, r) and (E', r') are complete, separable metric spaces, $\gamma : E' \mapsto E$ is continuous, and $E = \gamma(E')$, that is, the mapping γ is onto. For $f \in M(E, \overline{R})$, we define $\gamma f = f \circ \gamma \in M(E', \overline{R})$.

7.1. Viscosity solutions, definition and convergence

Recall that $\overline{R} = [-\infty, \infty]$, $M(E, \overline{R})$ denotes the collection of Borel measurable functions with values in $\overline{R}$ and $f(x) \in R$ for at least one $x \in E$, and $C(E, \overline{R}) \subset M(E, \overline{R})$ denotes the collection of functions that are continuous as mappings from E into $\overline{R}$ with the natural topology on $\overline{R}$. $M^u(E, \overline{R}) \subset M(E, \overline{R})$ and $C^u(E, \overline{R}) \subset C(E, \overline{R})$ denote the corresponding collections of functions that are bounded above (that is, $f \in M^u(E, \overline{R})$ implies $\sup_{x \in E} f(x) < \infty$), and $M^l(E, \overline{R}) \subset M(E, \overline{R})$ and $C^l(E, \overline{R}) \subset C(E, \overline{R})$ denote the collections of functions that are bounded below.

Let $H \subset M(E,\overline{R}) \times M(E',\overline{R})$, $h \in C_b(E)$, and $\alpha > 0$. Consider the equation

$$(I - \alpha H)f = h. \tag{7.1}$$

Since all the examples in which we are interested have the property that $(f,g) \in H$ implies $(f+c,g) \in H$ for all $c \in R$, we generalize the definition implied by Lemma 6.3 rather than Definition 6.1. (Note that for $c \in R$, we take $\infty + c = \infty$ and $-\infty + c = -\infty$.)

DEFINITION 7.1 (Viscosity Solution). Let E, E', H, h, α be as above. Let $f \in B(E)$, and define $g = \alpha^{-1}(f-h)$, that is, $f - \alpha g = h$. Then

a) f is a *viscosity subsolution* of (7.1) if and only if f is upper semicontinuous and for each $(f_0, g_0) \in H$ such that $\sup_x (f - f_0)(x) < \infty$ (that is f_0 is bounded below), there exist $\{y_n\} \subset E'$ satisfying

$$\lim_{n\to\infty} (\gamma f - \gamma f_0)(y_n) = \sup_x (f - f_0)(x) \tag{7.2}$$

and

$$\limsup_{n\to\infty} (\alpha^{-1}(\gamma f - \gamma h) - (g_0)^*)(y_n) \le 0. \tag{7.3}$$

b) f is a *viscosity supersolution* of (7.1) if and only if f is lower semicontinuous and for each $(f_0, g_0) \in H$ such that $\sup_x (f_0 - f)(x) < \infty$ (that is, f_0 is bounded above), there exist $\{y_n\} \subset E'$ satisfying

$$\lim_{n\to\infty} (\gamma f_0 - \gamma f)(y_n) = \sup_x (f_0 - f)(x) \tag{7.4}$$

and

$$\liminf_{n\to\infty} (\alpha^{-1}(\gamma f - \gamma h) - (g_0)_*)(y_n) \ge 0. \tag{7.5}$$

A function $f \in C_b(E)$ is a *viscosity solution* of (7.1), if it is both a subsolution and a supersolution.

f will be called a *strong* subsolution if for every $\{y_n\} \subset E'$ satisfying (7.2), there exists a sequence $\{\widehat{y}_n\} \subset E'$ satisfying (7.2) and (7.3) and $\liminf_{n\to\infty} r'(y_n, \widehat{y}_n) = 0$. Similarly, f will be called a *strong* supersolution if for every $\{y_n\} \subset E'$ satisfying (7.4), there exists a sequence $\{\widehat{y}_n\} \subset E'$ satisfying (7.4) and (7.5) and $\liminf_{n\to\infty} r'(y_n, \widehat{y}_n) = 0$.

REMARK 7.2. a) If E is compact, $E = E'$, $\gamma(x) = x$, $H \subset C(E) \times B(E)$, and $(f,g) \in H$ implies $(f+c,g) \in H$, $c \in R$, then the conditions of Definition 7.1 and Definition 6.1 are equivalent. If the conditions of Definition 6.1 are satisfied, then take $y_n = x_0$ to see that the conditions of Definition 7.1 are satisfied. Conversely, if $\{y_n\}$ satisfies (7.2) and (7.3) and $E = E'$ is compact, we can assume that $y_n \to x_0$. It follows from (7.2) and the upper semicontinuity of f, that $\lim_{n\to\infty} f(y_n) = f(x_0)$. Consequently, (7.3) implies $\alpha^{-1}(f(x_0) - h(x_0)) \le (g_0)^*(x_0)$. The definitions of strong sub- and supersolution are also equivalent.

b) If E is compact and $E \ne E'$, then f is a viscosity subsolution if and only if for each $(f_0, g_0) \in H$ such that $\sup_x (f - f_0)(x) < \infty$, there exist $x_0 \in E$ and $\{y_n\} \subset E'$ satisfying $\lim_{n\to\infty} \gamma(y_n) = x_0$, $\lim_{n\to\infty} \gamma f(y_n) = f(x_0)$,

$$f(x_0) - f_0(x_0) = \sup_x (f - f_0)(x_0) \tag{7.6}$$

and

$$\alpha^{-1}(f(x_0) - h(x_0)) \le \liminf_{n\to\infty} (g_0)^*(y_n). \tag{7.7}$$

f is a strong subsolution if for each x_0 satisfying (7.6), there exists a sequence y_n satisfying $\lim_{n\to\infty}\gamma(y_n) = x_0$ and (7.7).

We also generalize the definitions of the comparison principle and a viscosity extension.

DEFINITION 7.3 (Comparison Principle). Let $H_\dagger, H_\ddagger \subset M(E,\overline{R}) \times M(E',\overline{R})$, $h \in C_b(E)$ and $\alpha > 0$. Subsolutions of

$$(I - \alpha H_\dagger)f = h \tag{7.8}$$

and supersolutions of

$$(I - \alpha H_\ddagger)f = h \tag{7.9}$$

satisfy a *comparison principle*, if for every viscosity subsolution $\overline{f}$ of (7.8) and every viscosity supersolution $\underline{f}$ of (7.9) we have $\overline{f} \leq \underline{f}$ on E.

When $H_\dagger = H_\ddagger \equiv H$, we say that

$$(I - \alpha H)f = h$$

satisfies a *comparison principle.*

We have the following generalization of Lemma 6.5

LEMMA 7.4. *Let $H_\dagger = H_\ddagger = H$. If $h \in C_b(E)$ and there exist $(f_n, g_n) \in H \cap (C_b(E) \times B(E'))$ satisfying $\|\gamma f_n - \alpha g_n - \gamma h\| \to 0$, then the comparison principle holds for $f - \alpha Hf = h$.*

PROOF. Suppose $\overline{f}$ is a viscosity subsolution. Set $h_n = \gamma f_n - \alpha g_n$. For $\epsilon_n > 0$, $\epsilon_n \to 0$, there exist $y_n \in E'$ satisfying $\gamma \overline{f}(y_n) - \gamma f_n(y_n) \geq \sup_x(\overline{f}(x) - f_n(x)) - \epsilon_n$ and

$$\begin{aligned}
\sup_x(\overline{f}(x) - f_n(x)) &\leq \gamma\overline{f}(y_n) - \gamma f_n(y_n) + \epsilon_n \\
&\leq \gamma h(y_n) + \alpha g_n^*(y_n) - \gamma f_n(y_n) + 2\epsilon_n \\
&= \gamma h(y_n) - (h_n)_*(y_n) + 2\epsilon_n \\
&\to 0.
\end{aligned}$$

Similarly, if $\underline{f}$ is a supersolution of $f - \alpha Hf = h$,

$$\liminf_{n\to\infty} \inf_x(\underline{f}(x) - f_n(x)) \geq 0,$$

and it follows that $\overline{f} \leq \underline{f}$. □

DEFINITION 7.5. $\widehat{H} \subset M(E,\overline{R}) \times M(E',\overline{R})$ is a *viscosity subextension* of $H \subset M(E,\overline{R}) \times M(E',\overline{R})$, if $\widehat{H} \supset H$ and for each $h \in C_b(E)$, $\alpha > 0$, f is a viscosity subsolution of $(I - \alpha H)f = h$ if and only if f is a viscosity subsolution of $(I - \alpha\widehat{H})f = h$.

$\widehat{H} \subset M(E,\overline{R}) \times M(E',\overline{R})$ is a *viscosity superextension* of $H \subset M(E,\overline{R}) \times M(E',\overline{R})$, if $\widehat{H} \supset H$ and for each $h \in C_b(E)$, $\alpha > 0$, f is a viscosity supersolution of $(I - \alpha H)f = h$ if and only if f is a viscosity supersolution of $(I - \alpha\widehat{H})f = h$.

$\widehat{H}$ is a *viscosity extension* of H if it is both a viscosity subextension and a viscosity superextension.

LEMMA 7.6. *Let $H_\dagger \subset \widehat{H}_\dagger \subset M^l(E,\overline{R}) \times M(E',\overline{R})$. Suppose that for each $(f,g) \in \widehat{H}_\dagger$, there exists $\{(f_n, g_n)\} \subset H_\dagger$ such that for each $c, d \in R$,*

$$\lim_{n\to\infty} \|f_n \wedge c - f \wedge c\| = 0$$

and

$$\limsup_{n\to\infty} \sup_{z\in\{\gamma f_n \vee \gamma f \le c\}} (g_n^*(z) \vee d - g^*(z) \vee d) \le 0. \tag{7.10}$$

Then $\widehat{H}_\dagger$ is a viscosity subextension of $H_\dagger$.

Let $H_\ddagger \subset \widehat{H}_\ddagger \subset M^u(E,\overline{R}) \times M(E',\overline{R})$. Suppose that for each $(f,g) \in \widehat{H}_\ddagger$, there exists $\{(f_n, g_n)\} \subset H_\ddagger$ such that for each $c, d \in R$,

$$\lim_{n\to\infty} \|f_n \vee c - f \vee c\| = 0$$

and

$$\liminf_{n\to\infty} \inf_{z\in\{\gamma f_n \wedge \gamma f \ge c\}} ((g_n)_*(z) \wedge d - g_*(z) \wedge d) \ge 0. \tag{7.11}$$

Then $\widehat{H}_\ddagger$ is a viscosity superextension of $H_\ddagger$.

PROOF. Clearly, if f is a viscosity subsolution for $(I - \alpha\widehat{H}_\dagger)f = h$, then f is a viscosity subsolution for $(I - \alpha H_\dagger)f = h$. Suppose f is a viscosity subsolution for $(I - \alpha H_\dagger)f = h$, and let $(f_0, g_0) \in \widehat{H}_\dagger$. There exist $(f_n, g_n) \in H_\dagger$ such that for each $c, d \in R$, $\|f_0 \wedge c - f_n \wedge c\| \to 0$ and (7.10) holds with g replaced by g_0. Let $\epsilon_n > 0$ and $\epsilon_n \to 0$. There exist $z_n \in E'$ such that

$$\gamma f(z_n) - \gamma f_n(z_n) \ge \sup_x (f(x) - f_n(x)) - \epsilon_n$$

and $\alpha^{-1}(\gamma f(z_n) - \gamma h(z_n)) - g_n^*(z_n) \le \epsilon_n$. Since

$$\gamma f_n(z_n) \le \gamma f(z_n) + \epsilon_n - \inf_x (f_n(x) - f(x)),$$

$$\limsup_{n\to\infty} \gamma f_n(z_n) \le 2\|f\| - \inf_x f_0(x) < \infty,$$

we have

$$\begin{aligned}
\lim_{n\to\infty} \gamma f(z_n) - \gamma f_0(z_n) &= \lim_{n\to\infty} \gamma f(z_n) - \gamma f_n(z_n) \\
&= \lim_{n\to\infty} \sup_x (f(x) - f_n(x)) = \sup_x (f(x) - f_0(x)).
\end{aligned}$$

There exists $c < \infty$ such that $f_n(z_n) \vee f_0(z_n) < c$. Let $d < -\alpha^{-1}(\|f\| + \|h\|)$. Then for n sufficiently large, $g_n^*(z_n) > d$ and (by (7.10)) $g_0^*(z_n) > d$. Consequently,

$$\begin{aligned}
&\limsup_{n\to\infty} \alpha^{-1}(\gamma f(z_n) - \gamma h(z_n)) - g_0^*(z_n) \\
&\quad \le \limsup_{n\to\infty} \alpha^{-1}(\gamma f(z_n) - \gamma h(z_n)) - g_n^*(z_n) + (g_n^*(z_n) - g_0^*(z_n)) \\
&\quad \le \limsup_{n\to\infty} \alpha^{-1}(\gamma f(z_n) - \gamma h(z_n)) - g_n^*(z_n) \le 0,
\end{aligned}$$

verifying that f is a subsolution for $(I - \alpha\widehat{H}_\dagger)f = h$. The proof is similar for supersolutions. □

Lemma 7.6 determines a closure operation that preserves sub and supersolutions. The next lemma gives another.

LEMMA 7.7. *Let* $(f_0, g_0) \in M^l(E, \overline{R}) \times M(E, \overline{R})$. *Suppose that there exists* $\{(f_n, g_n)\} \subset H_\dagger$ *satisfying the following:*

a) f_n *is lower semicontinuous.*
b) $f_1 \le f_2 \le \cdots$ *and* $f_n \to f_0$ *pointwise.*
c) *If* $\{z_n\} \subset E'$ *satisfies* $\sup_n \gamma f_n(z_n) < \infty$ *and* $\inf_n g_n^*(z_n) > -\infty$, *then* $\{\gamma(z_n)\}$ *is relatively compact, and if* $\{\gamma(z_{n_k})\}$ *converges to* x_0,

$$\limsup_{n\to\infty} g_n^*(z_{n_k}) \le g_0^*(x_0). \tag{7.12}$$

Then $\widehat{H}_\dagger = H_\dagger \cup \{(f_0, \gamma g_0)\}$ *is a viscosity subextension of* $H_\dagger$.
The analogous result holds for superextensions.

PROOF. Suppose f is a viscosity subsolution for $(I - \alpha H_\dagger)f = h$.

Let $(f_n, g_n) \in H_\dagger$ satisfy (a) through (c), and let $\epsilon_n > 0$ and $\epsilon_n \to 0$. There exist $z_n \in E'$ such that

$$\gamma f(z_n) - \gamma f_n(z_n) \ge \sup_x (f(x) - f_n(x)) - \epsilon_n$$

and $\alpha^{-1}(\gamma f(z_n) - \gamma h(z_n)) - g_n^*(z_n) \le \epsilon_n$. Note that

$$\gamma f_n(z_n) \le 2\|f\| - \inf_y f_0(y) + \epsilon_n$$

and

$$g_n^*(z_n) \ge -\alpha^{-1}(\|f\| + \|h\|) - \epsilon_n.$$

It follows that $\{\gamma(z_n)\}$ is relatively compact, and by Lemma A.4, if x_0 is a limit point of $\{\gamma(z_n)\}$, then

$$\lim_{n\to\infty} \gamma f(z_n) - \gamma f_n(z_n) = f(x_0) - f_0(x_0) = \sup_x (f(x) - f_0(x)). \tag{7.13}$$

Since $f_n \le f_{n+1}$ and f_m is lower semicontinuous, for each $m = 1, 2, \ldots$ fixed,

$$\lim_{\delta\to 0+} \lim_{n\to\infty} \inf_{r(x,x_0)\le\delta} f_n(x) \ge \lim_{\delta\to 0+} \inf_{r(x,x_0)\le\delta} f_m(x) \ge f_m(x_0).$$

On the other hand, since $f_n \to f_0$ pointwise,

$$\lim_{\delta\to 0+} \lim_{n\to\infty} \inf_{r(x,x_0)\le\delta} f_n(x) \le \lim_{n\to\infty} f_n(x_0) = f_0(x_0).$$

Hence

$$f_0(x_0) = \lim_{\delta\to 0+} \lim_{n\to\infty} \inf_{r(x,x_0)\le\delta} f_n(x) = \sup_{\delta>0} \sup_n \inf_{r(x,x_0)\le\delta} f_n(x). \tag{7.14}$$

For simplicity, assume that $\gamma(z_n) \to x_0$. (Otherwise, consider a subsequence.) Then (7.13), (7.14), and (7.12) imply

$$\begin{aligned} & f(x_0) - f_0(x_0) - h(x_0) \\ & = \limsup_{n\to\infty} (\gamma f(z_n) - \gamma f_n(z_n) - \gamma h(z_n)) \le \limsup (\alpha g_n^*(z_n) + \alpha\epsilon_n - \gamma f_n(z_n)) \\ & \le \lim_{\delta\to 0+} \limsup_{n\to\infty} (\alpha(g_n^*(z_n) - \inf_{r(x,x_0)\le\delta} f_n(x)) \le \alpha g_0^*(x_0) - f_0(x_0). \end{aligned}$$

Consequently, (7.2) and (7.3) hold for any $y_n = y \in \gamma^{-1}(x_0)$. □

Next, we generalize Lemma 6.9 to the present setting. We will need the following simple technical lemma.

LEMMA 7.8. *Suppose* $f, g \in M(E, \overline{R})$, *and* $f - \epsilon g \in M(E, \overline{R})$ *for all* $0 < \epsilon < \epsilon_0$.

a) *If*

$$\sup_x f(x) \le \sup_x (f(x) - \epsilon g(x)) < \infty$$

for all $0 < \epsilon < \epsilon_0$*, then there exist* x_n *such that* $\lim_{n\to\infty} f(x_n) = \sup_x f(x)$ *and* $\limsup_{n\to\infty} g(x_n) \le 0$.

b) *If*

$$\inf_x f(x) \ge \inf_x (f(x) - \epsilon g(x)) > -\infty$$

for all $0 < \epsilon < \epsilon_0$*, then there exist* x_n *such that* $\lim_{n\to\infty} f(x_n) = \inf_x f(x)$ *and* $\liminf_{n\to\infty} g(x_n) \ge 0$.

PROOF. We prove Part (a). The proof for (b) is similar.
Note that $\inf_x g(x) > -\infty$. Let $\epsilon_n \to 0+$, and let x_n satisfy

$$\sup_x (f(x) - \epsilon_n g(x)) \le f(x_n) - \epsilon_n g(x_n) + \epsilon_n^2.$$

Combined with the condition $\sup_x f(x) \le \sup_x (f(x) - \epsilon g(x))$, then

$$\sup_x f(x) + \epsilon_n g(x_n) - \epsilon_n^2 \le f(x_n) \le f(x_n) - \epsilon_n g(x_n) + \epsilon_n^2$$

which implies

$$\limsup_{n\to\infty} g(x_n) \le \lim_{n\to\infty} \epsilon_n = 0$$

and

$$\lim_{n\to\infty} f(x_n) = \sup_x f(x).$$

□

We recall the notion of $\limsup_{n\to\infty} G_n$ and $\liminf_{n\to\infty} G_n$ for a sequence of subsets G_n in a topological space in Definition 2.4. In particular, we note that both limits are closed sets by definition. We require the following conditions on the approximating state spaces (cf. Definition 2.5).

CONDITION 7.9. (E_n, r_n)*,* $n = 2, 3, \ldots$*,* (E', r')*, and* (E, r) *are complete, separable metric spaces,* $\eta'_n : E_n \mapsto E'$ *is Borel measurable,* $\gamma : E' \mapsto E$ *is continuous and onto, and* $\eta_n = \gamma \circ \eta'_n$*. For* $n = 1, 2, \ldots$ *and* $q \in \mathcal{Q}$*,* $K_n^q \subset E_n$*. The* K_n^q *satisfy the following:*

a) *For* $q_1, q_2 \in \mathcal{Q}$*, there exists* $q_3 \in \mathcal{Q}$ *such that* $K_n^{q_1} \cup K_n^{q_2} \subset K_n^{q_3}$.

b) *For each* $x \in E$*, there exists* $q \in \mathcal{Q}$ *and* $z_n \in K_n^q$ *such that* $\lim_{n\to\infty} \eta_n(z_n) = x$.

c) *For each* $q \in \mathcal{Q}$*, there exist a compact* $\widehat{K}^q \subset E$ *such that*

$$\lim_{n\to\infty} \sup_{y\in K_n^q} \inf_{x\in\widehat{K}^q} r(\eta_n(y), x) = 0,$$

and a compact $\widehat{K}'^q \subset E'$ *such that*

$$\lim_{n\to\infty} \sup_{y\in K_n^q} \inf_{x\in\widehat{K}'^q} r'(\eta'_n(y), x) = 0. \tag{7.15}$$

d) *For each compact* $K \subset E$*, there exists* $q \in \mathcal{Q}$ *such that*

$$K \subset K^q \equiv \liminf_{n\to\infty} \eta_n(K_n^q). \tag{7.16}$$

(Note that by (c), $K^q \subset \widehat{K}^q$*, so* K^q *is compact.)*

REMARK 7.10. Note that if Condition 7.9(c) holds, then it holds with $\widehat{K}^q = \limsup_{n\to\infty} \eta_n(K_n^q)$ and $\widehat{K}'^q = \limsup_{n\to\infty} \eta'_n(K_n^q)$.

Define $\eta_n : M(E,\overline{R}) \to M(E_n,\overline{R})$ by $\eta_n f = f \circ \eta_n$, and define γf and $\eta'_n f$ similarly. Following Definition 2.5, for $f_n \in B(E_n)$ and $f \in B(E)$, $\text{LIM} f_n = f$, if $\sup_n \|f_n\| < \infty$ and for each $q \in \mathcal{Q}$,

$$\lim_{n\to\infty} \sup_{z\in K_n^q} |f_n(z) - \eta_n f(z)| = 0.$$

Somewhat abusing notation, for $g_n \in B(E_n)$ and $g \in B(E')$, we write $\text{LIM} g_n = g$, if $\sup_n \|g_n\| < \infty$ and for each $q \in \mathcal{Q}$,

$$\lim_{n\to\infty} \sup_{z\in K_n^q} |g_n(z) - \eta'_n g(z)| = 0.$$

We assume that A_n is the full generator for a Markov process Y_n with values in E_n and that

$$H_n = \{(f, \frac{1}{n} e^{-nf} g) : (e^{nf}, g) \in A_n\}, \tag{7.17}$$

or in the discrete time case, that T_n is a transition operator, $\epsilon_n > 0$, and

$$H_n f = \frac{1}{n\epsilon_n} \log e^{-nf} T_n e^{nf}, \qquad f \in B(E_n).$$

We will assume the following convergence condition.

CONDITION 7.11. (Convergence Condition)

Let $H_\dagger \subset C^l(E,\overline{R}) \times M^u(E',\overline{R})$ and $H_\ddagger \subset C^u(E,\overline{R}) \times M^l(E',\overline{R})$, and let $v_n \to \infty$. For each $(f_0, g_0) \in H_\dagger$, assume that there exist $(f_{n,0}, g_{n,0}) \in H_n$ such that

$$\text{LIM} f_{n,0} \wedge c = f_0 \wedge c, \text{ for each } c \in R, \tag{7.18}$$

$$\sup_n \frac{1}{v_n} \log \|g_{n,0}\| < \infty \text{ and } \sup_{n,y\in E_n} g_{n,0}(y) < \infty, \tag{7.19}$$

and that for each $q \in \mathcal{Q}$ and each sequence $z_n \in K_n^q$ satisfying $\eta'_n(z_n) \to y$ and $\lim_{n\to\infty} f_{n,0}(z_n) = \gamma f_0(y) < \infty$,

$$\limsup_{n\to\infty} g_{n,0}(z_n) \le (g_0)^*(y). \tag{7.20}$$

For each $(f_1, g_1) \in H_\ddagger$, assume that there exist $(f_{n,1}, g_{n,1}) \in H_n$ such that

$$\text{LIM} f_{n,1} \vee c = f_1 \vee c, \text{ for each } c \in R, \tag{7.21}$$

$$\sup_n \frac{1}{v_n} \log \|g_{n,1}\| < \infty \text{ and } \inf_{n,y\in E_n} g_{n,1}(y) > -\infty, \tag{7.22}$$

and that for each $q \in \mathcal{Q}$ and each sequence $z_n \in K_n^q$ satisfying $\eta'_n(z_n) \to y$ and $\lim_{n\to\infty} f_{n,1}(z_n) = \gamma f_1(y) > -\infty$,

$$\liminf_{n\to\infty} g_{n,1}(z_n) \ge (g_1)_*(y). \tag{7.23}$$

By definition, $H_n \subset B(E_n) \times B(E_n)$. Lyapunov function arguments (for example, Lemma 4.20) frequently require working with unbounded functions, and it may be useful to allow unbounded functions in the convergence condition. In the continuous time case, as in Section 4.5, define $H_{\dagger,n}$ to be the collection of pairs $(f,g) \in M^l(E_n,\overline{R}) \times M^u(E_n,\overline{R})$ such that

$$Z^n_{f,g}(t) = e^{n(f(Y_n(t)) - f(Y_n(0)) - \int_0^t ng(Y_n(s))ds} \tag{7.24}$$

is a supermartingale whenever $f(Y_n(0)) < \infty$ almost surely, and let $H_{\ddagger,n}$ be the collection of $(f,g) \in M^u(E_n,\overline{R}) \times M^l(E_n,\overline{R})$ such that (7.24) is a submartingale whenever $f(Y_n(0)) > -\infty$ almost surely. We have the following lemma.

LEMMA 7.12. *For $n = 1, 2, \ldots$, let A_n be the full generator for a Markov process and let H_n be given by (7.17). Let $(f_n, g_n) \in H_{\dagger,n}$, $(f_0, g_0) \in C^l(E,\overline{R}) \times M^u(E',\overline{R})$, and for each $\epsilon > 0$, let*

$$\sigma_n^\epsilon = \inf\{t : r'(\eta'(Y_n(t)), \eta'(Y_n(0))) > \epsilon\}.$$

Suppose that $\inf_{n,y\in E_n} f_n(y) > -\infty$, $\sup_{n,y\in E_n} g_n(y) < \infty$, and that there exist $c_n \to \infty$ and $\delta_n > 0$ satisfying

$$\lim_{n\to\infty} n\delta_n = 0, \qquad \lim_{n\to\infty} \frac{1}{v_n}(nc_n - \log(n\delta_n)) = 0 \tag{7.25}$$

such that for each $q \in \mathcal{Q}$,

$$\lim_{n\to\infty} \sup_{x\in K_n^q} |f_n(x) \wedge c_n - \eta_n f_0(x) \wedge c_n| = 0, \tag{7.26}$$

$$\lim_{n\to\infty} \sup_{x\in K_n^q} \sup_{s\le\delta_n} \left| E[e^{n(f_n(Y_n(s))\wedge c_n - f_n(x)\wedge c_n)}] - 1 \right| = 0, \tag{7.27}$$

and for each sequence $z_n \in K_n^q$ satisfying $\eta_n'(z_n) \to y$ and $\lim_{n\to\infty} f_n(z_n) = \gamma f_0(y) < \infty$,

$$\limsup_{n\to\infty} g_n(z_n) \le (g_0)^*(y). \tag{7.28}$$

In addition, assume that for each $0 < \epsilon < 1$

$$\lim_{n\to\infty} e^{nc_n} \sup_{x\in K_n^q} P\{\sigma_n^\epsilon \le \delta_n | Y_n(0) = x\} = 0. \tag{7.29}$$

Then there exist $(f_{n,0}, g_{n,0}) \in H_n$ satisfying (7.18), (7.19), and (7.20).

The analogous result holds for $(f_n, g_n) \in H_{\ddagger,n}$, $(f_1, g_1) \in C^u(E,\overline{R}) \times M^l(E,\overline{R})$, giving a sequence satisfying (7.21), (7.22), and (7.23).

PROOF. Define $f_{n,0}$ to satisfy

$$e^{nf_{n,0}(x)} = \frac{1}{\delta_n}\int_0^{\delta_n} E[e^{nf_n(Y_n(s))\wedge c_n}|Y_n(0) = x]ds,$$

so

$$\begin{aligned} g_{n,0}(x) &= H_n f_{n,0}(x) = \frac{1}{n} e^{-nf_{n,0}(x)} A_n e^{nf_{n,0}}(x) \\ &= \frac{1}{n} \frac{E[e^{nf_n(Y_n(\delta_n))\wedge c_n}|Y_n(0) = x] - e^{nf_n(x)\wedge c_n}}{\int_0^{\delta_n} E[e^{nf_n(Y_n(s))\wedge c_n}|Y_n(0) = x]ds}. \end{aligned}$$

Note that $b_f = \inf_{n,y\in E_n} f_{n,0}(y) \ge \inf_{n,y\in E_n} f_n(y) \wedge c_n > -\infty$, $f_{n,0} \le c_n$, and $\|g_{n,0}\| \le \frac{1}{n\delta_n} e^{n(c_n - c_n\wedge b_f)}$, and setting $\overline{C}_g = \sup_{n,y\in E_n} g_n(y)$, the supermartingale property implies

$$\sup_{n,y\in E_n} g_{n,0}(y) \le \sup_{n,y\in E_n} \frac{e^{n\delta_n \overline{C}_g} - 1}{n\delta_n \inf_{s\le\delta_n} E[e^{n(f_n(Y_n(s))\wedge c_n - f_n(y)\wedge c_n)}|Y_n(0) = y]},$$

and (7.19) holds by (7.25) and (7.27).

For each $q \in \mathcal{Q}$,

$$\begin{aligned}
&\sup_{x\in K_n^q} \left|e^{n(f_{n,0}(x)-f_n(x)\wedge c_n)} - 1\right| \\
\leq\ & \frac{1}{\delta_n}\int_0^{\delta_n} \sup_{x\in K_n^q} \left|E[e^{n(f_n(Y_n(s))\wedge c_n - f_n(x)\wedge c_n)}|Y_n(0)=x] - 1\right| ds \to 0,
\end{aligned}$$

so by (7.26) and (7.27) we have

$$\lim_{n\to\infty} \sup_{x\in K_n^q} |f_{n,0}(x) - \eta_n f_0(x)\wedge c_n| = 0,$$

and (7.18) follows.

Now suppose $z_n \in K_n^q$, $\eta_n'(z_n) \to y$, and $\lim_{n\to\infty} f_n(z_n) = \gamma f_0(y) < \infty$. Let

$$C_{n,\epsilon}(g_n) = \sup\{g_n(y) : r'(\eta'(y), \eta'(z_n)) \leq \epsilon\}.$$

Then

$$\begin{aligned}
&E[e^{n(f_n(Y_n(\delta_n))\wedge c_n - f_n(z_n)\wedge c_n)}|Y_n(0)=z_n] \\
&\quad\leq E[e^{n(f_n(Y_n(\delta_n))\wedge c_n - f_n(z_n)\wedge c_n) - \int_{\sigma_n^\epsilon\wedge\delta_n}^{\delta_n} n g_n(Y_n(s))ds + n(\delta_n - \delta_n\wedge\sigma_n^\epsilon)\overline{C}_g} \\
&\qquad\qquad \times \mathbf{1}_{\{f_n(Y_n(\delta_n\wedge\sigma_n^\epsilon))<c_n\}}|Y_n(0)=z_n] \\
&\qquad + E[e^{n(f_n(Y_n(\delta_n\wedge\sigma_n^\epsilon))\wedge c_n - f_n(z_n)\wedge c_n)}\mathbf{1}_{\{f_n(Y_n(\delta_n\wedge\sigma_n^\epsilon))\geq c_n\}}|Y_n(0)=z_n] \\
&\quad\leq E[e^{n(f_n(Y_n(\delta_n\wedge\sigma_n^\epsilon))\wedge c_n - f_n(z_n)\wedge c_n) + n(\delta_n - \delta_n\wedge\sigma_n^\epsilon)\overline{C}_g}|Y_n(0)=z_n] \\
&\quad\leq E[e^{n(f_n(Y_n(\delta_n\wedge\sigma_n^\epsilon))\wedge c_n - f_n(z_n)\wedge c_n)}|Y_n(0)=z_n] \\
&\qquad + e^{nc_n}(e^{n\delta_n\overline{C}_g} - 1)P\{\sigma_n^\epsilon < \delta_n|Y_n(0)=z_n\}
\end{aligned}$$

and for n large enough so that $f_n(z_n) \leq c_n$,

$$\begin{aligned}
&E[e^{n(f_n(Y_n(\delta_n\wedge\sigma_n^\epsilon))\wedge c_n - f_n(z_n))}|Y_n(0)=z_n] \\
&\quad\leq E[e^{n(f_n(Y_n(\delta_n\wedge\sigma_n^\epsilon))\wedge c_n - f_n(z_n)) - \int_0^{\delta_n\wedge\sigma_n^\epsilon} n g_n(Y_n(s))ds + n\delta_n C_{n,\epsilon}(g_n)}|Y_n(0)=z_n] \\
&\quad\leq e^{n\delta_n C_{n,\epsilon}(g_n)}.
\end{aligned}$$

Consequently,

$$\begin{aligned}
&\limsup_{n\to\infty} g_{n,0}(z_n)\\
&= \limsup_{n\to\infty} \frac{1}{n}\frac{E[e^{nf_n(Y_n(\delta_n))\wedge c_n}|Y_n(0)=z_n]-e^{nf_n(z_n)\wedge c_n}}{\int_0^{\delta_n} E[e^{nf_n(Y_n(s))\wedge c_n}|Y_n(0)=z_n]ds}\\
&= \limsup_{n\to\infty} \frac{1}{n\delta_n}\left(E[e^{n(f_n(Y_n(\delta_n))\wedge c_n-f_n(z_n)\wedge c_n)}|Y_n(0)=z_n]-1\right)\\
&= \limsup_{n\to\infty} \frac{1}{n\delta_n}\Big(E[e^{n(f_n(Y_n(\delta_n))\wedge c_n-f_n(z_n)\wedge c_n)}|Y_n(0)=z_n]\\
&\qquad -E[e^{n(f_n(Y_n(\delta_n\wedge\sigma_n^\epsilon))\wedge c_n-f_n(z_n)\wedge c_n)}|Y_n(0)=z_n]\Big)\\
&\quad +\limsup_{n\to\infty} \frac{1}{n\delta_n}\left(E[e^{n(f_n(Y_n(\delta_n\wedge\sigma_n^\epsilon))\wedge c_n-f_n(z_n)\wedge c_n)}|Y_n(0)=z_n]-1\right)\\
&\le \limsup_{n\to\infty} \frac{1}{n\delta_n}e^{nc_n}(e^{n\delta_n\overline{C}_g}-1)P\{\sigma_n^\epsilon<\delta_n|Y_n(0)=z_n\}\\
&\quad +\limsup_{n\to\infty} \frac{1}{n\delta_n}(e^{n\delta_n C_{n,\epsilon}(g_n)}-1)\\
&\le \limsup_{n\to\infty} C_{n,\epsilon}(g_n).
\end{aligned}$$

But $\lim_{\epsilon\to 0}\limsup_{n\to\infty} C_{n,\epsilon}(g_n)\le g_0^*(y)$, so (7.20) holds. □

The analogous lemma in discrete time is much simpler.

LEMMA 7.13. *For $n=1,2,\ldots$, let T_n be a transition operator on E_n and let $\epsilon_n>0$ with $\epsilon_n\to 0$. Suppose $(f_n,g_n)\in M^l(E_n,\overline{R})\times M^u(E_n,\overline{R})$ satisfy*

$$T_n e^{nf_n}\le e^{nf_n+\epsilon_n n g_n},$$

allowing $\infty\le\infty$, and $(f_0,g_0)\in C^l(E,\overline{R})\times M^u(E',\overline{R})$. Suppose that

$$\inf_{n,y\in E_n} f_n(y)>-\infty,\quad \sup_{n,y\in E_n} g_n(y)<\infty,$$

and that there exist $c_n\to\infty$ satisfying

$$\lim_{n\to\infty}\frac{1}{v_n}(c_n-\log\epsilon_n)=0 \tag{7.30}$$

such that for each $q\in\mathcal{Q}$,

$$\lim_{n\to\infty}\sup_{x\in K_n^q}|f_n(x)\wedge c_n-\eta_n f_0(x)\wedge c_n|=0, \tag{7.31}$$

and for each sequence $z_n\in K_n^q$ satisfying $\eta_n'(z_n)\to y$ and $\lim_{n\to\infty} f_n(z_n)=\gamma f_0(y)<\infty$,

$$\limsup_{n\to\infty} g_n(z_n)\le (g_0)^*(y). \tag{7.32}$$

Then setting $f_{n,0}=f_n\wedge c_n$ and

$$g_{n,0}=\frac{1}{n\epsilon_n}\log e^{-nf_{n,0}}T_n e^{nf_{n,0}},$$

(7.18), (7.19), and (7.20) are satisfied.

The analogous result holds for (f_n, g_n) satisfying

$$T_n e^{nf_n} \geq e^{nf_n + \epsilon_n n g_n}$$

and $(f_1, g_1) \in C^u(E, \overline{R}) \times M^l(E, \overline{R})$, giving a sequence satisfying (7.21), (7.22), and (7.23).

PROOF. The convergence in (7.18) follows from the lower bound on the f_n and (7.31). We have

$$\|g_{n,0}\| \leq \frac{1}{\epsilon_n}|c_n - \inf_{y \in E_n} f_n(y))|,$$

so the first part of (7.19) holds. If $f_n(y) > c_n$, then $g_{n,0}(y) \leq 0$, and if $f_n(y) \leq c_n$, then

$$g_{n,0}(y) \leq g_n(y),$$

so the second part of (7.19) holds and (7.20) is given by (7.32). □

LEMMA 7.14. *Let A_n be a bounded generator, that is, A_n is of the form*

$$A_n f(x) = \lambda_n(x) \int_{E_n} (f(z) - f(x)) q_n(x, dz), \tag{7.33}$$

with λ_n bounded, and let H_n be given by (7.17). Let Condition 7.9 hold and $\{Y_n\}$ satisfy Condition 2.8. Let $H_\dagger \subset C^l(E, \overline{R}) \times M^u(E', \overline{R})$ and $H_\ddagger \subset C^u(E, \overline{R}) \times M^l(E', \overline{R})$, and assume that the Convergence Condition 7.11 holds.

Let $\alpha > 0$ and $h \in C_b(E)$, and assume that the comparison principle holds for subsolutions of

$$(I - \alpha H_\dagger) f = h, \tag{7.34}$$

and supersolutions of

$$(I - \alpha H_\ddagger) f = h. \tag{7.35}$$

Let $(f_n, g_n) \in H_n$, and define $f_n - \alpha g_n = h_n$. If $h = \mathrm{LIM} h_n$, then there exists a unique $f \in B(E)$ that is a viscosity subsolution of (7.34) and a supersolution of (7.35), $f \in C_b(E)$, and

$$f = \mathrm{LIM} f_n.$$

For $(f_0, g_0) \in H_\dagger$,

$$\sup_{x \in E} (f - f_0)(x) \leq \sup_{y \in E'} (\gamma h - (\gamma f_0 - \alpha g_0))(y), \tag{7.36}$$

and for $(f_1, g_1) \in H_\ddagger$,

$$\inf_{x \in E} (f - f_1)(x) \geq \inf_{y \in E'} (\gamma h - (\gamma f_1 - \alpha g_1))(y). \tag{7.37}$$

REMARK 7.15. $\mathrm{LIM} g_{n,0} \vee c = g_0 \vee c$ for all $c \in R$ implies (7.20) and $\mathrm{LIM} g_{n,1} \wedge c = g_1 \wedge c$ implies (7.23).

PROOF. Note that $\|f_n\| \leq \|h_n\|$. By assumption, for each $x \in E$, there exists $q \in Q$ and $z_n \in K_n^q$ such that $\lim_{n \to \infty} \eta_n(z_n) = x$. Let

$$f_3(x) = \sup\{\limsup_{n \to \infty} f_n(z_n) : z_n \in K_n^q, \lim_{n \to \infty} \eta_n(z_n) = x, q \in Q\},$$

and

$$f_4(x) = \inf\{\liminf_{n \to \infty} f_n(z_n) : z_n \in K_n^q, \lim_{n \to \infty} \eta_n(z_n) = x, q \in Q\},$$

and define $\overline{f}, \underline{f} \in B(E)$ by $\overline{f} = f_3^*$ and $\underline{f} = (f_4)_*$. Then $\overline{f}$ is bounded and upper-semicontinuous, and $\underline{f}$ is bounded and lower-semicontinuous. We claim that they are, respectively, a subsolution of (7.34) and a supersolution of (7.35). We only prove this assertion for $\overline{f}$. The argument for $\underline{f}$ is similar.

Let $(f_0, g_0) \in H_\dagger$. By assumption, there exists $(f_{n,0}, g_{n,0}) \in H_n$ such that (7.18), (7.19) and (7.20) hold. Let $0 < \epsilon < \alpha$, and set $h_n^\epsilon = f_n - \epsilon \frac{f_n - h_n}{\alpha}$ and $h_{0,n}^\epsilon = f_{0,n} - \epsilon g_{0,n}$. Note that $\|h_n^\epsilon\| \le \|h_n\|$. As in the proof of Lemma 5.9,

$$f_n - f_{0,n} \le (I - \epsilon A_n^{f_n})^{-1}(h_n^\epsilon - h_{n,0}^\epsilon),$$

and by Condition 2.8, for each $q \in \mathcal{Q}$ and $\delta > 0$, there exists $\widehat{q} \in \mathcal{Q}$ depending on q, δ, $\sup_n \|h_n\|$, and $\inf_{n, x \in E_n} f_{0,n}(x)$ and $\sup_{n, x\in E_n} g_{0,n}(x)$ such that for any sequence $y_n \in K_n^q$ there is a sequence $z_n \in K_n^{\widehat{q}}$ satisfying

$$\begin{aligned}
(7.38)\quad \limsup_{n\to\infty}&(f_n(y_n) - f_{0,n}(y_n)) \\
&\le \limsup_{n\to\infty}(I - \epsilon A_n^{f_n})^{-1}(h_n^\epsilon - h_{0,n}^\epsilon)(y_n) \\
&\le \delta + \limsup_{n\to\infty} \sup_{y \in K_n^{\widehat{q}}} (h_n^\epsilon(y) - h_{0,n}^\epsilon(y)) \\
&= \delta + \limsup_{n\to\infty}((1-\epsilon\alpha^{-1})f_n(z_n) + \epsilon g_{0,n}(z_n) + \epsilon\alpha^{-1}h_n(z_n) - f_{0,n}(z_n)).
\end{aligned}$$

By Condition 7.9(c), there exists a subsequence along which the lim sup on the right is achieved and $\eta'_{n(k)}(z_{n(k)})$ converges to $z_0 \in \widehat{K}'^{\widehat{q}} \subset E'$. Consequently, we can assume that $\eta'_n(z_n)$ converges to z_0. If the left side of (7.38) is greater than $-\infty$, we must have $f_0(z_0) < \infty$ and hence $\limsup_{n\to\infty} f_{0,n}(z_n) = \lim_{n\to\infty} f_{0,n}(z_n) = f_0(z_0)$, and by the continuity of h and f_0, the definition of $\overline{f}$, and (7.20), we have

$$\begin{aligned}
(7.39)\qquad &\limsup_{n\to\infty}(f_n(y_n) - f_{0,n}(y_n)) \\
&\quad \le \delta + ((1-\epsilon\alpha^{-1})\gamma\overline{f}(z_0) + \epsilon g_0^*(z_0) + \epsilon\alpha^{-1}\gamma h(z_0) - \gamma f_0(z_0)).
\end{aligned}$$

Since δ is arbitrary, by the definition of f_3 and $\overline{f}$, we have

$$\begin{aligned}
\sup_{x\in E}(\overline{f}(x) - f_0(x)) &= \sup_{y\in E'}(\gamma\overline{f}(y) - \gamma f_0(y)) \\
(7.40)\qquad &\le \sup_{y\in E'}\left(\gamma\overline{f}(y) - \gamma f_0(y) - \epsilon(\frac{\gamma\overline{f}(y) - \gamma h(y)}{\alpha} - g_0^*(y))\right).
\end{aligned}$$

By Lemma 7.8, there exist $y_k \in E'$ such that

$$\sup_{y\in E'}(\gamma\overline{f}(y) - \gamma f_0(y)) = \lim_{k\to\infty}(\gamma\overline{f}(y_k) - \gamma f_0(y_k))$$

and

$$\limsup_{k\to\infty}(\frac{\gamma\overline{f}(y_k) - \gamma h(y_k)}{\alpha} - g_0^*(y_k)) \le 0.$$

Consequently, $\overline{f}$ is a subsolution of (7.34).

By the comparison principle and the construction of $\overline{f}$ and $\underline{f}$, $\overline{f} = \underline{f} \equiv f \in C_b(E)$. To see that $f = \mathrm{LIM} f_n$, it is enough to show that for each $q \in \mathcal{Q}$ and each sequence $y_n \in K_n^q$,

$$\lim_{n\to\infty}|f_n(y_n) - f(\eta_n(y_n))| = 0.$$

Since f is continuous, by Condition 7.9(c), it is enough to show that if $\eta_n(y_n) \to x_0$, then $f_n(y_n) \to f(x_0)$. By definition,

$$\underline{f}(x_0) \leq f_4(x_0) \leq \liminf_{n\to\infty} f_n(y_n) \leq \limsup_{n\to\infty} f_n(y_n) \leq f_3(x_0) \leq \overline{f}(x_0).$$

Since $\underline{f} = \overline{f}$, we have $f = \text{LIM} f_n$.

Finally, (7.36) follows from (7.40) by sending $\epsilon \to \alpha$, and (7.37) can be proved similarly. □

We generalize Lemma 6.12 and Theorem 6.13 next.

LEMMA 7.16. *Let A_n be a bounded generator. Recalling that in this case $\mathcal{R}(I - \alpha H_n) = B(E_n)$, define*

$$R_\alpha^{(n)} f = (I - \alpha H_n)^{-1} f, \quad f \in B(E_n),\ \alpha > 0.$$

Let Condition 7.9 hold and $\{Y_n\}$ satisfy Condition 2.8. Let $H_\dagger$ and $H_\ddagger$ be as in Lemma 7.14.

For $\alpha > 0$, let $D_\alpha \subset C_b(E)$. Assume that for each $h \in D_\alpha$, the comparison principle holds for subsolutions of

$$(I - \alpha H_\dagger) f = h, \tag{7.41}$$

and supersolutions of

$$(I - \alpha H_\ddagger) f = h. \tag{7.42}$$

Define

$$\mathcal{D}(R_\alpha) \equiv \{h \in C_b(E) : \text{LIM} R_\alpha^{(n)} \eta_n h \text{ exists and is continuous}\},$$

and $R_\alpha h = \text{LIM} R_\alpha^{(n)} \eta_n h$ for $h \in \mathcal{D}(R_\alpha)$.

Then

a) *$D_\alpha \subset \mathcal{D}(R_\alpha)$, and for each $h \in D_\alpha$, $R_\alpha h$ is the unique function in $C_b(E)$ that is both a viscosity subsolution of (7.41) and a supersolution of (7.42).*

b) *For $K \subset E$ compact, $\delta > 0$, and α, there exists another compact $\widehat{K} \subset E$, depending on K, δ, and α such that for $h_1, h_2 \in \mathcal{D}(R_\alpha)$,*

$$\sup_{x\in K} |R_\alpha h_1(x) - R_\alpha h_2(x)| \leq \delta \|h_1 - h_2\| + \sup_{x\in\widehat{K}} |h_1(x) - h_2(x)| \tag{7.43}$$

and hence

$$\|R_\alpha h_1 - R_\alpha h_2\| \leq \|h_1 - h_2\|. \tag{7.44}$$

c) *$\mathcal{D}(R_\alpha)$ is buc-closed.*

d) *For $0 < \alpha, \mu < \alpha_0$, if $g \in \mathcal{D}(R_\mu)$, then $R_\mu g - \alpha\mu^{-1}(R_\mu g - g) \in \mathcal{D}(R_\alpha)$ and*

$$R_\mu g = R_\alpha(R_\mu g - \alpha\mu^{-1}(R_\mu g - g)), \tag{7.45}$$

and hence

$$\|R_\alpha R_\mu g - R_\mu g\| \leq \alpha\mu^{-1}\|R_\mu g - g\|.$$

e) *Let $\alpha_0 > 0$. Suppose, in addition, that for $0 < \alpha < \alpha_0$, $C_b(E)$ is the buc-closure of D_α (so $\mathcal{D}(R_\alpha) = C_b(E)$). Then for $0 < \mu, \lambda, \alpha < \alpha_0$ and $h_1, h_2 \in C_b(E)$,*

$$\|R_\lambda h_1 - R_\mu h_2\| \leq \|R_\lambda h_1 - R_\mu h_2 - \alpha(\frac{R_\lambda h_1 - h_1}{\lambda} - \frac{R_\mu h_2 - h_2}{\mu})\|. \tag{7.46}$$

f) *For* $h \in \mathcal{D}(R_\alpha)$, $(f_0, g_0) \in H_\dagger$, *and* $(f_1, g_1) \in H_\ddagger$,

$$\sup_{x\in E}(R_\alpha h - f_0)(x) \le \sup_{y\in E'}(\gamma h - \gamma f_0 + \alpha g_0)(y) \tag{7.47}$$

and

$$\inf_{x\in E}(R_\alpha h - f_1)(x) \ge \inf_{y\in E'}(\gamma h - \gamma f_1 + \alpha g_1)(y), \tag{7.48}$$

and hence, if $h \in \overline{\mathcal{D}(H_\dagger)} \cap \overline{\mathcal{D}(H_\ddagger)} \cap_{\alpha<\alpha_0} \mathcal{D}(R_\alpha)$,

$$\lim_{\alpha\to 0} R_\alpha h = h. \tag{7.49}$$

PROOF. Let $h \in D_\alpha$. Define $h_n \equiv \eta_n h \in B(E_n) = \mathcal{R}(I - \alpha H_n)$. Then $h = \text{LIM} h_n$, and Part (a) follows by Lemma 7.14.

Let $h_1, h_2 \in \mathcal{D}(R_\alpha)$ and $K \subset E$ be compact. By Condition 7.9(d), there exists $q \in \mathcal{Q}$ such that $K \subset K^q$. By Lemma 5.9, for each $\delta > 0$, there exists $\widehat{q} \in \mathcal{Q}$, depending on q, δ, and α such that

$$\begin{aligned}
&\limsup_{n\to\infty} \sup_{z\in K_n^q} (\eta_n R_\alpha h_1(z) - \eta_n R_\alpha h_2(z)) \\
&\quad = \limsup_{n\to\infty} \sup_{z\in K_n^q} ((I - \alpha H_n)^{-1}\eta_n h_1(z) - (I - \alpha H_n)^{-1}\eta_n h_2(z)) \\
&\quad \le \delta\|h_1 - h_2\| + \limsup_{n\to\infty} \sup_{z\in K_n^{\widehat{q}}} (\eta_n h_1(z) - \eta_n h_2(z)).
\end{aligned}$$

By the continuity of h_i and $R_\alpha h_i$,

$$\sup_{z\in K^q} (R_\alpha h_1(z) - R_\alpha h_2(z) \le \delta\|h_1 - h_2\| + \sup_{x\in \widehat{K}^{\widehat{q}}} |h_1(x) - h_2(x)|, \tag{7.50}$$

where $K^q = \liminf_{n\to\infty} \eta_n(K_n^q)$ and $\widehat{K}^{\widehat{q}} = \limsup_{n\to\infty} \eta_n(K_n^{\widehat{q}})$. Since the roles of h_1 and h_2 can be interchanged and $\widehat{K}^{\widehat{q}}$ is compact by Condition 7.9(c), (7.43) follows.

Suppose h is *buc*-approximable by $\mathcal{D}(R_\alpha)$. (See Section A.2.) Then there exists a constant C_h and $h_{\epsilon,K} \in \mathcal{D}(R_\alpha)$, $\epsilon > 0$, $K \subset E$ compact, such that $\|h_{\epsilon,K}\| \le C_h$ and $\sup_{x\in K} |h(x) - h_{\epsilon,K}(x)| \le \epsilon$. By (7.43), for each $K \subset E$ compact and $\delta > 0$, there exists another compact $\widehat{K} \subset E$ depending on K, δ, and α such that

$$\sup_{x\in K} |R_\alpha h_{\epsilon_1,K_1}(x) - R_\alpha h_{\epsilon_2,K_2}(x)| \le 2\delta C_h + \sup_{x\in\widehat{K}} |(h_{\epsilon_1,K_1} - h_{\epsilon_2,K_2})(x)|. \tag{7.51}$$

Taking $K_1 \supset \widehat{K}$ and $K_2 \supset \widehat{K}$, the right side is bounded by $2\delta C_h + \epsilon_1 + \epsilon_2$. It follows that there exists a bounded function f such that for each compact $K \subset E$ and $\widehat{K} \subset E$ as above,

$$\sup_{x\in K} |f(x) - R_\alpha h_{\epsilon,K_2}(x)| \le 2\delta C_h + \sup_{x\in\widehat{K}} |(h - h_{\epsilon,K_2})(x)|, \tag{7.52}$$

where for $K_2 \supset \widehat{K}$, the right side is bounded by $2\delta C_h + \epsilon$. The continuity of $R_\alpha h_{\epsilon,K_2}$ implies the continuity of f.

By (5.7), for every $a, T > 0$ and $q \in \mathcal{Q}$, there exists $\widehat{q} = \widehat{q}(q, a, T) \in \mathcal{Q}$,

$$\begin{aligned}
&\sup_{y\in K_n^q} |R_\alpha^{(n)} \eta_n h(y) - R_\alpha^{(n)} \eta_n h_{\epsilon,K_2}(y)| \\
&\qquad \le (e^{-\alpha^{-1}T} + C(q, a, T, C_h)e^{-na})\|h - h_{\epsilon,K_2}\| \\
&\qquad\quad + \sup_{y\in K_n^{\widehat{q}(q,a,T)}} |\eta_n h(y) - \eta_n h_{\epsilon,K_2}(y)|,
\end{aligned} \tag{7.53}$$

and hence

$$\begin{aligned}(7.54)\qquad &\sup_{y\in K_n^q}|R_\alpha^{(n)}\eta_n h(y)-\eta_n f(y)| \\ &\le (e^{-\alpha^{-1}T}+C(q,a,T,C_h)e^{-na})\|h-h_{\epsilon,K_2}\| \\ &\quad + \sup_{y\in K_n^{\widehat{q}(q,a,T)}}|\eta_n h(y)-\eta_n h_{\epsilon,K_2}(y)| \\ &\quad + \sup_{y\in K_n^q}|\eta_n f(y)-\eta_n R_\alpha h_{\epsilon,K_2}(y)| \\ &\quad + \sup_{y\in K_n^q}|R_\alpha^{(n)}\eta_n h_{\epsilon,K_2}(y)-\eta_n R_\alpha h_{\epsilon,K_2}(y)|.\end{aligned}$$

Consequently,

$$\begin{aligned}(7.55)\qquad &\limsup_{n\to\infty}\sup_{y\in K_n^q}|R_\alpha^{(n)}\eta_n h(y)-\eta_n f(y)| \\ &\le 2C_h e^{-\alpha^{-1}T}+\sup_{x\in K^{\widehat{q}(q,a,T)}}|h(x)-h_{\epsilon,K_2}(x)| \\ &\quad + \sup_{x\in K^q}|f(x)-R_\alpha h_{\epsilon,K_2}(x)|,\end{aligned}$$

and (7.52) and the fact that T is arbitrary imply the left side is zero. It follows that $h\in\mathcal{D}(R_\alpha)$ and $R_\alpha h=f$.

By Lemma 2.11 of Miyadera [**86**],

$$R_\mu^{(n)}g=R_\alpha^{(n)}(R_\mu^{(n)}g-\alpha\frac{R_\mu^{(n)}g-g}{\mu}),$$

for $g\in B(E_n)$, $\alpha>0,\mu>0$. Part (d) follows by Lemma 5.9.

Considering Part (e), by Part (c), $\mathcal{D}(R_\alpha)=C_b(E)$, for all $\alpha<\alpha_0$. Let $h_1,h_2\in C_b(E)$. As in (5.7), for each $q\in\mathcal{Q}$ and $\delta>0$, there exists $\widehat{q}\in\mathcal{Q}$ such that

$$\begin{aligned}&\limsup_{n\to\infty}\sup_{y\in K_n^q}(R_\mu^{(n)}\eta_n h_1(y)-R_\lambda^{(n)}\eta_n h_2(y)) \\ &= \limsup_{n\to\infty}\sup_{y\in K_n^q}\Big(R_\alpha^{(n)}(R_\mu^{(n)}\eta_n h_1-\alpha\frac{R_\mu^{(n)}\eta_n h_1-\eta_n h_1}{\mu}) \\ &\qquad -R_\alpha^{(n)}(R_\lambda^{(n)}\eta_n h_2-\alpha\frac{R_\lambda^{(n)}\eta_n h_2-\eta_n h_2}{\lambda})\Big)(y) \\ &\le \delta+\limsup_{n\to\infty}\sup_{y\in K_n^{\widehat{q}}}\Big((R_\mu^{(n)}\eta_n h_1-\alpha\frac{R_\mu^{(n)}\eta_n h_1-\eta_n h_1}{\mu}) \\ &\qquad -(R_\lambda^{(n)}\eta_n h_2-\alpha\frac{R_\lambda^{(n)}\eta_n h_2-\eta_n h_2}{\lambda})\Big)(y).\end{aligned}$$

Consequently, for any $x_0\in E$,

$$\begin{aligned}&(R_\mu h_1-R_\lambda h_2)(x_0) \\ &\le \delta+\sup_x\Big((R_\mu h_1-\alpha\frac{R_\mu h_1-h_1}{\mu})-(R_\lambda h_2-\alpha\frac{R_\lambda h_2-h_2}{\lambda})\Big)(x),\end{aligned}$$

giving (7.46).

Finally, (7.47) can be obtained by replacing f_n in (7.38) by $R_\alpha^{(n)}\eta_n h$ and $\overline{f}$ by $R_\alpha h$ and then following the proof of (7.36). □

We now derive the main semigroup convergence theorem in this section. For $r > 0$, set $B_r = \{h \in C_b(E) : \|h\| \leq r\}$.

THEOREM 7.17. *For $n = 1, 2, \ldots$, let A_n be the full generator (not necessarily bounded) of a Markov process Y_n, and let H_n be given by (7.17). Let Condition 7.9 hold and $\{Y_n\}$ satisfy Condition 2.8. Let $H_\dagger \subset C^l(E, \overline{R}) \times M^u(E', \overline{R})$ and $H_\ddagger \subset C^u(E, \overline{R}) \times M^l(E', \overline{R})$, and assume that the Convergence Condition 7.11 holds.*

Let $\alpha_0 > 0$, and assume that for each $0 < \alpha < \alpha_0$, there exists $D_\alpha \subset C_b(E)$ such that for each $r > 0$, B_r is the buc-closure of $D_\alpha \cap B_r$ (see Theorem A.8), and for each $h \in D_\alpha$, the comparison principle holds for subsolutions of

$$(I - \alpha H_\dagger) f = h \tag{7.56}$$

and supersolutions of

$$(I - \alpha H_\ddagger) f = h. \tag{7.57}$$

Then

a) *For each $h \in D_\alpha$, $0 < \alpha < \alpha_0$, there exists a unique function in $C_b(E)$, denoted $R_\alpha h$, that is a viscosity subsolution of (7.56) and a supersolution of (7.57).*

b) *R_α has a unique extension to $C_b(E)$ that gives a continuous mapping of B_r into $C_b(E)$ under the buc-topology for each $r > 0$. (We will continue to denote the extension by R_α). In particular,*

$$R_\alpha f = buc\text{-}\lim_{m \to \infty} R_\alpha f_m \tag{7.58}$$

whenever buc-$\lim_{m\to\infty} f_m = f$.

c) *The operator*

$$\widehat{H} = \cup_{0<\alpha<\alpha_0} \{(R_\alpha h, \frac{R_\alpha h - h}{\alpha}) : h \in C_b(E)\} \tag{7.59}$$

is dissipative, and $\widehat{H}$ satisfies the range condition 5.1.

d) *$\widehat{H}$ defined in (7.59) generates a strongly continuous contraction semigroup $\{V(t)\}$ on $\overline{\mathcal{D}(\widehat{H})}$ given by*

$$V(t) f = \lim_{m \to \infty} R_{t/m}^m f. \tag{7.60}$$

$\overline{\mathcal{D}(\widehat{H})} \supset \mathcal{D}(H_\dagger) \cap \mathcal{D}(H_\ddagger)$.

e) *$V(t)$ has a unique extension to the buc-closure D of $\overline{\mathcal{D}(\widehat{H})}$ that is a continuous mapping of $D \cap B_r$ into $C_b(E)$ under the buc-topology for each $r > 0$. (We will continue to denote the extension by $V(t)$.) In particular,*

$$V(t) f = buc\text{-}\lim_{m \to \infty} V(t) f_m,$$

whenever $f_m \in D$ and buc-$\lim_{m\to\infty} f_m = f$.

f) *Define*

$$V_n(t) f(y) = \frac{1}{n} \log E[e^{nf(Y_n(t))} | Y_n(0) = y], \quad f \in B(E_n).$$

For each $f \in D$ and $f_n \in B(E_n)$ satisfying $f = \mathrm{LIM} f_n$ and $t_n \to t \geq 0$,

$$V(t) f = \mathrm{LIM} V_n(t_n) f_n. \tag{7.61}$$

PROOF. Let $\epsilon_n > 0$ and $\epsilon_n e^{nv_n} \to 0$, where $\{v_n\}$ is the sequence in Condition 7.11. (Note also the role of ϵ_n in (5.12).) Define

$$A_n^{\epsilon_n} f = A_n(I - \epsilon_n A_n)^{-1} f, \quad f \in B(E_n),$$

and $H_n^{\epsilon_n} f = \frac{1}{n} e^{-nf} A_n^{\epsilon_n} e^{nf}$. Then, as in the proof of Lemma 5.13, if $(f_n, g_n) \in H_n$ and $\lim_{n\to\infty} n\epsilon_n \|g_n\| = 0$, for n sufficiently large, $e^{nf_n}(1 - n\epsilon_n g_n) > 0$ and defining $\widehat{f}_n$ so that

$$e^{n\widehat{f}_n} = e^{nf_n}(1 - n\epsilon_n g_n), \tag{7.62}$$

we have

$$\widehat{g}_n = H_n^{\epsilon_n} \widehat{f}_n = e^{n(f_n - \widehat{f}_n)} g_n = \frac{g_n}{1 - n\epsilon_n g_n}. \tag{7.63}$$

It follows that

$$\lim_{n\to\infty} \|f_n - \widehat{f}_n\| = \lim_{n\to\infty} \|g_n - \widehat{g}_n\| = 0.$$

Suppose $(f_{n,0}, g_{n,0})$ satisfies (7.18), (7.19) and (7.20). Define $\widehat{f}_{n,0}$ and $\widehat{g}_{n,0}$ as in (7.62) and (7.63) with (f_n, g_n) replaced by $(f_{n,0}, g_{n,0})$. Then $\{(\widehat{f}_{n,0}, \widehat{g}_{n,0})\}$ satisfies (7.18), (7.19) and (7.20). Similarly, for $(f_{n,1}, g_{n,1})$ satisfying (7.21), (7.22) and (7.23), there exist $\widehat{f}_{n,1}$ and $\widehat{g}_{n,1} = H_n^{\epsilon_n} \widehat{f}_{n,1}$ satisfying (7.21), (7.22) and (7.23). $H_n^{\epsilon_n}$ then satisfies the conditions of Lemma 7.16, and Part (a) follows.

For each $r > 0$, by (7.43), the operator that arises as the limit of $(I - \alpha H_n^{\epsilon_n})^{-1}$ is a continuous mapping of B_r into $C_b(E)$ that agrees with the R_α of Part (a) on the dense set D_α. Let $\widehat{R}_\alpha$ be an extension of R_α to $C_b(E)$ that is continuous on B_r. Let $\widehat{D} = \{h \in C_b(E) : R_\alpha h = \widehat{R}_\alpha h\}$. Then by the continuity of R_α and $\widehat{R}_\alpha$ on B_r, $\widehat{D} \cap B_r$ must be closed, and hence, by Lemma A.7, $\widehat{D}$ is *buc*-closed. Since $R_\alpha = \widehat{R}_\alpha$ on D_α, it follows that $R_\alpha = \widehat{R}_\alpha$ on $C_b(E)$, so Part (b) follows.

By the definition of $\widehat{H}$, for each $0 < \alpha < \alpha_0$, $\mathcal{D}(\widehat{H}) \subset C_b(E) = \mathcal{R}(I - \alpha\widehat{H})$. Hence $\widehat{H}$ satisfies the range condition 5.1. The dissipativity of $\widehat{H}$ follows from Part (e) of Lemma 7.16.

The first statement in Part (d) follows from Proposition 5.3. The second statement follows by (7.49).

Considering Part (e), Part (c) of Lemma 7.16 implies that

$$\widehat{H} \subset ex\text{-}\mathrm{LIM} H_n^{\epsilon_n}.$$

Now, applying Lemmas 5.9, 5.12 and Part (b) of Proposition 5.5, for $f \in \overline{\mathcal{D}(\widehat{H})}$ and $f_n \in B(E_n)$ satisfying $\mathrm{LIM} f_n = f$ and $t_n \to t \geq 0$, we have

$$\mathrm{LIM} V_n^{\epsilon_n}(t_n) f_n = V(t) f. \tag{7.64}$$

If we define

$$\mathcal{D}(V(t)) = \{f \in C_b(E) : \mathrm{LIM} V_n^{\epsilon_n}(t_n)\eta_n f \text{ exists and is continuous}\},$$

then $\mathcal{D}(V(t))$ is *buc*-closed by essentially the same proof as for Part (c) of Lemma 7.16, using (5.13) in place of (5.7). By (5.13)

$$\sup_{y\in K} |V(t)h_1(y) - V(t)h_2(y)| \leq \sup_{y\in\widehat{K}(K,\|h_1\|\vee\|h_2\|,t)} |h_1(y) - h_2(y)|, \tag{7.65}$$

and it follows that $V(t)$ is *buc*-continuous on $\mathcal{D}(V(t)) \cap B_r$. Then $V(t)f$ agrees with (7.60) for $f \in \overline{\mathcal{D}(\widehat{H})}$. Let $\widehat{V}(t)$ be an extension of $V(t)|_{\overline{\mathcal{D}(\widehat{H})}}$ to D that is *buc*-continuous on $D \cap B_r$ for each $r > 0$. Let $\widehat{D} = \{f \in D : \widehat{V}(t)f = V(t)f\}$. Then $\widehat{D} \cap B_r$ is *buc*-closed for each r, so by Lemma A.7, $\widehat{D}$ is *buc*-closed. Hence $D = \widehat{D}$.

Part (f) follows by Part (e) and Lemma 5.11. □

7.2. Large deviation applications

Following the same arguments as in Theorem 6.14, we have the following large deviation theorem.

THEOREM 7.18. *Let Condition 7.9 hold and $\{Y_n\}$ satisfy Condition 2.8. Assume that one of the following holds:*

a) *(Continuous-time case.) For each $n = 1, 2, \ldots$, $A_n \subset B(E_n) \times B(E_n)$ and existence and uniqueness hold for the $D_{E_n}[0,\infty)$-martingale problem for (A_n, μ) for each initial distribution $\mu \in \mathcal{P}(E_n)$. Letting $P_y^n \in \mathcal{P}(D_{E_n}[0,\infty))$ denote the distribution of the solution of the martingale problem for A_n starting from $y \in E_n$, the mapping $y \mapsto P_y^n$ is Borel measurable taking the weak topology on $\mathcal{P}(D_{E_n}[0,\infty))$ (cf. Theorem 4.4.6, Ethier and Kurtz* [**36**]*).*

   Y_n is a solution of the martingale problem for A_n, and
$$H_n f = \frac{1}{n} e^{-nf} A_n e^{nf}, \quad e^{nf} \in \mathcal{D}(A_n),$$
   or if A_n is multi-valued,
$$H_n = \{(f, \frac{1}{n} e^{-nf} g) : (f,g) \in A_n\}.$$

b) *(Discrete-time case.) For each $n = 1, 2, \ldots$, T_n is a transition operator on $B(E_n)$ for a Markov chain, $\epsilon_n > 0$, Y_n is a discrete-time Markov chain with time-step ϵ_n and transition operator T_n, that is,*
$$E[f(Y_n(t))|Y_n(0) = y] = T_n^{[t/\epsilon_n]} f(y),$$
   and
$$H_n f = \frac{1}{n\epsilon_n} \log e^{-nf} T_n e^{nf}, \quad f \in B(E_n).$$
   The sequence $\epsilon_n \to 0$.

Define $\{V_n(t)\}$ on $B(E_n)$ by
$$V_n(t)f(y) = \frac{1}{n} \log E[e^{nf(Y_n(t))}|Y_n(0) = y].$$

Let $H_\dagger \subset C^l(E, \overline{R}) \times M^u(E', \overline{R})$ and $H_\ddagger \subset C^u(E, \overline{R}) \times M^l(E', \overline{R})$, and assume that the Convergence Condition 7.11 holds.

Let $\alpha_0 > 0$, and assume that for each $0 < \alpha < \alpha_0$, there exists a subset $D_\alpha \subset C_b(E)$ such that $C_b(E)$ is the buc-closure of D_α and that for each $h \in D_\alpha$, the comparison principle holds for subsolutions of
$$(I - \alpha H_\dagger) f = h,$$
and supersolutions of
$$(I - \alpha H_\ddagger) f = h.$$

Let $F \subset C_b(E)$ and $S \subset R$ satisfy Condition (c) of Corollary 4.19, and assume that for $f \in F$ and $\lambda \in S$, $\lambda f \in \mathcal{D}(H_\dagger)$.

Let $\widehat{H}$ be defined in (7.59), and let D be the buc-closure of $\overline{\mathcal{D}(\widehat{H})}$. Assume that D is closed under addition and contains a set that is bounded above and isolates points. (If $\cap_{0<\alpha<\alpha_0} D_\alpha$ or $\mathcal{D}(H_\dagger) \cap C_b(E)$ is buc-dense in $C_b(E)$, then $D = C_b(E)$ and this assumption is immediate. See Lemmas 7.19 and 7.20.)

Define $X_n = \eta_n(Y_n)$. Assume that $\{X_n\}$ satisfies the exponential compact containment condition (see (4.10)) and that $\{X_n(0)\}$ satisfies a large deviation principle with a good rate function I_0.

Then

a) *For $f \in D$, $f_n \in B(E_n)$, $\text{LIM} f_n = f$, and $t_n \to t$,*

$$V(t)f = \text{LIM} V_n(t_n) f_n,$$

where $V(t)$ is given in Theorem 7.17.

b) *$\{X_n\}$ is exponentially tight and satisfies the large deviation principle with rate function I defined by (5.30) and (5.31).*

PROOF. The theorem follows by Theorem 7.17 and Theorem 5.15. (See Remark 5.16.) □

LEMMA 7.19. *Under the assumptions of Theorem 7.18, for each $h_0 \in \cap_{0<\alpha<\alpha_0} D_\alpha$, $buc - \lim_{\alpha\to 0} R_\alpha h_0 = h_0$. Consequently, if $\cap_{0<\alpha<\alpha_0} D_\alpha \subset C_b(E)$ is buc-dense in $C_b(E)$, then $D = C_b(E)$.*

PROOF. For $0 < \alpha < \alpha_0$, D_α is *buc*-dense in $C_b(E)$, and since $\mathcal{D}(R_\alpha) \supset D_\alpha$ is *buc*-closed, $\mathcal{D}(R_\alpha) = C_b(E)$.

Let $\epsilon_n > 0$ satisfy $n\epsilon_n \to 0$, and let $A_n^{\epsilon_n}$ and $Y_n^{\epsilon_n}$ be as in Lemma 5.11.

First, we prove that, for each $q \in \mathcal{Q}$, $P\{Y_n^{\epsilon_n}(0) \in K_n^q\} = 1$, $n = 1, 2, \ldots$ implies $\{X_n = \eta_n(Y_n^{\epsilon_n})\}$ satisfies the exponential compact containment condition. Let $a, T > 0$. By Condition 2.8, there exists $K_n^{\widehat{q}}$ and constant $C > 0$ such that

$$\sup_{y \in K_n^q} \frac{1}{n} \log P(Y_n^{\epsilon_n}(t) \notin K_n^{\widehat{q}} \text{ for some } 0 \le t \le T | Y_n^{\epsilon_n}(0) = y) \le Ce^{-na}.$$

By Condition 7.9(c), there exists a compact $\widehat{K}^{\widehat{q}}$ such that for each $\delta > 0$

$$\limsup_{n\to\infty} \sup_{y \in K_n^q} \frac{1}{n} \log P(X_n(t) \notin (K^{\widehat{q}})^\delta \text{ for some } 0 \le t \le T | Y_n^{\epsilon_n}(0) = y) \le -a,$$

and if $P\{Y_n^{\epsilon_n}(0) \in K_n^q\} = 1$, $n = 1, 2, \ldots$, an argument similar to the proof of Lemma 3.3 gives the exponential compact containment condition.

Consequently, by Corollary 4.19, $P\{Y_n^{\epsilon_n}(0) \in K_n^q\} = 1$, $n = 1, 2, \ldots$ implies $\{\eta_n(Y_n^{\epsilon_n})\}$ is exponentially tight, so, for $q \in \mathcal{Q}$, there exists a compact $\mathcal{K}_q \subset D_E[0, \infty)$ and $C > 0$ such that

$$P\{\eta_n(Y_n^{\epsilon_n}) \notin \mathcal{K}_q | Y_n^{\epsilon_n}(0) = y\} \le Ce^{-n}, \quad y \in K_n^q.$$

By Lemma 5.7 (taking account of the factor n),

$$(7.66)\quad \begin{aligned}&(I-\alpha H_n^{\epsilon_n})^{-1}\eta_n h_0(y)-\eta_n h_0(y)\\ &=\sup_{g\in B(E_n)}\Big(I-\alpha(A_n^{\epsilon_n})^g\Big)^{-1}(\eta_n h_0-\alpha L_n^{\epsilon_n}g)(y)-\eta_n h_0(y)\\ &\geq (I-\alpha A_n^{\epsilon_n})^{-1}\eta_n h_0(y)-\eta_n h_0(y)\\ &=\int_0^\infty \alpha^{-1}e^{-\alpha^{-1}s}E[h_0(\eta_n(Y_n^{\epsilon_n}(s))-h_0(\eta_n(Y_n^{\epsilon_n}(0))|Y_n^{\epsilon_n}(0)=y]ds\\ &\geq \inf_{x\in\mathcal{K}_q}\int_0^\infty \alpha^{-1}e^{-\alpha^{-1}s}(h_0(x(s))-h_0(x(0)))ds\\ &\qquad -2\|h_0\|P\{Y_n^{\epsilon_n}\notin\mathcal{K}_q|Y_n^{\epsilon_n}(0)=y\}.\end{aligned}$$

Recall that

$$(7.67)\qquad R_\alpha h_0-h_0=\mathrm{LIM}\Big((I-\alpha H_n^{\epsilon_n})^{-1}\eta_n h_0-\eta_n h_0\Big).$$

For compact $K\subset E$ there exists $q\in\mathcal{Q}$ such that for any $x\in K$, there exist $y_n\in K_n^q$ such that

$$\eta_n(y_n)\to x.$$

It follows that

$$(7.68)\quad \inf_{x_0\in K}(R_\alpha h_0(x_0)-h_0(x_0))\geq \inf_{x\in\mathcal{K}_q}\int_0^\infty \alpha^{-1}e^{-\alpha^{-1}s}(h_0(x(s))-h_0(x(0)))ds.$$

By the compactness of $\mathcal{K}_q$, the right side goes to zero as α goes to zero.

We next show a corresponding upper bound. Denoting

$$f_n=(I-\alpha H_n^{\epsilon_n})^{-1}\eta_n h_0\in B(E_n),$$

by Lemma 5.7,

$$(7.69)\quad \|f_n\|\leq\|\eta_n h_0\|\leq\|h_0\|,\quad \|H_n^{\epsilon_n}f_n\|\leq\alpha^{-1}(\|f_n\|+\|\eta_n h_0\|)\leq\alpha^{-1}2\|h_0\|.$$

Moreover, by the same lemma,

$$H_n^{\epsilon_n}f_n=(A_n^{\epsilon_n})^{f_n}f_n-L_nf_n.$$

Hence

$$(I-\alpha H_n^{\epsilon_n})f_n=f_n-\alpha(A_n^{\epsilon_n})^{f_n}f_n+\alpha L_nf_n=\eta_n h_0.$$

Noting that $re^r-e^r+1\geq 0$, by the definition of L_n in Lemma 5.7, $L_nf_n\geq 0$. Therefore

$$\begin{aligned}(I-\alpha H_n^{\epsilon_n})^{-1}(\eta_n h_0) &= f_n=(I-\alpha(A_n^{\epsilon_n})^{f_n})^{-1}(\eta_n h_0-\alpha L_nf_n)\\ &\leq (I-\alpha(A_n^{\epsilon_n})^{f_n})^{-1}(\eta_n h_0),\end{aligned}$$

implying

$$\begin{aligned}&(I-\alpha H_n^{\epsilon_n})^{-1}\eta_n h_0(y)-\eta_n h_0(y)\\ &\leq (I-\alpha(A_n^{\epsilon_n})^{f_n})^{-1}(\eta_n h_0)(y)-(\eta_n h_0)(y)\\ &=\int_0^\infty \alpha^{-1}e^{-\alpha^{-1}s}E^{Q_n}[h_0(\eta_n(Y_n^{\epsilon_n}(s)))-h_0(\eta_n(Y_n^{\epsilon_n}(0)))|Y_n^{\epsilon_n}(0)=y]ds,\end{aligned}$$

where $Q_n\in\mathcal{P}(D_{E_n}[0,\infty))$ is the measure under which $Y_n^{\epsilon_n}$ has generator $(A_n^{\epsilon_n})^{f_n}$.

By Lemma 5.8, for each $T > 0$, Q_n satisfies

$$\begin{aligned}\frac{dQ_n}{dP}\Big|_{D_{E_n}[0,T]} &= \exp\{nf_n(Y_n^{\epsilon_n}(T)) - nf_n(Y_n^{\epsilon_n}(0)) - \int_0^T nH_n^{\epsilon_n} f_n(Y_n^{\epsilon_n}(s))ds\} \\ &\leq \exp\{2n\|f_n\| + nT\|H_n^{\epsilon_n} f_n\|\} \\ &\leq \exp\{2n(1+\alpha^{-1}T)\|h_0\|\},\end{aligned}$$

where the last inequality follows from (7.69).

By the exponential tightness of $\{\eta_n(Y_n^{\epsilon_n}) : n = 1,2,\ldots\}$ under P, for each $T_0 > 0$ and $q \in \mathcal{Q}$, there exists a compact $\mathcal{K}_{q,T_0} \subset D_E[0,\infty)$ and $C > 0$ such that

$$P\{\eta_n(Y_n^{\epsilon_n}) \notin \mathcal{K}_{q,T_0} | Y_n^{\epsilon_n}(0) = y\} \leq Ce^{-n-2n(1+T_0)\|h_0\|}, \quad y \in K_n^q.$$

Consequently

$$Q_n|_{D_{E_n}[0,\alpha T_0]}\{\eta_n(Y_n^{\epsilon_n}) \notin \mathcal{K}_{q,T_0} | Y_n^{\epsilon_n}(0) = y\} \leq Ce^{-n}, \quad y \in K_n^q.$$

Setting

$$G_{\alpha,T_0}(y) = \int_0^{\alpha T_0} \alpha^{-1}e^{-\alpha^{-1}s}E^{Q_n}[h_0(\eta_n(Y_n^{\epsilon_n}(s))) - h_0(\eta_n(Y_n^{\epsilon_n}(0)))|Y_n^{\epsilon_n}(0) = y]ds$$

and using the above estimate, for $y \in K_n^q$,

$$\begin{aligned}G_{\alpha,T_0}(y) &\leq \sup_{x\in\mathcal{K}_{q,T_0}} \int_0^{\alpha T_0} \alpha^{-1}e^{-\alpha^{-1}s}(h_0(x(s)) - h_0(x(0)))ds \\ &\qquad + 2\|h_0\|Q_n|_{D_{E_n}[0,\alpha T_0]}\{\eta_n(Y_n^{\epsilon_n}) \notin \mathcal{K}_{q,T_0}|Y_n^{\epsilon_n}(0) = y\} \\ &\leq \sup_{x\in\mathcal{K}_{q,T_0}} \int_0^{T_0} e^{-r}(h_0(x(\alpha r)) - h_0(x(0)))dr + 2C\|h_0\|e^{-n},\end{aligned}$$

implying

$$\limsup_{\alpha\to 0+}\limsup_{n\to\infty}\sup_{y\in K} G_{\alpha,T_0}(y) \leq 0.$$

Moreover,

$$\begin{aligned}&\int_{\alpha T_0}^{\infty} \alpha^{-1}e^{-\alpha^{-1}s}E^{Q_n}[h_0(\eta_n(Y_n^{\epsilon_n}(s))) - h_0(\eta_n(Y_n^{\epsilon_n}(0)))|Y_n^{\epsilon_n}(0) = y]ds \\ &\quad \leq \int_{\alpha T_0}^{\infty} \alpha^{-1}e^{-\alpha^{-1}s}ds2\|h_0\| = 2e^{-T_0}\|h_0\|.\end{aligned}$$

Combining the above estimates,

$$\limsup_{\alpha\to 0+}\limsup_{n\to\infty}\sup_{y\in K_n^q}\Big((I-\alpha H_n^{\epsilon_n})^{-1}\eta_n h_0(y) - \eta_n h_0(y)\Big) \leq 2e^{-T_0}\|h_0\|,$$

and by (7.67) and (7.68), we have

$$buc - \lim_{\alpha\to 0} R_\alpha h_0 = h_0.$$

□

LEMMA 7.20. *Under the assumptions of Theorem 7.18, for* $f_0 \in \mathcal{D}(H_\dagger)\cap C_b(E)$, $buc - \lim_{\alpha\to 0} R_\alpha f_0 = f_0$. *Consequently, if* $\mathcal{D}(H_\dagger)\cap C_b(E)$ *is buc-dense in* $C_b(E)$, *then* $D = C_b(E)$.

PROOF. Let $(f_0, g_0) \in H_\dagger$. For $0 < \alpha < \alpha_0$, D_α is *buc*-dense in $C_b(E)$, and since $\mathcal{D}(R_\alpha) \supset D_\alpha$ is *buc*-closed, $\mathcal{D}(R_\alpha) = C_b(E)$ and $f_0 \in \mathcal{D}(R_\alpha)$. Consequently, taking $h = f_0$ in (7.47),

$$\sup_{x\in E}(R_\alpha f_0 - f_0)(x) \leq \alpha \sup_{y\in E'} g_0(y). \tag{7.70}$$

The proof of the corresponding lower bound is the same as in Lemma 7.19. □

7.3. Convergence using projected operators

In a number of examples, it is not possible to select $\{K_n^q\}$ so that Condition 2.8 holds and Condition 7.9 holds for $\{\eta_n'(K_n^q)\}$. We can drop this requirement at the cost of relaxing the requirements on the solutions (and, hence, making the verification of the comparison principle more difficult). Define $\pi^*, \pi_* : M(E', \overline{R}) \to M(E, \overline{R})$ by

$$\pi^* g(x) = \lim_{\epsilon\to 0} \sup_{\gamma(y)\in B_\epsilon(x)} g(y) \qquad \pi_* g(x) = \lim_{\epsilon\to 0} \inf_{\gamma(y)\in B_\epsilon(x)} g(y).$$

For $H \subset C(E, \overline{R}) \times M(E', \overline{R})$, define $\pi^* H = \{(f, \pi^* g) : (f, g) \in H\} \subset C(E, \overline{R}) \times M(E, \overline{R})$ and $\pi_* H = \{(f, \pi_* g) : (f, g) \in H\}$. Let $h \in C_b(E)$. If f is a viscosity subsolution of (7.1), then f is a viscosity subsolution of

$$(I - \alpha\pi^* H)f = h,$$

and similarly, if f is a viscosity supersolution of (7.1), then f is a viscosity supersolution of

$$(I - \alpha\pi_* H)f = h.$$

In particular, if the comparison principle holds for subsolutions of $\pi^* H_\dagger$ and supersolutions of $\pi_* H_\ddagger$, then it holds for subsolutions of $H_\dagger$ and supersolutions of $H_\ddagger$.

The convergence condition then becomes

CONDITION 7.21. (Convergence Condition)
Let $H_\dagger \subset C^l(E, \overline{R}) \times M^u(E', \overline{R})$ and $H_\ddagger \subset C^u(E, \overline{R}) \times M^l(E', \overline{R})$, and let $v_n \to \infty$. For each $(f_0, g_0) \in H_\dagger$, assume that there exist $(f_{n,0}, g_{n,0}) \in H_n$ such that

$$\text{LIM} f_{n,0} \wedge c = f_0 \wedge c, \text{ for each } c \in R, \tag{7.71}$$

$$\sup_n \frac{1}{v_n} \log \|g_{n,0}\| < \infty \text{ and } \sup_{n, y\in E_n} g_{n,0}(y) < \infty, \tag{7.72}$$

and that for each $q \in \mathcal{Q}$ and each sequence $z_n \in K_n^q$ satisfying $\eta_n(z_n) \to x$

$$\limsup_{n\to\infty} g_{n,0}(z_n) \leq \pi^* g_0(x). \tag{7.73}$$

For each $(f_1, g_1) \in H_\ddagger$, assume that there exist $(f_{n,1}, g_{n,1}) \in H_n$ such that

$$\text{LIM} f_{n,1} \vee c = f_1 \vee c, \text{ for each } c \in R, \tag{7.74}$$

$$\sup_n \frac{1}{v_n} \log \|g_{n,1}\| < \infty \text{ and } \inf_{n, y\in E'} g_{n,1}(y) > -\infty, \tag{7.75}$$

and that for each $q \in \mathcal{Q}$ and each sequence $z_n \in K_n^q$ satisfying $\eta_n(z_n) \to yx$

$$\liminf_{n\to\infty} g_{n,1}(z_n) \geq \pi_* g_1(x). \tag{7.76}$$

We have the following variation of Lemma 7.14

LEMMA 7.22. *Let A_n be a bounded generator, that is, A_n is of the form*

$$A_n f(x) = \lambda_n(x) \int_{E_n} (f(z) - f(x)) q_n(x, dz), \tag{7.77}$$

with λ_n bounded, and let H_n be given by (7.17). Let Condition 7.9 hold without the requirement (7.15), and let $\{Y_n\}$ satisfy Condition 2.8. Let $H_\dagger \subset C^l(E, \overline{R}) \times M^u(E', \overline{R})$ and $H_\ddagger \subset C^u(E, \overline{R}) \times M^l(E', \overline{R})$, and assume that the Convergence Condition 7.21 holds.

Let $\alpha > 0$ and $h \in C_b(E)$, and assume that the comparison principle holds for subsolutions of

$$(I - \alpha\pi^* H_\dagger) f = h, \tag{7.78}$$

and supersolutions of

$$(I - \alpha\pi_* H_\ddagger) f = h. \tag{7.79}$$

Let $(f_n, g_n) \in H_n$, and define $f_n - \alpha g_n = h_n$. If $h = \mathrm{LIM} h_n$, then there exists a unique $f \in B(E)$ that is a viscosity subsolution of (7.78) and a supersolution of (7.79), $f \in C_b(E)$, and

$$f = \mathrm{LIM} f_n.$$

For $(f_0, g_0) \in H_\dagger$,

$$\sup_{x \in E} (f - f_0)(x) \le \sup_{x \in E} (h - (f_0 - \alpha\pi^* g_0))(x), \tag{7.80}$$

and for $(f_1, g_1) \in H_\ddagger$,

$$\inf_{x \in E} (f - f_1)(x) \ge \inf_{x \in E} (h - (f_1 - \alpha\pi_* g_1))(x). \tag{7.81}$$

REMARK 7.23. $\mathrm{LIM} g_{n,0} \vee c = g_0 \vee c$ for each $c \in R$ implies (7.73), and $\mathrm{LIM} g_{n,1} \wedge c = g_1 \wedge c$ for each $c \in R$ implies (7.76).

PROOF. Define $\overline{f}$ and $\underline{f}$ as in the proof of Lemma 7.14. We claim that they are, respectively, a subsolution of (7.78) and a supersolution of (7.79). We only prove this assertion for $\overline{f}$. The argument for $\underline{f}$ is similar.

Let $(f_0, g_0) \in H_\dagger$. By assumption, there exists $(f_{n,0}, g_{n,0}) \in H_n$ such that (7.71), (7.72) and (7.73) hold. Then, following the proof of Lemma 7.14, for any sequence $y_n \in K_n^q$ there is a sub-sequence $z_n \in K_n^{\widehat{q}}$ satisfying $\eta_n(z_n) \to x_0$, such that

$$\begin{aligned} &\limsup_{n\to\infty} (f_n(y_n) - f_{0,n}(y_n)) \\ &\quad \le \limsup_{n\to\infty} (I - \epsilon A_n^{f_n})^{-1} (h_n^\epsilon - h_{0,n}^\epsilon)(y_n) \\ &\quad \le \delta + \limsup_{n\to\infty} \sup_{y \in K_n^{\widehat{q}}} (h_n^\epsilon(y) - h_{0,n}^\epsilon(y)) \\ &\quad = \delta + \limsup_{n\to\infty} ((1 - \epsilon\alpha^{-1}) f_n(z_n) + \epsilon g_{0,n}(z_n) + \epsilon\alpha^{-1} h_n(z_n) - f_{0,n}(z_n)) \\ &\quad \le \delta + ((1 - \epsilon\alpha^{-1}) \overline{f}(x_0) + \epsilon\pi^* g_0(x_0) + \epsilon\alpha^{-1} h(x_0) - f_0(x_0)), \end{aligned} \tag{7.82}$$

and hence,

$$\sup_{x \in E} (\overline{f}(x) - f_0(x)) \le \sup_{x \in E} \left(\overline{f}(x) - f_0(x) - \epsilon\Big(\frac{\overline{f}(x) - h(x)}{\alpha} - \pi^* g_0(x)\Big) \right).$$

By Lemma 7.8, there exist $x_k \in E$ such that

$$\sup_{x \in E}(\overline{f}(x) - f_0(x)) = \lim_{k \to \infty}(\overline{f}(x_k) - f_0(x_k))$$

and

$$\limsup_{k \to \infty}\left(\frac{\overline{f}(x_k) - h(x_k)}{\alpha} - \pi^* g_0(x_k)\right) \le 0.$$

Consequently, $\overline{f}$ is a subsolution of (7.78).

The remainder of the proof is that same as for Lemma 7.14. □

Similar modifications for the remainder of the results of Section 7.1 hold. Consequently, we have the following theorem.

THEOREM 7.24. *Let Condition 7.9 hold without the requirement that (7.15) hold, and let $\{Y_n\}$ satisfy Condition 2.8. Assume that one of the following holds:*

a) *(Continuous-time case.) For each $n = 1, 2, \dots$, $A_n \subset B(E_n) \times B(E_n)$ and existence and uniqueness hold for the $D_{E_n}[0,\infty)$-martingale problem for (A_n, μ) for each initial distribution $\mu \in \mathcal{P}(E_n)$. Letting $P_y^n \in \mathcal{P}(D_{E_n}[0,\infty))$ denote the distribution of the solution of the martingale problem for A_n starting from $y \in E_n$, the mapping $y \mapsto P_y^n$ is Borel measurable taking the weak topology on $\mathcal{P}(D_{E_n}[0,\infty))$ (cf. Theorem 4.4.6, Ethier and Kurtz* [**36**]*).*

   Y_n is a solution of the martingale problem for A_n, and

$$H_n f = \frac{1}{n} e^{-nf} A_n e^{nf}, \quad e^{nf} \in \mathcal{D}(A_n),$$

   or if A_n is multi-valued,

$$H_n = \{(f, \frac{1}{n} e^{-nf} g) : (f, g) \in A_n\}.$$

b) *(Discrete-time case.) For each $n = 1, 2, \dots$, T_n is a transition operator on $B(E_n)$ for a Markov chain, $\epsilon_n > 0$, Y_n is a discrete-time Markov chain with time-step ϵ_n and transition operator T_n, that is,*

$$E[f(Y_n(t)) | Y_n(0) = y] = T_n^{[t/\epsilon_n]} f(y),$$

   and

$$H_n f = \frac{1}{n\epsilon_n} \log e^{-nf} T_n e^{nf}, \quad f \in B(E_n).$$

   The sequence $\epsilon_n \to 0$.

Define $\{V_n(t)\}$ on $B(E_n)$ by

$$V_n(t) f(y) = \frac{1}{n} \log E[e^{nf(Y_n(t))} | Y_n(0) = y].$$

Let $H_\dagger \subset C^l(E, \overline{R}) \times M^u(E', \overline{R})$ and $H_\ddagger \subset C^u(E, \overline{R}) \times M^l(E', \overline{R})$, and assume that the Convergence Condition 7.21 holds.

Let $\alpha_0 > 0$, and assume that for each $0 < \alpha < \alpha_0$, there exists a subset $D_\alpha \subset C_b(E)$ such that $C_b(E)$ is the buc-closure of D_α and that for each $h \in D_\alpha$, the comparison principle holds for subsolutions of

$$(I - \pi^* H_\dagger) f = h,$$

and supersolutions of

$$(I - \alpha \pi_* H_\ddagger) f = h.$$

Let $\widehat{H}$ be defined in (7.59), and let D be the buc-closure of $\overline{\mathcal{D}(\widehat{H})}$. Assume that D is closed under addition and contains a set that is bounded above and isolates points. (If $\cap_{0<\alpha<\alpha_0} D_\alpha$ or $\mathcal{D}(H_\dagger) \cap C_b(E)$ is buc-dense in $C_b(E)$, then $D = C_b(E)$ and this assumption is immediate. See Lemmas 7.19 and 7.20.)

Define $X_n = \eta_n(Y_n)$. Assume that $\{Y_n\}$ satisfies Condition (b) of Corollary 4.19, $F \subset C_b(E)$ and $S \subset R$ satisfy Condition (c) of Corollary 4.19, and that for $f \in F$ and $\lambda \in S$, $\lambda f \in \mathcal{D}(H_\dagger)$. Assume that $\{X_n(0)\}$ satisfies a large deviation principle with a good rate function.

Then

a) *If $f \in C_b(E)$, $f_n \in B(E_n)$, $\text{LIM} f_n = f$, and $t_n \to t$,*
$$V(t)f = \text{LIM} V_n(t_n) f_n,$$
where $V(t)$ is given in Theorem 7.17.

b) *$\{X_n\}$ is exponentially tight and satisfies the large deviation principle with rate function I defined by (5.30) and (5.31) with $D = C_b(E)$.*

CHAPTER 8

The Nisio semigroup and a control representation of the rate function

The maximum principle is a characteristic property for an operator A_n to correspond to a martingale problem. The maximum principle also implies a variational representation for H (8.39). (See the discussion in Section 8.6.1). Indeed, this representation has the form of the generator of a Nisio semigroup in control theory, and hence, at least formally, the semigroups $\{V(t)\}$ in the large deviation theorems can be interpreted as Nisio semigroups. By making this formal connection rigorous, we can determine a simpler representation of the rate function I. To make this approach practical to use and easier to present, we formulate the variational representation of H as a condition (see (8.6), (8.7), and (8.41)), rather than as a consequence of the maximum principle. We then discuss techniques for constructing the representation (Section 8.6.1).

Sections 8.1 and 8.2 formulate a control problem and give the necessary background for the corresponding Nisio semigroup. Section 8.3 gives the simplified form of the rate function, assuming that $\{V(t)\}$ is of the form described in Section 8.2. Section 8.4 develops properties and a priori estimates for certain Nisio semigroups and their resolvents. Section 8.5 uses these estimates to establish infinitesimal criteria for the Nisio semigroup representation of $\{V(t)\}$, and hence for the simplified representation of the rate function. Finally, Section 8.6 presents techniques for verifying the conditions given in Section 8.5.

8.1. Formulation of the control problem

DEFINITION 8.1 (A control equation). Let (U, q) and (E, r) be complete, separable, metric spaces. Let A be a single valued linear operator $A : \mathcal{D}(A) \subset B(E) \mapsto M(E \times U)$. Let $\mathcal{M}_m(U)$ be the space of Borel measures μ on $U \times [0, \infty)$ satisfying $\mu(U \times [0, t]) = t$ for all $t \geq 0$, and for $T > 0$, let $\mathcal{M}_m^T(U)$ denote the space of measures μ on $U \times [0, T]$ satisfying the same identity. We take the weak topology on $\mathcal{M}_m^T(U)$ and topologize $\mathcal{M}_m(U)$ with a variant of the weak topology so that $\mu_n \to \mu$ if and only if $\int_{U\times[0,T]} h(t, u)\mu_n(du \times ds) \to \int_{U\times[0,T]} h(t, u)\mu(du \times ds)$ for each $h \in C_b(E \times [0, \infty))$ and each $T > 0$. The pair $(x, \lambda) \in D_E[0, \infty) \times \mathcal{M}_m(U)$ satisfies the *relaxed control equation* for A, if and only if

$$\int_{U\times[0,t]} |Af(x(s), u)|\lambda(du \times ds) < \infty,$$

and

$$f(x(t)) - f(x(0)) - \int_{U\times[0,t]} Af(x(s), u)\lambda(du \times ds) = 0, \tag{8.1}$$

for each $f \in \mathcal{D}(A)$ and $t \geq 0$. We denote the collection of pairs satisfying the above identity by $\mathcal{J}$, and for $x_0 \in E$, set

$$\mathcal{J}_{x_0} = \{(x, \lambda) \in \mathcal{J} : x(0) = x_0\}.$$

For $T > 0$, let $\mathcal{J}^T$ denote the collection of pairs in $D_E[0,T] \times \mathcal{M}_m^T(U)$ satisfying (8.1) for $0 \leq t \leq T$. Set

$$\widehat{\mathcal{J}} = \{x \in D_E[0,\infty) : (x,\lambda) \in \mathcal{J}, \text{for some } \lambda \in \mathcal{M}_m(U)\},$$

and define $\widehat{\mathcal{J}}_{x_0}$ and $\widehat{\mathcal{J}}^T$ similarly.

EXAMPLE 8.2. Let $E = U = R^d$, and let $\sigma(x) : R^d \mapsto M_{d\times d}$ and $b(x) : R^d \mapsto R^d$ be bounded and continuous. Define

$$Af(x,u) = u \cdot (\sigma^T(x)\nabla f(x)) + b(x)\nabla f(x), \quad f \in C_c^\infty(R^d),$$

and suppose (x, λ) satisfies (8.1). Then

$$x(t) - x(0) = \int_{U\times[0,t]} \Big(u \cdot \sigma^T(x(s)) + b(x(s))\Big)\lambda(du \times ds),$$

provided $\int_{U\times[0,t]} |u|\lambda(du \times ds) < \infty$. Since $\lambda \in \mathcal{M}_m(U)$, we can write $\lambda(du \times ds) = \lambda_s(du)ds$ for some $\mathcal{P}(R^d)$-valued function $s \mapsto \lambda_s$. If we let $u(s) = \int_U u\lambda_s(du)$, then

$$\dot{x}(t) = u(t) \cdot \sigma^T(x(t)) + b(x(t)).$$

EXAMPLE 8.3. Let $(E, r) = (L^2(\mathcal{O}), \|\cdot\| = \|\cdot\|_{L^2(\mathcal{O})})$, where $\mathcal{O} = [0,1)^d$ with periodic boundary; $(U, q) = (H_{-1}(\mathcal{O}), \|\cdot\|_{-1})$, where H_{-1} is the completion of $L^2(\mathcal{O})$ under the norm $\|\rho\|_{-1} = \langle \rho, (I-\Delta)^{-1}\rho\rangle$ (see Section 13.1); and let $F \in C^2(R)$ have bounded second derivative. Let $\mathcal{D}(A)$ consist of functions of the form

$$f(\rho) = \varphi(\|\rho - \gamma\|_{-1}^2) + c, \qquad \varphi \in C_c^\infty(R), \gamma \in E, c \in R. \tag{8.2}$$

We define $\text{grad} f(\rho)$ as the unique element in $L^2(\mathcal{O})$ such that

$$\lim_{t\to 0} t^{-1}(f(\rho + tq) - f(\rho)) = \langle \text{grad} f(\rho), q\rangle, \quad \forall q \in L^2(\mathcal{O}).$$

Thus, setting $B = (I - \Delta)^{-1}$, for $f \in \mathcal{D}(A)$,

$$\text{grad} f(\rho) = 2\varphi'(\|\rho - \gamma\|_{-1}^2) B(\rho - \gamma) \in H_2(\mathcal{O}),$$

and we define

$$\begin{aligned} Af(\rho, u) &= \langle \Delta\rho - F'(\rho) + u, \text{grad} f(\rho)\rangle \\ &= 2\varphi'(\|\rho-\gamma\|_{-1}^2)\langle \Delta\rho - F'(\rho) + u, B(\rho - \gamma)\rangle \\ &= 2\varphi'(\|\rho-\gamma\|_{-1}^2)\langle (B - I)\rho - BF'(\rho) + Bu, \rho - \gamma\rangle. \end{aligned}$$

The last equality follows from the fact that $\Delta B = B - I$. Note that $Af(\rho, u)$ is continuous on $E \times U$. For $(\rho, \lambda) \in \mathcal{J}$,

$$\|\rho(t) - \gamma\|_{-1}^2 - \|\rho(0) - \gamma\|_{-1}^2 = \int_{U\times[0,t]} 2\langle \Delta\rho(s) - F'(\rho(s)) + u, B(\rho(s) - \gamma)\rangle \lambda(du \times ds)$$

for $\gamma \in L^2(\mathcal{O})$. Subtracting this identity with $\gamma = 0$ and dividing by -2 gives

$$\langle \rho(t), B\gamma\rangle - \langle \rho(0), B\gamma\rangle = \int_{U\times[0,t]} \langle \Delta\rho(s) - F'(\rho(s)) + u, B\gamma\rangle \lambda(du \times ds). \tag{8.3}$$

If the control satisfies

$$\int_{U\times[0,t]} \|u\|_{L^2(\mathcal{O})}\lambda(du \times ds) < \infty, \quad t \geq 0, \tag{8.4}$$

then writing $\lambda(du \times ds) = \lambda_s(du)ds$ and defining $u(s) = \int_U u\lambda_s(du) \in L^2(\mathcal{O})$, we see that ρ is a weak solution of

$$\frac{\partial}{\partial t}\rho = \Delta\rho - F'(\rho) + u.$$

Under (8.4), for $\rho(0) \in E$, classical theory guarantees that there is a unique solution to this partial differential equation satisfying $\int_0^t \|\rho(s)\|ds < \infty$, $t \geq 0$.

In some examples, it is necessary to restrict the controls that are allowed depending on the location of the state.

DEFINITION 8.4 (Admissible controls). Let $\Gamma \subset E \times U$ be closed, and for $z \in E$ define $\Gamma_z = \{u : (z, u) \in \Gamma\}$. The assumption that Γ is closed is equivalent to the assumption that the mapping $z \mapsto \Gamma_z$ is upper semicontinuous in the sense that $\limsup_{y\to z} \Gamma_y \subset \Gamma_z$, where

$$\limsup_{y\to z} \Gamma_y = \cap_{\epsilon>0} \mathrm{cl} \cup_{y\in B_\epsilon(z)} \Gamma_y,$$

$\mathrm{cl}D$ denoting the closure of the set D. Define

$$\mathcal{J}^\Gamma = \{(x, \lambda) \in \mathcal{J} : \int_{U\times[0,t]} I_\Gamma(x(s), u)\lambda(du \times ds) = t, t \geq 0\}. \tag{8.5}$$

Γ_z is the set of *admissible controls* for state z. $\mathcal{J}^\Gamma_{x_0}$, $\mathcal{J}^{\Gamma,t}$, $\widehat{\mathcal{J}}^\Gamma$, etc. are defined similarly to $\mathcal{J}_{x_0}$, $\mathcal{J}^t$, $\widehat{\mathcal{J}}$, etc.

The restriction in (8.5) is equivalent to the requirement that $\lambda_s(\Gamma_{x(s)}) = 1$ for almost every s.

REMARK 8.5. The measure λ is known as a relaxed control. For each $x \in \widehat{\mathcal{J}}$, there could exist more than one λ satisfying the above control equation.

A set $D \subset C_b(E)$ *separates points* if for any $x \neq y \in E$, there exists $f \in D$ such that $f(x) \neq f(y)$. A set $D \subset C_b(E)$ is *convergence determining* if for $\mu_n, \mu \in \mathcal{P}(E)$, $\int f d\mu_n \to \int f d\mu$ for all $f \in D$ implies μ_n converges to μ in the weak topology on on $\mathcal{P}(E)$. For example, the collection of bounded, uniformly continuous functions $C_b^u(E)$ is convergence determining as is $C_c^\infty(R^d)$ in the case $E = R^d$. It is easy to see that if $\mathcal{D}(A)$ separates points in E, $x \in \widehat{\mathcal{J}}$, and $\{x(s) : s \leq t\}$ is relatively compact for each $t > 0$, then $x \in C_E[0, \infty)$, and that if $\mathcal{D}(A)$ is convergence determining, every $x \in \widehat{\mathcal{J}}$ is continuous.

Let H be one of the operators arising as a limit as in the examples of Chapter 1, and suppose for the moment that H is single-valued. Frequently, we can find an operator A as in Definition 8.1 and a lower semicontinuous function L on $E \times U$ such that $Hf = \mathbf{H}f$, where

$$\mathbf{H}f \equiv \sup_{u\in U}(Af(x, u) - L(x, u)), \tag{8.6}$$

or if there is an admissibility condition,

$$\mathbf{H}f \equiv \sup_{u\in\Gamma_x}(Af(x, u) - L(x, u)). \tag{8.7}$$

In fact, this representation of H holds quite generally (see Section 8.6). We note that, by definition, $\mathbf{H}$ is single valued and $\mathcal{D}(\mathbf{H}) = \mathcal{D}(A)$. In general, H may be multivalued, and the relationship between H and $\mathbf{H}$ may be

$$\mathbf{H} \subset H,$$

or may be expressed implicitly through the resolvent by the requirement that *for a large class of* $h \in C_b(E)$ *and* $\alpha > 0$, *each viscosity solution of* $(I - \alpha H)f = h$ *is also a viscosity solution of* $(I - \alpha \mathbf{H})f = h$.

More generally, in the extended setup of Chapter 7, we may have two limit operators $H_\dagger$ and $H_\ddagger$ instead of H. In this case, we may find that

$$H_\ddagger f \leq \mathbf{H} f \leq H_\dagger f, \quad f \in \mathcal{D}(A), \tag{8.8}$$

as in Section 10.5, or the opposite inequalities,

$$H_\dagger f \leq \mathbf{H} f \leq H_\ddagger f, \quad f \in \mathcal{D}(A), \tag{8.9}$$

as in Chapter 11.

EXAMPLE 8.6. For Example 8.2, if we let $L(x,u) = \frac{1}{2}|u|^2$, then

$$\sup_{u \in R^d} (Af(x,u) - L(x,u)) = b(x) \cdot \nabla f(x) + \frac{1}{2}|\sigma^T(x)\nabla f(x)|^2 = Hf(x),$$

where the H coincides with (1.14).

Similarly, for Example 8.3, if we let $L(\rho, u) = 1/2\|u\|^2_{L^2(\mathcal{O})}$, then

$$\sup_{u \in U}(Af(\rho,u) - L(\rho,u)) = \langle \Delta\rho - F'(\rho), \frac{\delta f}{\delta \rho}\rangle + \frac{1}{2}\|\frac{\delta f}{\delta \rho}\|^2_{L^2(\mathcal{O})} = Hf(\rho),$$

where H coincides with an extended version of (1.32). (See Section 13.1.)

Let $\mathbf{V}(t)$ be the Nisio semigroup corresponding to the optimal control problem with dynamics determined by (8.1) and "reward" function $-L$, that is,

$$\mathbf{V}(t)g(x_0) = \sup_{(x,\lambda) \in \mathcal{J}^{\Gamma,t}_{x_0}} \{g(x(t)) - \int_{[0,t]\times U} L(x(s),u)\lambda(du \times ds)\}, \tag{8.10}$$

for each $x_0 \in E$. (The supremum of an empty set is defined to be $-\infty$.) Under the conditions discussed in the previous paragraph, we expect that $V(t) = \mathbf{V}(t)$ and that we will be able to simplify the form of the rate function I using a variational structure induced by A and L.

In the development that follows, we will apply results that were originally stated for the cost minimization problem

$$S(t)g(x_0) = \inf_{(x,\lambda) \in \mathcal{J}^{\Gamma,t}_{x_0}} \{g(x(t)) + \int_{[0,t]\times U} L(x(s),u)\lambda(du \times ds)\}, \tag{8.11}$$

(where inf over the empty set is ∞). Note that $S(t)g = -\mathbf{V}(t)(-g)$, so results on S translate immediately into results on $\mathbf{V}$.

DEFINITION 8.7 (Tightness Function). Let U be a complete, separable, metric space. A measurable function $\Phi : U \to (-\infty, \infty]$ is a *tightness function* if $\inf_{u \in U} \Phi(u) > -\infty$ and for each $M \in R$, $\{u : \Phi(u) \leq M\}$ is relatively compact in U.

If U is locally compact, then tightness functions are easy to construct. Take $\Phi = \psi^{-1}$ for any strictly positive $\psi \in \widehat{C}(U)$. Of course, if $U = R^d$, then $\Phi(u) = |u|^\beta$ for some $\beta > 0$ is a tightness function. If U is a Hilbert space, $\{v_i\}$ is a complete, orthonormal basis, and $\{a_i\}$ satisfies $a_i > 0$ and $a_i \to \infty$, then

$$\Phi(u) = \sum_i a_i \langle v_i, u\rangle^2$$

is a tightness function. More generally, if U is a Banach space, then one can construct tightness functions of the form

$$\Phi(u) = \sum_i a_i |\langle v_i, u\rangle|^\beta,$$

where $a_i > 0$, $\beta > 0$, and $v_i \in U^*$.

Tightness functions are introduced as a means of expressing compactness conditions on spaces of measures. In particular, we have the following lemma.

LEMMA 8.8. *Let S be a complete separable metric space, and suppose Φ is a tightness function on S. Then for each $M < \infty$,*

$$\mathcal{Q}_M = \{\mu \in \mathcal{P}(S) : \int_S \Phi(x)\mu(dx) \le M\}$$

is relatively compact in the weak topology.

PROOF. Adding a constant to Φ if necessary, we can assume $\Phi \ge 0$. For $0 < N < \infty$, let K_N be the closure of $\{x : \Phi(x) \le N\}$. Then K_N is compact and $x \in (K_N)^c$ implies $\Phi(x) > N$.

If $\mu \in \mathcal{Q}_M$, then

$$N\mu(K_N^c) \le \int_{K_N} \Phi d\mu + \int_{K_N^c} \Phi d\mu \le \int \Phi d\mu \le M.$$

Consequently

$$\mu(K_N^c) \le \frac{M}{N},$$

and, since N is arbitrary, it follows that $\mathcal{Q}_M$ is tight and, therefore, is relatively compact by Prohorov's theorem. □

Much of our development will be based on the following set of conditions.

CONDITION 8.9.

(1) *$A \subset C_b(E) \times C(E \times U)$ is single-valued and $\mathcal{D}(A)$ separates points.*

(2) *$\Gamma \subset E \times U$ is closed, and for each $x_0 \in E$, there exists $(x, \lambda) \in \mathcal{J}^\Gamma$ such that $x(0) = x_0$.*

(3) *$L(x, u) : E \times U \to [0, \infty]$ is a lower semicontinuous function, and for each $c \in [0, \infty)$ and compact $K \subset E$,*

$$\{(x, u) \in \Gamma : L(x, u) \le c\} \cap (K \times U)$$

is relatively compact.

(4) *For each compact $K \subset E$, $T > 0$, and $0 \le M < \infty$, there exists a compact $\widehat{K} \equiv \widehat{K}(K, T, M) \subset E$ such that $(x, \lambda) \in \mathcal{J}^\Gamma$, $x(0) \in K$, and*

$$\int_{U \times [0,T]} L(x(s), u)\lambda(du \times ds) \le M$$

imply

$$x(t) \in \widehat{K}, \quad 0 \le t \le T.$$

(5) *For each* $f \in \mathcal{D}(A)$ *and compact set* $K \subset E$*, there exists a right continuous, nondecreasing function* $\psi_{f,K} : [0,\infty) \to [0,\infty)$ *such that*

$$|Af(x,u)| \le \psi_{f,K}(L(x,u)), \quad \forall (x,u) \in \Gamma \cap (K \times U),$$

and

$$\lim_{r\to\infty} r^{-1}\psi_{f,K}(r) = 0.$$

CONDITION 8.10. *For each* $x_0 \in E$*, there exists* $(x,\lambda) \in \mathcal{J}^\Gamma$ *such that* $x(0) = x_0$ *and*

$$\int_{U\times[0,\infty)} L(x(s),u)\lambda(du \times ds) = 0. \tag{8.12}$$

We will need one additional assumption regarding existence of solutions of the control problem. Let $\mathbf{H}_\dagger \subset \mathcal{D}(A) \times M^u(E)$ and $\mathbf{H}_\ddagger \subset \mathcal{D}(A) \times M^l(E)$ satisfy

$$\mathbf{H}_\ddagger f(x) \le \mathbf{H}f(x), \forall f \in \mathcal{D}(\mathbf{H}_\ddagger), \qquad \mathbf{H}f(x) \le \mathbf{H}_\dagger f(x), \forall f \in \mathcal{D}(\mathbf{H}_\dagger). \tag{8.13}$$

We note that for $(x,\lambda) \in \mathcal{J}^\Gamma$ and $f \in \mathcal{D}(\mathbf{H}_\dagger)$,

$$\int_0^t (\mathbf{H}_\dagger f)(x(s))ds \ge \int_{U\times[0,t]} (Af(x(s),u) - L(x(s),u))\lambda(ds \times du), \quad t \ge 0 \tag{8.14}$$

and we will occasionally need to assume a similar inequality for $\mathbf{H}_\ddagger$.

CONDITION 8.11. *For each* $x_0 \in E$ *and each* $f \in \mathcal{D}(\mathbf{H}_\ddagger)$*, there exists* $(x,\lambda) \in \mathcal{J}^\Gamma_{x_0}$ *such that for every* $0 \le t_1 < t_2$,

$$\int_{t_1}^{t_2} (\mathbf{H}_\ddagger f)(x(s))ds \le \int_{U\times(t_1,t_2]} (Af(x(s),u) - L(x(s),u))\lambda(ds \times du). \tag{8.15}$$

REMARK 8.12.

a) Conditions 8.9.1 and 8.9.4 imply that each $x \in \widehat{\mathcal{J}}$ having finite cost on bounded time intervals is continuous.

b) Under Condition 8.9.2, each $(x,\lambda) \in \mathcal{J}^{\Gamma,t}$ can be extended to a solution in $\mathcal{J}^\Gamma$, so the supremum in (8.10) can be taken over $(x,\lambda) \in \mathcal{J}^\Gamma_{x_0}$.

c) Without loss of generality, we can assume that $r^{-1}\psi_{f,K}(r)$ is nonincreasing. (Otherwise, replace ψ by $\widetilde{\psi}_{f,K}(r) = r\sup_{s\ge r} s^{-1}\psi_{f,K}(s)$.)

d) Let V be the limit semigroup in Chapter 5 (or Chapters 6 or 7), then $V(t)1 \equiv 1$. Consequently, if (8.10) holds for A and L satisfying Condition 8.9 and $\mathbf{V} = V$, then by Proposition 8.13, for each $x_0 \in E$ there exists $(x,\lambda) \in \mathcal{J}^\Gamma$ such that $x(0) = x_0$ and (8.12) holds. Of course, Condition 8.10 implies the second part of Condition 8.9.2.

e) In the important special case $\mathbf{H}_\ddagger \equiv \mathbf{H} \equiv \mathbf{H}_\dagger$, (8.15) becomes

$$\int_0^t (\mathbf{H}f)(x(s))ds = \int_{U\times[0,t]} (Af(x(s),u) - L(x(s),u))\lambda(ds \times du), \quad t \ge 0. \tag{8.16}$$

Ordinarily, $\mathbf{H}_\ddagger 1 = \mathbf{H}_\dagger 1 = 0$, and Condition 8.11 implies Condition 8.10.

8.2. The Nisio semigroup

The following semigroup property is a consequence of the dynamic programming principle.

PROPOSITION 8.13. *Let Condition 8.9 be satisfied. Then*

a) *For each $M > 0$, compact $K \subset E$, and $T > 0$,*

$$\mathcal{K}_{M,K}^T = \{(x,\lambda) \in \mathcal{J}^{\Gamma,T} : \int_{U\times[0,T]} L(x(s),u)\lambda(du\times ds) \le M, x(0) \in K\}$$

is a compact set in $C_E[0,T] \times \mathcal{M}_m^T(U)$.

b) *$\{\mathbf{V}(t)\}$ defined in (8.10) forms a nonlinear contraction semigroup on the space of upper semicontinuous functions which are bounded above.*

PROOF. For $n = 1, 2, \ldots$, let $(x_n, \lambda_n) \in \mathcal{K}_{M,K}^T$. We must show that a subsequence converges to an element of $\mathcal{K}_{M,K}^T$. Without loss of generality, we may assume that $x_n(0) \to x_0$. By Condition 8.9.4, there exists a compact $\widehat{K} \subset E$ such that $x_n(t) \in \widehat{K}, 0 \le t \le T$, $n = 1, 2, \ldots$. Define the occupation measure on $E \times U \times [0,T]$ by

$$\mu_n(C \times [0,t]) = \int_{U\times[0,t]} I_C(x_n(s),u)\lambda_n(du \times ds).$$

Then μ_n is really concentrated on $(\Gamma \cap \widehat{K} \times U) \times [0,T]$. By Condition 8.9.3, with $K = \widehat{K}$, $\{\mu_n\}$ is relatively compact in $\mathcal{M}_m^T(E \times U \times [0,T])$, so we may as well assume that $\mu_n \to \mu$. This convergence implies that for each $c > 0$,

$$\begin{aligned}\lim_{n\to\infty} &\int_{E\times U\times[0,t]} \frac{cAf(x,u)}{c \vee |Af(x,u)|}\mu_n(dx \times du \times ds) \\ &= \int_{E\times U\times[0,t]} \frac{cAf(x,u)}{c \vee |Af(x,u)|}\mu(dx \times du \times ds),\end{aligned}$$

and Condition 8.9.5 implies

$$\begin{aligned}\Big|\int_{E\times U\times[0,t]} &\frac{cAf(x,u)}{c \vee |Af(x,u)|}\mu_n(dx \times du \times ds) \\ &- \int_{E\times U\times[0,t]} Af(x,u)\mu_n(dx \times du \times ds)\Big| \\ \le &\int_{E\times U\times[0,t]} I_{\{|Af(x,u)|>c\}}|Af(x,u)|\mu_n(dx \times du \times ds) \\ \le &\int_{E\times U\times[0,t]} I_{\{\psi_{f,\widehat{K}}(L(x,u))\ge c\}}\psi_{f,\widehat{K}}(L(x,u))\mu_n(dx \times du \times ds) \\ \le &\sup_{r\ge\psi_{f,\widehat{K}}^{-1}(c)} \frac{\psi_{f,\widehat{K}}(r)}{r}M,\end{aligned}$$

where $\psi^{-1}_{f,\widehat{K}}(c) = \inf\{r : \psi_{f,\widehat{K}}(r) \geq c\}$. A similar inequality holds with μ_n replaced by μ. Consequently,

$$\begin{aligned}\lim_{n\to\infty} \int_{U\times[0,t]} & Af(x_n(s),u)\lambda_n(du \times ds) \\ &= \lim_{n\to\infty} \int_{E\times U\times[0,t]} Af(x,u)\mu_n(dx \times du \times ds) \\ &= \int_{E\times U\times[0,t]} Af(x,u)\mu(dx \times du \times ds),\end{aligned}$$

for each $f \in \mathcal{D}(A)$ and each $0 \leq t \leq T$.

Since $x_n(t) \in \widehat{K}$, $0 \leq t \leq T$, $n = 1, 2, \ldots$, in any time interval $[a,b] \subset [0,T]$, there exists a subsequence along which t_n converges to $t \in [a,b]$ and $x_n(t_n)$ converges to a value $x(t) \in \widehat{K} \subset E$. Since $\mathcal{D}(A)$ separates points, the value of $x(t)$ is uniquely determined by the fact that

$$f(x(t)) = f(x_0) + \int_{E\times U\times[0,t]} Af(x,u)\mu(dx \times du \times ds), \tag{8.17}$$

for all $f \in \mathcal{D}(A)$, and it follows that for any sequence $s_n \to t$, $x_n(s_n) \to x(t)$. By this argument, we can construct $x(t) \in E$ for t in a countable dense subset $J \subset [0,T]$. The mapping $x : J \mapsto E$ is continuous and hence extends to a continuous mapping $x : [0,T] \mapsto E$ satisfying (8.17) for all $t \in [0,T]$ and all $f \in \mathcal{D}(A)$. Consequently, for any $t \in [0,T]$ and any sequence $s_n \to t$, $x_n(s_n) \to x(t)$ and it follows that

$$\lim_{n\to\infty} \sup_{s\leq T} r(x_n(s), x(s)) = 0,$$

and in turn, that μ must be of the form

$$\mu(dx \times du \times ds) = \delta_{x(s)}\lambda_s(du)ds.$$

Therefore, there exists $\lambda \in \mathcal{M}^T_m(U)$ such that (x, λ) satisfies (8.1).

The semigroup property follows as in Section 2 of Kurtz [**75**]. (Results for semigroups of the form $\{S(t)\}$ translate immediately into results for $\{\mathbf{V}(t)\}$.) In [**75**], the state and control spaces are assumed to be locally compact, but the extension to complete, separable metric spaces, under Condition 8.9.4, is straightforward. Also, [**75**] does not consider admissible controls, but $\mathcal{J}^\Gamma$ is a closed subset of $\mathcal{J}$, so the compactness result does not change. □

8.3. Control representation of the rate function

The primary result of this section is the representation of the rate function given by the following theorem.

THEOREM 8.14. *Suppose (E,r) and (U,q) are complete, separable, metric spaces. Let $A : \mathcal{D}(A) \subset C_b(E) \mapsto C(E \times U)$ and lower semicontinuous $L(x,u) \geq 0$ satisfy Conditions 8.9 and 8.10. Suppose that the conditions of Theorem 5.15 hold, and that V is the limit semigroup in that theorem. Define $\mathbf{V}$ according to (8.10), and suppose that $V(t) = \mathbf{V}(t)$ on $D \subset C_b(E)$, where D is the domain of V. Assume that D contains a set that is bounded above and isolates points. Then, for $x \in D_E[0,\infty)$,*

$$I(x) = I_0(x(0)) + \inf_{\lambda:(x,\lambda)\in\mathcal{J}^\Gamma_{x(0)}} \{\int_{U\times[0,\infty)} L(x(s),u)\lambda(du \times ds)\}, \tag{8.18}$$

where $\inf \emptyset = \infty$.

REMARK 8.15.

a) Note that x is fixed and that there may be more than one λ for which $(x,\lambda) \in \mathcal{J}^{\Gamma}_{x(0)}$.

b) If we restrict our attention to a bounded time interval $[0,T]$, then by Condition 8.10, for $x \in D_E[0,T]$

$$\begin{aligned} I^T(x) &= \inf\{I(z) : z \in D_E[0,\infty), z(s) = x(s), s \le T\} \\ &= \inf\{\int_{U\times[0,T]} L(y(s),u)\lambda(du \times ds) : (y,\lambda) \in \mathcal{J}^{\Gamma,T}, y = x\}. \end{aligned}$$

PROOF. For $f \in D$ and $x_0 \in E$,

$$\begin{aligned} V(t)f(x_0) &= \mathbf{V}(t)f(x_0) = \sup_{(x,\lambda)\in\mathcal{J}^{\Gamma}_{x_0}} \{f(x(t)) - \int_{U\times[0,t]} L(x(s),u)\lambda(du \times ds)\} \\ &= -\inf_{(x,\lambda)\in\mathcal{J}^{\Gamma}_{x_0}} \{\int_{U\times[0,t]} L(x(s),u)\lambda(du \times ds) - f(x(t))\}. \end{aligned}$$

With reference to Corollary 5.17, let $\{f_m\} \subset D$ satisfy the conditions of Corollary 3.21 with $S = E$ and x replaced by x_1. Then

$$\begin{aligned} &I_t(x_1|x_0) \\ &= \sup_{f\in D}(f(x_1) - \mathbf{V}(t)f(x_0)) \\ &= \sup_{f\in D} \inf_{(x,\lambda)\in\mathcal{J}^{\Gamma}_{x_0}} \{f(x_1) - f(x(t)) + \int_{U\times[0,t]} L(x(s),u)\lambda(du \times ds)\} \\ &= \lim_{m\to\infty} \inf_{(x,\lambda)\in\mathcal{J}^{\Gamma}_{x_0}} \{f_m(x_1) - f_m(x(t)) + \int_{U\times[0,t]} L(x(s),u)\lambda(du \times ds)\}. \end{aligned}$$

Either $I_t(x_1|x_0) = \infty$ or there exists a minimizing sequence (x_m, λ_m) such that $x_m(t) \to x_1$ and $\limsup_{m\to\infty} \int_{U\times[0,t]} L(x_m(s),u)\lambda_m(du \times ds) < \infty$. In the latter case, by Proposition 8.13, there exists a convergent subsequence whose limit $(x_\infty, \lambda_\infty)$ must satisfy $x_\infty(0) = x_0$, $x_\infty(t) = x_1$, and

$$\begin{aligned} I_t(x_1|x_0) &= \int_{U\times[0,t]} L(x_\infty(s),u)\lambda_\infty(du \times ds) \\ &= \inf\{\int_{U\times[0,t]} L(x(s),u)\lambda(du \times ds) : (x,\lambda) \in \mathcal{J}^{\Gamma}, x(0) = x_0, x(t) = x_1\}. \end{aligned}$$

Therefore, as in Corollary 5.17, (8.18) follows. □

Next, we derive infinitesimal conditions under which $V = \mathbf{V}$.

8.4. Properties of the control semigroup V

We study some properties of $\mathbf{V}$ under Conditions 8.9 and 8.10.

LEMMA 8.16. *Suppose that Conditions 8.9 and 8.10 hold. Then for each* $f \in C_b(E)$ *and each* $x_0 \in E$,

$$\inf_x f(x) \le \mathbf{V}(t)f(x_0) \le \sup_x f(x), \tag{8.19}$$

and $\mathbf{V}(t)f(x_0)$ *is a continuous function of* t.

PROOF. For (x,λ) satisfying (8.12), $\mathbf{V}(t)f(x_0)\geq f(x(t))$, giving the first inequality. The second inequality in (8.19) follows from the fact that $L\geq 0$.

To verify the continuity, let $(x^t,\lambda^t)\in\mathcal{J}^\Gamma$ satisfy $x^t(0)=x_0$,

$$\mathbf{V}(t)f(x_0)=f(x^t(t))-\int_{U\times[0,t]}L(x^t(s),u)\lambda^t(du\times ds),$$

and

$$\int_{U\times(t,\infty)}L(x^t(s),u)\lambda^t(du\times ds)=0.$$

Then for $r>t$

$$\begin{aligned}\mathbf{V}(r)f(x_0) &\geq f(x^t(r))-\int_{U\times[0,t]}L(x^t(s),u)\lambda^t(du\times ds)\\ &=f(x^t(r))-f(x^t(t))+\mathbf{V}(t)f(x_0).\end{aligned}$$

For $r<t$, the same inequality follows by the nonnegativity of L. It follows that

$$\liminf_{r\to t}V(r)f(x_0)\geq\mathbf{V}(t)f(x_0).$$

Note that

$$\int_{U\times[0,\infty)}L(x^t(s),u)\lambda^t(du\times ds)\leq 2\|f\|,$$

so by Proposition 8.13, $\{(x^t,\lambda^t):0\leq t\leq T\}$, with (x^t,λ^t) restricted to $[0,T]$, is relatively compact in $C_E[0,T]\times\mathcal{M}_m^T$. The upper semicontinuity of the mapping

$$(x,\lambda,t)\to f(x(t))-\int_{U\times[0,t]}L(x(s),u)\lambda(du\times ds) \tag{8.20}$$

implies that

$$\limsup_{r\to t}\mathbf{V}(r)f(x_0)\leq\mathbf{V}(t)f(x_0), \tag{8.21}$$

giving the desired continuity. □

Let h be measurable and bounded on compact sets, and let $\alpha>0$. Define $\mathbf{R}_\alpha h$ by

$$\begin{aligned}&\mathbf{R}_\alpha h(x_0) \qquad (8.22)\\ &=\limsup_{t\to\infty}\sup_{(x,\lambda)\in\mathcal{J}^\Gamma_{x_0}}\{\int_{U\times[0,t]}e^{-\alpha^{-1}s}(\frac{h(x(s))}{\alpha}-L(x(s),u))\lambda(du\times ds)\}\\ &=\limsup_{t\to\infty}\sup_{(x,\lambda)\in\mathcal{J}^\Gamma_{x_0}}\{\int_0^t\alpha^{-1}e^{-\alpha^{-1}s}h(x(s))ds\\ &\qquad-\int_0^\infty\alpha^{-1}e^{-\alpha^{-1}s}\int_{U\times[0,s\wedge t]}L(x(r),u)\lambda(du\times dr)ds\},\end{aligned}$$

where the second equality follows by

$$\begin{aligned}&\int_{U\times[0,t]}e^{-\alpha^{-1}s}L(x(s),u)\lambda(du\times ds)\\ &\quad=\int_{U\times[0,t]}\int_s^\infty\alpha^{-1}e^{-\alpha^{-1}r}dr L(x(s),u)\lambda(du\times ds)\\ &\quad=\int_0^\infty\int_{U\times[0,r\wedge t]}\alpha^{-1}e^{-\alpha^{-1}r}L(x(s),u)\lambda(du\times ds)dr,\end{aligned}$$

with the interchange justified by the fact that the integrand is nonnegative.

LEMMA 8.17. *Suppose $h \in B(E)$ is upper semicontinuous. Then under Condition 8.9, $\mathbf{R}_\alpha h$ is an upper semicontinuous function on E. Under Conditions 8.9 and 8.10, $\|\mathbf{R}_\alpha h\| \leq \|h\|$, and*

$$(8.23)\mathbf{R}_\alpha h(x_0)$$

$$= \sup_{(x,\lambda)\in\mathcal{J}^\Gamma_{x_0}} \{\int_{U\times[0,\infty)} e^{-\alpha^{-1}s}(\frac{h(x(s))}{\alpha} - L(x(s),u))\lambda(du\times ds)\}$$

$$= \sup_{(x,\lambda)\in\mathcal{J}^\Gamma_{x_0}} \{\int_0^\infty \alpha^{-1}e^{-\alpha^{-1}s}\Big(h(x(s)) - \int_{U\times[0,s]} L(x(r),u)\lambda(du\times dr)\Big)ds\},$$

PROOF. The boundedness of h and the nonnegativity of L implies $\mathbf{R}_\alpha h(x_0)$ is the decreasing limit of

$$\mathbf{R}^t_\alpha h(x_0) = \sup_{(x,\lambda)\in\mathcal{J}^\Gamma_{x_0}} \{e^{-\alpha^{-1}t}\|h\| + \int_0^t \alpha^{-1}e^{-\alpha^{-1}s}h(x(s))ds$$
$$- \int_0^\infty \alpha^{-1}e^{-\alpha^{-1}s}\int_{U\times[0,s\wedge t]} L(x(r),u)\lambda(du\times dr)ds\}.$$

Upper semicontinuity of $\mathbf{R}^t_\alpha h$ follows as in the proof of Proposition 8.13, so $\mathbf{R}_\alpha h$ is upper semicontinuous since it is the decreasing limit of a sequence of upper semicontinuous functions. Under Condition 8.10, the right side of (8.23) differs from $\mathbf{R}^t_\alpha h(x_0)$ by less than $2e^{-\alpha^{-1}t}\|h\|$, giving the desired equality. □

LEMMA 8.18. *Suppose Conditions 8.9 and 8.10 hold, and let $\{\mathbf{V}(t)\}$ be defined by (8.10). Then for each $f \in C_b(E)$ and each $x_0 \in E$,*

$$\mathbf{V}(t)f(x_0) = \lim_{n\to\infty} \mathbf{R}^{[nt]}_{n^{-1}} f(x_0)$$

PROOF. Note that

$$\int_0^\infty \alpha^{-1}e^{-\alpha^{-1}s}(h(x(s)) - \int_{U\times[0,s]} L(x(r),u))\lambda(du\times dr))ds$$
$$= E[h(x(\alpha\Delta)) - \int_{U\times[0,\alpha\Delta]} L(x(r),u))\lambda(du\times dr)],$$

where Δ is a unit exponential random variable, so

$$\mathbf{R}_\alpha h(x_0) = \sup_{(x,\lambda)\in\mathcal{J}^\Gamma_{x_0}} E[h(x(\alpha\Delta)) - \int_{U\times[0,\alpha\Delta]} L(x(r),u))\lambda(du\times dr)].$$

Fix $f \in C_b(E)$, $x_0 \in E$, and $t > 0$. It follows that

$$(8.24)\quad \mathbf{R}^{[nt]}_{n^{-1}} f(x_0) \geq \sup_{(x,\lambda)\in\mathcal{J}^\Gamma_{x_0}} E[f(x(\frac{\Delta_1+\cdots+\Delta_{[nt]}}{n}))$$
$$- \int_{U\times[0,\frac{\Delta_1+\cdots+\Delta_{[nt]}}{n}]} L(x(r),u)\lambda(du\times dr)],$$

and

$$\text{(8.25)}\quad \mathbf{R}_{n^{-1}}^{[nt]} f(x_0) \leq E[\sup_{(x,\lambda)\in\mathcal{J}_{x_0}^{\Gamma}} \Big(f(x(\frac{\Delta_1+\cdots+\Delta_{[nt]}}{n})) - \int_{U\times[0,\frac{\Delta_1+\cdots+\Delta_{[nt]}}{n}]} L(x(r),u)\lambda(du\times dr)\Big)],$$

where $\Delta_1, \Delta_2, \ldots$ are iid unit exponential random variables. Let $(x^t, \lambda^t) \in \mathcal{J}^{\Gamma}$ be as in the proof of Lemma8.16. Then by (8.24) we have

$$\mathbf{R}_{n^{-1}}^{[nt]} f(x_0) \geq E[f(x^t(\frac{\Delta_1+\cdots+\Delta_{[nt]}}{n})) - \int_{U\times[0,\frac{\Delta_1+\cdots+\Delta_{[nt]}}{n}]} L(x^t(r),u)\lambda^t(du\times dr)],$$

and hence, by the law of large numbers and the definition of (x^t, λ^t),

$$\liminf_{n\to\infty} \mathbf{R}_{n^{-1}}^{[nt]} f(x_0) \geq f(x^t(t)) - \int_{U\times[0,t]} L(x^t(s),u)\lambda^t(du\times ds) V(t) f(x_0).$$

By (8.25), we have

$$\mathbf{R}_{n^{-1}}^{[nt]} f(x_0) \leq E[V(\frac{\Delta_1+\cdots+\Delta_{[nt]}}{n}) f(x_0)],$$

and it follows by the law of large numbers and Lemma 8.16 that

$$\limsup_{n\to\infty} \mathbf{R}_{n^{-1}}^{[nt]} f(x_0) \leq \mathbf{V}(t) f(x_0).$$

□

LEMMA 8.19. *Let $\mathbf{H}_\ddagger$ and $\mathbf{H}_\dagger$ satisfy (8.13), and suppose Conditions 8.9 and 8.11 are satisfied. Then for each $x_0 \in E$, each $\alpha > 0$, and each $f \in \mathcal{D}(\mathbf{H}_\dagger)$,*

$$\begin{aligned}
&\mathbf{R}_\alpha(I - \alpha\mathbf{H}_\dagger) f(x_0) \\
&= \limsup_{t\to\infty} \sup_{(x,\lambda)\in\mathcal{J}_{x_0}^{\Gamma}} \int_{U\times[0,t]} e^{-\alpha^{-1}s}\Big(\frac{f(x(s)) - \alpha\mathbf{H}_\dagger f(x(s))}{\alpha} - L(x(s),u)\Big)\lambda(du\times ds) \\
&\leq f(x_0)
\end{aligned}$$

and for each $f \in \mathcal{D}(\mathbf{H}_\ddagger)$

$$\begin{aligned}
&\mathbf{R}_\alpha(I - \alpha\mathbf{H}_\ddagger) f(x_0) \\
&= \limsup_{t\to\infty} \sup_{(x,\lambda)\in\mathcal{J}_{x_0}^{\Gamma}} \int_{U\times[0,t]} e^{-\alpha^{-1}s}\Big(\frac{f(x(s)) - \alpha\mathbf{H}_\ddagger f(x(s))}{\alpha} - L(x(s),u)\Big)\lambda(du\times ds) \\
&\geq f(x_0).
\end{aligned}$$

PROOF. Let $(x,\lambda) \in \mathcal{J}_{x_0}^{\Gamma}$ and $f \in \mathcal{D}(\mathbf{H}_\dagger)$. Note that

$$\mathbf{H}_\dagger f(x) + L(x,u) \geq \mathbf{H} f(x) + L(x,u) \geq Af(x,u),$$

so allowing $-\infty \leq -\infty$,

$$\begin{aligned}
&\int_{U\times[0,t]} e^{-\alpha^{-1}s}\Big(\frac{f(x(s))-\alpha\mathbf{H}_\dagger f(x(s))}{\alpha} - L(x(s),u)\Big)\lambda(du\times ds)\\
\leq &\int_{U\times[0,t]} e^{-\alpha^{-1}s}\Big(\frac{f(x(s))-\alpha\mathbf{H}f(x(s))}{\alpha} - L(x(s),u)\Big)\lambda(du\times ds)\\
\leq &\int_{U\times[0,t]} \alpha^{-1}e^{-\alpha^{-1}s}(f(x(s))-\alpha Af(x(s),u))\lambda(du\times ds)\\
= &\int_0^t \alpha^{-1}e^{-\alpha^{-1}s}f(x(s))ds - \int_0^\infty \alpha^{-1}e^{-\alpha^{-1}s}\int_{U\times[0,s\wedge t]} Af(x(r),u)\lambda(du\times dr)ds\\
= &\; f(x_0) - e^{-\alpha^{-1}t}f(x(t)),
\end{aligned}$$

where the first equality comes from integration by parts and the second from the fact that $(x,\lambda)\in\mathcal{J}^\Gamma_{x_0}$. Letting $t\to\infty$, the first inequality of the lemma follows.

Now let $f\in\mathcal{D}(\mathbf{H}_\ddagger)$ and suppose that $(\widehat{x},\widehat{\lambda})\in\mathcal{J}^\Gamma_{x_0}$ satisfies (8.15). Then

$$\begin{aligned}
&\sup_{(x,\lambda)\in\mathcal{J}^\Gamma_{x_0}}\int_{U\times[0,t]} e^{-\alpha^{-1}s}\Big(\frac{f(x(s))-\alpha\mathbf{H}_\ddagger f(x(s))}{\alpha} - L(x(s),u)\Big)\lambda(du\times ds)\\
\geq &\int_{U\times[0,t]} e^{-\alpha^{-1}s}\Big(\frac{f(\widehat{x}(s))-\alpha\mathbf{H}_\ddagger f(\widehat{x}(s))}{\alpha} - L(\widehat{x}(s),u)\Big)\widehat{\lambda}(du\times ds)\\
\geq &\int_{U\times[0,t]} \alpha^{-1}e^{-\alpha^{-1}s}(f(\widehat{x}(s))-\alpha Af(\widehat{x}(s),u))\widehat{\lambda}(du\times ds)\\
= &\int_0^t \alpha^{-1}e^{-\alpha^{-1}s}f(\widehat{x}(s))ds - \int_0^\infty \alpha^{-1}e^{-\alpha^{-1}s}\int_{U\times[0,s\wedge t]} Af(\widehat{x}(r),u)\widehat{\lambda}(du\times dr)ds\\
= &\; f(x_0) - e^{-\alpha^{-1}t}f(\widehat{x}(t)),
\end{aligned}$$

proving the lemma. □

LEMMA 8.20. *Under Conditions 8.9 and 8.10, for each $h\in B(E)$ and $\alpha,\beta>0$,*

$$\mathbf{R}_\alpha h = \mathbf{R}_\beta\Big(\mathbf{R}_\alpha h - \beta\frac{\mathbf{R}_\alpha h - h}{\alpha}\Big).$$

PROOF. By Condition 8.10, in (8.22), we can take the supremum over solutions that satisfy $\int_{U\times[T,\infty)} L(x(s),u)\lambda(du\times ds) = 0$ for some $T>0$. In particular, we can work with solutions that satisfy

$$\int_{U\times[0,\infty)} e^{-\alpha^{-1}s}\Big|\frac{h(x(s))}{\alpha} - L(x(s),u)\Big|\lambda(du\times ds) < \infty$$

for all $\alpha>0$.

By definition,

$$
\begin{aligned}
&\mathbf{R}_\beta(\mathbf{R}_\alpha h - \beta\frac{\mathbf{R}_\alpha h - h}{\alpha})(x_0)\\
= &\sup_{(x,\lambda)\in\mathcal{J}^\Gamma_{x_0}}\{\int_{U\times[0,\infty)} e^{-\beta^{-1}s}\Big(\frac{1}{\beta}(\frac{\beta}{\alpha}h + \frac{\alpha-\beta}{\alpha}\mathbf{R}_\alpha h)(x(s)) - L(x(s),u)\Big)\lambda(du\times ds)\}\\
= &\sup_{(x,\lambda)\in\mathcal{J}^\Gamma_{x_0}}\{\int_{U\times[0,\infty)} e^{-\beta^{-1}s}\Big((\frac{h(x(s))}{\alpha} - L(x(s),u))\\
&\quad +\frac{\alpha-\beta}{\alpha\beta}\sup_{(\widehat{x},\widehat{\lambda})\in\mathcal{J}^\Gamma_{x(s)}}\int_{U\times[0,\infty)} e^{-\alpha^{-1}r}(\frac{h(\widehat{x}(r))}{\alpha}\\
&\qquad -L(\widehat{x}(r),u))\widehat{\lambda}(du\times dr))\Big)\lambda(du\times ds)\}\\
\geq &\int_{U\times[0,\infty)} e^{-\beta^{-1}s}\Big(\frac{h(x(s))}{\alpha} - L(x(s),u)\Big)\lambda(du\times ds)\\
&\quad +\int_0^\infty \frac{\alpha-\beta}{\alpha\beta}e^{-\beta^{-1}s}\Big(\int_{U\times[0,\infty)} e^{-\alpha^{-1}r}(\frac{h(x(s+r))}{\alpha}\\
&\qquad -L(x(s+r),u))\lambda(du\times dr))\Big)ds\\
\geq &\int_{U\times[0,\infty)} e^{-\beta^{-1}s}\Big(\frac{h(x(s))}{\alpha} - L(x(s),u)\Big)\lambda(du\times ds)\\
&\quad +\int_0^\infty \frac{\alpha-\beta}{\alpha\beta}e^{s(\frac{1}{\alpha}-\frac{1}{\beta})}\Big(\int_{U\times[s,\infty)} e^{-\alpha^{-1}r}(\frac{h(x(r))}{\alpha}\\
&\qquad -L(x(r),u))\lambda(du\times dr))\Big)ds\\
= &\int_{U\times[0,\infty)} e^{-\alpha^{-1}s}\Big(\frac{h(x(s))}{\alpha} - L(x(s),u)\Big)\lambda(du\times ds)
\end{aligned}
$$

for any $(x,\lambda)\in\mathcal{J}^\Gamma_{x_0}$ for which the integrals are all finite. (The last equality follows from integration by parts.) Therefore

$$\mathbf{R}_\beta(\mathbf{R}_\alpha h - \beta\frac{\mathbf{R}_\alpha h - h}{\alpha})(x_0) \geq \mathbf{R}_\alpha h(x_0).$$

Similarly, for each $\epsilon > 0$ and $s > 0$, there exist $(x_\epsilon,\lambda_\epsilon)\in\mathcal{J}^\Gamma_{x_0}$ and $(\widehat{x}_{\epsilon,s},\widehat{\lambda}_{\epsilon,s})\in\mathcal{J}^\Gamma_{x_\epsilon(s)}$ such that

$$
\begin{aligned}
(8.26)\qquad &\mathbf{R}_\beta(\mathbf{R}_\alpha h - \beta\frac{\mathbf{R}_\alpha h - h}{\alpha})(x_0)\\
&\leq \epsilon + \int_{U\times[0,\infty)} e^{-\beta^{-1}s}(\frac{h(x_\epsilon(s))}{\alpha} - L(x_\epsilon(s),u))\lambda_\epsilon(du\times ds)\\
&\quad +\frac{\alpha-\beta}{\alpha\beta}\int_0^\infty e^{-\beta^{-1}s}\Big(\int_{U\times[0,\infty)} e^{-\alpha^{-1}r}(\frac{h(\widehat{x}_{\epsilon,s}(r))}{\alpha}\\
&\qquad -L(\widehat{x}_{\epsilon,s}(r),u))\widehat{\lambda}_{\epsilon,s}(du\times dr)\Big)ds.
\end{aligned}
$$

For $s \geq 0$, define $(\widetilde{x}_{\epsilon,s},\widetilde{\lambda}_{\epsilon,s})\in\mathcal{J}^\Gamma_{x_0}$ by

$$\widetilde{x}_{\epsilon,s}(r) = \begin{cases} x_\epsilon(r), & 0\leq r\leq s\\ \widehat{x}_{\epsilon,s}(r), & s < r < \infty\end{cases}$$

and

$$\widetilde{\lambda}_{\epsilon,s}(du \times dr) = I_{U\times[0,s]}(u,r)\lambda_\epsilon(du \times dr) + I_{U\times(s,\infty)}(u,r)\widehat{\lambda}_{\epsilon,s}(du \times dr).$$

Then

$$\begin{aligned}
(8.27)\ &\mathbf{R}_\alpha h(x_0) \\
&\geq \int_{U\times[0,\infty)} e^{-\alpha^{-1}r}\Big(\frac{h(\widetilde{x}_{\epsilon,s}(r))}{\alpha} - L(\widetilde{x}_{\epsilon,s}(r),u)\Big)\widetilde{\lambda}_{\epsilon,s}(du \times dr) \\
&\geq \int_{U\times[0,s]} e^{-\alpha^{-1}r}\Big(\frac{h(x_\epsilon(r))}{\alpha} - L(x_\epsilon(r),u)\Big)\lambda_\epsilon(du \times dr) \\
&\qquad + e^{-\alpha^{-1}s}\int_{U\times[0,\infty)} e^{-\alpha^{-1}r}\Big(\frac{h(\widehat{x}_{\epsilon,s}(r))}{\alpha} - L(\widehat{x}_{\epsilon,s}(r),u)\Big)\widehat{\lambda}_{\epsilon,s}(du \times dr).
\end{aligned}$$

Multiplying both side of (8.27) by $e^{(\alpha^{-1}-\beta^{-1})s}$ and integrating over $s \in [0,\infty)$, we have

$$\begin{aligned}
(8.28)\ &\Big(\frac{\alpha-\beta}{\alpha\beta}\Big)^{-1}\mathbf{R}_\alpha h(x_0) \\
&\geq \int_0^\infty e^{(\alpha^{-1}-\beta^{-1})s}\int_{U\times[0,s]} e^{-\alpha^{-1}r}\Big(\frac{h(x_\epsilon(r))}{\alpha} - L(x_\epsilon(r),u)\Big)\lambda_\epsilon(du \times dr)ds \\
&\qquad + \int_0^\infty e^{-\beta^{-1}s}\int_{U\times[0,\infty)} e^{-\alpha^{-1}r}\Big(\frac{h(\widehat{x}_{\epsilon,s}(r))}{\alpha} - L(\widehat{x}_{\epsilon,s}(r),u)\Big)\widehat{\lambda}_{\epsilon,s}(du \times dr)ds \\
&= \Big(\frac{\alpha-\beta}{\alpha\beta}\Big)^{-1}\int_{U\times[0,\infty)} e^{-\beta^{-1}s}\Big(\frac{h(x_\epsilon(s))}{\alpha} - L(x_\epsilon(s),u)\Big)\lambda_\epsilon(du \times ds) \\
&\qquad + \int_0^\infty e^{-\beta^{-1}s}\int_{U\times[0,\infty)} e^{-\alpha^{-1}r}\Big(\frac{h(\widehat{x}_{\epsilon,s}(r))}{\alpha} - L(\widehat{x}_{\epsilon,s}(r),u)\Big)\widehat{\lambda}_{\epsilon,s}(du \times dr)ds.
\end{aligned}$$

the equality follows by interchanging the order of integration. Combining (8.26) and (8.28) gives.

$$\mathbf{R}_\alpha h(x_0) \geq \mathbf{R}_\beta\Big(\mathbf{R}_\alpha h - \beta\frac{\mathbf{R}_\alpha h - h}{\alpha}\Big)(x_0).$$

□

LEMMA 8.21. *a) Let $h_1 \in M(E)$ be bounded from above and $h_2 \in M(E)$ be bounded from below. Then, under Conditions 8.9 and 8.10,*

$$(8.29)\qquad \sup_x(\mathbf{R}_\alpha h_1(x) - \mathbf{R}_\alpha h_2(x)) \leq \sup_x(h_1(x) - h_2(x)),$$

with the possibility that the left side is $-\infty$. In particular, $\mathbf{R}_\alpha$ is a contraction on $B(E)$:

$$\|\mathbf{R}_\alpha h_1 - \mathbf{R}_\alpha h_2\| \leq \|h_1 - h_2\|, \quad h_1, h_2 \in B(E).$$

b) For each $\delta > 0$, $r > 0$, and compact $K \subset E$, there exist compact $K'(\delta, r, K) \subset E$, nondecreasing in δ^{-1}, r, and K such that

$$(8.30)\qquad \sup_{x\in K}|\mathbf{R}_\alpha h_1(x) - \mathbf{R}_\alpha h_2(x)| \leq \delta + \sup_{y\in K'(\delta,r,K)}|h_1(y) - h_2(y)|,$$

for any $h_1, h_2 \in B(E)$ satisfying $\|h_1\| \vee \|h_2\| \leq r$.

PROOF. Let h_1 be bounded from above and h_2 bounded from below. Letting $h_1^k = h_1 \vee k$ and $h_2^k = h_2 \wedge k$, by the definition of $\mathbf{R}_\alpha$ in (8.22),

$$\begin{aligned}(\mathbf{R}_\alpha h_1^k - \mathbf{R}_\alpha h_2^k)(x_0) &\leq \limsup_{t\to\infty} \sup_{(x,\lambda)\in\mathcal{J}_{x_0}^\Gamma} \int_0^t \alpha^{-1}e^{-\alpha^{-1}s}(h_1^k(x(s)) - h_2^k(x(s)))ds \\ &\leq \sup_x (h_1^k(x) - h_2^k(x)).\end{aligned}$$

Letting $k \to \infty$, (8.29) follows.

For δ, r, K as in Part (b), choose T large enough so that $4re^{-T/\alpha} < \delta$. Let $M = (\delta + 2r)e^{T/\alpha}$, and select the compact set $\widehat{K} = \widehat{K}(K, T, M)$ in Condition 8.9.4. Let $h_1, h_2 \in B(E)$ be such that $\|h_1\| \vee \|h_2\| \leq r$. Assume that there exists $x_0 = x_0(\delta, K, h_1, h_2) \in K$ satisfying

$$\sup_{x\in K} |\mathbf{R}_\alpha h_1(x) - \mathbf{R}_\alpha h_2(x)| \leq \frac{\delta}{4} + \mathbf{R}_\alpha h_1(x_0) - \mathbf{R}_\alpha h_2(x_0).$$

(Otherwise, interchange the roles of h_1 and h_2.) By the definition of $\mathbf{R}_\alpha$, there exists $(x, \lambda) \in \mathcal{J}_{x_0}^\Gamma$ such that

$$\mathbf{R}_\alpha h_1(x_0) \leq \frac{\delta}{4} + \int_{U\times[0,\infty)} e^{-\alpha^{-1}s}(\frac{h_1(x(s))}{\alpha} - L(x(s), u))\lambda(du \times ds), \tag{8.31}$$

and hence

$$\begin{aligned}&\sup_{x\in K} |\mathbf{R}_\alpha h_1(x) - \mathbf{R}_\alpha h_2(x)| \\ \leq\ & \frac{\delta}{2} + \int_0^\infty \alpha^{-1}e^{-\alpha^{-1}s}(h_1 - h_2)(x(s))ds \\ \leq\ & \frac{\delta}{2} + \int_0^T \alpha^{-1}e^{-\alpha^{-1}s}(h_1 - h_2)(x(s))ds + (\|h_1\| + \|h_2\|)e^{-\alpha^{-1}T} \\ \leq\ & \frac{\delta}{2} + \int_0^T \alpha^{-1}e^{-\alpha^{-1}s}(h_1 - h_2)(x(s))ds + \frac{\delta}{2}.\end{aligned}$$

By (8.31), we also have

$$\begin{aligned}&\int_{[0,T]\times U} L(x(s), u)\lambda(du \times ds) \\ \leq\ & e^{\alpha^{-1}T}\int_{U\times[0,T]} e^{-\alpha^{-1}s}L(x(s), u))\lambda(du \times ds) \\ \leq\ & e^{\alpha^{-1}T}(\frac{\delta}{4} + \int_0^\infty \alpha^{-1}e^{-\alpha^{-1}s}h_1(x(s))ds - \mathbf{R}_\alpha h_1(x_0)) \\ \leq\ & e^{\alpha^{-1}T}(\delta + 2\|h_1\|) \\ \leq\ & e^{\alpha^{-1}T}(\delta + 2r) = M,\end{aligned}$$

so

$$x(t) \in \widehat{K}, \quad t \in [0, T],$$

and

$$\sup_{x\in K} |\mathbf{R}_\alpha h_1(x) - \mathbf{R}_\alpha h_2(x)| \leq \delta + \sup_{y\in\widehat{K}} |h_1(y) - h_2(y)|.$$

□

Using essentially identical arguments as in Lemma 8.21, we can also prove the following property for $\mathbf{V}$:

LEMMA 8.22. *Under Conditions 8.9 and 8.10, $\mathbf{V}$ is a contraction on $B(E)$:*

$$\|\mathbf{V}(t)h_1 - \mathbf{V}(t)h_2\| \le \|h_1 - h_2\|, \quad h_1, h_2 \in B(E).$$

For each $\delta > 0$, $r > 0$ and a compact $K \subset E$, there exists compact $K'(\delta, r, K) \subset E$ that is nondecreasing in δ^{-1}, r and K such that

$$\sup_{x \in K} |\mathbf{V}(t)h_1(x) - \mathbf{V}(t)h_2(x)| \le \delta + \sup_{y \in \widehat{K}} |h_1(y) - h_2(y)|, \tag{8.32}$$

for every $h_1, h_2 \in B(E)$ satisfying $\|h_1\| \vee \|h_2\| \le r$.

8.5. Verification of semigroup representation

Lemma 8.19 suggests that $(I - \alpha\mathbf{H})^{-1}h = \mathbf{R}_\alpha h$. The following theorems demonstrate that this assertion is indeed valid in a certain sense. Consequently, Lemma 8.18 implies that $\mathbf{H}$ is related to $\mathbf{V}$ in the same sense that H (or $\widehat{H}$) in Chapters 5, 6, and 7 is related to V, and we can formulate conditions in terms of H and $\mathbf{H}$ that imply $V = \mathbf{V}$. The control problem then gives a more explicit representation of the rate function.

THEOREM 8.23. *Let (E, r), (U, q) be complete, separable, metric spaces. Suppose that $A \subset C_b(E) \times C(E \times U)$ and $L : E \times U \mapsto [0, \infty]$ satisfy Conditions 8.9, 8.10, and 8.11 with*

$$\mathbf{H}_\dagger f(x) = \mathbf{H}_\ddagger f(x) \equiv \mathbf{H}f(x) = \sup_{u \in \Gamma_x} (Af(x, u) - L(x, u)).$$

Suppose $\mathbf{H} \subset C_b(E) \times B(E)$ is dissipative and satisfies the range condition 5.1.

Then for each $0 < \alpha < \alpha_0$ (α_0 is the same constant as in the range condition 5.1),

$$\mathbf{R}_\alpha h = (I - \alpha\overline{\mathbf{H}})^{-1}h \in \overline{\mathcal{D}(\mathbf{H})} \subset C_b(E), \qquad \forall h \in \overline{\mathcal{D}(\mathbf{H})},$$

and $\mathbf{V}$ is the semigroup generated by $\mathbf{H}$ on $\overline{\mathcal{D}(\mathbf{H})}$, that is,

$$\begin{aligned} \mathbf{V}(t)f(x_0) &\equiv \sup_{(x,\lambda) \in \mathcal{J}_{x_0}^{\Gamma,t}} \{f(x(t)) - \int_{[0,t]\times U} L(x(s), u)\lambda(du \times ds)\} \\ &= \lim_{n \to \infty} (I - \frac{t}{n}\overline{\mathbf{H}})^{-n} f(x_0) \end{aligned}$$

for each $x_0 \in E$.

REMARK 8.24. If $H \subset C_b(E) \times B(E)$ is a dissipative operator satisfying $\mathbf{H} \subset \overline{H}$ and $\overline{\mathcal{D}(H)} = \overline{\mathcal{D}(\mathbf{H})}$ and $\mathbf{H}$ satisfies the conditions of Theorem 8.23, then H also satisfies the range condition 5.1 and

$$\mathbf{V}(t)f = \lim_{n \to \infty} (I - \frac{t}{n}\overline{H})^{-n} f$$

for each $f \in \overline{\mathcal{D}(H)}$.

PROOF. Since the range condition

$$\mathcal{D}(\mathbf{H}) \subset \overline{\mathcal{R}(I - \alpha\mathbf{H})}$$

is satisfied, for each $h \in \overline{\mathcal{D}(\mathbf{H})}$, there exist $h_n \in \mathcal{R}(I - \alpha\mathbf{H})$ such that $\|h_n - h\| \to 0$. Setting

$$f_n = (I - \alpha\mathbf{H})^{-1} h_n \in \mathcal{D}(\mathbf{H}) \subset \overline{C}(E),$$
$$\|f_n - f_m\| \leq \|h_n - h_m\|,$$

and, consequently, there exists $f \in \overline{\mathcal{D}(\mathbf{H})} \subset C_b(E)$ such that $\|f_n - f\| \to 0$. Let $g = (f - h)/\alpha \in C_b(E)$. Then $g \in \overline{\mathbf{H}} f$.

By Lemma 8.19,

$$f_n(x_0) = \mathbf{R}_\alpha (I - \alpha\mathbf{H}) f_n(x_0).$$

Letting $n \to \infty$, we have

$$f(x_0) = \mathbf{R}_\alpha (I - \alpha\overline{\mathbf{H}}) f(x_0).$$

Noting $f = (I - \alpha\overline{\mathbf{H}})^{-1} h$,

$$(I - \alpha\overline{\mathbf{H}})^{-1} h = \mathbf{R}_\alpha h.$$

By Lemma 8.18 and Proposition 5.3, the conclusion follows. □

Combining Corollary 5.19, 5.20 and Theorems 8.14 and 8.23 (including its remark), we arrive at the following:

COROLLARY 8.25 (Continuous-time case). *Assume that the conditions of Corollary 5.19 and Theorem 8.23 are satisfied, and that H of Corollary 5.19 and $\mathbf{H}$ of Theorem 8.23 satisfy $\mathbf{H} \subset \overline{H}$ and $\overline{D(\mathbf{H})} = \overline{D(H)}$. Then the sequence of stochastic processes $\{X_n\}$ is exponentially tight and satisfies the large deviation principle with good rate function I given by (8.18).*

COROLLARY 8.26 (Discrete-time case). *Assume that the conditions of Corollary 5.20 and Theorem 8.23 are satisfied, and that H of Corollary 5.20 and $\mathbf{H}$ of Theorem 8.23 satisfy $\mathbf{H} \subset \overline{H}$ and $\overline{D(\mathbf{H})} = \overline{D(H)}$. Then the sequence of stochastic processes $\{X_n\}$ is exponentially tight and satisfies the large deviation principle with good rate function I given by (8.18).*

As observed before, the range condition 5.1 is usually hard to verify. Next, we derive a viscosity solution version of Theorem 8.23 and Corollaries 8.25 and 8.26.

THEOREM 8.27. *Let (E, r) and (U, q) be complete, separable metric spaces. Suppose that $A \subset C_b(E) \times C(E \times U)$ and $L : E \times U \mapsto [0, \infty]$ satisfy Conditions 8.9, 8.10, and 8.11. Define*

$$\mathbf{H} f(x) = \sup_{u \in \Gamma_x} (A f(x, u) - L(x, u))$$

with $\mathcal{D}(\mathbf{H}) = \mathcal{D}(A)$. Let $\mathbf{H}_\dagger \subset \mathcal{D}(A) \times M^u(E)$ and $\mathbf{H}_\ddagger \subset \mathcal{D}(A) \times M^l(E)$, and assume

$$\mathbf{H}_\ddagger f(x) \leq \mathbf{H} f(x), \forall f \in \mathcal{D}(\mathbf{H}_\ddagger), \qquad \mathbf{H} f(x) \leq \mathbf{H}_\dagger f(x), \forall f \in \mathcal{D}(\mathbf{H}_\dagger). \tag{8.33}$$

Suppose that for each $0 < \alpha \leq \alpha_0$, there exists $D_\alpha \subset C_b(E)$ such that $C_b(E)$ is the buc-closure of D_α and that for each $h \in D_\alpha$, the comparison principle holds for subsolutions of

$$(I - \alpha\mathbf{H}_\dagger) f = h \tag{8.34}$$

and supersolutions of

$$(I - \alpha\mathbf{H}_\ddagger) f = h. \tag{8.35}$$

Then

a) *For each* $h \in D_\alpha$, $\mathbf{R}_\alpha h$ *given by (8.22) is continuous and is the unique function that is a subsolution of (8.34) and a supersolution of (8.35).*
b) $\mathbf{R}_\alpha : C_b(E) \mapsto C_b(E)$.
c) *The operator*

$$\widehat{\mathbf{H}} = \cup_{0<\alpha<\alpha_0}\{(\mathbf{R}_\alpha h, \frac{\mathbf{R}_\alpha h - h}{\alpha}) : h \in C_b(E)\},$$

is m-*dissipative, and hence generates a semigroup* $\widehat{\mathbf{V}}$ *on* $\overline{\mathcal{D}(\widehat{\mathbf{H}})} \subset C_b(E)$ *satisfying:*

$$\begin{aligned}(8.36)\quad \widehat{\mathbf{V}}(t)f(x_0) &\equiv \lim_{n\to\infty} \mathbf{R}_{n^{-1}}^{[nt]} f(x_0) \\ &= \sup_{(x,\lambda)\in\mathcal{J}_{x_0}^{\Gamma,t}} \{f(x(t)) - \int_{[0,t]\times U} L(x(s),u)\lambda(du\times ds)\} \\ &= \mathbf{V}(t)f(x_0).\end{aligned}$$

PROOF. For $h \in C_b(E)$, let $f = \mathbf{R}_\alpha h$ be given by (8.22). By Lemma 8.17, f is upper semicontinuous on E.

First, we prove that f is a subsolution of (8.34). For $f_0 \in \mathcal{D}(\mathbf{H}_\dagger)$ and $\beta > 0$, by Lemma 8.19,

$$\mathbf{R}_\beta(f_0 - \beta\mathbf{H}_\dagger f_0) \le f_0. \tag{8.37}$$

By Lemma 8.20, (8.37), and Lemma 8.21,

$$\begin{aligned}\sup_y(f - f_0)(y) &= \sup_y(\mathbf{R}_\alpha h - f_0)(y) \\ &\le \sup_y(\mathbf{R}_\beta(\mathbf{R}_\alpha h - \beta\alpha^{-1}(\mathbf{R}_\alpha h - h))(y) - \mathbf{R}_\beta(f_0 - \beta\mathbf{H}_\dagger f_0))(y)) \\ &\le \sup_y((\mathbf{R}_\alpha h(y) - \beta\alpha^{-1}(\mathbf{R}_\alpha h - h))(y) - f_0 + \beta\mathbf{H}_\dagger f_0)(y)),\end{aligned}$$

and hence,

$$\sup_y(f - f_0)(y) \le \sup_y\Big(f - f_0 - \beta(\frac{f-h}{\alpha} - \mathbf{H}_\dagger f_0)\Big)(y),$$

for all $\beta > 0$. By Lemma 7.8, there exist $x_n \in E$ such that

$$\limsup_{n\to\infty}(f - f_0)(x_n) = \sup_y(f - f_0)(y) \text{ and } \limsup_{n\to\infty}(\frac{f-h}{\alpha} - \mathbf{H}_\dagger f_0)(x_n) \le 0.$$

Therefore, f is a subsolution of (8.34).

Similarly, to see that f_* is a viscosity supersolution of (8.35), for $f_0 \in \mathcal{D}(\mathbf{H}_\ddagger)$, we have

$$\begin{aligned}\inf_y(f - f_0)(y) &= \inf_y(\mathbf{R}_\alpha h - f_0)(y) \\ &\ge \inf_y(\mathbf{R}_\beta(\mathbf{R}_\alpha h - \beta\alpha^{-1}(\mathbf{R}_\alpha h - h))(y) - \mathbf{R}_\beta(f_0 - \beta\mathbf{H}_\ddagger f_0))(y)) \\ &\ge \inf_y((\mathbf{R}_\alpha h(y) - \beta\alpha^{-1}(\mathbf{R}_\alpha h - h))(y) - f_0(y) + \beta\mathbf{H}_\ddagger f_0(y))\end{aligned}$$

Hence, for $\beta < \alpha$ (so that the expression on the right is $(1 - \beta\alpha^{-1})\mathbf{R}_\alpha h + \beta H_\ddagger f_0$ plus a continuous function),

$$\inf_y(f_* - f_0)(y) \ge \inf_y\Big(f_* - f_0 - \beta(\frac{f_* - h}{\alpha} - (\mathbf{H}_\ddagger f_0)_*)\Big)(y),$$

and by Lemma 7.8, there exist $x_n \in E$ such that

$$\liminf_{n\to\infty}(f_* - f_0)(x_n) = \inf_y (f_* - f_0)(y) \text{ and } \liminf_{n\to\infty}(\frac{f_* - h}{\alpha} - (\mathbf{H}_\ddagger f_0)_*)(x_n) \geq 0.$$

By the comparison principle, for each $h \in D_\alpha$, $f = \mathbf{R}_\alpha h \in C_b(E)$ and is the unique function that is a subsolution of (8.34) and a supersolution of (8.35), giving Part (a).

For Part (b), let

$$\mathcal{C}_\alpha \equiv \{f \in C_b(E) : \mathbf{R}_\alpha f \in C_b(E)\}.$$

Then $D_\alpha \subset \mathcal{C}_\alpha$, and by assumption, D_α is *buc*-dense in $C_b(E)$. Therefore, if $\mathcal{C}_\alpha$ is closed under the *buc*-topology, then $\mathcal{C}_\alpha = C_b(E)$. Suppose f is *buc*-approximable by $\mathcal{C}_\alpha$, that is, there exists a constant $C_{f,\mathcal{C}_\alpha}$, such that for any $\epsilon > 0$ and any compact set $\widehat{K} \subset E$, there is $f_{\epsilon,\widehat{K}} \in \mathcal{C}_\alpha$ such that

$$\|f_{\epsilon,\widehat{K}}\| \leq C_{f,\mathcal{C}_\alpha}, \quad \sup_{x\in\widehat{K}} |f_{\epsilon,\widehat{K}}(x) - f(x)| \leq \epsilon.$$

For $\delta > 0$ and compact $K \subset E$, by Lemma 8.21, there exists $K' \equiv K'(\delta, C_{f,\mathcal{C}}, K) \subset E$ such that

$$\sup_{x\in K} |\mathbf{R}_\alpha f(x) - \mathbf{R}_\alpha g(x)| \leq \delta + \sup_{x\in K'} |f(x) - g(x)|,$$

for every $\|g\| \leq C_{f,\mathcal{C}}$. In particular, taking $\widehat{K} = K'$ and $g = f_{\epsilon,\widehat{K}}$, we have

$$\sup_{x\in K} |\mathbf{R}_\alpha f(x) - \mathbf{R}_\alpha f_{\epsilon,K'}(x)| \leq \delta + \sup_{x\in K'} |f(x) - f_{\epsilon,K'}(x)| \leq \delta + \epsilon.$$

Since $f_{\epsilon,K'} \in D_\alpha$, $\mathbf{R}_\alpha f_{\epsilon,K'}$ is continuous and hence $\mathbf{R}_\alpha f \in C_b(E)$.

For Part (c), the range condition

$$\mathcal{D}(\widehat{\mathbf{H}}) \subset \mathcal{R}(I - \alpha\widehat{\mathbf{H}}) = C_b(E), \quad 0 < \alpha < \alpha_0,$$

follows directly from the definition of $\widehat{\mathbf{H}}$. To see that $\widehat{\mathbf{H}}$ is dissipative, let

$$(\mathbf{R}_\lambda h_1, \frac{\mathbf{R}_\lambda h_1 - h_1}{\lambda}), \quad (\mathbf{R}_\mu h_2, \frac{\mathbf{R}_\mu h_2 - h_2}{\mu}) \in \widehat{\mathbf{H}},$$

where $0 < \lambda, \mu < \alpha_0$, $h_1, h_2 \in C_b(E)$. Let $0 < \epsilon < \min\{\lambda, \mu\}$. By Lemma 8.20,

$$\mathbf{R}_\lambda h_1 = \mathbf{R}_\epsilon(\mathbf{R}_\lambda h_1 - \epsilon\frac{\mathbf{R}_\lambda h_1 - h_1}{\lambda}), \quad \mathbf{R}_\mu h_2 = \mathbf{R}_\epsilon(\mathbf{R}_\mu h_2 - \epsilon\frac{\mathbf{R}_\mu h_2 - h_2}{\mu}).$$

By the contractivity of $\mathbf{R}_\epsilon$,

$$(8.38) \|\mathbf{R}_\lambda h_1 - \mathbf{R}_\mu h_2\| \leq \|\mathbf{R}_\lambda h_1 - \mathbf{R}_\mu h_2 - \epsilon(\frac{\mathbf{R}_\lambda h_1 - h_1}{\lambda} - \frac{\mathbf{R}_\mu h_2 - h_2}{\mu})\|.$$

Since $(\|f - \alpha g\| - \|f\|)\alpha^{-1}$ is an increasing function of α, (8.38) holds for all $\epsilon > 0$ and the dissipativity of $\widehat{\mathbf{H}}$ follows. Finally, (8.36) follows by Proposition 5.3 and Lemma 8.18. □

The following two corollaries play key roles in the representation of rate functions in later examples.

COROLLARY 8.28. *Assume that the conditions of Theorems 7.18 and 8.27 are satisfied with the same* $D_\alpha, \alpha > 0$. *Let* $H_\dagger$ *and* $H_\ddagger$ *be as in Theorem 7.18 and* $\mathbf{H}_\dagger$ *and* $\mathbf{H}_\ddagger$ *as in Theorem 8.27. Suppose that one of the following holds:*

a) *For each* $h \in D_\alpha$, *every viscosity subsolution of* $(I - \alpha H_\dagger)f = h$ *is also a viscosity subsolution of* $(I - \alpha \mathbf{H}_\dagger)f = h$ *and every viscosity supersolution of* $(I - \alpha H_\ddagger)f = h$ *is also a viscosity supersolution of* $(I - \alpha \mathbf{H}_\ddagger)f = h$.
b) *For each* $h \in D_\alpha$, *every viscosity subsolution of* $(I - \alpha \mathbf{H}_\dagger)f = h$ *is also a viscosity subsolution of* $(I - \alpha H_\dagger)f = h$ *and every viscosity supersolution of* $(I - \alpha \mathbf{H}_\ddagger)f = h$ *is also a viscosity supersolution of* $(I - \alpha H_\ddagger)f = h$.

Then V *of Theorem 7.18 and* $\mathbf{V}$ *given by (8.10 satisfy* $Vf = \mathbf{V}f$, $f \in D$, *and the rate function for* $\{X_n\}$ *satisfies (8.18).*

PROOF. We follow the notation of Theorems 7.18 and 8.27. By assumption, $\mathbf{R}_\alpha h = R_\alpha h$ for $h \in D_\alpha$. Let

$$\mathcal{C}_\alpha = \{h \in C_b(E) : \mathbf{R}_\alpha h = R_\alpha h\}.$$

Then $\mathcal{C}_\alpha$ is *buc*-closed by (8.30), (7.43), and Lemma A.11. Consequently, $R_\alpha = \mathbf{R}_\alpha$ on $C_b(E)$, and

$$\widehat{H} = \widehat{\mathbf{H}}.$$

It follows that $\mathbf{V}(t) = V(t)$ on $\overline{\mathcal{D}(\widehat{H})}$, but again $\{f \in C_b(E) : V(t)f = \mathbf{V}(t)f\}$ is *buc*-closed, so $\mathbf{V}(t) = V(t)$ on D. The last statement follows by Theorem 8.14). □

COROLLARY 8.29 (Compact state space). *Assume that the conditions of Theorem 6.14 and Theorem 8.27 are satisfied with the same* $D_\alpha, \alpha > 0$. *Let* H *be as in Theorems 6.14 and* $\mathbf{H}_\dagger = \mathbf{H}_\ddagger = \mathbf{H}$ *in 8.27. Suppose that one of the following holds:*

a) *For each* $h \in D_\alpha$, *every viscosity solution of* $(I - \alpha H)f = h$ *is also a viscosity solution of* $(I - \alpha \mathbf{H})f = h$.
b) *For each* $h \in D_\alpha$, *every viscosity solution of* $(I - \alpha \mathbf{H})f = h$ *is also a viscosity solution of* $(I - \alpha H)f = h$.

Then V *of Theorem 6.14 and* $\mathbf{V}$ *given by (8.10) satisfy* $Vf = \mathbf{V}f$, $f \in C(E)$, *and the rate function for* $\{X_n\}$ *satisfies (8.18).*

8.6. Verifying the assumptions

Typically, the variational representation (8.7) holds and Conditions 8.10, 8.11 are satisfied. We describe some techniques for verifying these assumptions in this section. The discussion here is frequently heuristic; however, in the sections that follow, we will rigorously exploit these techniques in concrete examples.

8.6.1. Variational representation of H.

8.6.1.1. *Construction using operator duality.* Fleming [**44**], [**45**] and Sheu [**109**] observed that representation (8.7) holds quite generally. See also Feng [**41**] for a summary of sample path level large deviation results under the generalities discussed here. These papers present ways of constructing A and L based on the exponential martingale problem for H_n and discuss the probabilistic relationship to the Girsanov change of measure. We quote some relevant results from these papers without proof.

Let E be a compact metric space, and let $A \subset C(E) \times C(E)$ be the generator for a martingale problem. Assume that for each $x \in E$, there exists a solution of the martingale problem for A with sample paths in $D_E[0,\infty)$ and $X(0) = x$. (Note that A corresponds to the A_n that determine the Y_n, not the A in Definition 8.1.) Suppose that $D \equiv \mathcal{D}(A)$ is an algebra and that $f \in \mathcal{D}(A)$ implies $e^f \in \mathcal{D}(A)$. For simplicity, we assume A is single-valued. Define a new set of transformed operators under the above conditions. For $f, g \in D$,

$$\begin{aligned} Hf &= e^{-f}Ae^f \\ A^g f &= e^{-g}A(fe^g) - (e^{-g}f)Ae^g \\ Lg &= A^g g - Hg. \end{aligned}$$

The following natural *duality* between H and L was first observed by Fleming for diffusion processes and extended by Sheu [**108**] to more general Markov processes:

$$\begin{aligned} Hf(x) &= \sup_{g\in D}\{A^g f(x) - Lg(x)\} \\ Lg(x) &= \sup_{f\in D}\{A^g f(x) - Hf(x)\}, \end{aligned} \tag{8.39}$$

with both suprema attained when $f = g$. Note that this duality relationship is obtained for the H that appears in the exponential martingale problem. We are interested in a corresponding duality relationship for the rescaled limit H of a sequence $\{H_n\}$, where H_n is of the form

$$H_n f = \frac{1}{n} e^{-nf} A_n e^{nf}.$$

Define

$$\begin{aligned} A_n^g f &= e^{-ng}A_n(fe^{ng}) - (e^{-ng}f)A_n e^{ng} \\ L_n g &= A_n^g g - H_n g. \end{aligned}$$

Then

$$\begin{aligned} H_n f(x) &= \sup_{g\in D}\{A_n^g f(x) - L_n g(x)\} \\ L_n g(x) &= \sup_{f\in D}\{A_n^g f(x) - H_n f(x)\}. \end{aligned} \tag{8.40}$$

Let $H = \lim_{n\to\infty} H_n$. Feng [**41**] proved that (8.39) holds for the limiting operators under the assumption that the A_n are local operators. The locality seems nonessential. In applications, one can typically compute L_n and A_n^g and then directly verify that the desired duality holds in the limit.

The representation of Hf given by (8.39) is of the form (8.6) with $U = D$ and

$$L(x,u) = A^u u(x) - Hu(x).$$

One may view the duality formula (8.39) as an infinitesimal version of the usual proof of large deviation theorems using Girsanov change of measures.

To use D, a collection of functions, as a control space would be awkward. Feng [**41**] also discusses an embedding technique which may allow the replacement of D by a more reasonable control space.

Let U be a complete separable metric space, and let B map D to the space of U-valued, measurable functions (i.e. for each $g \in D$, $u = Bg$ is a measurable, U-valued function). Assume that an operator $A : D \mapsto C(E \times U)$ and a function $L : E \times U \mapsto [0,\infty)$ exist such that

$$Af(x, Bg(x)) = A^g f(x), \quad f, g \in D,$$

and

$$L(x, Bg(x)) = Lg(x), \quad g \in D.$$

Let Γ be the closure of $\{(x, Bg(x)) : g \in D\}$. Then we should have

$$Hf(x) = \sup_{u \in \Gamma_x} (Af(x,u) - L(x,u)). \tag{8.41}$$

8.6.1.2. *Construction using the Fenchel-Legendre transform.* When E is a convex subset of a topological vector space, the limiting operator H typically can be rewritten as $H(x, \nabla f(x))$, where $H(x,p)$ is convex in the second variable and ∇f is the gradient of f. In such cases, an alternative way of constructing a variational representation for the operator H is available through the Fenchel-Legendre transform. Let E^* be the dual of E. For each $q \in E^*$, define

$$L(x,q) = \sup_{p \in E}\{\langle p, q\rangle - H(x,p)\}. \tag{8.42}$$

Then with mild regularity on $H(x,\cdot)$, we also have

$$H(x,p) = \sup_{q \in E^*}\{\langle p, q\rangle - L(x,q)\},$$

which implies

$$Hf(x) = H(x, \nabla f(x)) = \sup_{q \in E^*}\{\langle \nabla f(x), q\rangle - L(x,q)\},$$

giving the desired representation.

The transform approach for constructing the rate function was employed by Freidlin and Wentzell and is currently popular in the literature. In general, the representation of H obtained through this approach differs from that of the operator duality approach. The alternative approach may be particularly useful when L defined by (8.42) is only determined implicitly while the L constructed using the operator duality has an explicit representation. The price paid for this explicit representation may be to complicate the control dynamics; however, the control dynamics are still given explicitly.

8.6.2. Verifying Condition 8.10. Constructing the variational representation through operator duality and the convex transformation lead to different approaches for the verification of Conditions 8.10 and 8.11.

8.6.2.1. *The operator duality approach.* Verifying Condition 8.10 is basically equivalent to finding a $g \in D$ such that $Lg(x) = 0$, for example, $g \equiv 0$. Note that, taking $g \equiv 0$,

$$A^0 = \lim_{n\to\infty} A_n.$$

Typically, A^0 will be a generator corresponding to a deterministic flow, and for each $x_0 \in E$, there will exist $x \in C_E[0,\infty)$ such that $x(0) = x_0$ and

$$f(x(t)) - f(x_0) - \int_0^t A^0 f(x(s))ds = 0, \quad t \geq 0, f \in D.$$

Defining $\lambda(du \times ds) = \delta_{\{Bg(x(s))\}}(du) \times ds$, we then have

$$f(x(t)) - f(x(0)) - \int_0^t Af(x(s),u)\lambda(du \times ds) = 0.$$

8.6.2.2. *The Fenchel-Legendre transform approach.* For each $x \in E$, $H(x,p)$ and $L(x,q)$ are typically convex functions in p and q respectively and $H(x,0) = 0$. In order to have a $q(x)$ such that $L(x,q(x)) = 0$, it is enough to prove that the maximum of $\langle p, q(x)\rangle - H(x,p)$, as a function of p, is attained at 0. Under certain regularity conditions, through the duality results in convex analysis, this is equivalent to $q(x) = (\partial_p H)(x,0)$. Therefore, we can define $q(x)$ as above for the given H and seek the smoothness of $q(x)$ in order to have existence of $x(t)$ satisfying

$$f(x(t)) - f(x(0)) - \int_0^t \langle q(x(s)), \nabla f(x(s))\rangle ds = 0, \forall f \in D, x(0) \in E.$$

If such an x exists, define $\lambda(du \times ds) = \delta_{\{q(x(s))\}}(du) \times ds$.

8.6.3. Verifying Condition 8.11.

8.6.3.1. *The operator duality approach.* Formally, at least, $Hf(x) = A^f f(x) - Lf(x) = Af(x, Bf(x)) - L(x, Bf(x))$. For each f, we expect A^f to be the generator of a flow, and hence, for each $x_0 \in E$ there should exist $x \in C_E[0,\infty)$ with $x(0) = x_0$ such that

$$g(x(t)) - g(x_0) - \int_0^t A^f g(x(s))ds = 0, \quad t \ge 0,\ g \in D,$$

and hence, for $g = f$ and $\lambda(du \times ds) = \delta_{\{Bf(x(s))\}}(du) \times ds$, we have (8.15).

8.6.3.2. *The Fenchel-Legendre transform approach.* We seek a function $q_f(x)$ such that $H(x, \nabla f(x)) = q_f(x)\nabla f(x) - L(x, q_f(x))$, where $q_f(x)$ is smooth enough so that for each $x_0 \in E$ there exists x satisfying $\dot{x}(t) = q_f(x(t))$. By a convexity argument, we should have $q_f(x) \in (\partial_p H)(x, \nabla f(x))$. Again, define $\lambda(du \times ds) = \delta_{\{q_f(x(s))\}}(du) \times ds$.

8.6.3.3. *A general approach.* Frequently, the variational representation of H enjoys stronger regularity than is expressed in Condition 8.9. Under these stronger conditions, Condition 8.11 always holds.

CONDITION 8.30.

(1) *E is compact, $\Gamma = E \times U$, $A \subset C(E) \times C(E \times U)$, $\mathcal{D}(A)$ is an algebra, and $\overline{\mathcal{D}(A)} = C(E)$. For each $u \in U$,*

$$A_u f \equiv Af(\cdot, u)$$

satisfies the positive maximum principle and

$$Af^2(x,u) = 2f(x)Af(x,u), \quad \forall f \in \mathcal{D}(A).$$

(2) *$L : E \times U \mapsto [0,\infty]$ is lower semicontinuous, and there exists a tightness function Φ on U such that $\Phi(u) \le L(x,u)$ for $(x,u) \in E \times U$.*
(3) *For each $f \in \mathcal{D}(A)$, $\mathbf{H}f \in C(E)$.*
(4) *For each $f \in \mathcal{D}(A)$, there exists a right continuous, nondecreasing function $\psi_f : [0,\infty) \mapsto [0,\infty)$ such that $|Af(x,u)| \le \psi_f(\Phi(u))$, for each $(x,u) \in E \times U$, and $\lim_{r\to\infty} r^{-1}\psi_f(r) = 0$.*

We give an estimate on $\mathbf{H}$ under Conditions 8.30 and 8.10.

LEMMA 8.31. *Suppose that Conditions 8.30 and 8.10 hold, and $\mathbf{H}$ is defined as in (8.7). Then for each $f \in \mathcal{D}(A)$, there exists $u_f : E \mapsto U$ such that $u_f(x) \in \Gamma_x$ and*

$$\mathbf{H}f(x) = Af(x, u_f(x)) - L(x, u_f(x)), \tag{8.43}$$

$$-\psi_f(0) \le \mathbf{H}f(x) \le c_f \equiv \sup_r(\psi_f(r) - r) < \infty, \tag{8.44}$$

and for every $u_f(x)$ satisfying (8.43),

$$L(x, u_f(x)) \le d_f, \tag{8.45}$$

where d_f is defined by

$$d_f \equiv \sup\{d : \psi_f(d) - d \ge -\psi_f(0)\} < \infty.$$

PROOF. By Condition 8.10, the nonnegativity of L, and the fact that each level set of L is compact, for each $x \in E$, there exists $u \in \Gamma_x$ such that $L(x, u) = 0$. Hence, by Condition 8.30.4, $Af(x, u) \ge -\psi_f(0)$ giving the left inequality in (8.44). Also by Condition 8.30.4, we have

$$\mathbf{H}f(x) \le \sup_{u \in \Gamma_x}(\psi_f(L(x, u)) - L(x, u)), \tag{8.46}$$

giving the right inequality. Finally, let $u_n \in \Gamma_x$ satisfy

$$\lim_{n\to\infty}(Af(x, u_n) - L(x, u_n)) = \mathbf{H}f(x).$$

Then (8.44) and Condition 8.30.4 imply

$$-\psi_f(0) \le \liminf_{n\to\infty}(\psi_f(L(x, u_n)) - L(x, u_n)),$$

and hence $\limsup_n L(x, u_n) \le d_f$. By Condition 8.30.2, we can assume that $u_n \to u_f(x) \in \Gamma_x$, and by the continuity of Af and the lower semicontinuity of L, we must have (8.43) and (8.45). Indeed, by (8.44) and Condition 8.30.4, for any $u_f(x)$ satisfying (8.43), (8.45) holds. □

LEMMA 8.32. *Suppose that Conditions 8.30 and 8.10 are satisfied with $\mathbf{H}_\ddagger = \mathbf{H}$. Then Conditions 8.9 and 8.11 hold.*

PROOF. Condition 8.9.2 is implied by the definition of Γ and Condition 8.10. The other parts of Condition 8.9 are immediate consequences of Condition 8.30. It remains to verify Condition 8.11.

To simplify notation, we establish (8.15) with $t = 1$. By Lemma 8.31, for each $x_0 \in E$ and $f \in \mathcal{D}(A)$, there exists $u_0 \in U$ such that

$$\mathbf{H}f(x_0) = Af(x_0, u_0) - L(x_0, u_0).$$

Since E is compact and A_{u_0} satisfies the positive maximum principle, by Theorem 5.4 in Chapter 4 of Ethier and Kurtz [**36**], there exists x_n (possibly random) such that $x_n(0) = x_0$ and for each $g \in \mathcal{D}(A)$,

$$M^g(t) = g(x_n(t)) - g(x_n(0)) - \int_0^t Ag(x_n(s), u_0)ds = M^g(t)$$

is a martingale on $0 \le t \le \frac{1}{n}$. Since $A_{u_0}g^2 = 2gA_{u_0}g$, the quadratic variation of M^g is zero, and hence $M^g(t) = 0$.

Define $u_n(s) = u_0$, for $0 \le s \le 1/n$, so that

$$g(x_n(t)) - g(x_n(0)) - \int_0^t Ag(x_n(s), u_n(s))ds = 0,$$

for $g \in \mathcal{D}(A)$ and $0 \le t \le 1/n$.

For $x_n(\frac{1}{n}) \in E$, there exists $u_n(\frac{1}{n}) \in U$ such that

$$\mathbf{H}f(x_n(\frac{1}{n})) = Af(x_n(\frac{1}{n}), u_n(\frac{1}{n})) - L(x_n(\frac{1}{n}), u_n(\frac{1}{n})).$$

Define $u_n(t) = u_n(\frac{1}{n})$ for $\frac{1}{n} < t \le \frac{2}{n}$. Continuing this construction, we have

$$g(x_n(t)) - g(x_n(0)) - \int_0^t Ag(x_n(s), u_n(s))ds = 0, \quad 0 \le t \le 1, \ \forall g \in \mathcal{D}(A),$$

and

$$\mathbf{H}f(x_n(\frac{k}{n})) = Af(x_n(\frac{k}{n}), u_n(\frac{k}{n})) - L(x_n(\frac{k}{n}), u_n(\frac{k}{n})), k = 0, 1, \ldots, n. \tag{8.47}$$

Define

$$\lambda_n(du \times ds) = \delta_{u_n(s)}(du) \times ds,$$

and $\widehat{x}_n(t) = x_n(\frac{[nt]}{n})$. Then $(x_n, \lambda_n) \in \mathcal{J}_{x_0}$ and

$$\int_0^1 \mathbf{H}f(\widehat{x}_n(s))ds = \int_{U\times[0,1]} (Af - L)(\widehat{x}_n(s), u)\lambda_n(du \times ds).$$

By Lemma 8.31,

$$\sup_n \int_{U\times[0,1]} L(\widehat{x}_n(s), u)\lambda_n(du \times ds) \le d_f < \infty,$$

and it follows by Condition 8.30.2 that λ_n is relatively compact.

For each $g \in \mathcal{D}(A)$ and $\epsilon > 0$,

$$\begin{aligned}
|g(x_n(t+\epsilon)) - g(x_n(t))| &= \Big|\int_{U\times[t,t+\epsilon]} Ag(x_n(s), u)\lambda_n(du \times ds)\Big| \\
&\le \int_{U\times[t,t+\epsilon]} \psi_g(L(x_n(s), u))\lambda_n(du \times ds) \\
&\le \psi_g(\epsilon^{-1})\epsilon + \sup_{r\ge\epsilon^{-1}} \frac{\psi(r)}{r} \int_{U\times[0,1]} L(x_n(s), u)\lambda_n(du \times ds) \\
&\equiv \omega_g(\epsilon),
\end{aligned}$$

and $\lim_{\epsilon\to 0} \omega_g(\epsilon) = 0$. Since $\mathcal{D}(A)$ is dense in $C(E)$, $\{x_n\}$ is relatively compact in $C_E[0,1]$. Consequently, there exists a modulus ω such that

$$\sup_n r(x_n(t+\epsilon), x_n(t)) \le \omega(\epsilon),$$

and it follows that

$$\sup_n r(\widehat{x}_n(t+\epsilon), \widehat{x}_n(t)) \le \omega(\frac{[n(t+\epsilon)]}{n} - \frac{[nt]}{n}) \le \omega(\epsilon).$$

Therefore $\{\widehat{x}_n\}$ is relatively compact as well (in the Skorohod topology). Select a subsequence so that $(\widehat{x}_n, x_n, \lambda_n)$ converges to (x, x, λ).

Define occupation measures on $E \times U \times [0,1]$

$$\mu_n(C \times [0,t]) = \int_0^t I_C(x_n(s), u)\lambda_n(du \times ds).$$

Since $(x_n, \lambda_n) \to (x, \lambda)$, $\mu_n \to \mu$ in the weak topology and

$$\mu(C \times [0,t]) = \int_0^t I_C(x(s), u)\lambda(du \times ds).$$

Since

$$|Ag(x_n(s), u_n(s))| \le \psi_g(\Phi(u_n(s))) \le \psi_g(L(\widehat{x}_n(s), u_n(s))), \forall g \in \mathcal{D}(A),$$

we have

$$\sup_n \int_{E\times U\times[0,1]} \psi_g^{-1}(|Ag(x,u)|)\mu_n(dx\times du\times dt)$$
$$= \sup_n \int_0^1 \psi_g^{-1}(|Ag(x_n(s),u_n(s))|)ds \le d_g < \infty.$$

Therefore, by uniform integrability,

$$\begin{aligned}\lim_{n\to\infty}\int_{U\times[0,t]} Ag(x_n(s),u)\lambda_n(du\times ds) &= \lim_{n\to\infty}\int_{E\times U\times[0,t]} Ag(x,u)\mu_n(dx\times du\times ds)\\ &= \int_{U\times[0,t]} Ag(x,u)\lambda(du\times ds),\end{aligned}$$

for each $g\in\mathcal{D}(A)$ and $0\le t\le 1$. Therefore

$$g(x(t)) - g(x_0) - \int_{U\times[0,t]} Ag(x(s),u)\lambda(du\times ds) = 0,$$

that is, $(x,\lambda)\in\mathcal{J}_{x_0}$. In addition, by (8.47) and the upper semicontinuity of $Af-L$,

$$\begin{aligned}\int_0^1 \mathbf{H}f(x(s))ds &= \lim_{n\to\infty}\int_0^1 \mathbf{H}f(\widehat{x}_n(s))ds\\ &= \lim_{n\to\infty}\int_0^1 (Af-L)(\widehat{x}_n(s),u)\lambda_n(du\times ds)\\ &\le \int_0^1 (Af-L)(x(s),u)\lambda(du\times ds).\end{aligned}$$

Noting that $\mathbf{H}f(x)\ge (Af-L)(x,u)$, we conclude

$$\int_0^1 \mathbf{H}f(x(s))ds = \int_0^1 (Af-L)(x(s),u)\lambda(du\times ds),$$

and $(x,\lambda)\in\mathcal{J}_{x_0}$. □

Part 3

Examples of large deviations and the comparison principle

CHAPTER 9

The comparison principle

At the beginning of Chapter 2, we outlined a four-step program for proving large deviation results. The technically most difficult step in the program is usually the verification of the comparison principle. For $h_1, h_2 \in B(E)$, we consider subsolutions of

$$(I - \alpha H_\dagger)f = h_1 \tag{9.1}$$

and supersolutions of

$$(I - \alpha H_\ddagger)f = h_2. \tag{9.2}$$

For $h_1 = h_2 = h$, the comparison principle states that if $\overline{f}$ is a viscosity subsolution of (9.1) and $\underline{f}$ is a viscosity supersolution (9.2), then $\overline{f} \le \underline{f}$. Consequently, to verify the comparison principle, we must be able to bound $\overline{f}$ from above and $\underline{f}$ from below.

The techniques in this chapter are mainly extensions of those in Crandall, Ishii and Lions [**17**]. However, these generalizations are non-trivial. The most significant improvement is the treatment for equations with infinite dimensional state space. The method here does not rely on Ekeland's perturbed optimization principle or its variants. See Section 9.4. In the past, Ekeland's principle has been a basic point of departure for Hamilton-Jacobi theory in infinite dimensions. See the available theory developed by Crandall and Lions [**21**], [**20**] and Tataru [**117**]. This work applies in a restricted class of Banach spaces for equations with sufficiently regular coefficients. These requirements exclude equations arising for large deviations of interacting particle systems, such as Example 1.14. The method in Section 9.4 allows us to treat a large class of problems in this type of situation. It is based on some techniques in Feng and Katsoulakis [**43**].

Throughout this chapter, we assume that $H_\dagger \subset M^l(E,\overline{R}) \times M^u(E',\overline{R})$, $H_\ddagger \subset M^u(E,\overline{R}) \times M^l(E',\overline{R})$, and $H = H_\dagger \cap H_\ddagger$. It follows that $H \subset B(E) \times B(E)$. Note that if the comparison principle holds for H, then it holds for the pair $(H_\dagger, H_\ddagger)$. We assume that $(f,g) \in H_\dagger$ and $c \in R$ imply that $(f+c, g) \in H_\dagger$ and similarly for $H_\ddagger$. Recall that by Remark 7.2, if $E = E'$ and E is compact, then Definition 7.1 is equivalent to Definition 6.1.

9.1. General estimates

The definitions of subsolution and supersolution suggest an approach to the estimates needed on $\overline{f}$ and $\underline{f}$. Suppose that E is compact. If $\overline{f}$ is a subsolution of (9.1) and $(f_0, g_0) \in H_\dagger$, then there exists $x_0 \in E$ such that

$$\overline{f}(x_0) - f_0(x_0) = \sup_x \overline{f}(x) - f_0(x) \tag{9.3}$$

and

$$\overline{f}(x_0) \leq \alpha(g_0)^*(x_0) + h_1(x_0), \tag{9.4}$$

and if $\underline{f}$ is a supersolution of (9.2) and $(f_1, g_1) \in H_\ddagger$, then there exists $x_1 \in E$ such that

$$f_1(x_1) - \underline{f}(x_1) = \sup_x f_1(x) - \underline{f}(x), \tag{9.5}$$

and

$$\underline{f}(x_1) \geq \alpha(g_1)_*(x_1) + h_2(x_1). \tag{9.6}$$

If $x_1 = x_0$ and (9.4) and (9.6) both hold, we have

$$\overline{f}(x_0) - \underline{f}(x_0) \leq \alpha((g_0)^*(x_0) - (g_1)_*(x_0)) + h_1(x_0) - h_2(x_0). \tag{9.7}$$

If in addition

$$\overline{f}(x_0) - \underline{f}(x_0) = \sup_x(\overline{f}(x) - \underline{f}(x)) \tag{9.8}$$

and the right side of (9.7) is less than or equal to zero, we have $\overline{f} \leq \underline{f}$. In other words, for x_0 satisfying (9.8), we would like to find $(f_0, g_0) \in H_\dagger$, $(f_1, g_1) \in H_\ddagger$ satisfying (9.3), (9.4), (9.5), and (9.6) (with $x_1 = x_0$) such that

$$(g_0)^*(x_0) - (g_1)_*(x_0) \leq 0.$$

Then $\sup_x(\overline{f}(x) - \underline{f}(x)) \leq \sup_x(h_1(x) - h_2(x))$, so if $h_1 = h_2 = h$, $\overline{f} \leq \underline{f}$.

In the absence of knowledge of the regularity of $\overline{f}$ and $\underline{f}$, however, selecting f_0 and f_1 so that (9.3) and (9.5) hold with $x_0 = x_1$ is likely to be impossible. The following lemma formalizes an approximate approach to the same end. For a subsolution $\overline{f}$ and $(f_0, g_0) \in H_\dagger$, let $\overline{V}(\overline{f}, f_0, g_0) \subset E$ be the collection of x_0 satisfying (9.3) and (9.4). For a supersolution $\underline{f}$ and $(f_1, g_1) \in H_\ddagger$, let $\underline{V}(\underline{f}, f_1, g_1)$ be the collection of x_1 satisfying (9.5) and (9.6). The lemma does not assume that E is compact; however, it does assume that the appropriate extrema are achieved which may be difficult to verify for non-compact E.

LEMMA 9.1. *Let $\overline{f}$ be a viscosity subsolution of (9.1) and $\underline{f}$ a viscosity supersolution of (9.2). Suppose that there exist $\{(f_0^m, g_0^m)\} \subset H_\dagger$, $\{(f_1^m, g_1^m)\} \subset H_\ddagger$, $x_0^m \in \overline{V}(\overline{f}, f_0^m, g_0^m)$, and $x_1^m \in \underline{V}(\underline{f}, f_1^m, g_1^m)$ such that*

$$\lim_{m\to\infty} \overline{f}(x_0^m) - \underline{f}(x_1^m) = \sup_x \overline{f}(x) - \underline{f}(x)$$

$$\lim_{m\to\infty} h_1(x_0^m) - h_2(x_1^m) = 0$$

and

$$\limsup_{m\to\infty} (g_0^m)^*(x_0^m) - (g_1^m)_*(x_1^m) \leq 0.$$

Then $\overline{f} \leq \underline{f}$.

PROOF. Note that (9.4) and (9.6) imply

$$\overline{f}(x_0^m) - \underline{f}(x_1^m) \leq \alpha((g_0^m)^*(x_0^m) - (g_1^m)_*(x_1^m)) + h_1(x_0^m) - h_2(x_1^m), \tag{9.9}$$

and hence passing to the limit, we have

$$\sup_x(\overline{f}(x) - \underline{f}(x)) \leq 0.$$

□

The following purely analytic lemma is useful in constructing the sequences required in the previous lemma. It is a straightforward adaptation to metric spaces of Proposition 3.7 in Crandall, Ishii and Lions [**17**].

LEMMA 9.2. *Let E be a metric space. Suppose that $\Psi : E \times E \to [0,\infty]$ is lower semicontinuous, $\Psi \geq 0$, and for each $\mu > 0$, $\Phi_\mu : E \times E \mapsto [-\infty,\infty)$ is upper semicontinuous. Suppose that there exists an upper semicontinuous $\Phi : E \times E \mapsto [-\infty,\infty)$, $\Phi \not\equiv -\infty$, such that*

$$\lim_{\mu\to\infty} \sup_{x,y\in E} |\Phi(x,y) \vee C - \Phi_\mu(x,y) \vee C| = 0, \qquad \forall C \in R.$$

Define

$$M_\mu = \sup_{(x,y)\in E\times E} (\Phi_\mu(x,y) - \mu\Psi(x,y)).$$

Suppose that $M_\mu < \infty$ for some $\mu > 0$ and that $M_\infty \equiv \lim_{\mu\to\infty} M_\mu > -\infty$. Select $(x_\mu, y_\mu) \in E \times E$ so that

$$\lim_{\mu\to\infty} (M_\mu - (\Phi(x_\mu,y_\mu) - \mu\Psi(x_\mu,y_\mu))) = 0.$$

Then the following hold:

a) $\lim_{\mu\to\infty} \mu\Psi(x_\mu,y_\mu) = 0$.

b) $\lim_{\mu\to\infty} \Phi_\mu(x_\mu,y_\mu) = \lim_{\mu\to\infty} M_\mu \equiv M_\infty \geq \sup_{\{(x,y):\Psi(x,y)=0\}} \Phi(x,y)$.

c) *If $(\widehat{x},\widehat{y})$ is a limit point of (x_μ,y_μ) as $\mu\to\infty$, then $\Psi(\widehat{x},\widehat{y}) = 0$ and*

$$\lim_{\mu\to\infty} M_\mu = \lim_{\mu\to\infty} \Phi_\mu(x_\mu,y_\mu) = \Phi(\widehat{x},\widehat{y}) = \sup_{\{(x,y):\Psi(x,y)=0\}} \Phi(x,y). \tag{9.10}$$

PROOF. Note that the hypotheses will still hold if we replace Φ_μ by Φ and that the conclusions with Φ in place of Φ_μ imply the original conclusions. Consequently, we assume that $\Phi_\mu = \Phi$. Then M_μ is decreasing and for $\mu > \mu_0$,

$$\limsup_{\mu\to\infty} (\mu - \mu_0)\Psi(x_\mu,y_\mu) \leq M_{\mu_0} - M_\infty. \tag{9.11}$$

Since the right side of (9.11) can be made arbitrarily small, Part (a) follows, which in turn implies Part (b). The fact that $\Psi(\widehat{x},\widehat{y}) = 0$ follows from the lower semicontinuity of Ψ. The first equality in (9.10) follows from Part (a) and the second and third from the upper semicontinuity of Φ and the definition of (x_μ,y_μ), □

In the setting of Chapter 6, the natural domain of H will usually be closed under affine transformations, that is, if $f \in \mathcal{D}(H)$, then for constants a, b, $af+b \in \mathcal{D}(H)$ and $H(af+b) = H(af)$, or more precisely, we will say that H is *closed under affine transformations* if $f \in \mathcal{D}(H)$ implies that $af \in \mathcal{D}(H)$ and $(af,g) \in H$ and $b \in R$ imply that $(af+b,g) \in H$. In the setting of Chapter 7, $\mathcal{D}(H_\dagger)$ and $\mathcal{D}(H_\ddagger)$ will usually be closed under multiplication by positive constants. (As noted, we can always assume that the domains are closed under addition of a constant.)

In order to simplify the notation, in the next lemma, for an operator H, we write $H^*f(x) = (Hf)^*(x)$ and $H_*f(x) = (Hf)_*(x)$.

LEMMA 9.3. *Suppose that $\mathcal{D}(H_\dagger)$ and $\mathcal{D}(H_\ddagger)$ are closed under addition of arbitrary constants with $H_\dagger(f+c) = H_\dagger f, H_\ddagger(f+c) = H_\ddagger f$. Let $h_1, h_2 \in C_b(E)$ and $\alpha > 0$, and let $\overline{f}$ be a viscosity subsolution of (9.1) and $\underline{f}$ a viscosity supersolution of (9.2).*

Suppose that $V : E \mapsto \overline{R}$ is lower semicontinuous, $V \geq 0$, and $V \not\equiv \infty$, and that Ψ is lower semicontinuous on $E \times E$, $\Psi \geq 0$, $\Psi(x,y) > 0$ when $x \neq y$, and

$\{x \in E : \Psi(x,x) = 0\}$ *is nonempty. Suppose that there exist* $\mu_0 > 0$, $\delta_0 > 0$, *and* $\epsilon_0 > 0$ *such that for* $\mu > \mu_0$, $0 < \delta < \delta_0$, *and* $|\lambda - 1| \leq \epsilon_0$, *and each* $x \in E$, $-\mu\Psi(x,\cdot) - \delta V \in \mathcal{D}(H_\ddagger)$ *and* $\lambda^{-1}\mu\Psi(\cdot,x) + \delta V \in \mathcal{D}(H_\dagger)$. *Let* $|\lambda_m - 1| \leq \epsilon_0$, $\mu_m > \mu_0$, $\lim_{m\to\infty}\lambda_m \to \lambda > 0$, *and* $\lim_{m\to\infty}\mu_m = \infty$.

Then for δ *sufficiently small, there exist* x_0^m, x_1^m, $\widehat{x}_0^m$ *and* $\widehat{x}_1^m$ *such that*

$$\begin{aligned}(9.12)\quad & \lambda_m\overline{f}(x_0^m) - \underline{f}(x_1^m) - \mu_m\Psi(x_0^m,x_1^m) - \lambda_m\delta V(x_0^m) - \delta V(x_1^m) \\ &\geq \sup_{x,y\in E}(\lambda_m\overline{f}(x) - \underline{f}(y) - \mu_m\Psi(x,y) - \lambda_m\delta V(x) - \delta V(y)) - \frac{1}{2m},\end{aligned}$$

$$\begin{aligned}(9.13)\quad & \lambda_m\overline{f}(\widehat{x}_0^m) - \underline{f}(x_1^m) - \mu_m\Psi(\widehat{x}_0^m,x_1^m) - \lambda_m\delta V(\widehat{x}_0^m) - \delta V(x_1^m) \\ &\geq \sup_{x,y\in E}(\lambda_m\overline{f}(x) - \underline{f}(y) - \mu_m\Psi(x,y) - \lambda_m\delta V(x) - \delta V(y)) - \frac{1}{m},\end{aligned}$$

$$\begin{aligned}(9.14)\quad & \lambda_m\overline{f}(x_0^m) - \underline{f}(\widehat{x}_1^m) - \mu_m\Psi(x_0^m,\widehat{x}_1^m) - \lambda_m\delta V(x_0^m) - \delta V(\widehat{x}_1^m) \\ &\geq \sup_{x,y\in E}(\lambda_m\overline{f}(x) - \underline{f}(y) - \mu_m\Psi(x,y) - \lambda_m\delta V(x) - \delta V(y)) - \frac{1}{m},\end{aligned}$$

$$\begin{aligned}(9.15)\quad & \sup_{\{x:\Psi(x,x)=0\}}(\lambda\overline{f}(x) - \underline{f}(x) - (1+\lambda)\delta V(x)) \\ &\leq \liminf_{m\to\infty}\alpha\Big(\lambda_m(H_\dagger)^*(\lambda_m^{-1}\mu_m\Psi(\cdot,x_1^m) + \delta V)(\widehat{x}_0^m) \\ &\qquad - (H_\ddagger)_*(-\mu_m\Psi(x_0^m,\cdot) - \delta V)(\widehat{x}_1^m)\Big) \\ &\qquad + \limsup_{m\to\infty}(\lambda h_1(\widehat{x}_0^m) - h_2(\widehat{x}_1^m)),\end{aligned}$$

$$(9.16)\quad \limsup_{m\to\infty}(H_\ddagger)_*(-\mu_m\Psi(x_0^m,\cdot) - \delta V)(\widehat{x}_1^m) \leq \alpha^{-1}(\|\underline{f}\| + \|h_2\|) < \infty,$$

$$(9.17)\quad \liminf_{m\to\infty}(H_\dagger)^*(\lambda_m^{-1}\mu_m\Psi(\cdot,x_1^m) + \delta V)(\widehat{x}_0^m) \geq -\alpha^{-1}(\|\overline{f}\| + \|h_1\|) > -\infty,$$

and

$$(9.18)\quad \lim_{m\to\infty}\mu_m(\Psi(x_0^m,x_1^m) + \Psi(\widehat{x}_0^m,x_1^m) + \Psi(x_0^m,\widehat{x}_1^m)) = 0.$$

If the level sets $\{(x,y) : \mu\Psi(x,y) + \lambda\delta V(x) + \delta V(y) \leq c\}$ *are relatively compact for some* $\mu > 0$ *(and hence for all* $\mu > 0$*), then the sequence* $\{(x_0^m, x_1^m, \widehat{x}_0^m, \widehat{x}_1^m)\}$ *is relatively compact in* $E \times E \times E \times E$, *any limit point will be of the form* (x,x,x,x) *with* $\Psi(x,x) = 0$, *and the right side of (9.15) is bounded by*

$$\begin{aligned}(9.19)\quad & \liminf_{m\to\infty}\alpha\Big(\lambda_m(H_\dagger)^*(\lambda_m^{-1}\mu_m\Psi(\cdot,x_1^m) + \delta V)(\widehat{x}_0^m) \\ &\qquad - (H_\ddagger)_*(-\mu_m\Psi(x_0^m,\cdot) - \delta V)(\widehat{x}_1^m)\Big) \\ &\qquad + \sup_{\{x:\Psi(x,x)=0\}}(\lambda h_1(x) - h_2(x)).\end{aligned}$$

Suppose that x_0^m *and* x_1^m *satisfy*

$$\begin{aligned}(9.20)\quad & \lambda_m\overline{f}(x_0^m) - \underline{f}(x_1^m) - \mu_m\Psi(x_0^m,x_1^m) - \lambda_m\delta V(x_0^m) - \delta V(x_1^m) \\ &= \sup_{x,y\in E}(\lambda_m\overline{f}(x) - \underline{f}(y) - \mu_m\Psi(x,y) - \lambda_m\delta V(x) - \delta V(y)).\end{aligned}$$

(If the level sets of $\mu_m\Psi(x,y)+\lambda_m\delta V(x)+\delta V(y)$ are relatively compact, then x_0^m, x_1^m will exist by the semicontinuity assumptions). If $\overline{f}$ is a strong subsolution and $\underline{f}$ is a strong supersolution, then the $\widehat{x}_0^m$ and $\widehat{x}_1^m$ can be chosen so that $\widehat{x}_0^m = x_0^m$ and $\widehat{x}_1^m = x_1^m$.

REMARK 9.4. The lemma will be applied in situations in which the first expression in (9.19) is bounded by a function $\omega(\lambda)$ satisfying $\lim_{\lambda\to 1}\omega(\lambda)=0$. Observe that $V(x)$ may be infinite, so this estimate only implies

$$\sup_{\{x:\Psi(x,x)=0,V(x)<\infty\}}(\overline{f}(x)-\underline{f}(x)) \leq \sup_{\{x:\Psi(x,x)=0\}}(h_1(x)-h_2(x)), \tag{9.21}$$

and an additional argument is needed to ensure $\overline{f}(x)-\underline{f}(x)\leq 0$ for all x.

PROOF. Select (x_0^m, x_1^m) satisfying (9.12). Apply Lemma 9.2 with $\Phi_{\mu_m}(x,y) = \lambda_m\overline{f}(x)-\underline{f}(y)-\lambda_m\delta V(x)-\delta V(y)$. The lemma ensures that

$$\begin{aligned}\liminf_{m\to\infty}&(\lambda_m\overline{f}(x_0^m)-\underline{f}(x_1^m)-\lambda_m\delta V(x_0^m)-\delta V(x_1^m))\\ &\geq \sup_{x\in E,\Psi(x,x)=0}(\lambda\overline{f}(x)-\underline{f}(x)-(1+\lambda)\delta V(x)),\end{aligned} \tag{9.22}$$

since $\lim_{m\to\infty}\mu_m\Psi(x_0^m,x_1^m)=0$. Since

$$f_0=\lambda_m^{-1}(\underline{f}(x_1^m)+\mu_m\Psi(\cdot,x_1^m)+\lambda_m\delta V(\cdot)+\delta V(x_1^m))\in\mathcal{D}(H_\dagger),$$

by Definition 7.1, there exist $\widehat{x}_0^m$ satisfying

$$\begin{aligned}\overline{f}(\widehat{x}_0^m)-\lambda_m^{-1}&(\underline{f}(x_1^m)+\mu_m\Psi(\widehat{x}_0^m,x_1^m)+\lambda_m\delta V(\widehat{x}_0^m)+\delta V(x_1^m))\\ &\geq \sup_x(\overline{f}(x)-f_0(x))-\frac{1}{2m},\end{aligned} \tag{9.23}$$

and hence (9.13), and

$$\overline{f}(\widehat{x}_0^m)\leq \alpha(H_\dagger)^*(\lambda_m^{-1}\mu_m\Psi(\cdot,x_1^m)+\delta V)(\widehat{x}_0^m)+h_1(\widehat{x}_0^m)+\frac{1}{m}, \tag{9.24}$$

which implies (9.17). Similarly, $\lambda_m\overline{f}(x_0^m)-\mu_m\Psi(x_0^m,\cdot)-\lambda_m\delta V(x_0^m)-\delta V(\cdot)\in\mathcal{D}(H_\ddagger)$ and there exist $\widehat{x}_1^m$ satisfying (9.14) and

$$\underline{f}(\widehat{x}_1^m)\geq\alpha(H_\ddagger)_*(-\mu_m\Psi(x_0^m,\cdot)-\delta V(\cdot))(\widehat{x}_1^m)+h_2(\widehat{x}_1^m)-\frac{1}{m}, \tag{9.25}$$

which gives (9.16). In addition, (9.24) and (9.25) give

$$\begin{aligned}\lambda_m\overline{f}(\widehat{x}_0^m)&-\underline{f}(\widehat{x}_1^m)-\lambda_m\delta V(\widehat{x}_0^m)-\delta V(\widehat{x}_1^m)\\ &\leq\lambda_m\overline{f}(\widehat{x}_0^m)-\underline{f}(\widehat{x}_1^m)\\ &\leq\lambda_m\alpha(H_\dagger)^*(\lambda_m^{-1}\mu_m\Psi(\cdot,x_1^m)+\delta V)(\widehat{x}_0^m)+\lambda_m h_1(\widehat{x}_0^m)+\frac{\lambda_m}{m}\\ &\quad-\alpha(H_\ddagger)_*(-\mu_m\Psi(x_0^m,\cdot)-\delta V(\cdot))(\widehat{x}_1^m)-h_2(\widehat{x}_1^m)+\frac{1}{m}.\end{aligned} \tag{9.26}$$

By Lemma 9.2

$$\lim_{m\to\infty}\mu_m\Psi(\widehat{x}_0^m,x_1^m)=\lim_{m\to\infty}\mu_m\Psi(x_0^m,x_1^m)=0, \tag{9.27}$$

giving (9.18).

By (9.12) and (9.13),

$$\begin{aligned}|\lambda_m\overline{f}(x_0^m) - \lambda_m\delta V(x_0^m) - \lambda_m\overline{f}(\widehat{x}_0^m) + \lambda_m\delta V(\widehat{x}_0^m) \\ +\mu_m\Psi(\widehat{x}_0^m, x_1^m) - \mu_m\Psi(x_0^m, x_1^m)| \;\leq\; \frac{3}{2m},\end{aligned}$$

so

$$\overline{f}(\widehat{x}_0^m) - \delta V(\widehat{x}_0^m) - \overline{f}(x_0^m) + \delta V(x_0^m) \to 0,$$

and similarly,

$$\underline{f}(\widehat{x}_1^m) + \delta V(\widehat{x}_1^m) - \underline{f}(x_1^m) - \delta V(x_1^m) \to 0.$$

Consequently, (9.22) gives (9.15).

Under the assumption that the level sets of $\mu\Psi(x,y) + \lambda\delta V(x) + \delta V(y)$ are compact, the relative compactness of the the sequence $\{(x_0^m, x_1^m, \widehat{x}_0^m, \widehat{x}_1^m)\}$ follows from the definitions of the components and the boundedness of $\overline{f}$ and $\underline{f}$. Since, by (9.18) and the lower semicontinuity of Ψ, any limit point $(x_0, x_1, \widehat{x}_0, \widehat{x}_1)$ must satisfy

$$\Psi(x_0, x_1) = \Psi(\widehat{x}_0, x_1) = \Psi(x_0, \widehat{x}_1) = 0,$$

it follows that all four components are the same. This observation then implies (9.19) bounds (9.15).

If $\overline{f}$ is a strong subsolution and (9.20) holds, then by definition, there exists $\lim_{k\to\infty} \widehat{x}_0^{m,k} = x_0^m$ such that (9.23) and (9.24) hold with the $\widehat{x}_0^m$ replaced by $\widehat{x}_0^{m,k}$ and the $1/m$ replaced by $1/k$. Similar inequalities hold in the supersolution case as well. Letting $k \to \infty$, by the semicontinuity assumptions, we see that x_0^m and x_1^m satisfy the requirements for $\widehat{x}_0^m$ and $\widehat{x}_1^m$ respectively. □

The following condition is slightly weaker than assuming that all sub- and supersolutions are strong, but it still allows us to take $\widehat{x}_0^m = x_0^m$ and $\widehat{x}_1^m = x_1^m$ in applications of Lemma 9.3.

CONDITION 9.5. *There exist $\mu_0, \delta_0, \epsilon_0 > 0$ such that for each $x_0, y_0 \in E$ satisfying $V(x_0) + V(y_0) < \infty$, $\mu > \mu_0$, $0 < \delta < \delta_0$, and $|\lambda - \lambda_0| \leq \epsilon_0$, the following hold:*

(1) *There exists a lower semicontinuous g_0 such that $\mu\Psi(\cdot, y_0) + \delta V(\cdot) + g_0 \in \mathcal{D}(H_\dagger)$, $g_0(x_0) = 0$, and $g_0(x) > 0$ for $x \neq x_0$, and*

$$H_\dagger^*\left(\lambda^{-1}\mu\Psi(\cdot, y_0) + \delta V(\cdot) + g_0\right)(x_0) \leq H_\dagger^*\left(\lambda^{-1}\mu\Psi(\cdot, y_0) + \delta V(\cdot)\right)(x_0).$$

(2) *There exists a lower semicontinuous g_1 such that $-\mu\Psi(x_0, \cdot) - \delta V(\cdot) - g_1 \in \mathcal{D}(H_\ddagger)$, $g_1(y_0) = 0$ and $g_1(y) > 0$ for $y \neq y_0$, and*

$$(H_\ddagger)_*(-\mu\Psi(x_0, \cdot) - \delta V(\cdot) - g_1)(y_0) \geq (H_\ddagger)_*(-\mu\Psi(x_0, \cdot) - \delta V(\cdot))(y_0).$$

LEMMA 9.6. *Let Condition 9.5 hold in Lemma 9.3. Suppose that x_0^m and x_1^m satisfy (9.20). Then $\widehat{x}_0^m$ and $\widehat{x}_1^m$ can be chosen so that $\widehat{x}_0^m = x_0^m$ and $\widehat{x}_1^m = x_1^m$.*

PROOF. Let (9.20) be satisfied. Then there exists g_0 satisfying Condition 9.5 with x_0 replaced by x_0^m such that for $x \neq x_0^m$,

$$\begin{aligned}&\lambda_m\overline{f}(x_0^m) - (\underline{f}(x_1^m) + \mu_m\Psi(\cdot, x_1^m) + \lambda_m\delta V(\cdot) + g_0 + \delta V(x_1^m))(x_0^m) \\ &\quad > \left(\lambda_m\overline{f}(x) - (\underline{f}(x_1^m) + \mu_m\Psi(\cdot, x_1^m) + \lambda_m\delta V(\cdot) + g_0 + \delta V(x_1^m)\right)(x).\end{aligned}$$

Since x_0^m is the unique maximum,

$$\begin{aligned}\overline{f}(x_0^m) &\le \alpha H_\dagger^*\left(\lambda_m^{-1}\mu_m\Psi(\cdot,x_1^m)+\delta V+g_0\right)(x_0^m)+h_1(x_0^m)\\ &\le \alpha H_\dagger^*\left(\lambda_m^{-1}\mu_m\Psi(\cdot,x_1^m)+\delta V\right)(x_0^m)+h_1(x_0^m),\end{aligned}$$

that is, $\widehat{x}_0^m = x_0^m$. Similarly, one can select $\widehat{x}_1^m = x_1^m$. □

A special case of Lemma 9.3 is of particular interest.

LEMMA 9.7. *In addition to the assumptions of Lemma 9.3, let $K = \{x : \Psi(x,x) = 0\}$ be compact and assume that for each open $U \supset K$,*

$$\inf_{(x,y)\in(U\times U)^c} \Psi(x,y) > 0.$$

Then, for x_0^m, x_1^m, $\widehat{x}_0^m$, and $\widehat{x}_1^m$ as in Lemma 9.3,

$$\begin{aligned}(9.28)\quad &\sup_{x\in K}(\lambda\overline{f}(x)-\underline{f}(x)-(1+\lambda)\delta V(x))\\ &\le \sup_{x\in K}(\lambda h_1(x)-h_2(x))\\ &\quad+\liminf_{m\to\infty}\alpha(\lambda_m(H_\dagger)^*(\lambda_m^{-1}\mu_m\Psi(\cdot,x_1^m)+\delta V(\cdot))(\widehat{x}_0^m)\\ &\qquad -(H_\ddagger)_*(-\mu_m\Psi(x_0^m,\cdot)-\delta V(\cdot))(\widehat{x}_1^m)).\end{aligned}$$

If one of the four sequences, $\{x_0^m\}$, $\{\widehat{x}_0^m\}$, $\{x_1^m\}$ or $\{\widehat{x}_1^m\}$, converges, all four converge to the same point.

PROOF. Since $\lim_{m\to\infty}\mu_m\Psi(x_0^m,x_1^m)=0$, by the assumptions on Ψ, $\{x_0^m\}$, $\{x_1^m\}$ are both relatively compact. Similarly, $\{\widehat{x}_0^m\}$ and $\{\widehat{x}_1^m\}$ are relatively compact. Therefore (9.28) follows from (9.15) and the continuity of h_1 and h_2. □

As (9.15) and (9.28) suggest, we can verify the comparison principle by estimating

$$\lambda H_\dagger(\lambda^{-1}f_1)(x) - H_\ddagger f_2(y)$$

for some $\lambda^{-1}f_1 \in \mathcal{D}(H_\dagger), f_2 \in \mathcal{D}(H_\ddagger)$, where the x,y are close to each other. In the remainder of this section, we discuss concrete techniques for estimating this difference for a variety of operators $H_\dagger$ and $H_\ddagger$ with special structures.

The first technique we discuss deals with a simple but general situation where $H_\dagger = H_\ddagger = H$ and H admits a variational representation (9.29) of the form considered in Chapter 8. Note that in the following lemma, the role of λ above is played by $(1+\epsilon)$.

LEMMA 9.8. *Let (E,r) and (U,q) be metric spaces, $\Gamma \subset E\times U$ and $\Gamma_x = \{u : (x,u)\in\Gamma\}$. Suppose $H \subset C_b(E)\times B(E)$ satisfies*

$$(9.29)\qquad Hf(x) = \sup_{u\in\Gamma_x}\{Af(x,u)-L(x,u)\},$$

where A is linear and L is a function defined on Γ. Let $f_1, f_2 \in \mathcal{D}(H)$, $x,y\in E$, $C>0$, and $\lambda>1$ be given. Suppose that for each $u\in\Gamma_x$, there exists $v\in\Gamma_y$ such that

$$\begin{aligned}(9.30)\qquad Af_1(x,u)-Af_2(y,v) &\le \frac{\epsilon}{2}(C+L(x,u)),\\ L(y,v)-L(x,u) &\le \frac{\epsilon}{2}(C+L(x,u)).\end{aligned}$$

Then

$$(1+\epsilon)H(\frac{f_1}{1+\epsilon})(x) - Hf_2(y) \le C\epsilon.$$

PROOF. There exists $u_n \in \Gamma_x$ such that

$$(1+\epsilon)H(\frac{f_1}{1+\epsilon})(x) \le Af_1(x,u_n) - (1+\epsilon)L(x,u_n) + \frac{1}{n}.$$

For each n, choose $v_n \in \Gamma_y$ satisfying (9.30) for $u = u_n$. Then

$$\begin{aligned}
&(1+\epsilon)H(\frac{f_1}{1+\epsilon})(x) - Hf_2(y)\\
&\le \frac{1}{n} + Af_1(x,u_n) - Af_2(y,v_n) + L(y,v_n) - (1+\epsilon)L(x,u_n)\\
&\le \frac{1}{n} + \frac{\epsilon}{2}(C + L(x,u_n)) + \frac{\epsilon}{2}(C + L(x,u_n)) - \epsilon L(x,u_n) = \frac{1}{n} + C\epsilon.
\end{aligned}$$

Sending $n \to \infty$, the result follows. □

9.2. General conditions in R^d

Application of Lemma 9.7 is simpler if every viscosity subsolution (supersolution) is a strong subsolution (supersolution). If the natural state space is R^d, we can frequently replace R^d by the one-point compactification $E = R^d \cup \{\infty\}$. For many examples,

$$\mathcal{D}(H) \subset \{f \in C(E) : \nabla f \text{ is continuous}, \nabla f(\infty) \equiv \lim_{|x|\to\infty} \nabla f(x) = 0\},$$

and the operator H can be written as

$$Hf(x) = H(x, \nabla f(x)), \tag{9.31}$$

for some function $H(x,p)$ (or similarly for $H_\dagger$ and $H_\ddagger$).

For the function $H(x,p)$, H^* denotes the upper semicontinuous regularization and H_* denotes the lower semicontinuous regularization, that is,

$$H^*(x,p) = \lim_{\epsilon\to 0} \sup_{(y,q)\in B_\epsilon(x,p)} H(y,q),$$

with sup replaced by inf in the definition of H_*. For the operator H,

$$H^*f(x) = \lim_{\epsilon\to 0} \sup_{y\in B_\epsilon(x)} H(y, \nabla f(y)).$$

In general, $H^*f(x)$ need not equal $H^*(x, \nabla f(x))$ (we always have

$$H^*f(x) \le H^*(x, \nabla f(x)));$$

however, if H satisfies

$$\lim_{\delta\to 0} \sup_{q\in B_\delta(p)} \sup_{x\in\Gamma} |H(x,q) - H(x,p)| = 0,$$

for each compact $\Gamma \subset R^d$, then for $x \in R^d$

$$H^*(x,p) = \lim_{\epsilon\to 0} \sup_{y\in B_\epsilon(x)} H(y,p).$$

Consequently, we will assume that

$$(H_\dagger f)^*(x) = H_\dagger^*(x, \nabla f(x)), \quad f \in \mathcal{D}(H_\dagger) \tag{9.32}$$

and, similarly, that

$$(H_\ddagger f)_*(x) = (H_\ddagger)_*(x, \nabla f(x)), \quad f \in \mathcal{D}(H_\ddagger). \tag{9.33}$$

For operators having this kind of local form, one typically has equivalence of viscosity and strong viscosity solutions.

LEMMA 9.9. *Let $E = R^d \cup \{\infty\}$. Let $H(x,p)$ be measurable,*

$$\mathcal{D}(H_\dagger) \subset \{f \in C(E) : \nabla f \text{ is continuous}, \nabla f(\infty) \equiv \lim_{|x|\to\infty} \nabla f(x) = 0\},$$

and $H_\dagger f(x) = H^*(x, \nabla f(x))$. Assume that $\mathcal{D}(H_\dagger)$ is closed under addition and that for each $x_0 \in E$, there exists $f \in \mathcal{D}(H_\dagger)$ such that $f(x_0) = 0$ and $f(x) > 0$ for $x \neq x_0$. Let $h \in C(E)$. If $\overline{f}$ is a viscosity subsolution of

$$(I - \alpha H_\dagger) f = h,$$

then $\overline{f}$ is a strong subsolution.

Similarly, let

$$\mathcal{D}(H_\ddagger) \subset \{f \in C(E) : \nabla f \text{ is continuous}, \nabla f(\infty) \equiv \lim_{|x|\to\infty} \nabla f(x) = 0\}$$

and $H_\ddagger f(x) = H_*(x, \nabla f(x))$. Assume that $\mathcal{D}(H_\ddagger)$ is closed under addition and that for each $x_0 \in E$, there exists $f \in \mathcal{D}(H_\ddagger)$ such that $f(x_0) = 0$ and $f(x) < 0$ for $x \neq x_0$. Let $h \in C(E)$. If $\underline{f}$ is a viscosity supersolution of

$$(I - \alpha H_\ddagger) f = h,$$

then $\underline{f}$ is a strong supersolution.

PROOF. Since E is compact, it is sufficient to show that if $f_0 \in \mathcal{D}(H_\dagger)$ and

$$\overline{f}(x_0) - f_0(x_0) = \sup_x (\overline{f}(x) - f_0(x)),$$

then

$$\alpha^{-1}(\overline{f}(x_0) - h(x_0)) \leq H^*(x_0, \nabla f_0(x_0)). \tag{9.34}$$

Let $f_1 \in \mathcal{D}(H_\dagger)$, $f_1(x_0) = 0$, and $f_1(x) > 0$ for $x \neq x_0$. Then $f_0 + f_1 \in \mathcal{D}(H_\dagger)$ and x_0 is the unique point in E such that

$$\overline{f}(x_0) - f_0(x_0) - f_1(x_0) = \sup_x (\overline{f}(x) - f_0(x) - f_1(x)).$$

Consequently,

$$\alpha^{-1}(\overline{f}(x_0) - h(x_0)) \leq H_\dagger^*(x_0, \nabla f_0(x_0) + \nabla f_1(x_0)) = H_\dagger^*(x_0, \nabla f_0(x_0)). \tag{9.35}$$

The proof of the supersolution result is similar. □

We will prove the comparison principle under the following conditions on $H_\dagger$ and $H_\ddagger$.

CONDITION 9.10. *$H_\dagger$ satisfies (9.32) and $H_\ddagger$ satisfies (9.33) and the following hold:*

(1) *For each compact* $\Gamma \subset R^d$*, there exist* $\mu_m \to \infty$ *and* $\omega : (0,\infty) \to [0,\infty]$ *with*

$$\lim_{\epsilon \to 0} \inf_{|\lambda - 1| \le \epsilon} \omega(\lambda) = 0 \tag{9.36}$$

such that for each $\lambda > 0$ *and each sequence* $\{(x_m, y_m)\} \subset \Gamma \times \Gamma$ *satisfying* $\mu_m |x_m - y_m|^2 \to 0$,

$$\sup_m (H_\ddagger)_*(y_m, \mu_m(x_m - y_m)) < \infty, \tag{9.37}$$

and

$$\inf_m H_\dagger^*(x_m, \frac{\mu_m(x_m - y_m)}{\lambda}) > -\infty, \tag{9.38}$$

we have

$$\liminf_{m\to\infty} [\lambda H_\dagger^*(x_m, \frac{\mu_m(x_m - y_m)}{\lambda}) - (H_\ddagger)_*(y_m, \mu_m(x_m - y_m))] \le \omega(\lambda). \tag{9.39}$$

(2) *If* $x_m \to \infty$ *and* $|x_m||p_m| \to 0$*, then*

$$\limsup_{m\to\infty} H_\dagger(x_m, p_m) \le 0 \text{ and } \liminf_{m\to\infty} H_\ddagger(x_m, p_m) \ge 0.$$

Occasionally, we will use a stronger version of Condition 9.10.1.

CONDITION 9.11. *If* $\lambda > 1$, $\mu_m \to \infty$, $\sup_m(|x_m| + |y_m|) < \infty$, *and* $\mu_m |x_m - y_m|^2 \to 0$*, then*

$$\liminf_{m\to\infty} [\lambda H_\dagger^*(x_m, \frac{\mu_m(x_m - y_m)}{\lambda}) - (H_\ddagger)_*(y_m, \mu_m(x_m - y_m))] \le 0. \tag{9.40}$$

The following lemmas are immediate.

LEMMA 9.12. *If Condition 9.11 holds, then Condition 9.10.1 holds with*

$$\omega(\lambda) = \begin{cases} 0 & \lambda > 1 \\ \infty & \lambda \le 1. \end{cases}$$

LEMMA 9.13. *Suppose* $(H_\dagger^{(1)}, H_\ddagger^{(1)}), \ldots, (H_\dagger^{(k)}, H_\ddagger^{(k)})$ *satisfies Condition 9.11 and* $a_1, \ldots, a_k \ge 0$*. Then* $(H_\dagger, H_\ddagger) = (a_1 H_\dagger^{(1)} + \cdots + a_k H_\ddagger^{(k)}, a_1 H_\ddagger^{(1)} + \cdots + a_k H_\ddagger^{(k)})$ *satisfies Condition 9.11.*

REMARK 9.14. Although Condition 9.10 has no obvious intuitive interpretation, we will see that it covers many examples. In particular, it gives the classical results in the Freidlin-Wentzell theory (Freidlin and Wentzell [**52**]) and many generalization. Freidlin and Wentzell placed regularity conditions on the Fenchel-Legendre transform L of H $(= H_\dagger = H_\ddagger)$ (see Remark 10.16), whereas Condition 9.10 places regularity conditions on H (or $H_\dagger$ and $H_\ddagger$). For a given problem, regularity of H is typically easier to check. A number of sufficient conditions will be given in lemmas below.

LEMMA 9.15. *Let* $E = R^d \cup \{\infty\}$ *and* $\mathcal{D}(H_\dagger) = \mathcal{D}(H_\ddagger) = D_d$ *given by*

$$D_d = \{f \in C(E) : f|_{R^d} - f(\infty) \in C_c^2(R^d)\}. \tag{9.41}$$

Define $\nabla f(\infty) = 0$*, and assume that* $H_\dagger$ *and* $H_\ddagger$ *have the form*

$$H_\dagger f(x) = H_\dagger(x, \nabla f(x)), \quad H_\ddagger f(x) = H_\ddagger(x, \nabla f(x)), \tag{9.42}$$

where $H_\dagger(x,p)$ and $H_\ddagger(x,p)$ satisfy Condition 9.10. Then for $h_1 = h_2 = h \in C(E)$ and $\alpha > 0$, the comparison principle holds for subsolutions of (9.1) and supersolutions of (9.2).

PROOF. By Lemma 7.6, under Condition 9.10.2, we can extend the domains to contain the subspace of functions that are continuously differentiable on R^d and satisfy

$$\lim_{|x|\to\infty} |x||\nabla f(x)| = 0. \tag{9.43}$$

By Lemma 9.9, every subsolution of (9.1) is a strong subsolution and every supersolution of (9.2) is a strong supersolution, so we can apply Lemma 9.7 with $\widehat{x}_i^m = x_i^m$. Let $\mu_m = m$, $\lambda = 1$,

$$\Psi(x,y) = \frac{1}{1+|x|^3} + \frac{1}{1+|y|^3}$$

and $K = \{\infty\}$. Then by (9.28),

$$\overline{f}(\infty) - \underline{f}(\infty) \le \liminf_{m\to\infty} \alpha(H_\dagger^*(x_0^m, -m\frac{3|x_0^m|x_0^m}{(1+|x_0^m|^3)^2}) - H_{\ddagger*}(x_1^m, m\frac{3|x_1^m|x_1^m}{(1+|x_1^m|^3)^2})),$$

and since $\lim_{m\to\infty} m\Psi(x_0^m, x_1^m) = 0$, the right side is less than or equal to zero by Condition 9.10.2.

Suppose $\delta \equiv \sup_x(\overline{f}(x) - \underline{f}(x)) > 0$. By upper semicontinuity, there exists $\kappa > 0$ and $\epsilon > 0$ such that

$$\sup_{|\lambda-1|<\epsilon} \sup_{|x|>\kappa} (\lambda\overline{f}(x) - \underline{f}(x)) \le \frac{\delta}{3}$$

and

$$\inf_{|\lambda-1|<\epsilon} \sup_x (\lambda\overline{f}(x) - \underline{f}(x)) \ge \frac{2\delta}{3}.$$

Let $\Gamma = \{x : |x| \le 2\kappa\}$, and let μ_m and ω be as in Condition 9.10.1.

Let $\varphi_i : [0,\infty) \to [0,\infty)$, $i = 1,2$, be nondecreasing and continuously differentiable, $\varphi_1(r) = \frac{1}{2}r$, $0 \le r \le 1$, $\varphi_1(r) = 1$, $r > 2$, $\varphi_2(r) = 0$, $0 \le r \le 2\kappa^2$, $\varphi_2(r) > 0$, $r > 2\kappa^2$, and $\varphi_2(r) = 1$, $r > 4\kappa^2$. Now apply Lemma 9.7 with

$$\Psi(x,y) = \begin{cases} \varphi_1(|x-y|^2) + \varphi_2(|x|^2) + \varphi_2(|y|^2) & |x|, |y| < \infty \\ 2 + \varphi_2(|x|^2) & |x| < \infty, y = \infty \\ 2 + \varphi_2(|y|^2) & |y| < \infty, x = \infty \\ 2 & x = y = \infty \end{cases}$$

and $K = \{x : |x| \le 2\kappa^2\}$. Let $\lambda_m = \lambda$ and x_0^m and x_1^m satisfy

$$\lambda\overline{f}(x_0^m) - \underline{f}(x_1^m) - \mu_m\Psi(x_0^m, x_1^m) = \sup_{x,y\in E} (\lambda\overline{f}(x) - \underline{f}(y) - \mu_m\Psi(x,y)). \tag{9.44}$$

For m sufficiently large, we must have $|x_0^m|, |x_1^m| < \kappa$ and $|x_0^m - x_1^m| < 1$, so the right side of (9.28) becomes

$$\liminf_{m\to\infty} \Big(\alpha(\lambda H_\dagger^*(x_0^m, \frac{\mu_m(x_0^m - x_1^m)}{\lambda}) - H_{\ddagger*}(x_1^m, -\mu_m(x_1^m - x_0^m))) \\ +\lambda h(x_0^m) - h(x_1^m)\Big),$$

which is less than or equal to

$$\omega(\lambda) + \sup_{|x|\le 2\kappa^2} (\lambda - 1)h(x),$$

by Condition 9.10.1. Then

$$\lim_{\epsilon\to 0}\inf_{|\lambda-1|\le\epsilon}(\omega(\lambda)+\sup_{|x|\le 2\kappa^2}(\lambda-1)h(x))\le 0,$$

contradicting the assumption that $\sup_x(\overline{f}(x)-\underline{f}(x))>0$. □

The following lemmas give sufficient conditions for Condition 9.10 to hold. The first treats H ($H_\dagger$ and $H_\ddagger$) that are nondegenerate in a sense that will be clear from specific examples.

LEMMA 9.16. *Suppose $H_\dagger$ is upper semicontinuous and $H_\ddagger$ is lower semicontinuous on $R^d\times R^d$ and $H_\dagger(x,p)\le H_\ddagger(x,p)$, $x,p\in R^d$. Suppose that for each compact $\Gamma\subset R^d$,*

$$\lim_{|p|\to\infty}\inf_{x\in\Gamma}H_\ddagger(x,p)=\infty. \tag{9.45}$$

Then Condition 9.10.1 holds with

$$\omega(\lambda)=\begin{cases}0 & \lambda=1\\ \infty & \lambda\ne 1.\end{cases}$$

If $H_\ddagger$ is convex in p and for each $x,p\in R^d$,

$$\lim_{r\to\infty}H_\ddagger(x,rp)=\infty, \tag{9.46}$$

then (9.45) holds.

PROOF. Suppose that $\lambda=1$, $\mu_m\to\infty$, $x_m,y_m\in\Gamma$, $\mu_m|x_m-y_m|^2\to 0$, and

$$\sup_m H_{\ddagger *}(y_m,\mu_m(x_m-y_m))<\infty.$$

By (9.45), $\sup_m\mu_m|x_m-y_m|<\infty$. Select a subsequence along which $(x_m,y_m,\mu_m(x_m-y_m))$ converges to some (x_0,x_0,p_0). Then by the semicontinuity assumptions

$$\begin{aligned}&\liminf_{m\to\infty}[H_\dagger^*(x_m,\mu_m(x_m-y_m))-(H_\ddagger)_*(y_m,\mu_m(x_m-y_m))]\\&\le H_\dagger^*(x_0,p_0)-(H_\ddagger)_*(x_0,p_0)]\le 0.\end{aligned}$$

and (9.39) holds for $\omega(1)=0$.

To see that (9.46) implies (9.45), it is sufficient to consider sequences $p_n\to\infty$ and $x_n\to x\in\Gamma$. Without loss of generality, we can assume that $\theta_n=p_n/|p_n|\to\theta$. By lower semicontinuity,

$$\liminf_{n\to\infty}(H_\ddagger(x_n,r\theta_n)-H_\ddagger(x_n,0))\ge H_\ddagger(x,r\theta)-H_\ddagger^*(x,0),$$

and, since for n sufficiently large, $|p_n|>r$, by convexity,

$$\begin{aligned}\liminf_{n\to\infty}\frac{H_\ddagger(x_n,p_n)-H_\ddagger(x_n,0)}{|p_n|}&\ge\lim_{n\to\infty}\frac{H_\ddagger(x_n,r\theta_n)-H_\ddagger(x_n,0)}{r}\\&\ge\frac{H_\ddagger(x,r\theta)-H_\ddagger^*(x,0)}{r}.\end{aligned}$$

For r sufficiently large, the right side is positive, so $H_\ddagger(x_n,p_n)\to\infty$. □

The next lemmas are applicable under Lipschitz assumptions on the generator.

LEMMA 9.17. *Suppose there exists a continuous function $K : R \mapsto [0,\infty)$ such that*

$$H_\dagger(x,p) - H_\ddagger(y,p) \le K(H_\ddagger(y,p))(1+|p|)|x-y|. \tag{9.47}$$

Then Condition 9.10.1 holds with

$$\omega(\lambda) = \begin{cases} 0 & \lambda = 1 \\ \infty & \lambda \ne 1. \end{cases}$$

PROOF. For μ_m, x_m, and y_m as in Condition 9.10.1, set $p_m = \mu_m(x_m - y_m)$. Replacing x, y, and p in (9.47) by x_m, y_m, and p_m, the right side goes to zero and the condition follows. □

The conditions of Lemma 9.17 are easy to state but do not appear to cover a variety of examples for which some kind of Lipschitz continuity holds. The next set of conditions is more generally applicable. For simplicity, we state the results for $H_\dagger = H_\ddagger = H$.

CONDITION 9.18. *H is continuous and*

$$H(x,p) = \int_V \beta(x,v)\phi(c(x,v)\cdot p; v)\gamma(dv),$$

where

(1) *γ is a σ-finite measure on a measurable space $(V, \mathcal{V})$.*
(2) *$\phi : R^m \times V \mapsto R$ is measurable and $\phi(q;v)$ is convex in q for each fixed v.*
(3) *$c : R^d \times V \mapsto M^{m\times d}$ (the $m \times d$ matrices) is measurable.*
(4) *There exists $\Phi : R_+ \times V \mapsto R_+$ such that for each $v \in V$, $\Phi(0;v) = 0$ and $\Phi(\cdot; v)$ is nondecreasing and*

$$\phi(q;v) \le \Phi(|q|;v), \quad (q,v) \in R_+ \times V.$$

(5) *For each $r > 0$, there exists a $\delta_r > 0$ such that for $\delta < \delta_r$,*

$$\kappa_r(\delta) \equiv \sup\{\int_V \beta(x,v)\Phi(\delta\frac{|c(x,v)-c(y,v)|}{|x-y|}; v)\gamma(dv) : |x|, |y| < r, x \ne y\} < \infty.$$

LEMMA 9.19. *Suppose that Condition 9.18 holds with $\beta \equiv 1$. Then Condition 9.11 holds.*

PROOF. Let $\lambda > 1$. By convexity,

$$\begin{aligned} &\phi(\frac{1}{\lambda}c(x,v)\cdot p; v) \\ &\quad = \phi(\frac{\lambda-1}{\lambda}\frac{(c(x,v)-c(y,v))\cdot p}{\lambda-1} + \frac{1}{\lambda}c(y,v)\cdot p; v) \\ &\quad \le (1-\frac{1}{\lambda})\phi(\frac{(c(x,v)-c(y,v))\cdot p}{\lambda-1}; v) + \frac{1}{\lambda}\phi(c(y,v)\cdot p; v). \end{aligned}$$

Hence,

$$\begin{aligned} &\lambda\phi(\frac{1}{\lambda}c(x,v)\cdot p; v) - \phi(c(y,v)\cdot p; v) \\ &\quad \le (\lambda-1)\phi(\frac{(c(x,v)-c(y,v))\cdot p}{\lambda-1}; v) \\ &\quad \le (\lambda-1)\Phi(\frac{1}{\lambda-1}\frac{|c(x,v)-c(y,v)|}{|x-y|}|p||x-y|; v). \end{aligned}$$

Integrating by γ and taking appropriate limits in x and y to obtain H^* and H_*, this inequality implies

$$\lambda H^*(x, \frac{1}{\lambda}p) - H_*(y,p) \le (\lambda - 1)\kappa_r((\lambda-1)^{-1}|p||x-y|),$$

for x, y, p satisfying $|x|, |y| < r$ and $(\lambda-1)^{-1}|p||x-y| < \delta_r$.

Note that by the dominated convergence theorem, $\lim_{\delta\to 0}\kappa_r(\delta) = 0$.

Let $0 < r < \infty$, $|x_m|, |y_m| < r$, $\mu_m \to \infty$ and $\mu_m|x_m - y_m|^2 \to 0$. For $\lambda > 1$, there exists m_0 such that $m > m_0$ implies

$$0 < \frac{1}{\lambda - 1}\mu_m|x_m - y_m|^2 < \delta_r.$$

Consequently,

$$\begin{aligned}\liminf_{m\to\infty} \lambda H^*(x_m, \frac{\mu_m(x_m - y_m)}{\lambda}) &- H_*(y_m, \mu_m(x_m - y_m)) \\ &\le \lim_{m\to\infty}(\lambda-1)\kappa_r((\lambda-1)^{-1}\mu_m|x_m - y_m|^2) \\ &= 0.\end{aligned}$$

□

LEMMA 9.20. *Suppose that H_0 satisfies Condition 9.18 with $\phi \ge 0$ and β such that*

$$\beta(x,v) - \beta(y,v) \le K(x,y)\beta(y,v),$$

where K is continuous and $K(x,x) = 0$, for all $x \in R^d$. Then H_0 satisfies Condition 9.10.1 with

$$\omega(\lambda) = \begin{cases} 0 & \lambda > 1 \\ \infty & \lambda \le 1. \end{cases}$$

Let $b : R^d \to R^d$ be Lipschitz and define

$$H(x,p) = H_0(x,p) + b(x)\cdot p.$$

If K satisfies

$$K(x,y) \le \overline{K}|x-y| \tag{9.48}$$

for some $\overline{K} < \infty$, then H satisfies Condition 9.10.1 with the same ω.

PROOF. As in the previous lemma, by convexity,

$$\begin{aligned}&\phi(\frac{1}{\lambda}c(x,v)\cdot p; v) \\ &\quad = \phi(\frac{\lambda-1}{\lambda}\frac{(c(x,v)-c(y,v))\cdot p}{\lambda-1} + \frac{1}{\lambda}c(y,v)\cdot p; v) \\ &\quad \le (1-\frac{1}{\lambda})\phi(\frac{(c(x,v)-c(y,v))\cdot p}{\lambda-1}; v) + \frac{1}{\lambda}\phi(c(y,v)\cdot p; v).\end{aligned}$$

Hence,

$$\begin{aligned}
&\lambda\beta(x,v)\phi(\frac{1}{\lambda}c(x,v)\cdot p;v)-\beta(y,v)\phi(c(y,v)\cdot p;v)\\
&\le(\lambda-1)\beta(x,v)\phi(\frac{(c(x,v)-c(y,v))\cdot p}{\lambda-1};v)\\
&\qquad+(\beta(x,v)-\beta(y,v))\phi(c(y,v)\cdot p;v)\\
&\le(\lambda-1)\beta(x,v)\Phi(\frac{1}{\lambda-1}\frac{|(c(x,v)-c(y,v))|}{|x-y|}|p||x-y|;v)\\
&\qquad+K(x,y)\beta(y,v))\phi(c(y,v)\cdot p;v).
\end{aligned}$$

As before, we have

$$\lambda(H_0)^*(x,\frac{1}{\lambda}p)-(H_0)_*(y,p)\le(\lambda-1)\kappa_r((\lambda-1)^{-1}|p||x-y|)+K(x,y)(H_0)_*(y,p),$$

which implies

$$\begin{aligned}
\lambda H^*(x,\frac{1}{\lambda}p)-H_*(y,p)&\le(\lambda-1)\kappa_r((\lambda-1)^{-1}|p||x-y|)+K(x,y)H_*(y,p)\\
&\qquad+(b(x)-b(y)-K(x,y)b(y))\cdot p,
\end{aligned}\tag{9.49}$$

Let $0<r<\infty$, $|x_m|,|y_m|<r$, $\mu_m\to\infty$, $\mu_m|x_m-y_m|^2\to 0$ and

$$\sup_m H_*(y_m,\mu_m(x_m-y_m))<\infty.$$

If $b=0$ or if b is Lipschitz and K satisfies (9.48), the last two terms on the right of (9.49) with x and y replaced by x_m and y_m will go to zero. For $\lambda>1$, there exists m_0 such that $m>m_0$ implies

$$0<\frac{1}{\lambda-1}\mu_m|x_m-y_m|^2<\delta_r.$$

Consequently,

$$\begin{aligned}
&\liminf_{m\to\infty}\lambda H^*(x_m,\frac{\mu_m(x_m-y_m)}{\lambda})-H_*(y_m,\mu_m(x_m-y_m))\\
&\le(\lambda-1)\liminf_{m\to\infty}\Big[(\lambda-1)\kappa_r((\lambda-1)^{-1}\mu_m|x_m-y_m|^2)\\
&\qquad+K(x_m,y_m)H_*(y_m,\mu_m(x_m-y_m)\\
&\qquad+(b(x_m)-b(y_m)-K(x_m,y_m)b(y_m))\cdot\mu_m(x_m-y_m)\Big]\\
&=0,
\end{aligned}$$

and the lemma follows. □

9.3. Bounded smooth domains in R^d with (possibly oblique) reflection

Let $\varphi\in C^2(R^d)$ and suppose that $E\equiv\{x:\varphi(x)\ge 0\}$ is compact, E is the closure of $E^o=\{x:\varphi(x)>0\}$, and for $x\in\partial E=\{x:\varphi(x)=0\}$, $\nabla\varphi(x)\neq 0$. We assume that $\nu:R^d\mapsto R^d$, $|\nu(x)|=1$, and

$$\kappa_0=\inf_{x\in\partial E}\nu(x)\cdot\nabla\varphi(x)>0.\tag{9.50}$$

With reference to (9.1) and (9.2), and the reflecting diffusion example in Section 10.5, we let $\widetilde{H},H_b:R^d\times R^d\mapsto R$ be measurable and $H_b(x,p)\ge 0$. Define

$H_\dagger f(x) = H_\dagger(x, \nabla f(x))$ and for $f \in C^1(E)$, $H_\ddagger f(x) = H_\ddagger(x, \nabla f(x))$, where

$$H_\dagger(x,p) = \begin{cases} \widetilde{H}(x,p) + H_b(x,p)(0 \vee \nu(x)\cdot p) & x \in \partial E \\ \widetilde{H}(x,p) & x \in E^o \end{cases}$$

and

$$H_\ddagger(x,p) = \begin{cases} \widetilde{H}(x,p) + H_b(x,p)(0 \wedge \nu(x)\cdot p) & x \in \partial E \\ \widetilde{H}(x,p) & x \in E^o \,. \end{cases}$$

Let $n(x) = \nabla\varphi(x)/|\nabla\varphi(x)|$. Then for $x \in \partial E$, $n(x)$ is the unit interior normal at x, and n is C^1 in a neighborhood of ∂E. The smoothness of the boundary also implies that there exists a constant $C_1 > 0$ such that

$$n(x)\cdot(x-y) < C_1|x-y|^2, \forall x \in \partial E, y \in E.$$

Our analysis is an adaptation of Lions and Sznitman [**81**] and Lions [**80**]. In particular, the following lemma is essentially Lemma 4.1 of [**81**].

LEMMA 9.21. *Let φ and E be as above, and let ν be the restriction to ∂E of a C^1 function (which we will also denote by ν). Then there exists a $d\times d$ symmetric-matrix-valued function $m(x) = ((m_{ij}(x)\ ))$ such that m is positive definite and C^1 and*

$$m(x)\nu(x) = n(x), \quad x \in \partial E.$$

PROOF. Let $\rho : [0,\infty) \to [0,\infty)$ be smooth, $\rho(0) = 1$ and $\rho(r) = 0$, $r \geq 1$. Define the matrix $p(x)$ by

$$p(x) = I - \rho(\varphi(x)/\delta)(\nu(x)\cdot n(x))^{-1}\nu(x)n^T(x),$$

for $\delta > 0$ small enough so that ν and n are differentiable and $\nu(x)\cdot n(x)$ is bounded away from zero on $\{x : 0 \leq \varphi(x) \leq \delta\}$. Note that for $x \in \partial E$, $p(x)\nu(x) = 0$. Then

$$m(x) = p^T(x)p(x) + \rho(\varphi(x)/\delta)(\nu(x)\cdot n(x))^{-1}n(x)n^T(x)$$

satisfies the requirements of the lemma. □

Let $\overline{f}$ be a subsolution of (9.1) and $\underline{f}$ a supersolution of (9.2) with $h_1 = h_2 = h \in C(E)$. Define

$$\overline{f}_\epsilon = \overline{f} + \epsilon\varphi, \quad \underline{f}_\epsilon = \underline{f} - \epsilon\varphi,$$

and let

$$H_{\dagger,\epsilon}(x,p) = H_\dagger(x, p - \epsilon\nabla\varphi(x)), \quad H_{\ddagger,\epsilon}(x,p) = H_\ddagger(x, p + \epsilon\nabla\varphi(x)),$$

and

$$H_{\dagger,\epsilon}f(x) \equiv H_{\dagger,\epsilon}(x, \nabla f(x)), \quad H_{\ddagger,\epsilon}f(x) \equiv H_{\ddagger,\epsilon}(x, \nabla f(x)).$$

Then the following lemma is immediate from the definitions.

LEMMA 9.22. *$\overline{f}_\epsilon$ is a subsolution of*

$$(I - \alpha H_{\dagger,\epsilon})f = h + \epsilon\varphi, \tag{9.51}$$

and $\underline{f}_\epsilon$ is a supersolution of

$$(I - \alpha H_{\ddagger,\epsilon})f = h - \epsilon\varphi. \tag{9.52}$$

We will compare $\overline{f}_\epsilon$ and $\underline{f}_\epsilon$ by applying (9.28) of Lemma 9.7. Instead of $|x-y|^2$, we take

$$\Psi(x,y) = (x-y)^T \frac{m(x)+m(y)}{2}(x-y). \tag{9.53}$$

Define

$$\nabla_1\Psi(x,y) \equiv \nabla_x\Psi(x,y) \qquad \nabla_2\Psi(x,y) \equiv \nabla_y\Psi(x,y).$$

Note that in the following lemma, the various sequences depend on ϵ, but to simplify notation, we do not specifically indicate this fact.

LEMMA 9.23. *Suppose $h \in C(E)$, $\alpha > 0$, and $\epsilon > 0$. Then for $\lambda_m \to \lambda > 0$ and $0 < \mu_m \to \infty$, there exist $x_0^m, x_1^m \in E$ such that*

$$\begin{aligned}\sup_{x\in E}(\lambda\overline{f}_\epsilon(x) - \underline{f}_\epsilon(x)) &\le |\lambda-1|\|h\| + \epsilon(1+\lambda)\|\varphi\| \\ &+\alpha \liminf_{m\to\infty}\{\lambda_m \widetilde{H}^*\Big(x_0^m, \frac{p_m}{\lambda_m} - \epsilon\nabla\varphi(x_0^m)\Big) \\ &\qquad -\widetilde{H}_*\Big(x_1^m, p_m + q_m + \epsilon\nabla\varphi(x_1^m)\Big)\},\end{aligned} \tag{9.54}$$

where

$$p_m = \mu_m \nabla_1\Psi(x_0^m, x_1^m)$$

and

$$\begin{aligned} q_m &= -\mu_m(\nabla_2\Psi(x_0^m,x_1^m) + \nabla_1\Psi(x_0^m,x_1^m)) \\ &= -\mu_m\Big((x_0^m - x_1^m)^T \frac{\partial_1 m(x_0^m) + \partial_1 m(x_1^m)}{2}(x_0^m - x_1^m), \\ &\qquad \ldots, (x_0^m - x_1^m)^T\frac{\partial_d m(x_0^m) + \partial_d m(x_1^m)}{2}(x_0^m - x_1^m)\Big).\end{aligned}$$

In addition, $\lim_{m\to\infty}\mu_m\Psi(x_0^m,x_1^m) = 0$, implying $\mu_m|x_0^m - x_1^m|^2 \to 0$ and $|q_m| \to 0$.

PROOF. As in Lemma 9.9, any viscosity subsolution is a strong subsolution and any supersolution is a strong supersolution. Applying Lemma 9.7, there exist $x_0^m, x_1^m \in E$ such that

$$\begin{aligned}\sup_{x\in E}(\lambda\overline{f}_\epsilon(x) - \underline{f}_\epsilon(x)) &\le |\lambda-1|\|h\| + \epsilon(1+\lambda)\|\varphi\| \\ &+\alpha \liminf_{m\to\infty}\{\lambda_m H^*_{\dagger,\epsilon}\Big(x_0^m, \frac{\mu_m}{\lambda_m}\nabla_1\Psi(x_0^m,x_1^m)\Big) \\ &\qquad -(H_{\ddagger,\epsilon})_*\Big(x_1^m, -\mu_m\nabla_2\Psi(x_0^m,x_1^m)\Big)\}.\end{aligned} \tag{9.55}$$

Noting that

$$\begin{aligned}\nabla_x\Psi(x,y) &= m(x)(x-y) + m(y)(x-y) \\ &\quad +\frac{1}{2}\Big((x-y)^T\partial_1 m(x)(x-y), \ldots, (x-y)^T\partial_d m(x)(x-y)\Big)\end{aligned}$$

and setting $\nu(x) = (\nu_1(x), \dots, \nu_d(x))^T$, there is a constant C_2 such that

$$\begin{aligned}
&\nu(x) \cdot \nabla_x \Psi(x,y) \\
&= \nu(x)^T (m(x) + m(y)) (x-y) + \frac{1}{2} \sum_{k=1}^{d} \nu_k(x)(x-y)^T \partial_k m(x)(x-y) \\
&\le 2n(x) \cdot (x-y) + |m(y) - m(x)||x-y| + \frac{1}{2} \sum_{k=1}^{d} |\partial_k m(x)||x-y|^2 \\
&\le C_2 |x-y|^2.
\end{aligned}$$

By Lemma 9.2, $\lim_{m\to\infty} \mu_m \Psi(x_0^m, x_1^m) = 0$, so $\mu_m |x_0^m - x_1^m|^2 \to 0$. Therefore, when $x_0^m \in \partial E$, for m sufficiently large, we have

$$\nu(x_0^m) \cdot (\frac{\mu_m}{\lambda_m} \nabla_1 \Psi(x_0^m, x_1^m) - \epsilon \nabla \varphi(x_0^m)) \le \frac{C_2}{\lambda_m} \mu_m |x_0^m - x_1^m|^2 - \epsilon \kappa_0 < 0.$$

Consequently

$$H^*_{\dagger,\epsilon}(x_0^m, \frac{\mu_m}{\lambda_m} \nabla_1 \Psi(x_0^m, x_1^m)) \le \widetilde{H}^*(x_0^m, \frac{\mu_m}{\lambda_m} \nabla_1 \Psi(x_0^m, x_1^m) - \epsilon \nabla \varphi(x_0^m)).$$

Similarly

$$(H_{\ddagger,\epsilon})_*(x_1^m, -\mu_m \nabla_2 \Psi(x_0^m, x_1^m)) \ge \widetilde{H}_*(x_1^m, -\mu_m \nabla_2 \Psi(x_0^m, x_1^m) + \epsilon \nabla \varphi(x_1^m)).$$

The conclusion follows from these inequalities and (9.55). □

The following lemma extends the comparison principle for nondegenerate equations, Lemma 9.16, to equations with reflecting boundary conditions.

LEMMA 9.24. *Let $h_1 = h_2 = h \in C(E)$ and $\alpha > 0$. Suppose that $\widetilde{H}$ is continuous and satisfies the nondegeneracy condition (9.45). Then the comparison principle holds for subsolutions of (9.1) and supersolutions of (9.2).*

PROOF. Let $\lambda_m \equiv 1$ in (9.54), and let $\{x_0^m\}$ and $\{x_1^m\}$ be as in Lemma 9.23. Then (9.16) gives

$$\begin{aligned}
&\limsup_{m\to\infty} \widetilde{H}(x_1^m, p_m + q_m + \epsilon \nabla \varphi(x_1^m)) \\
&\qquad \le \limsup_{m\to\infty} (H_{\ddagger,\epsilon})_*(x_1^m, p_m + q_m + \epsilon \nabla \varphi(x_1^m)) \\
&\qquad < \alpha^{-1}(\|\underline{f}_\epsilon\| + \|h\|),
\end{aligned}$$

implying $\sup_m |p_m + q_m| < \infty$ with a bound uniform in $0 < \epsilon < 1$. Since $|q_m| \to 0$, $|x_0^m - x_1^m| \to 0$, and $\widetilde{H}$ is continuous, taking a subsequence if necessary, we have $\lim_{m\to\infty} x_0^m = \lim_{m\to\infty} x_1^m = x^\epsilon$ and p^ϵ satisfying

$$\sup_{x\in E} (\overline{f}_\epsilon(x) - \underline{f}_\epsilon(x)) \le 2\epsilon \|\varphi\| + \alpha(\Big(H(x^\epsilon, p^\epsilon - \epsilon \nabla \varphi(x^\epsilon)) - \widetilde{H}(x^\epsilon, p^\epsilon + \epsilon \nabla \varphi(x^\epsilon))\Big).$$

Noting that p^ϵ is uniformly bounded in ϵ and letting $\epsilon \to 0$, we have the result. □

In specific cases, the nondegeneracy assumption in the above lemma can be replaced by Lipschitz continuity in the coefficients. As an example, we consider

$$\widetilde{H}(x,p) = \frac{1}{2} |\sigma(x)^T p|^2 + b(x)p,$$

where $\sigma(x)$ is a $d \times d$ matrix and $b(x) \in R^d$.

LEMMA 9.25. *Suppose $h_1 = h_2 = h \in C(E)$ and $\alpha > 0$. Let σ, b be Lipschitz continuous. Then the comparison principle holds for subsolutions of (9.1) and supersolutions of (9.2).*

PROOF. Let $\lambda > 1$, $x, y \in E$, $p, q, r, s \in R^d$ and $|r|, |s| \le \delta$. Then

$$
\begin{aligned}
(9.56)\qquad & \lambda \widetilde{H}(x, \frac{p}{\lambda} - r) - \widetilde{H}(y, p+q+s) \\
&\le \frac{1}{2}\lambda^{-1}|(\sigma(x) - \sigma(y) + \sigma(y))^T(p - \lambda r)|^2 - \frac{1}{2}|\sigma^T(y)(p+q+s)|^2 \\
&\quad + |b(x) - b(y)||p| + \sup_{z\in E}|b(z)|(|q| + (1+\lambda)\delta) \\
&\le (2\lambda)^{-1}\Big(L_\sigma^2|(x-y)p|^2 + \lambda^2 L_\sigma^2|x-y|^2\delta^2 \\
&\qquad + 2\lambda L_\sigma^2|(x-y)p||x-y|\delta + |\sigma^T(y)p|^2 \\
&\qquad + \lambda^2 \sup_z |\sigma(z)|^2\delta^2 \\
&\qquad + 2\lambda|\sigma^T(y)p|\sup_z|\sigma^T(z)|\delta + 2L_\sigma|(x-y)p||\sigma^T(y)p|\Big) \\
&\quad - \frac{1}{2}|\sigma^T(y)p|^2 + |\sigma^T(y)p||\sigma^T(y)(q+s)| \\
&\quad + L_b|x-y||p| + \sup_{z\in E}|b(z)|(|q| + (1+\lambda)\delta),
\end{aligned}
$$

where L_σ, L_b are the Lipschitz constants for σ and b. We next use the above inequality to estimate the right side of (9.54) with $\lambda_m = \lambda$. Note that for each $\epsilon > 0$, $|x_0^m - x_1^m| \to 0$, $q_m \to 0$ and $|x_0^m - x_1^m||p_m| \to 0$, so we have

$$
\begin{aligned}
&\sup_{x\in E}(\lambda\overline{f}(x) - \underline{f}(x)) - |\lambda - 1|\|h\| \\
&\le \alpha^{-1}\limsup_{\epsilon\to 0+}\limsup_{m\to\infty}\Big(\frac{1}{2\lambda}|\sigma^T(x_1^m)p_m|^2 \\
&\qquad + |\sigma^T(x_1^m)p_m|\sup_z|\sigma^T(z)|\epsilon\sup_z|\nabla\varphi(z)| \\
&\qquad + \frac{L_\sigma}{\lambda}|(x_0^m - x_1^m)p_m||\sigma^T(x_1^m)p_m| \\
&\qquad - \frac{1}{2}|\sigma^T(x_1^m)p_m|^2 + |\sigma^T(x_1^m)p_m||\sigma^T(x_1^m)(q_m + \epsilon\nabla\varphi(x_1^m))|\Big) \\
&= \alpha^{-1}\limsup_{\epsilon\to 0+}\limsup_{m\to\infty}\Big(-\frac{1}{2}\frac{\lambda-1}{\lambda}|\sigma^T(x_1^m)p_m|^2 \\
&\qquad + |\sigma^T(x_1^m)p_m|\{\sup_z|\sigma(z)|\epsilon\sup_z|\nabla\varphi(z)| \\
&\qquad + \frac{L_\sigma}{\lambda}|(x_0^m - x_1^m)p_m| + \sigma^T(x_1^m)(q_m + \epsilon\varphi(x_1^m))\}\Big) \\
&\le \alpha^{-1}\frac{\lambda}{2(\lambda-1)}\limsup_{\epsilon\to 0+}\limsup_{m\to\infty}\Big(\sup_z|\sigma(z)|\epsilon\sup_z|\nabla\varphi(z)| \\
&\qquad + \frac{L_\sigma}{\lambda}|(x_0^m - x_1^m)p_m| + \sigma^T(x_1^m)(q_m + \epsilon\varphi(x_1^m))\Big)^2 = 0.
\end{aligned}
$$

The last inequality follows from the fact that

$$
-au^2 + 2bu \le \frac{b^2}{a}, \quad a > 0.
$$

Sending $\lambda \to 1$, we obtain the result. □

9.4. Conditions for infinite dimensional state space

In Examples 1.12, 1.13 and 1.14, the limit operators are first order differential operators in infinite dimensions. Next, we develop estimates useful for applying Lemma 9.3 to these situations.

In the examples, E is either a subspace of some function space, or a space of probability measures. Therefore, we use ρ, γ to denote a typical element in E and let x denote an element of the set on which the functions are defined. More motivation for the examples considered here can be found in Feng and Katsoulakis [**43**]. A *distance function* on E is a metric defined on $E \times E$ that is not necessarily equivalent to the original metric r. "Convergence" always refers to convergence in the original metric r unless otherwise noted.

CONDITION 9.26.

(1) *There exists a nonnegative, lower semicontinuous function $\mathcal{E} \in M(E)$, $\mathcal{E} \not\equiv \infty$, that has compact level sets, a lower semicontinuous distance function $d(\rho, \gamma)$, and $a_0, b_0 > 0$ such that for $\rho_0, \gamma_0 \in E$ satisfying $\mathcal{E}(\rho_0) + \mathcal{E}(\gamma_0) < \infty$, $a > a_0$, and $0 < b < b_0$,*

$$ad^2(\cdot, \gamma_0) + b\mathcal{E}(\cdot) \in \mathcal{D}(H_\dagger) \tag{9.57}$$

and

$$-ad^2(\rho_0, \cdot) - b\mathcal{E}(\cdot) \in \mathcal{D}(H_\ddagger). \tag{9.58}$$

(2) *For $0 < \kappa < b_0$, there exist $\omega_\kappa : [0, \infty) \mapsto [0, \infty]$ satisfying*

$$\liminf_{\kappa \to 0+} \liminf_{r \to 0+} \omega_\kappa(r) = 0,$$

and for $m(1-\kappa) > a_0$ and $\rho, \gamma \in E$,

$$\begin{aligned} \Big\{\frac{1}{1-\kappa}\Big(H_\dagger(m(1-\kappa)d^2(\cdot,\gamma) + \kappa\mathcal{E})\Big)^*(\rho) \\ -\frac{1}{1+\kappa}\Big(H_\ddagger(-m(1+\kappa)d^2(\rho,\cdot) - \kappa\mathcal{E}(\cdot))\Big)_*(\gamma)\Big\} \\ \leq \omega_\kappa(md^2(\rho,\gamma)). \end{aligned} \tag{9.59}$$

CONDITION 9.27.

(1) *$\overline{f}$ and $\underline{f}$ are d-continuous in the sense that $\lim_{n\to\infty} d(\rho_n, \rho_0) = 0$ implies*

$$\lim_{n\to\infty} \overline{f}(\rho_n) = \overline{f}(\rho_0) \text{ and } \lim_{n\to\infty} \underline{f}(\rho_n) = \underline{f}(\rho_0).$$

(2) *For each $\rho \in E$, there exist $\rho_n \in E$ with $\mathcal{E}(\rho_n) < \infty$ such that*

$$\lim_{n\to\infty} d(\rho_n, \rho) = 0,$$

that is, $\{\rho : \mathcal{E}(\rho) < \infty\}$ is dense in E in the topology determined by d.

THEOREM 9.28. *Assume that Conditions 9.26 and 9.5 hold with $V = \mathcal{E}$. For $\alpha > 0$ and $h_1, h_2 \in C_b(E)$, let $\overline{f}$ be a subsolution of (9.1) and $\underline{f}$ be a supersolution of (9.2). Then*

$$\sup_{\rho\in\{\rho:\mathcal{E}(\rho)<\infty\}} (\overline{f}(\rho) - \underline{f}(\rho)) \leq \sup_{\rho\in E}(h_1(\rho) - h_2(\rho)). \tag{9.60}$$

If, in addition, Condition 9.27 holds, then

$$\sup_{\rho\in E}(\overline{f}(\rho) - \underline{f}(\rho)) \le \sup_{\rho\in E}(h_1(\rho) - h_2(\rho)). \tag{9.61}$$

PROOF. The theorem follows from Lemma 9.3 and Lemma 9.6, taking $\mu_m = m(1+\kappa)$, $\lambda_m = (1+\kappa)/(1-\kappa)$, $\Psi = d^2$, $\delta = \kappa$, and $V = \mathcal{E}$. The compactness of the level sets of $\mathcal{E}$ allows us to assume x_0^m and x_1^m satisfy (9.20), and Lemma 9.6 allows us to take $\widehat{x}_0^m = x_0^m$ and $\widehat{x}_0^m = x_1^m$ in (9.19). By (9.18), (9.59) implies (9.60). (See Remark 9.4.) Condition 9.27 then gives (9.61). □

We apply Theorem 9.28 to the operators arising in Examples 1.12, 1.13 and 1.14. In these examples, it is useful to formally view the state space E as a manifold. Then H can be written in the form

$$\begin{aligned} Hf(\rho) &= \Big[-\operatorname{grad}\mathcal{E}(\rho), \operatorname{grad}f(\rho)\Big]_\rho + \frac{1}{2}\Big[\operatorname{grad}f(\rho), \operatorname{grad}f(\rho)\Big]_\rho, \\ &= \frac{1}{2}(\|\operatorname{grad}f(\rho) - \operatorname{grad}\mathcal{E}(\rho)\|_\rho^2 - \|\operatorname{grad}\mathcal{E}(\rho)\|_\rho^2) \end{aligned} \tag{9.62}$$

where $[\cdot,\cdot]_\rho$ is an inner product determined by the local differential structure of E at $\rho \in E$, $\|h\|_\rho^2 = [h,h]_\rho$, $\mathcal{E} : E \mapsto [0,\infty]$ is a functional, which in certain physical contexts is called a *free energy*, grad is a natural notion of gradient that respects the constraints satisfied by the state variables (e.g. $\int_O \rho(x)dx = 0$ in Example 1.13, and $\rho > 0$, $\int_O \rho(dx) = 1$ in Example 1.14).

For consistency of notation in these examples, we only use $H_\dagger, H_\ddagger$ to denote limit operators arising in the application of Theorem 7.17. The comparison principles we prove in the rest of this section, however, are not for these $H_\dagger, H_\ddagger$. $\widehat{H}_\dagger$ and $\widehat{H}_\ddagger$ are modifications of $H_\dagger$ and $H_\ddagger$ for which the comparison principle is easier to prove. Chapter 13 will discuss the large deviation principle for these examples, and we will show that the comparison principle for $\widehat{H}_\dagger$ and $\widehat{H}_\ddagger$ implies the comparison principle for the corresponding $H_\dagger$ and $H_\ddagger$.

EXAMPLE 9.29. (Reaction-diffusion equation, Example 1.12.)

Let $E = H_0(O) = L^2(O)$, where $O = [0,1)^d$ with periodic boundary. For $\beta = 1, 2, \ldots$, let $H_\beta(O)$ be the completion of $C^\infty(O)$ under the norm $\|\rho\|_\beta^2 = \langle (I-\Delta)^\beta \rho, \rho\rangle$ and $H_{-\beta}(O)$ be the completion of $H_0(O)$ under the norm $\|\rho\|_{-\beta}^2 = \langle (I-\Delta)^{-\beta}\rho, \rho\rangle$. The inner product $\langle \rho, \gamma\rangle$ extends consistently to $\rho \in H_{-\beta}(O)$ and $\gamma \in H_\beta(O)$, and

$$|\langle \rho, \gamma\rangle| \le \|\rho\|_{-\beta}\|\gamma\|_\beta.$$

For each $\rho \in E$, we take

$$\Big[u, v\Big]_\rho = \langle u, v\rangle, \quad u \in H_{-\beta}(O), v \in H_\beta(O).$$

The intuition is that at each point $\rho \in E$, we take $L^2(O)$ as the tangent space and allow the possibility that the gradient only exists in "smooth directions." Let $\|\cdot\| = \|\cdot\|_0 = \|\cdot\|_{L^2(O)}$.

DEFINITION 9.30 (Gradient). Let $f : E \mapsto \overline{R}$ and $\rho \in E$. Then, if it exists, $\text{grad}f(\rho)$ is the unique element in $\cup_\beta H_{-\beta}(\mathcal{O})$ such that for each $p \in C^\infty(\mathcal{O})$

$$\lim_{t\to 0}\frac{1}{t}\Big(f(\rho+tp)-f(\rho)\Big) = \langle \text{grad}f(\rho_0), p\rangle.$$

With this definition of grad, H given by (1.32) is of the form

$$Hf(\rho) = \langle \Delta\rho - F'(\rho), \text{grad}f(\rho)\rangle + \frac{1}{2}\|\text{grad}f(\rho)\|^2. \tag{9.63}$$

Define

$$\mathcal{E}(\rho) = \int_{\mathcal{O}} \{\frac{1}{2}|\nabla\rho(x)|^2 + F(\rho(x))\}dx. \tag{9.64}$$

Then

$$\text{grad}\mathcal{E}(\rho) = -\Delta\rho + F'(\rho), \quad \forall \rho, \mathcal{E}(\rho) < \infty,$$

and we see that (9.63) has the form (9.62).

Let $\{e_1, e_2, \ldots\} \subset C^\infty(\mathcal{O})$ be the orthonormal basis of eigenfunctions for $-\Delta$ in $L^2(\mathcal{O})$ (see Appendix C). Define

$$G(\rho,\gamma) = \sum_{k=1}^\infty 2^{-k}\langle \rho-\gamma, e_k\rangle^2. \tag{9.65}$$

Then fixing γ_0,

$$\text{grad}G(\rho,\gamma_0) = \sum_{k=1}^\infty 2^{-k+1}\langle \rho-\gamma_0, e_k\rangle e_k,$$

and $\|\text{grad}G(\rho,\gamma_0)\| < \infty$.

Let $d(\rho,\gamma) = \|\rho-\gamma\|$, $a, c \geq 0$, $0 < b < 1$, and $C \in R$. We extend the definition of H in (9.63), according to (9.62), to test functions of the form

$$f_0(\rho) = ad^2(\rho,\gamma_0) + b\mathcal{E}(\rho) + cG(\rho,\rho_0) + C \tag{9.66}$$

for subsolutions, and to test functions of the form

$$f_1(\gamma) = -ad^2(\rho_0,\gamma) - b\mathcal{E}(\gamma) - cG(\gamma,\gamma_0) + C,$$

for supersolutions.

We define

$$\widehat{H}_\dagger f_0(\rho) = \begin{cases} 2a(1-b)\langle\Delta\rho - F'(\rho), \rho-\gamma_0\rangle + 2a^2\|\rho-\gamma_0\|^2 \\ \quad +(\frac{1}{2}b^2-b)\|\Delta\rho - F'(\rho)\|^2 \\ \quad +c(1-b)\langle \text{grad}G(\rho,\rho_0), \Delta\rho - F'(\rho)\rangle \\ \quad +2^{-1}c^2\|\text{grad}G(\rho,\rho_0)\|^2 \\ \quad +2ac\langle\text{grad}G(\rho,\rho_0), \rho-\gamma_0\rangle, & \Delta\rho - F'(\rho) \in L^2(\mathcal{O}) \\ -\infty & \text{otherwise.} \end{cases} \tag{9.67}$$

and

$$\widehat{H}_\ddagger f_1(\gamma) = \begin{cases} -2a(1+b)\langle\Delta\gamma - F'(\gamma), \gamma-\rho_0\rangle + 2a^2\|\gamma-\rho_0\|^2 \\ \quad +(\frac{1}{2}b^2+b)\|\Delta\gamma - F'(\gamma)\|^2 \\ \quad -c(1+b)\langle \text{grad}G(\gamma,\gamma_0), \Delta\gamma - F'(\gamma)\rangle \\ \quad +2^{-1}c^2\|\text{grad}G(\gamma,\gamma_0)\|^2 \\ \quad +2ac\langle\text{grad}G(\gamma,\gamma_0), \gamma-\rho_0\rangle, & \Delta\gamma - F'(\gamma) \in L^2(\mathcal{O}) \\ \infty & \text{otherwise.} \end{cases} \tag{9.68}$$

The "free energy" $\mathcal{E}$ introduces an important higher order term $\|\Delta\rho - F'(\rho)\|^2$ that ensures $\widehat{H}_\dagger f_0$ is upper semicontinuous and $\widehat{H}_\ddagger f_1$ is lower semicontinuous.

THEOREM 9.31. *Let $h_1, h_2 \in C_b(E)$ and $\alpha > 0$. Suppose $\overline{f}$ is a subsolution of $(I - \alpha\widehat{H}_\dagger)f = h_1$ and $\underline{f}$ is a supersolution of $(I - \alpha\widehat{H}_\ddagger)f = h_2$. Then*

$$\sup_{\rho\in\{\rho:\mathcal{E}(\rho)<\infty\}} (\overline{f}(\rho) - \underline{f}(\rho)) \leq \sup_{\rho\in E}(h_1(\rho) - h_2(\rho)). \tag{9.69}$$

If, in addition, $\overline{f}$ and $\underline{f}$ are continuous, then

$$\sup_{\rho\in E}(\overline{f}(\rho) - \underline{f}(\rho)) \leq \sup_{\rho\in E}(h_1(\rho) - h_2(\rho)). \tag{9.70}$$

PROOF. We apply Theorem 9.28. G provides the family of functions needed to verify Condition 9.5.

If either $\Delta\gamma - F'(\gamma) \notin L^2(\mathcal{O})$ or $\Delta\rho - F'(\rho) \notin L^2(\mathcal{O})$, then (9.59) follows trivially. Otherwise, for $0 < \kappa < 1$ and $m > 0$, we have

$$\begin{aligned}
&\frac{1}{1-\kappa}\Big(\widehat{H}_\dagger(m(1-\kappa)d^2(\cdot,\gamma) + \kappa\mathcal{E})\Big)^*(\rho) \\
&\quad -\frac{1}{1+\kappa}\Big(\widehat{H}_\ddagger(-m(1+\kappa)d^2(\rho,\cdot) - \kappa\mathcal{E}(\cdot))\Big)_*(\gamma) \\
&= 2m\langle(\Delta\rho - F'(\rho)) - (\Delta\gamma - F'(\gamma)), \rho - \gamma\rangle \\
&\quad -2\kappa m\langle(\Delta\rho - F'(\rho)) + (\Delta\gamma - F'(\gamma)), \rho - \gamma\rangle - 4\kappa m^2\|\gamma - \rho\|^2 \\
&\quad -\frac{\kappa(2-\kappa)}{2(1-\kappa)}\|\Delta\rho - F'(\rho)\|^2 - \frac{\kappa(2+\kappa)}{2(1+\kappa)}\|\Delta\gamma - F'(\gamma)\|^2 \\
&\leq -2m\langle F'(\rho)) - F'(\gamma), \rho - \gamma\rangle \\
&\quad -\frac{3\kappa}{8}\|\Delta\rho - F'(\rho)\|^2 - \frac{3\kappa}{8}\|\Delta\gamma - F'(\gamma)\|^2 \\
&\leq 2mL_F\|\rho - \gamma\|^2,
\end{aligned} \tag{9.71}$$

where the first inequality follows from the inequalities $\langle\Delta(\rho - \gamma), \rho - \gamma\rangle \leq 0$ and

$$-\kappa m\langle q, \rho - \gamma\rangle - 2\kappa m^2\|\rho - \gamma\|^2 \leq \frac{\kappa}{8}\|q\|^2$$

(expand $\|\sqrt{2}m(\rho-\gamma) + 2^{-3/2}q\|^2$), and the second inequality from $\|F'(\rho) - F'(\gamma)\| \leq L_F\|\rho - \gamma\|$. Condition 9.26 follows.

Finally, since $\{\rho : \mathcal{E}(\rho) < \infty\}$ is dense in E, if $\overline{f}$ and $\underline{f}$ are continuous, (9.69) implies (9.70). □

EXAMPLE 9.32. (Discrete Cahn-Hilliard equation, Example 1.13.)

In this example, $\sum_{x\in\Lambda_m} Y_n(t, x)$ is constant in time. Without loss of generality, we can assume that this conserved quantity is zero. The limiting state space then becomes

$$E = H^0(\mathcal{O}) = \{\rho \in L^2(\mathcal{O}) : \int_{\mathcal{O}} \rho(x)dx = 0\}.$$

For $\beta = 1, 2, \ldots$, define $H^\beta(\mathcal{O})$ to be the completion of $H^0(\mathcal{O}) \cap C^\infty(\mathcal{O})$ under the norm $\|\rho\|_\beta = (-1)^\beta\langle\Delta^\beta\rho, \rho\rangle_0$. Note that

$$\langle\rho, \gamma\rangle_0 = \langle\rho, \gamma\rangle = \int_{\mathcal{O}} \rho(x)\gamma(x)dx,$$

but we use the subscript to emphasize the differences among the inner products.

The inverse Laplacian is well-defined on E, and we define $H^{-\beta}(\mathcal{O})$ to be the completion of E under the norm given by

$$\|\rho\|_{-\beta}^2 = (-1)^\beta \langle \rho, \Delta^{-\beta}\rho\rangle_0,$$

that is,

$$\langle \rho, \gamma\rangle_{-\beta} = (-1)^\beta \langle \Delta^{-\beta}\rho, \gamma\rangle_0 = (-1)^\beta \langle \rho, \Delta^{-\beta}\gamma\rangle_0, \tag{9.72}$$

if $\rho, \gamma \in H^0(\mathcal{O})$. The norm on $H^{-1}(\mathcal{O})$ can also be represented as

$$\|u\|_{-1}^2 = \sup_{p\in C^\infty(\mathcal{O})} \{2\int_{\mathcal{O}} up - \int_{\mathcal{O}} |\nabla p|^2 dx\}.$$

The additional fact that we will need regarding these spaces is that for $\rho, \gamma \in H^0(\mathcal{O})$,

$$|\langle \rho, \gamma\rangle_0| \le \|\rho\|_{-\beta}\|\gamma\|_\beta,$$

where the right side is infinite if $\gamma \notin H^\beta(\mathcal{O})$.

We view E as a manifold for which the tangent space at $\rho \in E$ is given by $H^{-1}(\mathcal{O})$ with

$$\Big[u, v\Big]_\rho = \langle u, v\rangle_{-1}.$$

We will use the following definition of gradient:

DEFINITION 9.33 (Gradient). Let $f : E \mapsto \overline{R}$ and $\rho \in E$. Then, if it exists, grad$f(\rho)$ is the unique element in $\cup_\beta H^{-\beta}(\mathcal{O})$ such that for each $p \in C^\infty(\mathcal{O})$

$$\lim_{t\to 0} \frac{1}{t}\Big(f(\rho - t\Delta p) - f(\rho)\Big) = \langle \text{grad} f(\rho_0), p\rangle.$$

Then for $f(\rho) = \|\rho - \gamma\|^2$, grad$f(\rho) = -2\Delta(\rho - \gamma)$; for $f(\rho) = d^2(\rho, \gamma) \equiv \|\rho - \gamma\|_{-1}^2$, grad$f(\rho) = 2(\rho - \gamma)$ and $\langle \text{grad} d^2(\rho, \gamma), \text{grad} d^2(\rho, \gamma)\rangle_{-1} = 4\|\rho - \gamma\|_{-1}^2$; for $\mathcal{E}$ given by (9.64) and ρ satisfying $\mathcal{E}(\rho) < \infty$,

$$\text{grad}\mathcal{E}(\rho) = -\nabla\cdot(\nabla(-\Delta\rho + F'(\rho))) = \Delta^2\rho - \Delta F'(\rho), \tag{9.73}$$

and $\langle \text{grad}\mathcal{E}(\rho), \text{grad}\mathcal{E}(\rho)\rangle_{-1} = \|\Delta(\Delta\rho - F'(\rho))\|_{-1}^2$; and for f of the form (1.30),

$$\text{grad} f(\rho) = -\nabla\cdot(\nabla\frac{\delta f}{\delta\rho}) = -\Delta\frac{\delta f}{\delta\rho}.$$

Consequently, H in (1.41) can be written in the form (9.62):

$$\begin{aligned} Hf(\rho) &= \langle -\text{grad}\mathcal{E}(\rho), \text{grad} f(\rho)\rangle_{-1} + \frac{1}{2}\|\text{grad} f(\rho)\|_{-1}^2 \\ &= \langle \Delta\rho - F'(\rho), \text{grad} f(\rho)\rangle + \frac{1}{2}\|\text{grad} f(\rho)\|_{-1}^2, \end{aligned} \tag{9.74}$$

where the second expression follows by (9.72).

We next introduce $\widehat{H}_\dagger$ and $\widehat{H}_\ddagger$ which can be viewed as extending H to a new class of test functions.

Taking $d(\rho, \gamma) = \|\rho - \gamma\|_{-1}$ in Condition 9.26, and $m > 0, 0 < \kappa < 1$. Let

$$f_0(\rho) = m(1-\kappa)\|\rho - \gamma\|_{-1}^2 + \kappa\mathcal{E}(\rho) \tag{9.75}$$

and

$$f_1(\gamma) = -m(1+\kappa)\|\rho - \gamma\|_{-1}^2 - \kappa\mathcal{E}(\gamma). \tag{9.76}$$

If ρ, γ satisfy $\mathcal{E}(\rho) < \infty, \mathcal{E}(\gamma) < \infty$,

$$\begin{aligned} \text{grad} f_0(\rho) &= 2(1-\kappa)m(\rho-\gamma) + \kappa\Delta(\Delta\rho - F'(\rho)) \\ \text{grad} f_1(\gamma) &= -2(1+\kappa)m(\gamma-\rho) - \kappa\Delta(\Delta\gamma - F'(\gamma)). \end{aligned}$$

Therefore, we define

$$\begin{aligned} (9.77)\ \widehat{H}_\dagger f_0(\rho) &= -2m(1-\kappa)^2\langle \Delta(\Delta\rho - F'(\rho)), \rho-\gamma\rangle_{-1} \\ &\quad +2(1-\kappa)^2 m^2\|\rho-\gamma\|_{-1}^2 - (\kappa - \frac{\kappa^2}{2})\|\Delta(\Delta\rho - F'(\rho))\|_{-1}^2 \end{aligned}$$

if $\|\Delta(\Delta\rho - F'(\rho))\|_{-1}^2 < \infty$, and define $\widehat{H}_\dagger f_0(\rho) = -\infty$ otherwise. We also define

$$\begin{aligned} (9.78)\ \widehat{H}_\ddagger f_1(\gamma) &= -2m(1+\kappa)^2\langle \Delta(\Delta\gamma - F'(\gamma)), \rho-\gamma\rangle_{-1} \\ &\quad +2(1+\kappa)^2 m^2\|\rho-\gamma\|_{-1}^2 + (\kappa + \frac{\kappa^2}{2})\|\Delta(\Delta\gamma - F'(\gamma))\|_{-1}^2 \end{aligned}$$

if $\|-\Delta(-\Delta\gamma + F'(\gamma))\|_{-1}^2 < \infty$, and define $H_\ddagger f_1(\gamma) = \infty$ otherwise. Because of the higher order perturbation term, for any $0 < \kappa \le 1$, $\widehat{H}_\dagger f_0$ is upper semicontinuous and $\widehat{H}_\ddagger f_1$ is lower semicontinuous in E.

As in the previous example, we can introduce the G in (9.65) into the test functions f_0, f_1, and modify the definition of $\widehat{H}_\dagger f_0$ and $\widehat{H}_\ddagger f_1$ correspondingly.

THEOREM 9.34. *Let $h_1, h_2 \in C_b(E)$ and $\alpha > 0$. Suppose $\overline{f}$ is a subsolution of $(I - \alpha\widehat{H}_\dagger)f = h$ and $\underline{f}$ is a supersolution of $(I - \alpha\widehat{H}_\ddagger)f = h$. Then*

$$\sup_{\rho\in\{\rho:\mathcal{E}(\rho)<\infty\}} (\overline{f}(\rho) - \underline{f}(\rho)) \le \sup_{\rho\in E}(h_1(\rho) - h_2(\rho)). \tag{9.79}$$

If, in addition, $\overline{f}$ and $\underline{f}$ are continuous, then

$$\sup_{\rho\in E}(\overline{f}(\rho) - \underline{f}(\rho)) \le \sup_{\rho\in E}(h_1(\rho) - h_2(\rho)). \tag{9.80}$$

PROOF. Condition 9.5 follows as in the previous example. If either $\|\Delta(\Delta\rho - F'(\rho))\|_{-1}^2 = \infty$ or $\|\Delta(\Delta\gamma - F'(\gamma))\|_{-1}^2 = \infty$, then (9.59) follows trivially. Otherwise, for $0 < \kappa < 1$ and $m > 0$, we have

$$\begin{aligned} (9.81)\ &\frac{1}{1-\kappa}\Big(\widehat{H}_\dagger(m(1-\kappa)d^2(\cdot,\gamma) + \kappa\mathcal{E})\Big)^*(\rho) \\ &\quad -\frac{1}{1+\kappa}\Big(\widehat{H}_\ddagger(-m(1+\kappa)d^2(\rho,\cdot) - \kappa\mathcal{E}(\cdot))\Big)_*(\gamma) \\ &= -2m\langle \Delta(\Delta\rho - F'(\rho)) - \Delta(\Delta\gamma - F'(\gamma)), \rho-\gamma\rangle_{-1} \\ &\quad +2\kappa m\langle \Delta(\Delta\rho - F'(\rho)) + \Delta(\Delta\gamma - F'(\gamma)), \rho-\gamma\rangle_{-1} - 4\kappa m^2\|\gamma-\rho\|_{-1}^2 \\ &\quad -\frac{\kappa(2-\kappa)}{2(1-\kappa)}\|\Delta^2\rho - \Delta F'(\rho)\|_{-1}^2 - \frac{\kappa(2+\kappa)}{2(1+\kappa)}\|\Delta^2\gamma - \Delta F'(\gamma)\|_{-1}^2 \\ &\le 2m(\langle \Delta(F'(\rho) - F'(\gamma)), \rho-\gamma\rangle_{-1} - \|\Delta(\rho-\gamma)\|_{-1}^2) \\ &\quad -\frac{3\kappa}{8}\|\Delta^2\rho - \Delta F'(\rho)\|_{-1}^2 - \frac{3\kappa}{8}\|\Delta^2\gamma - \Delta F'(\gamma)\|_{-1}^2 \\ &\le 2m(L_F\|\rho-\gamma\|_0^2 - \|\rho-\gamma\|_1^2) \\ &\le 2m(L_F\|\rho-\gamma\|_{-1}\|\rho-\gamma\|_1 - \|\rho-\gamma\|_1^2) \\ &\le \frac{1}{2}L_F^2 m\|\rho-\gamma\|_{-1}^2, \end{aligned}$$

where the first inequality follows from the inequality

$$\kappa m\langle q,\rho-\gamma\rangle_{-1}-2\kappa m^2\|\rho-\gamma\|_{-1}^2\le\frac{\kappa}{8}\|q\|_{-1}^2$$

(expand $\|\sqrt{2}m(\rho-\gamma)-2^{-3/2}q\|_{-1}^2$), and the second inequality from

$$\begin{aligned}|\langle\Delta(F'(\rho)-F'(\gamma)),\rho-\gamma\rangle_{-1}| &= |\langle F'(\rho)-F'(\gamma),\rho-\gamma\rangle_0|\\ &\le \|F'(\rho)-F'(\gamma)\|_0\|\rho-\gamma\|_0\le L_F\|\rho-\gamma\|_0^2.\end{aligned}$$

Finally, since $\{\rho:\mathcal{E}(\rho)<\infty\}$ is dense in E, under the continuity assumption, (9.79) implies (9.80). □

EXAMPLE 9.35. (Weakly Interacting Diffusions, Example 1.14)

Let $E=\mathcal{P}_2(R^d)$ be the set of all Borel probability measures on R^d with finite second moment. Let $r=d$ be the 2-Wasserstein metric on E given in Definition D.13 by

$$d^2(\rho,\gamma)=\inf_{\pi\in\mathbf{\Pi}(\rho,\gamma)}\int|x-y|^2\pi(dx,dy) \tag{9.82}$$

where

$\mathbf{\Pi}(\rho,\gamma)=\{\pi\in\mathcal{P}(R^d\times R^d)$ satisfying $\pi(dx\times R^d)=\rho(dx),\pi(R^d\times dy)=\gamma(dy)\}$.

The infimum in (9.82) is achieved, and we let $\mathbf{\Pi}^{opt}(\rho,\gamma)$ denote the subset of $\pi\in\mathbf{\Pi}(\rho,\gamma)$ such that $d^2(\rho,\gamma)=\int|x-y|^2\pi(dx,dy)$. See Appendix D.

(E,r) is complete and separable (Lemma D.17). Throughout this example, with a slight abuse of notation, we write $\rho(dx)=\rho(x)dx$ if the Lebesgue density exists.

Following Otto [**92**], we formally view E as a manifold. With reference to (9.62), the form of H in (1.48) suggests a form for the inner product on the tangent space at $\rho_0\in E$ and the definition of the gradient of a function on E (Definition 9.36). $p\in C_c^\infty(R^d)$ determines a "smooth" direction in E in the following way. Let $z_p(t,x)$ satisfy $\dot z_p=\nabla p(z_p)$, with $z_p(0,x)=x$, and for $\rho_0\in E$, define $\rho^p(t)$ by

$$\langle\rho^p(t),g\rangle=\int_{R^d}g(x)\rho^p(t,dx)=\int_{R^d}g(z_p(t,x))\rho_0(dx),\quad\forall g\in B(R^d) \tag{9.83}$$

that is, ρ^p is a weak solution of

$$\rho^p+\nabla\cdot(\rho^p\nabla p)=0,\quad\rho^p(0)=\rho_0. \tag{9.84}$$

DEFINITION 9.36 (Gradient). Let $f:E\mapsto\overline{R}$ and $\rho_0\in E$. The gradient of f at ρ_0, $\text{grad}f(\rho_0)$, exists, if and only if it can be identified as the unique element in the Schwartz space $\mathcal{D}'(R^d)$ (see Section D.1) such that for each $p\in C_c^\infty(R^d)$ and $\rho^p(t)$ satisfying (9.84),

$$\lim_{t\to0}\frac{f(\rho^p(t))-f(\rho_0)}{t}=\langle\text{grad}f(\rho_0),p\rangle.$$

For example, let $p_1,\ldots,p_l\in C_c^2(R^d)$, and define

$$\mathbf{p}=(p_1,\ldots,p_l),\qquad\langle\mathbf{p},\rho\rangle=(\int_{R^d}p_1d\rho,\ldots,\int_{R^d}p_ld\rho).$$

Then for

$$f(\rho)=\varphi(\langle\mathbf{p},\rho\rangle),\qquad\varphi\in C^2(R^l), \tag{9.85}$$

$$\begin{aligned}\langle \text{grad} f(\rho_0), p\rangle &= \sum_{i=1}^{l} \partial_i \varphi(\langle \mathbf{p}, \rho_0\rangle) \int_{R^d} \nabla p_i(x) \cdot \nabla p(x) \rho_0(dx) \\ &= \int_{R^d} \nabla \frac{\delta f}{\delta \rho}(x) \cdot \nabla p(x) \rho_0(dx) \\ &= \langle \frac{\delta f}{\delta \rho}, p\rangle_{1,\rho_0},\end{aligned}$$

where, as before,

$$\frac{\delta f}{\delta \rho} = \sum_{i=1}^{l} \partial_i \varphi(\langle \mathbf{p}, \rho\rangle) p_i.$$

With this definition of gradient, we consider the tangent space at a given point $\rho \in E$ to be a weighted Sobolev space

$$H_\rho^{-1}(R^d) \equiv \{u \in \mathcal{D}'(R^d) : \|u\|_{-1,\rho}^2 \equiv \sup_{p \in C_c^\infty(R^d)} \Big(2\langle u, p\rangle - \int_{R^d} |\nabla p|^2 d\rho\Big) < \infty\}.$$

Appendix D.5 discusses the properties of this space. $H_\rho^{-1}(R^d)$ is a Hilbert space, and we denote its inner product by $\langle \cdot, \cdot \rangle_{-1,\rho}$. Note that for f given by (9.85), we have

$$\begin{aligned}\|\text{grad} f(\rho)\|_{-1,\rho}^2 &= \sup_{p \in C_c^\infty(R^d)} \int_{R^d} (2\nabla \frac{\delta f}{\delta \rho}(x) - \nabla p(x)) \cdot \nabla p(x) \rho(dx) \\ &= \int_{R^d} |\nabla \frac{\delta f}{\delta \rho}|^2 d\rho.\end{aligned} \tag{9.86}$$

Next, we introduce a number of functionals on E and identify their gradients.

LEMMA 9.37. *For $\gamma, \rho \in E$, let $f(\rho) = d^2(\rho, \gamma)$. Then for $\rho(t)$ satisfying (9.84),*

$$\lim_{t \to 0} \frac{f(\rho(t)) - f(\rho_0)}{t} = 2 \inf_{\pi \in \mathbf{\Pi}^{opt}(\rho_0, \gamma)} \int \nabla p(x) \cdot (x - y) \pi(dx, dy). \tag{9.87}$$

REMARK 9.38. If ρ_0 is absolutely continuous, then $\pi \in \mathbf{\Pi}^{opt}(\rho_0, \gamma)$ is unique and satisfies

$$\pi_{\rho_0,\gamma}(dx, dy) = \rho_0(x) dx \delta_{T_{\rho_0,\gamma}(x)}(dy),$$

where $T_{\rho_0,\gamma}$ is defined in (D.29). In this case, the right side of (9.87) is linear in p and determines an element of $\mathcal{D}'(R^d)$, and

$$\langle \text{grad} f(\rho_0), p\rangle = 2 \int \nabla p(x) \cdot (x - T_{\rho_0,\gamma}(x)) \rho_0(x) dx. \tag{9.88}$$

See Theorem D.25.

PROOF. Since $x \mapsto z_p(t, x)$ is invertible,

$$\begin{aligned}&t^{-1}(d^2(\rho(t), \gamma) - d^2(\rho_0, \gamma)) \\ &= t^{-1}\Big(\inf_{\pi \in \mathbf{\Pi}(\rho_0,\gamma)} \int |z_p(t, x) - y|^2 \pi(dx, dy) - \inf_{\pi \in \mathbf{\Pi}(\rho_0,\gamma)} \int |x - y|^2 \pi(dx, dy)\Big) \\ &\geq t^{-1} \int |z_p(t, x) - x|^2 \rho_0(dx) + 2 \int t^{-1}(z_p(t, x) - x) \cdot (x - y) \pi_t(dx, dy),\end{aligned}$$

where π_t is an element of $\mathbf{\Pi}(\rho,\gamma)$ that achieves the first infimum. Since all limit points of π_t, $t\to 0$, will be in $\mathbf{\Pi}^{opt}(\rho_0,\gamma)$, we have

$$\lim_{t\to 0} t^{-1}(d^2(\rho(t),\gamma)-d^2(\rho_0,\gamma)) \geq 2 \inf_{\pi\in\mathbf{\Pi}^{opt}(\rho_0,\gamma)} \int \nabla p(x)\cdot(x-y)\pi(dx,dy). \tag{9.89}$$

If π_0 is the element of $\mathbf{\Pi}^{opt}(\rho,\gamma)$ that achieves the infimum in (9.89), then

$$\begin{aligned} &t^{-1}(d^2(\rho(t),\gamma)-d^2(\rho_0,t)) \\ &\quad \leq t^{-1}\int |z_p(t,x)-x|^2\rho_0(dx) + 2\int t^{-1}(z_p(t,x)-x)\cdot(x-y)\pi_0(dx,dy), \end{aligned}$$

and it follows that equality holds in (9.89). □

Throughout this section, we assume $\Psi,\Phi\in C^2(R^d)$ are semiconvex in the sense that there exist constants λ_Ψ and λ_Φ such that

$$\begin{aligned} (\nabla\Psi(x)-\nabla\Psi(y))\cdot(x-y) &\geq \lambda_\Psi|x-y|^2 \\ (\nabla\Phi(x)-\nabla\Phi(y))\cdot(x-y) &\geq \lambda_\Phi|x-y|^2. \end{aligned} \tag{9.90}$$

We also assume

$$\lim_{|x|\to\infty}\frac{\Psi(x)}{|x|^2}=\infty \tag{9.91}$$

and $\inf_z\Phi(z)>-\infty$. The main large deviation example in Dawson and Gärtner [**22**] is the Curie-Weiss model with double well potential, where $\Psi(x)=a|x|^4-b|x|^2$, $a>0$, $b\in R$, which satisfies these conditions.

Setting

$$\mu_\infty(dx)\equiv Z^{-1}e^{-2\Psi(x)}dx, \tag{9.92}$$

we define the *relative entropy* with respect μ_∞ by

$$\mathcal{R}(\rho\|\mu_\infty)=\int_{R^d}\frac{d\rho}{d\mu_\infty}\log\frac{d\rho}{d\mu_\infty}d\mu_\infty \tag{9.93}$$

and the *free energy* functional by

$$\mathcal{E}(\rho)=\frac{1}{2}\mathcal{R}(\rho\|\mu_\infty)+\frac{1}{2}\int_{R^d\times R^d}\Phi(x-y)\rho(dx)\rho(dy). \tag{9.94}$$

By a variational representation due to Donsker and Varadhan (e.g. Lemma 1.4.3 of [**35**]),

$$\begin{aligned} \mathcal{R}(\rho\|\mu_\infty) &= \sup_{p\in B(R^d)}\{\int p d\rho-\log\int e^p d\mu_\infty\} \\ &= \sup_{p\in C_c^\infty(R^d)}\{\int p d\rho-\log\int e^p d\mu_\infty\}. \end{aligned} \tag{9.95}$$

This representation and the assumption that Φ is bounded below ensure that $\mathcal{E}$ is lower semicontinuous in the weak topology on $\mathcal{P}(R^d)$ and hence also in the 2-Wasserstein topology on $\mathcal{P}_2(R^d)$. Note that if ρ is not absolutely continuous with respect to μ_∞, $\mathcal{E}(\rho)=\infty$. Taking $p\equiv 0$ in (9.95), we see that $\mathcal{R}(\rho\|\mu_\infty)\geq 0$ and hence $\mathcal{E}(\rho)\geq\inf_z\Phi(z)$.

For $\rho \in E$ such that $\rho(dx) = \rho(x)dx$, we can write

$$\mathcal{E}(\rho) = \frac{1}{2}\log Z + \frac{1}{2}\int_{R^d} \rho(x)\log\rho(x)dx + \int_{R^d} \Psi(x)\rho(x)dx + \frac{1}{2}\int_{R^d\times R^d} \Phi(x-y)\rho(x)\rho(y)dxdy. \tag{9.96}$$

Note that by Jensen's inequality,

$$\begin{aligned} \int_{R^d} \rho(x)\log\rho(x)dx &= -\int_{R^d} \rho(x)\log\rho(x)^{-1}e^{-|x|^2}dx - \int_{R^d} |x|^2\rho(x)dx \\ &\geq -\log\int_{R^d} e^{-|x|^2}dx - \int_{R^d} |x|^2\rho(x)dx \\ &> -\infty. \end{aligned} \tag{9.97}$$

By (9.91) and (9.97), the sum of the second and third terms in (9.96) is bounded below, as are the other terms.

Let p, ρ_0, and $\rho(t)$ be defined as in (9.83), and assume that $\mathcal{E}(\rho_0) < \infty$. Following calculations in [**63**],

$$\int \rho(t,x)\log\rho(t,x)dx = \int \rho_0(x)\log\frac{\rho_0(x)}{|\det\nabla z_p(t,x)|}dx\ ,$$

so

$$\frac{\partial}{\partial t}\int \rho(t,x)\log\rho(t,x)dx_{|t=0} = -\int \rho_0(x)\Delta p(x)dx.$$

Furthermore,

$$\begin{aligned} \frac{\partial}{\partial t}\int \Psi(x)\rho(t,dx) &= \frac{\partial}{\partial t}\int \Psi(z_p(t,x))\rho_0(dx) \\ &= \int \nabla\Psi(z_p(t,x))\nabla p(z_p(t,x))\rho_0(dx) \end{aligned}$$

and

$$\begin{aligned} &\frac{\partial}{\partial t}\int \Phi(x-y)\rho(t,dx)\rho(t,dy) \\ &= \frac{\partial}{\partial t}\int \Phi(z_p(t,x)-z_p(t,y))\rho_0(dx)\rho_0(dy) \\ &= \int \nabla\Phi(z_p(t,x)-z_p(t,y))(\nabla p(z_p(t,x)) - \nabla p(z_p(t,y))\rho_0(dx)\rho_0(dy) \\ &= 2\int \nabla\Phi(z_p(t,x)-z_p(t,y))\nabla p(z_p(t,x))\rho_0(dx)\rho_0(dy), \end{aligned}$$

where the last equality follows from the symmetry of Φ. Consequently,

$$\begin{aligned} \langle \text{grad}\mathcal{E}(\rho_0), p\rangle &= \langle \rho_0, \nabla\Psi\cdot\nabla p + \nabla\Phi(\cdot)*\rho_0\cdot\nabla p - \frac{1}{2}\Delta p\rangle \\ &= -\langle\frac{1}{2}\Delta\rho_0 + \nabla\cdot(\rho_0(\nabla\Psi + \nabla\Phi*\rho_0)), p\rangle, \end{aligned} \tag{9.98}$$

so $\text{grad}\mathcal{E}(\rho_0) = -(\frac{1}{2}\Delta\rho_0 + \nabla\cdot(\rho_0(\nabla\Psi + \nabla\Phi*\rho_0)))$ in $\mathcal{D}'(R^d)$.

$$\|\text{grad}\mathcal{E}(\rho_0)\|_{-1,\rho_0} = \|\frac{1}{2}\Delta\rho_0 + \nabla\cdot(\rho_0(\nabla\Psi + \nabla\Phi*\rho_0))\|_{-1,\rho_0}. \tag{9.99}$$

By Theorem D.45,

$$\|\mathrm{grad}\mathcal{E}(\rho_0)\|_{-1,\rho_0} = \int_{R^d} \frac{|\nabla\rho_0(x) + \rho_0(x)\nabla(2\Psi + 2\rho_0 * \Phi)|^2}{\rho_0(x)} dx \tag{9.100}$$

whenever ρ_0 has a Lebesgue density with $\nabla\rho_0 \in L^1_{loc}(R^d)$, and $\|\mathrm{grad}\mathcal{E}(\rho_0)\|_{-1,\rho_0} = \infty$ otherwise.

For f of the form (9.85) and ρ satisfying $\mathcal{E}(\rho) < \infty$ and $\nabla\rho \in L^1_{loc}(R^d)$, by (9.86) and (9.98),

$$\begin{aligned}
&\frac{1}{2}(\|\mathrm{grad}(f(\rho) - \mathcal{E}(\rho))\|^2_{-1,\rho} - \|\mathrm{grad}\mathcal{E}(\rho)\|^2_{-1,\rho}) \\
&\quad = \frac{1}{2}(\int |\nabla\frac{\delta f}{\delta\rho} - \frac{\nabla\rho}{2\rho} - (\nabla\Psi + \nabla\Phi(\cdot) * \rho)|^2 d\rho \\
&\qquad\qquad - \int |\frac{\nabla\rho}{2\rho} + (\nabla\Psi + \nabla\Phi(\cdot) * \rho)|^2 d\rho) \\
&\quad = -\int \nabla\frac{\delta f}{\delta\rho}(\frac{\nabla\rho}{2\rho} + (\nabla\Psi + \nabla\Phi(\cdot) * \rho))d\rho + \frac{1}{2}\int |\nabla\frac{\delta f}{\delta\rho}|^2 d\rho,
\end{aligned}$$

so H in (1.48) can be written as

$$Hf(\rho) = \langle -\mathrm{grad}\mathcal{E}(\rho), \mathrm{grad}f(\rho)\rangle_{-1,\rho} + \frac{1}{2}\|\mathrm{grad}f(\rho)\|^2_{-1,\rho}, \tag{9.101}$$

and (9.62) holds.

As in the previous two examples, this observation motivates us to extend H by including test functions of the forms (9.57) and (9.58), where the d in (9.57) and (9.58) is the 2-Wasserstein metric. With Condition 9.26 in mind, we assume conditions that imply the level sets of $\mathcal{E}$ are compact.

LEMMA 9.39. *Assuming (9.91), for each $C \in R$,*

$$\{\rho \in E : \mathcal{E}(\rho) \le C\}$$

is compact in (E, r).

PROOF. There exists $G \ge 0$ such that $\lim_{r\to\infty} r^{-1}G(r) = \infty$ and

$$\int e^{G(|x|^2) - 2\Psi(x)} dx < \infty,$$

and for each $M > 0$, (9.95) implies

$$\int G(|x|^2) \wedge M d\rho \le \mathcal{R}(\rho\|\mu_\infty) + \log\int e^{G(|x|^2)\wedge M} d\mu_\infty$$

(take $p = G(|x|^2) \wedge M$), and hence

$$\sup_{\rho\in E:\mathcal{E}(\rho)\le C} \int G(|x|^2) d\rho < \infty.$$

Therefore, $\{\rho \in E : \mathcal{E}(\rho) \le C\}$ is relatively compact in (E, r). □

We will express our extended operators in terms of the following *Fisher information* functionals: for each $\rho \in \mathcal{P}(R^d)$, let

(9.102)

$$I(\rho) = \begin{cases} \int_{R^d} \frac{|\nabla\rho(x) + 2\rho(x)\nabla\Psi(x)|^2}{\rho(x)} dx & \text{when } \rho(dx) = \rho(x)dx, \nabla\rho \in L^1_{loc}(R^d) \\ +\infty & \text{otherwise} \end{cases}$$

and
(9.103)

$$I^{\#}(\rho) \equiv \begin{cases} \int_{R^d} \frac{|\nabla\rho(x)+\rho(x)\nabla(2\Psi+2\rho*\Phi)|^2}{\rho(x)}dx & \text{when } \rho(dx)=\rho(x)dx, \nabla\rho \in L^1_{loc}(R^d) \\ +\infty & \text{otherwise,} \end{cases}$$

where we follow the convention that $0/0 = 0$. By Lemma D.44,

$$I(\rho) \equiv \begin{cases} \int_{R^d} \frac{|\nabla \frac{d\rho}{d\mu_\infty}|^2}{\frac{d\rho}{d\mu_\infty}} d\mu_\infty & \text{when } \rho(dx)=\rho(x)dx, \nabla \frac{d\rho}{d\mu_\infty} \in L^1_{loc}(R^d), \\ +\infty & \text{otherwise.} \end{cases}$$

A similar representation for $I^{\#}$ also holds.

We introduce a family of measures indexed by a state variable $\gamma \in E$:

$$\mu_{\infty,\gamma}(dx) = Z_\gamma^{-1} e^{-2(\Psi+\gamma*\Phi)}dx, \quad Z_\gamma = \int_{R^d} e^{-2(\Psi+\gamma*\Phi)}dx, \tag{9.104}$$

and consider the quantity

$$I(\rho|\mu_{\infty,\gamma}) \equiv \int_{R^d} \frac{|\nabla \frac{d\rho}{d\mu_{\infty,\gamma}}|^2}{\frac{d\rho}{d\mu_{\infty,\gamma}}} d\mu_{\infty,\gamma}. \tag{9.105}$$

Then

$$I^{\#}(\rho) = I(\rho|\mu_{\infty,\rho}).$$

Let $\mathcal{E}(\rho) < \infty$ and $\nabla\rho \in L^1_{loc}(R^d)$. Then by (9.100)

$$\|\text{grad}\mathcal{E}(\rho)\|^2_{-1,\rho} = \frac{1}{4} I^{\#}(\rho).$$

If ρ has Lebesgue density, by (9.88),

$$\|\text{grad}_\rho d^2(\rho,\gamma)\|^2_{-1,\rho} = 4\int_{R^d} |x - T_{\rho,\gamma}(x)|^2 \rho(x)dx = 4d^2(\rho,\gamma).$$

Let $p_{\rho,\gamma}$ be given in Theorem D.25 so that

$$T_{\rho,\gamma}(x) = x - \nabla p_{\rho,\gamma}(x). \tag{9.106}$$

If $I^{\#}(\rho) < \infty$, then

$$\langle \text{grad}\mathcal{E}(\rho), \text{grad}_\rho \frac{1}{2} d^2(\rho,\gamma)\rangle_{-1,\rho} = \int_{R^d} (\frac{1}{2} \frac{\nabla\rho + \rho\nabla(2\Psi + 2\Phi*\rho)}{\rho}) \nabla p_{\rho,\gamma} d\rho.$$

Given the above discussion, we introduce further extensions $\widehat{H}_\dagger$ and $\widehat{H}_\ddagger$ given by the right side of (9.101) on appropriate domains. We prove a comparison principle for this pair of operators here, but leave the rigorous justification of $\widehat{H}_\dagger$ and $\widehat{H}_\ddagger$ being viscosity sub- and super-extensions of H until Section 13.3.

For $0 < \kappa < 1$ and $m > 0$, let

$$f_0(\rho) = m(1-\kappa)\frac{1}{2}d^2(\rho,\gamma) + \kappa\mathcal{E}(\rho),$$

and

$$f_1(\gamma) = -m(1+\kappa)\frac{1}{2}d^2(\rho,\gamma) - \kappa\mathcal{E}(\gamma).$$

Applying (9.101), we define

(9.107)

$$\widehat{H}_\dagger f_0(\rho) \equiv \begin{cases} -m(1-\kappa)^2 \int_{R^d} (\frac{1}{2}\frac{\nabla\rho}{\rho} + \nabla\Psi + \nabla(\Phi * \rho))\nabla p_{\rho,\gamma} d\rho \\ \quad +(1-\kappa)^2 \frac{m^2}{2} d^2(\rho,\gamma) - (\kappa - \frac{\kappa^2}{2})\frac{1}{4} I^\#(\rho), & \text{if } I^\#(\rho) < \infty \\ -\infty & \text{otherwise;} \end{cases}$$

and

(9.108)

$$\widehat{H}_\ddagger f_1(\gamma) \equiv \begin{cases} -m(1+\kappa)^2 \int_{R^d} (\frac{1}{2}\frac{\nabla\gamma}{\gamma} + \nabla\Psi + \nabla(\Phi * \gamma))\nabla p_{\gamma,\rho} d\gamma \\ \quad +(1+\kappa)^2 \frac{m^2}{2} d^2(\rho,\gamma) + (\kappa + \frac{\kappa^2}{2})\frac{1}{4} I^\#(\gamma), & \text{if } I^\#(\gamma) < \infty \\ +\infty & \text{otherwise,} \end{cases}$$

where $p_{\rho,\gamma}$ is defined in (9.106) and $p_{\gamma,\rho}$ is defined similarly with the roles of γ and ρ interchanged.

If $\mathcal{E}(\rho) < \infty$ and $I^\#(\rho) < \infty$,

$$\begin{aligned} & \widehat{H}_\dagger f_0(\rho) \\ = \; & \frac{1}{2}(\|\text{grad}((1-\kappa)\mathcal{E}(\rho) - (1-\kappa)m\frac{1}{2}d^2(\rho,\gamma))\|^2_{-1,\rho} - \|\text{grad}\mathcal{E}(\rho)\|^2_{-1,\rho}) \\ = \; & (1-\kappa)^2\frac{1}{2}\|\text{grad}(\mathcal{E}(\rho) - m\frac{1}{2}d^2(\rho,\gamma))\|^2_{-1,\rho} - \frac{1}{2}\|\text{grad}\mathcal{E}(\rho)\|^2_{-1,\rho} \\ = \; & -\frac{(1-\kappa)^2}{2}\langle \text{grad}(\mathcal{E}(\rho), \text{grad} d^2(\rho,\gamma)\rangle_{-1,\rho} + \frac{(1-\kappa)^2 m^2}{2} d^2(\rho,\gamma) \\ & \qquad - \frac{1}{2}(2\kappa - \kappa^2)\|\text{grad}\mathcal{E}(\rho)\|^2_{-1,\rho} \\ \leq \; & (1-\kappa)\frac{1}{2}(\|\text{grad}(\mathcal{E}(\rho) - m\frac{1}{2}d^2(\rho,\gamma)\|^2_{-1,\rho} - \|\text{grad}\mathcal{E}(\rho)\|^2_{-1,\rho}) - \frac{\kappa}{2}\|\text{grad}\mathcal{E}(\rho)\|^2_{-1,\rho}, \end{aligned}$$

and we have

$$\widehat{H}_\dagger f_0(\rho) \leq (1-\kappa)\Big(m\langle -\text{grad}\mathcal{E}(\rho), \text{grad}_\rho \frac{1}{2}d^2(\rho,\gamma)\rangle_{-1,\rho} + \frac{m^2}{2}d^2(\rho,\gamma)\Big) - \frac{\kappa}{8}I^\#(\rho).$$

Similarly, if $\mathcal{E}(\gamma) < \infty$ and $I^\#(\gamma) < \infty$,

$$\widehat{H}_\ddagger f_1(\gamma) \geq (1+\kappa)\Big(m\langle \text{grad}\mathcal{E}(\gamma), \text{grad}_\gamma \frac{1}{2}d^2(\rho,\gamma)\rangle_{-1,\gamma} + \frac{m^2}{2}d^2(\rho,\gamma)\Big) + \frac{\kappa}{8}I^\#(\gamma).$$

Consequently, by (D.78),

$$(1-\kappa)^{-1}\widehat{H}_\dagger f_0(\rho) - (1+\kappa)^{-1}\widehat{H}_\ddagger f_1(\gamma) \leq m(|\lambda_\Psi| + 2|\lambda_\Phi|)d^2(\rho,\gamma)), \tag{9.109}$$

where λ_Ψ and λ_Φ are the constants in (9.90).

The small perturbations $\kappa\mathcal{E}$ introduce the $I^\#$ term in $\widehat{H}_\dagger f_0$ and $\widehat{H}_\ddagger f_1$ which ensures the appropriate semicontinuity of $\widehat{H}_\dagger f_0$ and $\widehat{H}_\ddagger f_1$.

LEMMA 9.40. *$\widehat{H}_\dagger f_0$ in (9.107) is upper semicontinuous, and $\widehat{H}_\ddagger f_1$ in (9.108) is lower semicontinuous.*

PROOF. The conclusion follows by Lemma D.48 of Appendix D. □

Finally, to verify condition (9.5), let $\{p_k : k = 1, 2, \ldots\} \subset C_c^\infty(R^d)$ be buc-dense in $C_b(R^d)$. Define

$$G(\rho,\gamma) = \sum_{k=1}^\infty \frac{1}{2^k(1+m_k^2)}\langle \rho - \gamma, p_k\rangle^2, \tag{9.110}$$

where

$$m_k = \sup_x(|p_k(x)| + |\Delta p_k(x)| + |\nabla p_k(x)|).$$

Then, as in the previous examples, we can further enlarge the domains of $\widehat{H}_\dagger$ and $\widehat{H}_\ddagger$ to include test functions

$$f_0(\rho) = ad^2(\rho,\gamma_0) + b\mathcal{E}(\rho) + cG(\rho,\rho_0) + C \tag{9.111}$$

and

$$f_1(\gamma) = -ad^2(\rho_0,\gamma) - b\mathcal{E}(\gamma) - cG(\gamma,\gamma_0) + C, \tag{9.112}$$

with $a, c > 0, 0 < b < 1$, $\rho_0, \gamma_0 \in E$ and $C \in R$. $\widehat{H}_\dagger f_0$ and $\widehat{H}_\ddagger f_1$ are still well defined and are respectively upper- and lower- semicontinuous.

THEOREM 9.41. *Suppose that $h_0, h_1 \in C_b(E)$ and that $\overline{f} \in C_b(E)$ is a subsolution of $(I - \alpha\widehat{H}_\dagger)f = h$ and $\underline{f} \in C_b(E)$ is a supersolution of $(I - \alpha\widehat{H}_\ddagger)f = h$. Then*

$$\sup_{\rho\in E}(\overline{f}(\rho) - \underline{f}(\rho)) \le \sup_{\rho\in E}(h_0(\rho) - h_1(\rho)).$$

PROOF. We apply Theorem 9.28. As in the previous two examples, G provides the family of functions needed to verify Condition 9.5.

(9.59) trivially holds if $I^\#(\rho)+I^\#(\gamma) = \infty$ (it reduces to $-\infty \le \omega_\kappa(md^2(\rho,\gamma))$). Otherwise, it follows from (9.109).

By Theorem 9.28, the conclusion follows. □

CHAPTER 10

Nearly deterministic processes in R^d

The fundamental problem for large deviations for processes introduced by Freidlin and Wentzell (see [**52**]) and, in particular cases, by Varadhan [**119**], considers processes with generators of the form

$$\begin{aligned}(10.1)\quad A_n f(x) &= n\int_{R^d}(f(x+\frac{1}{n}z)-f(x)-\frac{1}{n}z\cdot\nabla f(x))\eta(x,dz)\\ &\qquad +b(x)\cdot\nabla f(x)+\frac{1}{2n}\sum_{i,j}a_{ij}(x)\partial_i\partial_j f(x).\end{aligned}$$

For large values of n, the corresponding Markov process is essentially a solution of the ordinary differential equation

$$\dot{X}=b(X),$$

and the corresponding large deviation theory is concerned with the probabilities of sample paths significantly different from the solution of this equation. We begin by considering processes having stationary, independent increments, that is, when the coefficients in (10.1) do not depend on x. The spatial homogeneity makes verifying the comparison principle simple. (The function H in Lemma 9.15 does not depend on x.) In Section 10.1, Theorem 10.1 generalizes a result on R^d-valued Lévy processes by Lynch and Sethuraman [**82**]; Section 10.1.6 discusses another type of scaling of Lévy processes considered by Mogulskii [**88**]; and Section 10.1.7 derives the results for Lévy processes in separable Banach spaces due to de Acosta [**25**].

In Section 10.2, we treat the discrete-time analogues of the results in Section 10.1. Results of this type have been given by Mogulskii [**87**] and Borovkov [**11**].

In Section 10.3, we consider results for more general Markov processes with generators of the form (10.1), obtaining the results of Freidlin and Wentzell [**52**] as well as various generalizations that have been given by a number of authors. In particular, Theorem 10.2.6 of Dupuis and Ellis [**35**] follows from Lemmas 10.5, 10.21 and Remark 10.16, Theorem 5.6.7 of Dembo and Zeitouni [**29**] and Theorem 3.1 of de Acosta [**26**] follows from Lemmas 10.10, 10.21 and Remark 10.11.

Discrete time analogs of these results have been considered by Gulinsky and Veretennikov [**57**], Dupuis and Ellis [**35**], Chapter 6, de Acosta [**27**]. Results of this type are discussed in Section 10.4.

Finally, in Section 10.5, we consider processes with reflection, extending results of Anderson and Orey [**4**] and Doss and Priouret [**34**].

10.1. Processes with independent increments

We begin by considering processes with stationary, independent increments in R^d. For this class of problems, the comparison principle is trivial, since H does not depend on x and Condition 9.10 is immediate.

As a first example, we consider Schilder's theorem, that is the large deviation principle for

$$X_n = \frac{1}{\sqrt{n}} W,$$

where W is standard Brownian motion. Note that it would be distributionally equivalent to define X_n by

$$X_n(t) = \frac{1}{n} W(nt).$$

We have

$$A_n f = \frac{1}{2n} \Delta f$$

and

$$H_n f = \frac{1}{2n} \Delta f + \frac{1}{2} |\nabla f|^2.$$

It follows that $Hf = \frac{1}{2}|\nabla f|^2$.

Letting $E = R^d \cup \infty$, exponential tightness is immediate by Corollary 4.17. As noted above, the comparison principle follows by Lemma 9.15, since $H(x,p) = |p|^2$ does not depend on x. Theorem 6.14 then gives that the large deviation principle holds for $\{X_n\}$.

By a simple calculation

$$Hf(x) = H(x, \nabla f(x)) = \mathbf{H}f(x) \equiv \sup_u (u \cdot \nabla f(x) - \frac{1}{2}|u|^2),$$

and any solution of

$$f(x(t)) = f(0) + \int_0^t \int_{R^d} u \cdot \nabla f(x(s)) \lambda(du \times ds) = f(0) + \int_0^t \int_{R^d} u \lambda_s(du) \cdot \nabla f(x(s)) ds$$

satisfies

$$\dot{x}(t) = \int_{R^d} u \lambda_t(du), \quad a.e.$$

Consequently, if x is absolutely continuous,

$$I(x) = \inf\{\int_0^\infty \int_{R^d} \frac{1}{2}|u|^2 \lambda_s(du) : \int_{R^d} u \lambda_s(du) = \dot{x}(s), s \geq 0\} = \int_0^\infty \frac{1}{2}|\dot{x}(s)|^2 ds,$$

where the last equality follows by Jensen's inequality, that is

$$\int_{R^d} |u|^2 \lambda_s(du) \geq \left| \int_{R^d} u \lambda_s(du) \right|^2,$$

and $I(x) = \infty$ otherwise.

We now consider more general Levy processes with generators of the form

$$Ag(x) = \int_{R^d} (g(x+z) - g(x) - z \cdot \nabla g(x)) \eta(dz) + \frac{1}{2} \sum_{i,j} a_{ij} \partial_i \partial_j g(x) + b \cdot \nabla g(x). \tag{10.2}$$

If X corresponds to A, then X_n defined by $X_n(t) = \frac{1}{n} X(nt)$ has generator

$$A_n g(x) = n \int_{R^d} (g(x + \frac{1}{n} z) - g(x) - \frac{1}{n} z \cdot \nabla g(x)) \eta(dz) + \frac{1}{2n} \sum_{i,j} a_{ij} \partial_i \partial_j g(x) + b \cdot \nabla g(x). \tag{10.3}$$

10.1.1. Convergence of H_n. H_n is given by

$$\begin{aligned}H_nf(x) &= \int_{R^d}(e^{n(f(x+\frac{1}{n}z)-f(x))}-1-z\cdot\nabla f(x))\eta(dz)\\ &\quad+\frac{1}{2n}\sum_{i,j}a_{ij}\partial_i\partial_jf(x)+\frac{1}{2}\sum_{i,j}a_{ij}\partial_if(x)\partial_jf(x)+b\cdot\nabla f(x),\end{aligned}$$

and, assuming

$$\int_{R^d}(e^{\alpha\cdot z}-1-\alpha\cdot z)\eta(dz)<\infty, \tag{10.4}$$

for all $\alpha\in R^d$, for $f\in D_d$ defined as in (9.41) to be

$$D_d=\{f\in C(E): f|_{R^d}-f(\infty)\in C_c^2(R^d)\}, \tag{10.5}$$

$Hf=\lim_{n\to\infty}H_nf$ is given by

$$\begin{aligned}Hf(x) &= \int_{R^d}(e^{\nabla f(x)\cdot z}-1-z\cdot\nabla f(x))\eta(dz)+\frac{1}{2}\sum_{i,j}a_{ij}\partial_if(x)\partial_jf(x)+b\cdot\nabla f(x)\\ &= H(\nabla f(x)),\end{aligned}$$

where

$$H(p)=\int_{R^d}(e^{p\cdot z}-1-z\cdot p)\eta(dz)+\frac{1}{2}p^Tap+b\cdot p.$$

10.1.2. Exponential tightness. The exponential compact containment condition follows by Example 4.23, and exponential tightness follows by Corollary 4.17.

10.1.3. The comparison principle. As noted above, the comparison principle follows by Lemma 9.15, since $H(p)$ does not depend on x.

10.1.4. The large deviation theorem.

THEOREM 10.1. *Let A_n be given by (10.3) for $g\in D_d$. Suppose that X_n is a solution of the martingale problem for A_n and that the large deviation principle holds for $X_n(0)$ in R^d with a good rate function I_0 and rate transform Λ_0. Then*

a) *For each $\alpha>0$ and $f\in C(E)$, there exists a unique $R_\alpha f\in C(E)$ such that*

$$(I-\alpha H)R_\alpha f=f$$

in the viscosity sense, and

$$V(t)f=\lim_n R_{t/n}^nf$$

defines a semigroup $V(t): C(E)\mapsto C(E)$.

b) *The large deviation principle holds for $\{X_n\}$ in $D_E[0,\infty)$ with good rate function:*

$$I(x)=\sup_{\{t_i\}\subset\Delta_x^c}I_{t_1,\ldots,t_m}(x(t_1),\ldots,x(t_m)),$$

where

$$\begin{aligned}&I_{t_1,\ldots,t_m}(x_1,\ldots,x_m)\\ &\quad=\sup_{f_1,\ldots,f_k\in D_d}\{f_1(x_1)+\ldots+f_k(x_k)\\ &\qquad-\Lambda_0(V(t_1)(f_1+V(t_1-t_1)(f_2+\ldots V(t_k-t_{k-1})f_k)\ldots)))\}.\end{aligned}$$

c) *If* $x \in D_E[0,\infty) - C_{R^d}[0,\infty)$, *then* $I(x) = \infty$.

REMARK 10.2. Particular cases of this theorem were considered by Borovkov [**11**] and Lynch and Sethuraman [**82**]. Mogulskii [**88**] gives a more general version, including results under the assumption that (10.4) only holds for α in a ball containing the origin.

PROOF. Since the comparison principle holds for $(I - \alpha H)f = h$, parts (a) and (b) follow by Theorem 6.14.

The compact containment condition implies that if $x \in D_E[0,\infty) - D_{R^d}[0,\infty)$, then $I(x) = \infty$. (See Example 4.23.)

Since η determines the Poisson rate of jumps of a particular size, it follows that

$$\begin{aligned} P\{\sup_{s \le T} |X_n(s) - X_n(s-)| \ge \epsilon\} &= 1 - e^{-nT\eta\{z:|z|\ge n\epsilon\}} \\ &\le nT\eta\{z : |z| \ge n\epsilon\} \\ &\le \frac{nT\int_{\{z:|z|\ge n\epsilon\}} e^{k|z|}\eta(dz)}{e^{kn\epsilon}}, \end{aligned}$$

and hence,

$$\limsup_{n\to\infty} P\{\sup_{s\le T} |X_n(s) - X_n(s-)| \ge \epsilon\} \le -kn\epsilon.$$

Since k is arbitrary, the left side is $-\infty$, and by Theorem 4.13, $x \in D_{R^d}[0,\infty) - C_{R^d}[0,\infty)$ implies $I(x) = \infty$. □

10.1.5. Variational representation of the rate function. Since H is convex, setting

$$L(q) = \sup_{p\in R^d} (p \cdot q - H(p)),$$

we have

$$H(p) = \sup_{q\in R^d} (p \cdot q - L(q)).$$

(See [**104**], Chapter 12.) It follows that

$$Hf(x) = H(\nabla f(x)) = \mathbf{H}f(x) \equiv \sup_{q\in R^d} (q \cdot \nabla f(x) - L(q)),$$

and hence by Theorem 8.14, if x is absolutely continuous,

$$\begin{aligned} I(x) &= I_0(x(0)) + \inf\{\int_0^\infty \int_{R^d} L(u)\lambda_s(du) : \int u\lambda_s(du) = \dot{x}(s)\} \\ &= I_0(x(0)) + \int_0^\infty L(\dot{x}(s))ds, \end{aligned}$$

where the last equality follows by the convexity of L. $I(x) = \infty$ otherwise.

10.1.6. Other scalings. Mogulskii [**88**] also considers different scalings of X. In our notation, let

$$X_n(t) = \frac{1}{\sqrt{n\beta_n}}(X(\beta_n t) - \beta_n bt),$$

where $n/\beta_n \to 0$. A_n then becomes

$$(10.6) A_n g(x) = \beta_n \int_{R^d} (g(x + \frac{1}{\sqrt{n\beta_n}} z) - g(x) - \frac{1}{\sqrt{n\beta_n}} z \cdot \nabla g(x))\eta(dz) + \frac{1}{2n} \sum_{i,j} a_{ij}\partial_i\partial_j g(x),$$

giving

$$\begin{aligned} H_n f(x) &= \frac{\beta_n}{n} \int_{R^d} (e^{n(f(x+\frac{1}{\sqrt{n\beta_n}}z)-f(x))} - 1 - \sqrt{\frac{n}{\beta_n}} z \cdot \nabla f(x))\eta(dz) \\ &\quad + \frac{1}{2n} \sum_{i,j} a_{ij}\partial_i\partial_j f(x) + \frac{1}{2} \sum_{i,j} a_{ij}\partial_i f(x)\partial_j f(x) \\ &= \int_{R^d} \beta_n n (f(x + \frac{1}{\sqrt{n\beta_n}} z) - f(x))^2 \\ &\quad \times \frac{e^{n(f(x+\frac{1}{\sqrt{n\beta_n}}z)-f(x))} - 1 - n(f(x+\frac{1}{\sqrt{n\beta_n}}z) - f(x))}{\left(n(f(x+\frac{1}{\sqrt{n\beta_n}}z) - f(x))\right)^2} \eta(dz) \\ &\quad + \frac{\beta_n}{n} \int_{R^d} \left(n(f(x + \frac{1}{\sqrt{n\beta_n}} z) - f(x)) - \sqrt{\frac{n}{\beta_n}} z \cdot \nabla f(x)\right) \eta(dz) \\ &\quad + \frac{1}{2n} \sum_{i,j} a_{ij}\partial_i\partial_j f(x) + \frac{1}{2} \sum_{i,j} a_{ij}\partial_i f(x)\partial_j f(x). \end{aligned}$$

If

$$(10.7) \qquad \int_{R^d} |z|^2 e^{\alpha|z|} \eta(dz) < \infty, \quad \text{some } \alpha > 0,$$

then H_n converges to H given by

$$Hf(x) = \frac{1}{2} \int_{R^d} (z \cdot \nabla f(x))^2 \eta(dz) + \frac{1}{2} \sum_{i,j} a_{ij}\partial_i f(x)\partial_j f(x).$$

Setting

$$\widehat{a}_{ij} = \int_{R^d} z_i z_j \eta(dz) + a_{ij},$$

we have

$$Hf(x) = \frac{1}{2} \sum_{i,j} \widehat{a}_{ij}\partial_i f(x)\partial_j f(x),$$

which is the same limiting operator as would be obtained if we took $\eta = 0$ and replaced a_{ij} by $\widehat{a}_{ij}$ in the original generator. In other words, speeding up time by more than n allows the central limit theorem to "beat" the large deviation principle, so that the large deviation behavior is the same as for a Brownian motion.

The moment condition (10.7) can be weakened under more restrictive assumptions about the behavior of β_n. Assuming, at a minimum, that $\int_{R^d} |z|^2 \eta(dz) < \infty$ and observing that the exponent is bounded on the set $|z| \le \sqrt{\beta_n/n}$, to obtain the desired limit, it is enough to show

$$\lim_{n\to\infty} \frac{\beta_n}{n} \int_{\{|z|>\sqrt{\beta_n/n}\}} e^{n(f(x+\frac{1}{\sqrt{n\beta_n}}z)-f(x))} \eta(dz) = 0.$$

Consequently, it is sufficient to select β_n so that for each $K > 0$

$$\lim_{n\to\infty} \frac{\beta_n}{n} \int_{\{|z|>\sqrt{\beta_n/n}\}} e^{K(n\wedge \frac{|z|}{\sqrt{n\beta_n}})} \eta(dz) = 0. \tag{10.8}$$

For example (cf. [**87**]), if

$$\int_{R^d} |z|^2 e^{\epsilon |z|^\gamma} \eta(dz) < \infty,$$

for some $\epsilon > 0$ and $0 < \gamma \leq 1$, then it is sufficient to have

$$\lim_{n\to\infty} \frac{n^{2/\gamma}}{n^3 \beta_n} = 0. \tag{10.9}$$

Since we are already assuming $n/\beta_n \to 0$, (10.9) places no additional restrictions on β_n unless $\gamma < 1/2$. To see that (10.9) implies (10.8), note that the expression in the limit is bounded by

$$\int_{\{|z|>\sqrt{\beta_n/n}\}} |z|^2 e^{K(n\wedge \frac{|z|}{\sqrt{n\beta_n}})} \eta(dz) = \int_{\{|z|>\sqrt{\beta_n/n}\}} |z|^2 e^{|z|^\gamma K(\frac{n}{|z|^\gamma} \wedge \frac{|z|^{1-\gamma}}{\sqrt{n\beta_n}})} \eta(dz),$$

and

$$\frac{n}{|z|^\gamma} \wedge \frac{|z|^{1-\gamma}}{\sqrt{n\beta_n}} \leq \frac{n}{(n\sqrt{n\beta_n})^\gamma},$$

which tends to zero by (10.9).

The compact containment condition follows by applying Lemma 4.22 with

$$f_n(x) = \log(\frac{1+|x|^2}{1+\delta|x|^2}),$$

for $\delta > 0$ sufficiently small.

If $\widehat{a}$ is nonsingular, then $L(q) = \frac{1}{2} q^T \widehat{a}^{-1} q$, and assuming for simplicity that $X_n(0) = 0$,

$$I(x) = \frac{1}{2} \int_0^\infty \dot{x}(s)^T \widehat{a}^{-1} \dot{x}(s) ds,$$

for absolutely continuous x, and $I(x) = \infty$ otherwise. If $\widehat{a}$ has rank $m < d$, then there exists a $d \times m$ matrix σ such that $\widehat{a} = \sigma\sigma^T$. Let W be a standard R^m-valued Brownian motion. Then the rate function for $\{X_n\}$ is the same as the rate function for

$$Y_n = \frac{1}{\sqrt{n}} \sigma W.$$

Consequently, by Schilder's theorem and the contraction principle,

$$I(x) = \inf\{\int_0^\infty \frac{1}{2} |u(t)|^2 dt : x(t) = \sigma \int_0^t u(s) ds, t \geq 0\},$$

where $\inf \emptyset = \infty$.

10.1.7. Extension to infinite dimensions. de Acosta [25] considers the analog of Theorem 10.1 for Banach space-valued processes. Let E be a separable Banach space and E^* its dual, and let $\widehat{E} \subset E^*$ be a separable subspace of E^* such that for each $z \in E$, $\|z\| = \sup_{\xi \in \widehat{E}} \langle z, \xi \rangle / \|\xi\|$.

Let η be a σ-finite measure on E satisfying

$$\int_E \|z\|^2 e^{\alpha \|z\|} \eta(dz) < \infty,$$

for each $\alpha > 0$. Let Ψ be a mean zero, Gaussian, E-valued random variable, and for $\xi_1, \xi_2 \in E^*$, define

$$a(\xi_1, \xi_2) = E[\langle \Psi, \xi_1 \rangle \langle \Psi, \xi_2 \rangle].$$

Let $\mathcal{D}(A)$ be the collection of functions of the form

$$f(x) = g(\langle x, \xi_1 \rangle, \ldots, \langle x, \xi_m \rangle), \tag{10.10}$$

for $\xi_1, \ldots, \xi_m \in \widehat{E}$ and $g \in D_m = \{g : g - c \in C_c^2(R^m) \text{ some } c \in R\}$, $m = 1, 2, \ldots$. To simplify notation, we write $\langle x, \xi \rangle = (\langle x, \xi_1 \rangle, \ldots, \langle x, \xi_m \rangle)$ and

$$a(\xi) = ((a(\xi_i, \xi_j)\ )).$$

Let X be the E-valued process with generator

$$\begin{aligned} Af(x) &= \frac{1}{2} \sum_{i,j} a(\xi_i, \xi_j) \partial_i \partial_j g(\langle x, \xi \rangle) + \sum_i \langle b, \xi_i \rangle \partial_i g(\langle x, \xi \rangle) \\ &\quad + \int_E (g(\langle x + z, \xi \rangle) - g(\langle x, \xi \rangle) - \sum_i \langle z, \xi_i \rangle \partial_i g(\langle x, \xi \rangle)) \eta(dz), \end{aligned}$$

and define $X_n(t) = \frac{1}{n} X(nt)$.

Since $X(nt) - X(0)$ can be represented as a sum of n independent copies of $X(t) - X(0)$, for each $t > 0$, exponential tightness for $\{X_n(t)\}$ follows as in Example 3.4. Since X has stationary, independent increments, for $\lambda > 0$, we have

$$\begin{aligned} E[e^{n\lambda \|X_n(t+u) - X_n(t)\| \wedge 1} | \mathcal{F}_t^n] &= E[e^{n\lambda \|X_n(u) - X_n(0)\| \wedge 1}] \\ &= E[e^{\lambda \|X(nu) - X(0)\| \wedge 1}] \\ &\le E[e^{\lambda \|X(u) - X(0)\| \wedge 1}]^n. \end{aligned}$$

Consequently, in Theorem 4.1, we can take

$$\gamma_n(\delta, \lambda, T) = n \sup_{0 \le u \le \delta} \log E[e^{\lambda \|X(u) - X(0)\| \wedge 1}].$$

Since

$$\lim_{u \to 0} \limsup_{n \to \infty} \frac{1}{n} \log E[e^{\gamma_n(\delta, \lambda, T)}] = \lim_{u \to 0} \log \sup_{0 \le u \le \delta} E[e^{\lambda \|X(u) - X(0)\| \wedge 1}] = 0,$$

exponential tightness for $\{X_n\}$ follows.

Assume for the moment, that $X_n(0) = 0$. Then $x_0 + X_n$ is the process with generator A_n and initial value x_0. By the finite-dimensional results, for $\xi_1, \ldots, \xi_m \in E^*$, the large deviation principle holds for $\langle X_n, \xi \rangle$. Let V^ξ denote the corresponding semigroup and I^ξ denote the corresponding rate function. It follows that for $f(x) = g(\langle x,\xi\rangle)$ and $x_0^n \to x_0$,

$$\begin{aligned} V(t)f(x_0) &= \lim_{n\to\infty} V_n(t)f(x_0^n) \\ &= \lim_{n\to\infty} \frac{1}{n}\log E[e^{nf(x_0^n+X_n(t))}] \\ &= \lim_{n\to\infty} \frac{1}{n}\log E[e^{ng(\langle x_0^n,\xi\rangle+\langle X_n(t),\xi\rangle)}] \\ &= V^\xi(t)g(\langle x_0,\xi\rangle) \end{aligned}$$

exists, and hence the convergence of $V_n(t)f$ to $V(t)f$ will be uniform on compact sets.

Suppose that $X_n(0)$ satisfies the large deviation principle with rate function I_0 and rate transform Λ_0. Observing that $\mathcal{D}(A)$ contains a subset that is bounded above and isolates points, by Theorem 5.15, $\{X_n\}$ satisfies the large deviation principle with

$$\begin{aligned} &I_{t_1,\ldots,t_k}(x_1,\ldots,x_k) \\ (10.11)\quad &= \sup_{f_1,\ldots,f_k\in\mathcal{D}(A)} \{f_1(x_1)+\ldots+f_k(x_k) \\ &\qquad -\Lambda_0(V(t_1)(f_1+V(t_2-t_1)(f_2+\ldots V(t_k-t_{k-1})f_k)\ldots))\} \\ &= \sup_m \sup_{\xi\in\widehat{E}^m} \sup_{g_1,\ldots,g_k\in D_m} \{g_1(\langle x_1,\xi\rangle)+\ldots+g_k(\langle x_k,\xi\rangle) \\ &\qquad -\Lambda_0^\xi(V^\xi(t_1)(g_1+V^\xi(t_2-t_1)(g_2+\ldots V^\xi(t_k-t_{k-1})g_k)\ldots))\} \\ &= \sup_m \sup_{\xi\in\widehat{E}^m} I^\xi_{t_1,\ldots,t_k}(\langle x_1,\xi\rangle,\ldots,\langle x_k,\xi\rangle). \end{aligned}$$

It follows that

$$\begin{aligned} I(x) &= \sup_m \sup_{\xi\in\widehat{E}^m} I^\xi(\langle x,\xi\rangle) \\ &= \sup_m \sup_{\xi\in\widehat{E}^m} \left(I_0^\xi(\langle x(0),\xi\rangle)+\int_0^\infty L^\xi(\frac{d}{ds}\langle x(s),\xi\rangle)ds\right) \\ &= \sup_m \sup_{\xi\in\widehat{E}^m} \left(I_0^\xi(\langle x(0),\xi\rangle)+\int_0^\infty \sup_{p\in R^m}(p\cdot\frac{d}{ds}\langle x(s),\xi\rangle-H^\xi(p))ds\right) \\ &= \lim_{m\to\infty} \sup_{\xi\in\widehat{E}^m} \left(I_0^\xi(\langle x(0),\xi\rangle)+\int_0^\infty \sup_{\zeta\in S(\xi)}(\frac{d}{ds}\langle x(s),\zeta\rangle-H(\zeta))ds\right) \\ &= \lim_{m\to\infty} \left(I_0^{\xi^m}(\langle x(0),\xi^m\rangle)+\int_0^\infty \sup_{\zeta\in S(\xi^m)}(\frac{d}{ds}\langle x(s),\zeta\rangle-H(\zeta))ds\right) \\ &= \left(I_0(x(0))+\int_0^\infty \sup_{\zeta\in\widehat{E}}(\frac{d}{ds}\langle x(s),\zeta\rangle-H(\zeta))ds\right), \end{aligned}$$

where for $\xi\in\widehat{E}^m$ and $p\in R^m$, $S(\xi)$ denotes the linear subspace of $\widehat{E}$ spanned by $\xi_1,\ldots,\xi_m$,

$$H^\xi(p)=\int_{R^d}(e^{p\cdot\langle z,\xi\rangle}-1-\langle z,\xi\rangle\cdot p)\eta(dz)+\frac{1}{2}p^Ta(\xi)p+\langle b,\xi\rangle\cdot p,$$

for $\zeta \in \widehat{E}$,

$$H(\zeta) = \int_{R^d} (e^{\langle z,\zeta\rangle} - 1 - \langle z,\zeta\rangle)\eta(dz) + \frac{1}{2}a(\zeta) + \langle b,\zeta\rangle,$$

and $\xi^m = (\xi_1, \ldots, \xi_m)$ for some sequence $\xi_1, \xi_2, \ldots$ that is dense in $\widehat{E}$. Note that $H^\xi(p) = H(\sum_{i=1}^m p_i\xi_i)$.

10.2. Random walks

The discrete-time analogues of the results in Section 10.1 are results on random walks considered by Borovkov [**11**] and Mogulskii [**87**]. Again, the comparison principle is trivial, since H does not depend on x and Condition 9.10 is immediate.

Let $\xi_1, \xi_2, \ldots$ be independent and identically distributed R^d-valued random variables with distribution ν. Define

$$X_n(t) = \beta_n^{-1} \sum_{k=1}^{[t/\epsilon_n]} \xi_k.$$

Then,

$$A_n f(x) = \epsilon_n^{-1}(\int_{R^d} f(x + \beta_n^{-1} z)\nu(dz) - f(x))$$

and

$$H_n f(x) = \frac{1}{n\epsilon_n} \log \int_{R^d} e^{n(f(x+\beta_n^{-1}z) - f(x))}\nu(dz).$$

10.2.1. Convergence of H_n. If $\int_{R^d} e^{\alpha\cdot z}\nu(dz) < \infty$, $\alpha \in R^d$, $\epsilon_n = n^{-1}$, and $\beta_n = n$, then $H_n f \to Hf$ given by

$$Hf(x) = \log \int_{R^d} e^{\nabla f(x)\cdot z}\nu(dz).$$

If $\int_{R^d} z\nu(dz) = 0$, $\int_{R^d} e^{\alpha|z|}\nu(dz) < \infty$ for some $\alpha > 0$, $\epsilon_n\beta_n^2 = n$, and $\epsilon_n n \to 0$, then $H_n f \to Hf$ given by

$$Hf(x) = \frac{1}{2}\int_{R^d} (\nabla f(x) \cdot z)^2 \nu(dz).$$

The moment condition can be relaxed as in Section 10.1.6.

The large deviation theorem and rate function representation are essentially the same as in Section 10.1.

10.3. Markov processes

Let $E = R^d \cup \infty$ denote the one-point compactification of R^d. Let $a = \sigma\sigma^T$, and for $f \in D_d$ given by (9.41), define

$$(10.12)\quad A_n f(x) = n\int_{R^d} (f(x + \frac{1}{n}z) - f(x) - \frac{1}{n}z \cdot \nabla f(x))\eta(x, dz) + b(x) \cdot \nabla f(x) + \frac{1}{2n}\sum_{i,j} a_{ij}(x)\partial_i\partial_j f(x),$$

for $x \in R^d$, and $A_n f(\infty) = 0$. We will require σ, b, and η to satisfy the following condition.

CONDITION 10.3. *$\sigma : R^d \mapsto M^{d\times d}$ and $b : R^d \mapsto R^d$ are locally bounded and measurable, $\psi : R^d \mapsto [1,\infty)$ and $\eta : R^d \mapsto \mathcal{M}(R^d)$, the space of σ-finite measures on R^d, satisfies*

$$\sup_x \int_{R^d} \left|\frac{z}{\psi(x)}\right|^2 e^{\alpha|z|/\psi(x)}\eta(x,dz) < \infty, \tag{10.13}$$

for each $\alpha > 0$.

If a is bounded and continuous, b is bounded, for each $x \in R^d$, $a(x)$ is positive definite, and for each $\Gamma \in \mathcal{B}(R^d)$, the mapping

$$x \to \int_\Gamma |z|^2(1+|z|^2)^{-1}\eta(x,dz)$$

is bounded and continuous, then for any initial distribution $\nu \in \mathcal{P}(R^d)$, existence and uniqueness will hold for the martingale problem for (A_n,ν) by Theorem 4.3 of Stroock [**114**] (Theorem 8.3.3 of Ethier and Kurtz [**36**]). Existence and uniqueness can also be obtained by verifying existence and uniqueness for a stochastic equation of the form

$$\begin{aligned} X_n(t) &= X_n(0) + \frac{1}{\sqrt{n}}\int_0^t \sigma(X_n(s))dW(s) + \int_0^t b(X_n(s))ds \\ &\quad + \frac{1}{n}\int_{[0,\infty)\times U\times[0,t]} I_{[0,n\beta(X_n(s-),u)]}(v)g(X_n(s-),u)\widetilde{\xi}(dv\times du\times ds), \end{aligned} \tag{10.14}$$

where $(U,\mathcal{U})$ is a measurable space, ξ is a Poisson random measure on $[0,\infty)\times U\times[0,\infty)$ with mean measure $m\times\gamma\times m$ (m denotes Lebesgue measure on $[0,\infty)$), γ is a σ-finite measure on U, and $\widetilde{\xi} = \xi - m\times\gamma\times m$. X_n given by (10.14) is a solution of the martingale problem for A_n with

$$\eta(x,\Gamma) = \int_U \beta(x,u)I_\Gamma(g(x,u))\gamma(du).$$

Conditions for uniqueness are given in Graham [**58**], Theorem 1.2, and Kurtz and Protter [**76**], Theorem 8.3.

More generally, we will simply assume that A_n has an extension $\widehat{A}_n$ that is the full generator for a well-posed martingale problem, that X_n is a solution of the martingale problem for $\widehat{A}_n$, and that the large deviation principle holds for $X_n(0)$ with a good rate function I_0. If $A_n \subset C(E)\times C(E)$, then such an extension exists by Krylov's theorem (see Theorems 4.5.11 and 4.5.19 of [**36**]), and the distribution of the solution, $P_x^n \in \mathcal{P}(D_E[0,\infty))$ is a measurable function of the initial position x.

10.3.1. Convergence of H_n. The operator $H_nf = \frac{1}{n}e^{-nf}A_ne^{nf}$ is given by

$$\begin{aligned} H_nf(x) &= \int_{R^d}(e^{n(f(x+\frac{1}{n}z)-f(x))} - 1 - z\cdot\nabla f(x))\eta(x,dz) \\ &\quad + \frac{1}{2n}\sum_{i,j} a_{ij}(x)\partial_i\partial_j f(x) \\ &\quad + \frac{1}{2}\sum_{i,j} a_{ij}(x)\partial_i f(x)\partial_j f(x) + b(x)\cdot\nabla f(x), \end{aligned}$$

and if Condition 10.3 holds with $\psi(x) = 1 + |x|$, for each $f \in D_d$, $H_n f$ converges, uniformly in x, to Hf given by

$$Hf(x) = \int_{R^d} (e^{\nabla f(x)\cdot z} - 1 - z \cdot \nabla f(x))\eta(x, dz) \\ + \frac{1}{2}\sum_{i,j} a_{ij}(x)\partial_i f(x)\partial_j f(x) + b(x) \cdot \nabla f(x).$$

Uniform convergence on compact subsets of R^d follows easily from the exponential moment conditions. To see that convergence is uniform on all of R^d, suppose the support of $f - f(\infty)$ is contained in $B_K(0)$ for some $K > 0$. Assume that $|x| > 2K + 1$. Then

$$|f(x + \frac{1}{n}z) - f(x)| \le \frac{2n^{-1}|z|\|f\|}{1 + |x|},$$

since the left side is zero unless $|x + n^{-1}z| < K$ and hence, $n^{-1}|z| > |x| - K \ge \frac{1}{2}(|x| + 1)$. It follows that

$$|H_n f(x)| \le \int_{\{|z|/(1+|x|)>n/2\}} (e^{2\|f\|\frac{|z|}{1+|x|}} - 1)\eta(x, dz). \tag{10.15}$$

Taking $\psi(x) = (1 + |x|)$ in Condition 10.3, the right side of (10.15) converges to zero uniformly on R^d, so the left side converges uniformly on $\{x : |x| > 2K + 1\}$.

Defining $H(x,p) : R^d \times R^d \mapsto R$ by

$$H(x,p) = \frac{1}{2}|\sigma^T(x)p|^2 + b(x) \cdot p + \int_{R^d} (e^{p\cdot z} - 1 - p \cdot z)\eta(x, dz), \tag{10.16}$$

we have

$$Hf(x) = H(x, \nabla f(x)). \tag{10.17}$$

Note that if H is continuous, then b is continuous, but that there is an interplay between σ and η.

10.3.2. Exponential tightness. The compact containment condition can be verified as in Example 4.23. Alternatively, we can consider the process in the one-point compactification of R^d, $E = R^d \cup \{\infty\}$. In either case, exponential tightness for $\{X_n\}$ follows by Corollary 4.17.

Since Theorem 6.14 is proved under a compactness assumption on E, we follow the latter alternative with $E = R^d \cup \{\infty\}$. We still have $Hf(x) = H(x, \nabla f(x))$, where $\nabla f(\infty)$ is understood to be 0.

10.3.3. The comparison principle. By Lemma 6.8, we can consider $\overline{H}$, the closure of H, rather than H. Under Condition 10.3, if for some $K > 0$, $|\sigma(x)| + |b(x)| \le K\psi(x)$, then $\mathcal{D}(\overline{H})$ contains all functions that are continuously differentiable on R^d and satisfy $\lim_{|x|\to\infty} f(x) = f(\infty)$ and $\lim_{|x|\to\infty} \psi(x)|\nabla f(x)| = 0$.

The next lemmas verify Condition 9.10 for most of the standard large deviation theorems for R^d-valued Markov processes. We say that σ and b have *at most linear growth* if there exist constants K_1 and K_2 such that

$$|\sigma(x)| + |b(x)| \le K_1 + K_2|x|. \tag{10.18}$$

LEMMA 10.4. *If σ and b have at most linear growth and and there exists $\delta > 0$ such that*

$$\sup_{|p|<\delta}\sup_x \int_{R^d} (e^{p\cdot z/(1+|x|)} - 1 - \frac{p\cdot z}{1+|x|})\eta(x,dz) < \infty, \tag{10.19}$$

then Condition 9.10.2 is satisfied.

PROOF. If $|x_m| \to \infty$ and $|x_m|p_m \to 0$, then

$$|b(x_m)\cdot p_m| + |\sigma^T(x_m)p_m| \leq (K_1 + K_2|x_m|)|p_m| \to 0,$$

and

$$\begin{aligned}&\int_{R^d}(e^{p_m\cdot z} - 1 - p_m\cdot z)\eta(x_m,dz)\\ &\quad = \int_{R^d}(e^{(1+|x_m|)p_m\cdot z/(1+|x_m|)} - 1 - \frac{(1+|x_m|)p_m\cdot z}{(1+|x_m|)})\eta(x_m,dz) \to 0,\end{aligned}$$

and it follows that $H(x_m,p_m) \to 0$. □

LEMMA 10.5. *Suppose H is continuous, and for each $x, p \in R^d$,*

$$\lim_{r\to\infty} H(x,rp) = \infty. \tag{10.20}$$

Suppose σ and b have at most linear growth and Condition 10.3 holds with $\psi(x) = 1+|x|$. Then Condition 9.10 and the comparison principle holds for H.

REMARK 10.6. Roughly speaking, (10.20) implies that the process can move in any direction from any point. In particular, (9.46) is satisfied if $a(x) = \sigma(x)\sigma^T(x)$ is positive definite or if for each p and x, $\eta(x,\{z : z\cdot p > 0\}) > 0$. If Conditions 10.2.2 and 10.2.4 of Dupuis and Ellis [**35**] hold, then the conditions of the lemma hold.

PROOF. Condition 9.10.1 follows by Lemma 9.16. Condition 9.10.2 holds by Lemma 10.4 which in turn implies that $\mathcal{D}(\overline{H})$ contains all continuously differentiable f satisfying (9.43). The comparison principle then follows by Lemma 9.15. □

The next lemmas verify Condition 9.11, which implies Condition 9.10.1 by Lemma 9.12. Condition 9.11 is preferable to Condition 9.10.1, since the collection of H satisfying Condition 9.11 is closed under positive linear combinations.

LEMMA 10.7. *Let $(U,\mathcal{U})$ be a measurable space, and let $\sigma : R^d\times U \mapsto R^d$ be locally Lipschitz in the sense that for each $r > 0$, there exists C_r such that*

$$|\sigma(x,u) - \sigma(y,u)| < C_r|x-y|, \quad |x|,|y| \leq r, u \in U.$$

Suppose ν is a finite measure on U, $\int_U |\sigma(x,u)|^2\nu(du) < \infty$, $x \in R^d$, and

$$H(x,p) = \frac{1}{2}\int_U |\sigma^T(x,u)\cdot p|^2\nu(du).$$

Then H satisfies Condition 9.11.

REMARK 10.8. Note that this lemma covers the case $H(x,p) = |\sigma^T(x)p|^2$ for Lipschitz $\sigma : R^d \to M^{d\times d}$.

PROOF. Fix $r > 0$. Let $\lambda > 1$, $\mu > 0$, and define $\delta = \mu|x-y|^2$. Then

$$\begin{aligned}
&\lambda H(x, \frac{\mu(x-y)}{\lambda}) - H(y, \mu(x-y)) \\
&= \frac{\mu^2}{2\lambda}\int_U |\sigma^T(x,u)\cdot(x-y)|^2\nu(du) - \frac{\mu^2}{2}\int_U |\sigma^T(y,u)\cdot(x-y)|^2\nu(du) \\
&= \frac{\mu^2}{2\lambda}\int_U |(\sigma(x,u)-\sigma(y,u))^T\cdot(x-y)|^2\nu(du) \\
&\qquad +\frac{\mu^2}{\lambda}\int_U (\sigma(y,u)\cdot(x-y))(\sigma(x,u)-\sigma(y,u))^T\cdot(x-y)\nu(du) \\
&\qquad -(1-\lambda^{-1})\frac{\mu^2}{2}\int_U |\sigma^T(y,u)\cdot(x-y)|^2\nu(du) \\
&\le 2^{-1}\lambda^{-1}C_r^2\delta^2\nu(U) + \lambda^{-1}\delta C_r\mu\int_U |\sigma^T(y,u)\cdot(x-y)|\nu(du) \\
&\qquad -2^{-1}(1-\lambda^{-1})\mu^2\int_U |\sigma^T(y,u)\cdot(x-y)|^2\nu(du) \\
&\le \left(2^{-1}\lambda^{-1}C_r^2\delta^2 + \frac{(\delta C_r)^2}{2\lambda(\lambda-1)}\right)\nu(U),
\end{aligned}$$

where the last inequality follows from the fact that

$$au - \frac{1}{2}bu^2 \le \frac{a^2}{2b}.$$

Consequently, if $\sup_m(|x_m| + |y_m|) \le r$ and $\delta_m \equiv \mu_m|x_m - y_m|^2 \to 0$, (9.40) follows. □

LEMMA 10.9. *Let $b : R^d \mapsto R^d$ be locally Lipschitz and define $H(x,p) = b(x)\cdot p$. Then H satisfies Condition 9.11.*

PROOF. Fix $r > 0$, and let C satisfy

$$|b(x) - b(y)| < C|x-y|, \quad |x|, |y| \le r.$$

Let $\lambda > 1$, $\mu > 0$. Then

$$\begin{aligned}
&\lambda H(x, \frac{\mu(x-y)}{\lambda}) - H(y, \mu(x-y)) \\
&= b(x)\cdot\mu(x-y) - b(y)\cdot\mu(x-y) \\
&= \mu(b(x)-b(y))\cdot(x-y) \\
&\le C\mu|x-y|^2,
\end{aligned}$$

and the lemma follows. □

LEMMA 10.10. *Suppose that σ, b, and η satisfy Condition 10.3 with $\psi(x) = 1+|x|$ and H is given by (10.16). Suppose that σ and b are Lipschitz and that η is of the form*

$$\eta(x,\Gamma) = \int_V I_\Gamma(g(x,v))\gamma(dv), \tag{10.21}$$

where $g : R^d \times V \mapsto R^d$. Let $\Phi(u) = u^2 e^{|u|}$, and assume that for each $r > 0$ there exists a $\delta_r > 0$ such that

$$\sup_{|x|,|y|\le r, x\neq y} \int_V \Phi(\delta_r \frac{|g(x,v)-g(y,v)|}{|x-y|})\gamma(dv) < \infty. \tag{10.22}$$

Then H satisfies Condition 9.11, Condition 9.10, and the comparison principle.

REMARK 10.11. With $\eta = 0$, large deviation results for Lipschitz σ and b are given by Theorem 5.6.7 of Dembo and Zeitouni [**29**]. de Acosta [**26**] gives results for Lipschitz σ and b and η of the form (10.21). The condition in (10.22) is essentially Condition 3.5 of his result.

PROOF. Let

$$H_0(x,p) = \int_{R^d} (e^{p\cdot z} - 1 - p\cdot z)\eta(x,dz) = \int_V (e^{g(x,v)\cdot p} - 1 - g(x,v)\cdot p)\gamma(dv).$$

Then H_0 satisfies Condition 9.18 with $\beta \equiv 1$, and hence, by Lemma 9.19, Condition 9.11. By Lemmas 10.7 and 10.9 and Lemma 9.13, H satisfies Condition 9.11 and (9.39) follows with

$$\omega(\lambda) = \begin{cases} 0 & \lambda > 1 \\ \infty & \lambda \le 1. \end{cases}$$

By the Lipschitz continuity of σ and b, σ and b have at most linear growth. Condition 10.3 with $\psi(x) = 1 + |x|$ implies (10.19). Consequently, Condition 9.10.2 follows by Lemma 10.4.

The lemma then follows by Lemma 9.15 □

LEMMA 10.12. *Suppose that σ, b, and η satisfy Condition 10.3 with $\psi(x) = 1 + |x|$ and H is given by (10.16). Suppose that σ and b are Lipschitz and that η is of the form*

$$\eta(x,\Gamma) = \int_U \beta(x,u) I_\Gamma(g(x,u))\gamma(du), \tag{10.23}$$

where $g : R^d \times U \to R^d$ and $\beta \ge 0$. Assume that there exists $\overline{K}$ such that

$$\beta(x,u) - \beta(y,u) \le \overline{K}|x-y|\beta(y,u).$$

Let $\Phi(z) = z^2 e^{|z|}$, and assume that for each $r > 0$ there exists a $\delta_r > 0$ such that

$$\sup_{|x|,|y|\le r, x\neq y} \int_U \beta(x,u)\Phi(\delta_r \frac{|g(x,u)-g(y,u)|}{|x-y|})\gamma(du) < \infty. \tag{10.24}$$

Then H satisfies Condition 9.10, and the comparison principle.

REMARK 10.13. This lemma covers processes with jump terms of the form

$$n\int_U \beta(x,u)(f(x + \frac{1}{n}g(x,u)) - f(x) - \frac{1}{n}g(x,u)\cdot\nabla f(x))\gamma(du).$$

For example, taking $U = R^d$ and $g(x,u) = u$, we have

$$n\int_{R^d} \beta(x,u)(f(x + \frac{1}{n}u) - f(x) - \frac{1}{n}u\cdot\nabla f(x))\gamma(du).$$

PROOF. Let

$$\begin{aligned} H_0(x,p) &= \frac{1}{2}|\sigma(x)^T p|^2 + \int_{R^d} (e^{p\cdot z} - 1 - p\cdot z)\eta(x,dz) \\ &= \frac{1}{2}|\sigma(x)^T p|^2 + \int_U \beta(x,u)(e^{g(x,u)\cdot p} - 1 - g(x,u)\cdot p)\gamma(du). \end{aligned}$$

Noting that

$$|\sigma(x)^T p|^2 = \sum_{i=1}^d |\sigma_i(x)\cdot p|^2,$$

where $\sigma_i(x)$ is the ith column of $\sigma(x)$, we can take $V = U \cup \{1,\ldots,d\}$ and extend γ to a measure on V by setting $\gamma(\{i\}) = 1$, extend β by defining $\beta(x,i) = 1$, and extend g by defining $g(x,i) = \sigma_i(x)$. Then H_0 satisfies the conditions of Lemma 9.20. Consequently, H satisfies Condition 9.10.1 with

$$\omega(\lambda) = \begin{cases} 0 & \lambda > 1 \\ \infty & \lambda \le 1. \end{cases}$$

The remainder of the proof is the same as for Lemma 10.10 □

Occasionally, the convex dual L is of simpler form than H. The following lemma gives conditions on L that can be useful in verifying the comparison principle.

LEMMA 10.14. *Suppose* $L : R^d \times R^d \mapsto [0,\infty)$ *and*

$$H(x,p) = \sup_{q\in R^d}\{p\cdot q - L(x,q)\}, \quad x,p \in R^d. \tag{10.25}$$

Then Condition 9.10.1 holds if L *satisfies the following:*

CONDITION 10.15. *For each compact set* $\Gamma \subset R^d$ *and* $\epsilon > 0$ *there exist* $\delta > 0$ *and* $C > 0$ *such that if* $x,y \in \Gamma$ *satisfies* $|x-y| < \delta$ *and* $q \in R^d$, *then there exists* $\widehat{q} \in R^d$ *such that*

$$\begin{aligned} L(y,\widehat{q}) - L(x,q) &\le \frac{\epsilon}{2}(1 + L(x,q)) \\ |\widehat{q} - q| &\le C|x-y|(1 + L(x,q)). \end{aligned} \tag{10.26}$$

If, in addition, Condition 9.10.2 holds, then the comparison principle holds.

REMARK 10.16. Condition 10.15 is Condition 10.2.5 of Dupuis and Ellis [**35**]. This condition is implied by

$$\lim_{\delta\to\infty} \sup_{x,y\in R^d, |x-y|<\delta} \sup_{q, L(y,q)<\infty} \frac{L(y,q) - L(x,q)}{1 + L(x,q)} = 0,$$

which is Condition III on page 146 of Freidlin and Wentzell [**52**]. (Take $\widehat{q} = q$.)

PROOF. For $\lambda > 1$, let $\epsilon = \lambda - 1$ and suppose that $C\mu|x-y|^2 < \frac{\epsilon}{2}$. For $q \in R^d$, let $\widehat{q}$ satisfy (10.26). Then, setting $p = \mu(x-y)$,

$$\begin{aligned}
&\lambda H(x, \frac{p}{\lambda}) - H(y,p) \\
&\quad = \sup_q (p \cdot q - \lambda L(x,q)) - \sup_q (p \cdot q - L(y,q)) \\
&\quad = \sup_q \left(\inf_{q'} (L(y,q') - p \cdot q') + p \cdot q - \lambda L(x,q)) \right) \\
&\quad \leq \sup_q (L(y,\widehat{q}) - \lambda L(x,q) + p \cdot (q - \widehat{q})) \\
&\quad \leq \sup_q (-(\lambda - 1) L(x,q) + \frac{\epsilon}{2}(1 + L(x,q)) + C\mu|x-y|^2(1 + L(x,q)) \\
&\quad \leq \epsilon,
\end{aligned}$$

and the lemma follows. □

10.3.4. The large deviation theorem.

THEOREM 10.17. *Assume that A_n given by (10.12) has an extension $\widehat{A}_n$ that is the full generator for a well-posed martingale problem whose distribution on $D_E[0,\infty)$ is a measurable function of the initial value. Suppose that X_n is a solution of the martingale problem for $\widehat{A}_n$ and that the large deviation principle holds for $X_n(0)$ in R^d with a good rate function I_0.*

Assume that σ and b have at most linear growth, that Condition 10.3 holds with $\psi(x) = 1 + |x|$, and that 9.10.1 holds. Then

a) *For each $\alpha > 0$ and $f \in C(E)$, there exists a unique $R_\alpha f \in C(E)$ such that $(I - \alpha H) R_\alpha f = f$ in the viscosity sense, and*

$$V(t)f = \lim_n R^n_{t/n} f$$

defines a semigroup $V(t) : C(E) \mapsto C(E)$.

b) *The large deviation principle holds for $\{X_n\}$ in $D_E[0,\infty)$ with good rate function:*

$$I(x) = \sup_{\{t_i\} \subset \Delta_x^c} I_{t_1,\ldots,t_m}(x(t_1), \ldots, x(t_m)),$$

where

$$\begin{aligned}
&I_{t_1,\ldots,t_m}(x_1,\ldots,x_m) \\
&\quad = \inf_{x_0} \{ I_0(x_0) + \sup_{f_1,\ldots,f_k \in D_d} \{ f_1(x_1) + \ldots + f_k(x_k) \\
&\qquad\qquad - V(t_1)(f_1 + V(t_1 - t_1)(f_2 + \ldots V(t_k - t_{k-1}) f_k) \ldots))(x_0) \}\}.
\end{aligned}$$

c) *If $x \in D_E[0,\infty) - D_{R^d}[0,\infty)$, then $I(x) = \infty$.*

REMARK 10.18. a) Condition 9.10.2 holds by Lemma 10.4.

b) The critical assumption is that Condition 9.10.1 holds. The results of Section 10.3.3 identify a number of settings in which this condition is valid. In particular, the condition holds under any of the following conditions:

(1) H is continuous and nondegenerate in the sense of (10.20). For example, if $\sigma(x)$ is nonsingular for each x, then the nondegeneracy condition holds.

See Lemma 10.5. These conditions combined with the variational conditions covered by Lemma 10.14 extend Theorem 10.2.6 of Dupuis and Ellis [**35**].

(2) σ and b are Lipschitz and $\eta = 0$. See Lemmas 10.7 and 10.9 and recall that Condition 9.11 implies Condition 9.10.1. These conditions give Theorem 5.6.7 of Dembo and Zeitouni [**25**].

(3) σ and b are Lipschitz and η satisfies the conditions of Lemma 10.10 or Lemma 10.12. Under the conditions of Lemma 10.10, the theorem extends Theorem 3.1 of de Acosta [**26**].

PROOF. Let $H_n f = \frac{1}{n} e^{-nf} A_n e^{nf}$ and H be defined as in (10.17). Let D_d be given by (9.41). Then $\lim_n \|H_n f - Hf\| = 0$ for each $f \in D_d$.

The comparison principle for $(I - \alpha H)f = h$ is given by Lemma 9.15, and parts (a) and (b) follow by Theorem 6.14.

Part (c) follows by the compact containment condition. (See Example 4.23.) □

10.3.5. Variational representation of the rate function.

LEMMA 10.19. *Assume Condition 10.3 holds. Then for each $x \in R^d$, $H(x, \cdot)$ is convex and finite (hence continuous) on R^d.*

Define $L(x, \cdot)$ as the Fenchel-Legendre transform of $H(x, \cdot)$, that is,

$$L(x,q) = \sup_{p \in R^d} \{p \cdot q - H(x,p)\}, \quad x, q \in R^d. \tag{10.27}$$

Then $L(x, q)$ is convex and lower semicontinuous in $q \in R^d$ and

$$H(x,p) = \sup_{q \in R^d} \{p \cdot q - L(x,q)\}. \tag{10.28}$$

REMARK 10.20. In the diffusion case, $H(x,p) = \frac{1}{2}|\sigma(x)^T p|^2 + b(x) \cdot p$. If $a(x) = \sigma(x)\sigma(x)^T$ is invertible for all x,

$$L(x,q) = \frac{1}{2}(q - b(x))^T a(x)^{-1}(q - b(x)).$$

PROOF. The convexity and finiteness of $H(x, \cdot)$ follows by direct calculation. The Fenchel and Moreau theorem in convex analysis gives (10.28). The convexity and lower semicontinuity of $L(x, \cdot)$ are basic properties of the transform (10.27). (See [**104**], Chapter 12.) □

Note that if we define

$$L_\psi(x,q) = \sup_{p \in R^d} (p \cdot q - H(x, \frac{p}{\psi(x)}),$$

then $L_\psi(x,q) = L(x, q\psi(x))$ and

$$H(x,p) = \sup_{q \in R^d} (\psi(x) p \cdot q - L_\psi(x,q)).$$

LEMMA 10.21 (Variational representation of H). *Assume that Condition 10.3 holds, that ψ and H are continuous, and that there exists $K > 0$ such that $|\sigma(x)| + |b(x)| \le K\psi(x)$, $x \in R^d$. Let $U = R^d$. Define*

$$Af(x,q) = \psi(x) q \cdot \nabla f(x), \quad q \in U, f \in D_d,$$

where $\nabla f(\infty)$ *is defined to be* 0. *Extend the definition of* L_ψ *by:*

$$L_\psi(\infty, q) = \liminf_{|x|\to\infty, q'\to q} L_\psi(x, q'). \tag{10.29}$$

Then $L_\psi \geq 0$ *is lower semicontinuous on* $E \times U$,

$$Hf(x) = \mathbf{H}f(x) \equiv \sup_{q\in U}((Af - L_\psi)(x,q)), \quad x \in E, f \in D_d, \tag{10.30}$$

and Conditions 8.9, 8.10, and 8.11 are satisfied.

PROOF. $L_\psi(x,q) \geq 0$ follows from $H(x, 0) = 0$. By the continuity of H, L_ψ as extended in (10.29) is lower semicontinuous. The representation (10.30) follows from (10.28).

Next consider Condition 8.9. $A \subset C_b(E) \times C(E \times U)$ and $\overline{D}_d = C_b(E)$ are immediate giving Condition 8.9.1; $\Gamma = E \times U$ and for each $x_0 \in E$, $x(t) \equiv x_0$ and $\lambda(dq \times ds) = \delta_{\{0\}}(dq) \times ds$ defines a pair such that $(x, \lambda) \in \mathcal{J}^\Gamma$ giving 8.9.2.

By Condition 10.3,

$$\overline{H}(c) \equiv \sup_{|p|\leq c} \sup_{x\in R^d} H(x, p/\psi(x)) < \infty,$$

and it follows that $\lim_{|q|\to\infty} \inf_x L_\psi(x,q) = \infty$ implying Condition 8.9.3. Condition 8.9.4 is immediate from the compactness of E.

To obtain the estimate in Condition 8.9.5, define, which is finite and note that

$$\frac{L_\psi(x,q)}{|q|} \geq \sup_{|p|=c} p \cdot \frac{q}{|q|} - \frac{\overline{H}(|c|)}{|q|} = c - \frac{\overline{H}(c)}{|q|}.$$

It follows that

$$\lim_{N\to\infty} \inf_{x\in R^d} \inf_{|q|=N} \frac{L_\psi(x,q)}{|q|} = \infty. \tag{10.31}$$

Define

$$\varphi(s) = s \inf_{x\in R^d} \inf_{|q|\geq s} \frac{L_\psi(x,q)}{|q|}, \quad s \geq 0.$$

Then φ is a strictly increasing function. For each $f \in D_d$, there exists $0 < C_f < \infty$ such that $|Af(x,q)| = |\psi(x)q \cdot \nabla f(x)| \leq C_f|q|$. By (10.31), $r^{-1}\varphi(r) \to +\infty$ as $r \to \infty$. Let $\psi_f(r) = C_f\varphi^{-1}(r)$. Then $r^{-1}\psi_f(r) \to 0$, and since $\varphi(C_f^{-1}|Af(x,q)|) \leq \varphi(|q|) \leq L_\psi(x,q)$, we have $|Af(x,q)| \leq \psi_f(L_\psi(x,q))$.

Since $r^{-1}\varphi(r) \to +\infty$ as $r \to +\infty$, φ is a tightness function, and Condition 8.9.3 follows.

To obtain Condition 8.10, observe that for $x \in R^d$

$$\frac{b(x)}{\psi(x)} = \frac{\partial}{\partial p} H(x, \frac{p}{\psi(x)}) \mid_{p=0} = (\partial_2 H)(x, 0)/\psi(x). \tag{10.32}$$

Since b is continuous, for each $x_0 \in R^d$, there exists a (local) solution of the differential equation $\dot{x} = b(x)$ satisfying $x(0) = x_0$. Note that x can be constructed so that either $\sup_{s\leq t} |x(s)| < \infty$ for all t or there exists t_0 such that $\lim_{t\to t_0} |x(t)| = \infty$. Define $x(t) = \infty$ for $t \geq t_0$.

Observe that $p \cdot b(x)/\psi(x) - H(x, p/\psi(x))$ attains its supremum in $p \in R^d$ at $p = 0$, and hence $L_\psi(x, b(x)/\psi(x)) = 0b(x)/\psi(x) - H(x, 0) = 0$. Note also that since b/ψ is bounded and L_ψ is lower semicontinuous, there exists $q_0 \in U$ such that $L_\psi(\infty, q_0) = 0$. Let $\lambda(du \times ds) = \delta_{\{b(x(s))/\psi(x(s))\}}(du) \times ds$ for $s < t_0$ and $\lambda(du \times ds) = \delta_{\{q_0\}}(du) \times ds$ for $s \geq t_0$. Then (x, λ) satisfies (8.12).

For $x_0 = \infty$, let $x(t) = \infty$ and $\lambda(du \times ds) = \delta_{\{q_0\}}(du) \times ds$. Then (x, λ) satisfies (8.12).

Finally, consider Condition 8.11. For each $f \in D_d$. let

$$q_f(x) = (\partial_2 H)(x, \nabla f(x)), \quad x \in R^d.$$

Then q_f is continuous and there exists a solution of $\dot{x}(t) = q_f(x(t))$ and $x(0) = x_0$ on an interval $[0, t_0)$ such that either $t_0 = \infty$ or $\lim_{t \to t_0} x(t) = \infty$. By the convex duality between $L_\psi(x, \cdot)$ and $H(x, \cdot/\psi(x))$, the finiteness of $H(x, p/\psi(x))$ and continuity of $H(x, p/\psi(x))$, applying Theorem 23.5 in Rockafellar [**104**], $q_f(x) = (\partial_p H)(x, \nabla f(x))$ implies that $\psi(x) q \nabla f(x) - L_\psi(x, q)$ attains its supremum in $q \in R^d$ at $q = q_f(x)/\psi(x)$. Therefore $Hf(x) = q_f(x)\nabla f(x) - L(x, q_f(x))$. Taking $\lambda(du \times ds) = \delta_{\{q_f(x(s))/\psi(x(s))\}}(du) \times ds$ for $s < t_0$ and $\lambda(du \times ds) = \delta_{\{q_0\}}(du) \times ds$ for $s \geq t_0$, $(x, \lambda) \in \mathcal{J}^\Gamma_{x_0}$ and satisfies (8.15).

For $x_0 = \infty$, let $x(t) \equiv \infty$ and $\lambda(du \times ds) = \delta_{\{q_0\}}(du) \times ds$. Then $(x, \lambda) \in \mathcal{J}^\Gamma_{x_0}$ and, since $Hf(\infty) = 0$, (8.15) holds as well. □

We have the following variation of the classical Freidlin-Wentzell theorem. By Lemmas 10.5, 10.21 and Remark 10.16, the result extends Theorem 10.2.6 of [**35**]; by Lemmas 10.10 and 10.21, it extends Theorem 3.1 of [**26**]

THEOREM 10.22. *Assume that the conditions of Theorem 10.17 hold and that H is continuous. Then the large deviation principle holds for $\{X_n\}$ in $D_{R^d}[0, \infty)$ with good rate function*

$$I(x) = \begin{cases} I_0(x(0)) + \int_0^\infty L(x(s), \dot{x}(s)) ds & \text{if } x \text{ is absolutely continuous} \\ +\infty & \text{otherwise} \end{cases}$$

PROOF. By Theorem 10.17, the large deviation principle holds for $\{X_n\}$ in $D_E[0, \infty)$ with some good rate function I. By Lemma 10.21, the conditions of Corollary 8.29 hold and the rate function satisfies

$$I(x) = I_0(x(0)) + \inf\{\int_{U \times [0,\infty)} L(x(s), \psi(X(s))u)\lambda(du \times ds) : (x, \lambda) \in \mathcal{J}\},$$

where $(x, \lambda) \in \mathcal{J}$ if

$$f(x(t)) - f(x(0)) = \int_0^t \psi(X(s)) u \nabla f(x(s)) \lambda(du \times ds), \quad f \in D_d.$$

For each $(x, \lambda) \in \mathcal{J}$, there exists a $\mathcal{P}(U)$-valued process $\mu(t)$ such that $\lambda(du \times dt) = \mu_t(du) \times dt$. If $\int_{U \times [0,\infty)} L(x(s), \psi(x(s))u)\lambda(du \times ds) < \infty$, then by (10.31), $\int_0^\infty |u| \lambda(du \times ds) < \infty$, $u(t) = \int_U u \mu_t(du)$ is integrable in t, and

$$x(t) = x(0) + \int_0^t \psi(x(s)) u(s) ds.$$

Consequently, x is absolutely continuous with $\dot{x}(t) = \psi(x(t))u(t)$. Since $L(x, q) = \sup_{p \in R^d}\{p \cdot q - H(x, p)\}$ is convex in q,

$$\begin{aligned} \int_0^\infty L(x(s), \psi(x(s))u(s)) ds &\leq \int_0^\infty L(x(s), \psi(x(s))u) \mu_s(du) \times ds \\ &= \int_{U \times [0,\infty)} L(x(s), \psi(x(s))u) \lambda(du \times ds), \end{aligned}$$

and it follows that

$$I(x) = I_0(x(0)) + \int_0^\infty L(x(s), \dot{x}(s))ds.$$

□

The following result is essentially Theorem 5.6.7 of Dembo and Zeitouni [**29**]. It provides an alternative representation of the rate function for diffusion processes.

LEMMA 10.23 (Variational Representation of H). *Let σ and b be bounded and continuous and $\eta = 0$. Define $U = R^d$,*

$$Af(x, u) = u \cdot (\sigma^T(x)\nabla f(x)) + b(x)\nabla f(x),$$

and $L(x, u) = |u|^2/2$ for any $(x, u) \in E \times U$. Then

$$Hf(x) = \mathbf{H}f(x) \equiv \sup_{u\in U}((Af - L)(x, u)),$$

and Conditions 8.9, 8.10, and 8.11 are satisfied.

PROOF. Verification of Condition 8.9 is straightforward. Condition 8.10 follows by taking $u \equiv 0$. Then for $x_0 \in R^d$, a solution of $\dot{x} = b(x)$, $x(0) = x_0$ exists by the continuity of b. For $x_0 = \infty$, $x(t) \equiv \infty$.

Finally, consider Condition 8.11. For $x_0 \in R^d$ and $f \in D_d$, let $u(x) = \sigma(x)\nabla f(x)$, and solve

$$\dot{x}(t) = \sigma(x(t))\sigma^T(x(t))\nabla f(x(t)) + b(x(t)), \quad x(0) = x_0.$$

Again, the solution exists by the continuity of σ and b.

Since $Hf(x) = Af(x, u(x)) - L(x, u(x))$, taking $\lambda(du \times ds) = \delta_{u(x(s))}(du) \times ds$, Condition 8.11 is verified. If $x_0 = \infty$, take $\lambda(du \times ds) = \delta_0(du) \times ds$ and $x(t) \equiv \infty$. □

THEOREM 10.24. *Assume that the conditions of Theorem 10.17 and Lemma 10.23 hold. Then the large deviation principle holds for $\{X_n\}$ with good rate function*

$$\begin{aligned} I(x) = \quad & I_0(x(0)) + \frac{1}{2}\inf\{\int_0^\infty |u(s)|^2 ds : u \in L^2[0,\infty) \text{ satisfies} \\ & \qquad x(t) = x(0) + \int_0^t b(x(s))ds + \int_0^t \sigma^T(x(s))u(s)ds\}. \end{aligned}$$

PROOF. Condition 9.10 holds by Lemma 10.10. By Theorem 10.17, the large deviation principle in $D_E[0,\infty)$ holds for X_n with a good rate function. By Lemma 10.23 and Corollary 8.29,

$$I(x) = I_0(x(0)) + \inf\{\int_{R^d\times[0,\infty)} \frac{|u|^2}{2}\lambda(du \times ds) : (x, \lambda) \in \mathcal{J}\},$$

where $(x, \lambda) \in \mathcal{J}$ means that

$$f(x(t)) - f(x(0)) - \int_{R^d\times[0,t]} [u(\sigma^T(x(s))\nabla f(x(s))) + b(x(s))\nabla f(x(s))]\lambda(du \times ds) = 0.$$

Decompose $\lambda(du \times ds) = \mu_s(du) \times ds$ as in the previous example, and define $u(t) = \int_U u\mu_t(du)$. Then

$$x(t) - x(0) = \int_0^t \sigma^T(x(s))u(s)ds + \int_0^t b(x(s))ds,$$

and the result follows as in Theorem 10.22. □

10.4. Nearly deterministic Markov chains

Let $\xi_1, \xi_2, \ldots$ be independent and identically distributed U-valued random variables with distribution ν. Consider the Markov chain in R^d satisfying

$$X_n(t+\epsilon_n) = X_n(t) + \epsilon_n F(X_n(t), \xi_{[t/\epsilon_n]+1}). \tag{10.33}$$

Then

$$T_n f(x) = \int_U f(x + \epsilon_n F(x,u))\nu(du)$$

and

$$H_n f(x) = \frac{1}{n\epsilon_n} \log \int_U e^{n(f(x+\epsilon_n F(x,u)) - f(x))}\nu(du).$$

We consider the case $n\epsilon_n \to 1$, although results can also be obtained under the assumption $n\epsilon_n \to 0$. (See Section 11.1.)

10.4.1. Convergence of H_n. Let $\psi(x) = 1 + |x|$ and assume that for each $c > 0$,

$$\sup_x \int_U e^{c|F(x,u)|/\psi(x)}\nu(du) < \infty. \tag{10.34}$$

Then for $f \in D_d$, $H_n f$ converges uniformly to

$$Hf(x) = \log \int e^{F(x,u)\cdot\nabla f(x)}\nu(du),$$

by an estimate similar to (10.15). We define

$$H(x,p) = \log \int e^{F(x,u)\cdot p}\nu(du).$$

10.4.2. Exponential tightness. Taking $E = R^d \cup \{\infty\}$, exponential tightness follows by Corollary 4.17.

10.4.3. The comparison principle.

LEMMA 10.25. *If (10.34) holds, then Condition 9.10.2 holds.*

PROOF. Let $x_m, p_m \in R^d$ be such that $|x_m| \to \infty$ and $|x_m||p_m| \to 0$. Then

$$|E[e^{F(x_m,\xi)\cdot p_m}] - 1| \le |E[e^{(1+|x_m|)|p_m||F(x_m,\xi)|/\psi(x_m)}] - 1| \to 0$$

by the uniform integrability implied by (10.34), and hence

$$\lim_{m\to+\infty} H(x_m,p_m) = \log E[\lim_{m\to+\infty} e^{F(x_m,\xi)\cdot p_m}] = 0.$$

□

LEMMA 10.26. *Let $h \in C_b(E)$ and $\alpha > 0$. Suppose $H(x,p)$ is continuous in x, p, (10.34) holds for all $c > 0$, and (9.45) holds. Then the comparison principle holds for*

$$(I - \alpha H)f = h.$$

PROOF. Lemma 9.16 gives Condition 9.10.1, and Lemma 10.25 gives Condition 9.10.2, so the comparison principle follows by Lemma 9.15 (taking $H_\dagger = H_\ddagger = H$). □

We also have the comparison principle under a Lipschitz assumption.

CONDITION 10.27. *For each $r > 0$, there exists a $\delta_r > 0$ such that for $0 < \delta < \delta_r$,*

$$\kappa_r(\delta) \equiv \sup\{E[\exp\{\delta\frac{|F(x,z) - F(y,z)|}{|x-y|}\}] : x \neq y, |x| + |y| \leq r\} < \infty. \tag{10.35}$$

LEMMA 10.28. *Let $h \in C_b(E)$ and $\alpha > 0$. Suppose $H(x,p)$ is continuous in x, p, (10.34) holds for all $c > 0$, and Condition 10.27 holds. Then the comparison principle holds for*

$$(I - \alpha H)f = h.$$

PROOF. Let $\lambda > 1$, and ξ have distribution ν. By the Hölder inequality

$$E[e^{\frac{1}{\lambda}F(x,\xi)\cdot p}] = E[e^{\frac{\lambda-1}{\lambda}\frac{(F(x,\xi)-F(y,\xi))\cdot p}{\lambda-1} + \frac{1}{\lambda}F(y,\xi)\cdot p}] \leq E^{\frac{\lambda-1}{\lambda}}[e^{\frac{(F(x,\xi)-F(y,\xi))\cdot p}{\lambda-1}}]E^{\frac{1}{\lambda}}[e^{F(y,\xi)\cdot p}].$$

That is,

$$\begin{aligned}
\lambda H(x, \frac{p}{\lambda}) - H(y,p) &\leq (\lambda - 1)\log E[e^{\frac{F(x,\xi)-F(y,\xi)\cdot p}{\lambda-1}}] \\
&\leq (\lambda - 1)\log E[e^{\frac{1}{\lambda-1}\frac{|F(x,\xi)-F(y,\xi)|}{|x-y|}|p||x-y|}].
\end{aligned}$$

Therefore, by (10.35), Condition 9.10.1 is satisfied.

Again, by Lemma 10.25, Condition 9.10.2 holds, so the comparison principle holds by Lemma 9.15. □

10.4.4. The large deviation theorem.

THEOREM 10.29. *Let X_n be given by (10.33), and suppose that the large deviation principle holds for $X_n(0)$ in R^d with a good rate function I_0.*

Assume that (10.34) holds with $\psi(x) = 1 + |x|$, that $H(x,p)$ is continuous, and that either Condition 10.27 or (9.45) holds. Then

a) *For each $\alpha > 0$ and $f \in C(E)$, there exists a unique $R_\alpha f \in C(E)$ such that $(I - \alpha H)R_\alpha f = f$ in the viscosity sense, and*

$$V(t)f = \lim_n R^n_{t/n} f$$

defines a semigroup $V(t) : C(E) \mapsto C(E)$.

b) *The large deviation principle holds for $\{X_n\}$ in $D_E[0,\infty)$ with good rate function:*

$$I(x) = \sup_{\{t_i\}\subset\Delta_x^c} I_{t_1,\ldots,t_m}(x(t_1),\ldots,x(t_m)),$$

where

$$\begin{aligned}
&I_{t_1,\ldots,t_m}(x_1,\ldots,x_m) \\
&= \inf_{x_0}\{I_0(x_0) + \sup_{f_1,\ldots,f_k\in D_d}\{f_1(x_1) + \ldots + f_k(x_k) \\
&\qquad -V(t_1)(f_1 + V(t_1 - t_1)(f_2 + \ldots V(t_k - t_{k-1})f_k)\ldots))(x_0)\}\}.
\end{aligned}$$

c) *If $x \in D_E[0,\infty) - D_{R^d}[0,\infty)$, then $I(x) = \infty$.*

PROOF. The comparison principle for $(I - \alpha H)f = h$ is given by Lemma 10.26 or Lemma 10.28, and Parts (a) and (b) follow by Theorem 6.14.

The exponential compact containment condition can be verified by an argument similar to Example 4.23 and Part (c) follows by Theorem 4.11(a). □

10.4.5. Variational representation of the rate function. As in Theorem 10.22, the rate function can be written

$$I(x) = \begin{cases} I_0(x(0)) + \int_0^\infty L(x(s), \dot{x}(s))ds & \text{if } x \text{ is absolutely continuous} \\ \infty & \text{otherwise} \end{cases}$$

where

$$L(x,q) = \sup_{p \in R^d} \{p \cdot q - H(x,p)\}, \; x, q \in R^d.$$

10.5. Diffusion processes with reflecting boundaries

We next consider reflecting diffusions in a smooth domain. Specifically, let $\varphi \in C^2(R^d)$ and suppose that $E \equiv \{x : \varphi(x) \geq 0\}$ is compact, E is the closure of $E^o = \{x : \varphi(x) > 0\}$, and for $x \in \partial E = \{x : \varphi(x) = 0\}$, $\nabla \varphi(x) \neq 0$.

Let X_n satisfy

$$\begin{aligned} X_n(t) &= X_n(0) + \frac{1}{\sqrt{n}} \int_0^t \sigma(X_n(s))dW(s) + \int_0^t b(X_n(s))ds \\ &\quad + \int_0^t \nu(X_n(s))d\lambda_n(s), \end{aligned} \tag{10.36}$$

where $X_n(t) \in E$ for all t a.s. and λ_n is nondecreasing and increases only when $X_n(t) \in \partial E$. We assume that σ, b, and ν are bounded and continuous and that

$$\kappa_0 = \inf_{x \in \partial E} \nu(x) \cdot \nabla \varphi(x) > 0. \tag{10.37}$$

If we define $\widetilde{A}_n$ by

$$\widetilde{A}_n f(x) = b(x) \cdot \nabla f(x) + \frac{1}{2n} \sum_{i,j} a_{ij}(x) \partial_i \partial_j f(x)$$

and $Bf(x) = \nu(x) \cdot \nabla f(x)$, Itô's formula gives

$$\begin{aligned} f(X_n(t)) - f(X_n(0)) - \int_0^t \widetilde{A}_n f(X_n(s))ds - \int_0^t Bf(X_n(s))d\lambda_n(s) \\ = \frac{1}{\sqrt{n}} \int_0^t \nabla f(X_n(s)) \cdot \sigma(X_n(s))dW(s), \end{aligned} \tag{10.38}$$

so the left side of (10.38) is a martingale. If we define A_n to be the restriction of $\widetilde{A}_n$ to $D_{E,\nu} = \{f \in C^2(E) : Bf(x) = 0, x \in \partial E\}$, then any solution of the stochastic differential equation will be a solution of the martingale problem for A_n. Alternatively, we can formulate a martingale problem in terms of $\widetilde{A}_n$ and B. In either case, under the continuity assumptions on σ, b, and ν, and (10.37), we can apply Krylov's selection theorem (Theorem 4.5.19 of [**36**]) to obtain a Markov process with full generator A_n extending A_n restricted to $D_{E,\nu}$, and the process can be constructed so that it is a weak solution of (10.36). In general, A_n (and the corresponding H_n) with domain $D_{E,\nu}$ may not be large enough either to characterize the process (hence the application of the selection theorem) or to obtain a limiting H satisfying the comparison principle; however, if $\{T_n(t)\}$ denotes the semigroup given by the selection theorem, any function of the form

$$h_n = \frac{1}{\epsilon_n} \int_0^{\epsilon_n} T_n(t)h dt \tag{10.39}$$

will be in the domain of the full generator A_n and

$$A_n h_n = \frac{T_n(\epsilon_n)h - h}{\epsilon_n}. \tag{10.40}$$

10.5.1. Convergence of H_n. With (10.39) and (10.40) in mind, for $f \in C^2(E)$, define f_n by

$$e^{nf_n} = \frac{1}{\epsilon_n}\int_0^{\epsilon_n} T_n(t)e^{nf}dt.$$

Then, letting E_x denote the expectation under the assumption that $X_n(0) = x$, and setting $\widetilde{H}_n f = \frac{1}{n}e^{-nf}\widetilde{A}_n e^{nf}$, for $f \in C^2(E)$, we have

$$\begin{aligned} A_n e^{nf_n(x)} &= \epsilon_n^{-1}E_x[e^{nf(X_n(\epsilon_n))} - e^{nf(x)}] \\ &= \epsilon_n^{-1}nE_x[\int_0^{\epsilon_n} e^{nf(X_n(s))}\widetilde{H}_n f(X_n(s))ds \\ &\qquad + \int_0^{\epsilon_n} e^{nf(X_n(s))}Bf(X_n(s))d\lambda_n(s)]. \end{aligned}$$

It follows that $H_n f_n$ is given by

$$\begin{aligned} &\Big(E_x[\int_0^{\epsilon_n} e^{n(f(X_n(s))-f(x))}\widetilde{H}_n f(X_n(s))ds \\ &\qquad + \int_0^{\epsilon_n} e^{n(f(X_n(s))-f(x))}Bf(X_n(s))d\lambda_n(s)]\Big)/E_x[\int_0^{\epsilon_n} e^{n(f(X_n(s))-f(x))}ds] \\ &= \Big(E_x[\int_0^1 e^{n(f(X_n(\epsilon_n s))-f(x))}\widetilde{H}_n f(X_n(\epsilon_n s))ds \\ &\qquad + \int_0^1 e^{n(f(X_n(\epsilon_n s))-f(x))}Bf(X_n(\epsilon_n s))d\widehat{\lambda}_n(s)]\Big)/E_x[\int_0^1 e^{n(f(X_n(\epsilon_n s))-f(x))}ds], \end{aligned}$$

where $\widehat{\lambda}_n(t) = \epsilon_n^{-1}\lambda_n(\epsilon_n t)$. Taking $\epsilon_n = n^{-1}$ and defining $U_n(t) = n(X_n(n^{-1}t) - x)$, we have

$$\begin{aligned} U_n(t) &= \int_0^t \sigma(x + \frac{1}{n}U_n(s))dW_n(s) + \int_0^t b(x + \frac{1}{n}U_n(s))ds \\ &\qquad + \int_0^t \nu(x + \frac{1}{n}U_n(s))d\widehat{\lambda}_n(s), \end{aligned} \tag{10.41}$$

where $W_n(t) = \sqrt{n}W(n^{-1}t)$ is a standard Brownian motion. To estimate $\widehat{\lambda}_n$, note that

$$\begin{aligned} Z_n(t) \equiv n\varphi(X_n(n^{-1}t)) &= n\varphi(X_n(0)) + \int_0^t \widetilde{A}_n\varphi(X_n(n^{-1}s))ds \\ &\qquad + \int_0^t \nabla\varphi(X(n^{-1}s))\cdot\sigma(X_n(n^{-1}s))dW_n(s) \\ &\qquad + \int_0^t \nu(X_n(n^{-1}s))\cdot\nabla\varphi(X_n(n^{-1}s))d\widehat{\lambda}_n(s), \end{aligned}$$

where $Z_n \geq 0$ and, by (10.37), the last term is nondecreasing and increases only when Z_n is zero. Let

$$R_n(t) = \int_0^t \widetilde{A}_n\varphi(X_n(n^{-1}s))ds + \int_0^t \nabla\varphi(X(n^{-1}s))\cdot\sigma(X_n(n^{-1}s))dW_n(s).$$

From properties of the Skorohod reflection map (see, for example, Theorem 6.1 of [**13**]),

$$\begin{aligned}\kappa_0\widehat{\lambda}_n(t) &\leq \int_0^t \nu(X_n(n^{-1}s)\nabla\varphi(X_n(n^{-1}s))d\widehat{\lambda}_n(s)\\ &= 0\vee\sup_{r\leq t}\{-n\varphi(X_n(0))-R_n(r)\}\\ &\leq \sup_{r\leq t}(-R_n(r)).\end{aligned}$$

Noting that there exist constants C and δ independent of x such that

$$P\{\sup_{r\leq 1}(-R_n(r))\geq a\}\leq Ce^{-\delta a^2}.$$

This inequality and (10.41) imply a similar inequality for U_n, and consequently, it is easy to check that $\lim_{n\to\infty}\|f_n-f\|=0$ and $\sup_n\|H_nf_n\|<\infty$. If $x_n\to x$ and $n\varphi(x_n)\to\infty$, $\widehat{\lambda}_n(t)\Rightarrow 0$ and

$$\lim_{n\to\infty}H_nf_n(x_n)=\lim_{n\to\infty}\widetilde{H}_nf_n(x_n)=\widetilde{H}(x,\nabla f(x)),$$

where $\widetilde{H}(x,p)=\frac{1}{2}|\sigma^T(x)p|^2+b(x)\cdot p$. If $x_n\to x$ and $n\varphi(x_n)\to z$, then $Z_n\Rightarrow Z$ satisfying

$$Z(t)=z+b(x)\cdot\nabla f(x)t+\nabla\varphi(x)\cdot\sigma(x)W(t)+\nu(x)\cdot\nabla\varphi(x)\widehat{\lambda}_{x,z}(t),$$

where

$$\begin{aligned}\widehat{\lambda}_{x,z}(t)&=\frac{0\vee\sup_{r\leq t}\{-z-R(r)\}}{\nu(x)\cdot\nabla\varphi(x)},\\ R(t)&=b(x)\cdot\nabla f(x)t+\nabla\varphi(x)\cdot\sigma(x)W(t),\end{aligned}$$

and $U_n\Rightarrow U_{x,z}$ satisfying

$$U_{x,z}(t)=\sigma(x)W(t)+b(x)t+\nu(x)\widehat{\lambda}_{x,z}(t).$$

Consequently,

$$\lim_{n\to\infty}H_nf_n(x_n)=\widetilde{H}(x,\nabla f(x))+\frac{E[\int_0^1 e^{\nabla f(x)\cdot U_{x,z}(s)}d\widehat{\lambda}_{x,z}(s)]}{E[\int_0^1 e^{\nabla f(x)\cdot U_{x,z}(s)}ds]}Bf(x).$$

Defining

$$\begin{aligned}H_b(x,p)&=\sup_{z\geq 0}\frac{E[\int_0^1 e^{p\cdot U_{x,z}(s)}d\widehat{\lambda}_{x,z}(s)]}{E[\int_0^1 e^{p\cdot U_{x,z}(s)}ds]},\\ \widetilde{H}(x,p)&=\frac{1}{2}|\sigma^T(x)p|^2+b(x)\cdot p,\end{aligned}$$

$$H_\dagger(x,p)=\begin{cases}\widetilde{H}(x,p)+H_b(x,p)(0\vee\nu(x)\cdot p) & x\in\partial E\\ \widetilde{H}(x,p) & x\in E^o\end{cases}$$

and

$$H_\ddagger(x,p)=\begin{cases}\widetilde{H}(x,p)+H_b(x,p)(0\wedge\nu(x)\cdot p) & x\in\partial E\\ \widetilde{H}(x,p) & x\in E^o\end{cases}$$

and setting $H_\dagger f(x)=H_\dagger(x,\nabla f(x))$ and $H_\ddagger f(x)=H_\ddagger(x,\nabla f(x))$, for any sequence $x_n\to x$,

$$\limsup_{n\to\infty}H_nf_n(x_n)\leq H_\dagger f(x)$$

and

$$\liminf_{n\to\infty} H_n f_n(x_n) \geq H_\ddagger f(x).$$

10.5.2. Exponential tightness. Since we are assuming E is compact and $\sup_n \|H_n f_n\| < \infty$ for the sequences $\{f_n\}$ defined above, exponential tightness follows by Corollary 4.17.

10.5.3. The Comparison Principle.

LEMMA 10.30. *Suppose $h \in C(E)$ and $\alpha > 0$. If σ and b are continuous and*

$$\inf_{x\in E, |\xi|=1} |\sigma(x)\xi| > 0, \tag{10.42}$$

or if σ, b are Lipschitz, then the comparison principle holds for subsolutions of

$$(I - \alpha H_\dagger)f = h$$

and supersolutions of

$$(I - \alpha H_\ddagger)f = h.$$

PROOF. Assuming the σ and b are continuous and (10.42) holds, then the result follows by Lemma 9.24.

If σ and b are Lipschitz, then the result follows by Lemma 9.25. □

10.5.4. The large deviation theorem. Having verified the convergence of H_n and the corresponding comparison principle, we have the following large deviation theorem.

THEOREM 10.31. *Let $E \subset R^d$ be defined as above, let ν be C^1 on E and satisfy (10.37), and let σ and b satisfy the conditions of Lemma 10.30. If $\{X_n(0)\}$ satisfies a large deviation principle with good rate function I_0, then $\{X_n\}$ satisfies a large deviation principle with good rate function defined as in Theorem 10.17.*

10.5.5. Variational representation of the rate function. As in the unconstrained case, there are two possible choices for the control problem. Assuming that (10.42) holds and a and b are continuous, $a(x) = \sigma(x)\sigma(x)^T$ is invertible for all x and setting $u = (q, r) \in R^d \times R$, we can take $\mathcal{D}(A) = C^1(E)$ and

$$Af(x,u) = (q + r\nu(x)) \cdot \nabla f(x).$$

Setting

$$\beta(x,q) = -\frac{\nabla\varphi(x)\cdot q}{\nabla\varphi(x)\cdot\nu(x)} \wedge 0,$$

define

$$L(x,u) \equiv L(x,q,r) = \frac{1}{2}(q - b(x))^T a(x)^{-1}(q - b(x)) + r\,(\beta(x,q) - r)$$

and

$$\Gamma = \{(x,q,0) : x \in E, q \in R^d\} \cup \{(x,q,r) : x \in \partial E, q \in R^d, 0 \leq r \leq \beta(x,q)\}.$$

Solutions of the control problem satisfy

$$x(t) = x(0) + \int_0^t (q(s) + r(s)\nu(x(s)))ds,$$

and since $r(t) = 0$ unless $x(t) \in \partial E$, for almost every t with $r(t) > 0$,

$$\nabla\varphi(x(t)) \cdot \dot{x}(t) = \nabla\varphi(x(t)) \cdot (q(t) + r(t)\nu(x(t))) = 0.$$

It follows that

$$r(t) = -\frac{\nabla\varphi(x(t))\cdot q(t)}{\nabla\varphi(x(t))\cdot\nu(x(t))} = \beta(x(t),q(t)), \tag{10.43}$$

for almost every t in $\{t : \varphi(x(t)) = 0\}$. Consequently, the boundary control adds nothing to the cost of the solution.

LEMMA 10.32. *Let* $\mathbf{H}$ *be given by (8.7), that is,*

$$\mathbf{H}f(x) = \sup_{(q,r)\in\Gamma_x} \left((q + r\nu(x))\cdot\nabla f(x) - L(x,q,r)\right).$$

Then for $x \in E^o$, $\mathbf{H}f(x) = \widetilde{H}(x,\nabla f(x))$. *If* $x \in \partial E$ *and* $\nu(x)\cdot\nabla f(x) \le 0$, *then*

$$\mathbf{H}f(x) = \widetilde{H}(x,\nabla f(x)). \tag{10.44}$$

If $\nu(x)\cdot\nabla f(x) > 0$, *then*

$$\widetilde{H}(x,\nabla f(x)) \le \mathbf{H}f(x) \le \widetilde{H}(x,\nabla f(x)) + G_b(x,\nabla f(x))\nu(x)\cdot\nabla f(x) \tag{10.45}$$

where

$$G_b(x,p) = \beta(x,\widehat{q}(x,p))$$

for $\widehat{q}(x,p)$ *maximizing*

$$q\cdot p - \frac{\nabla\varphi(x)\cdot q}{\nabla\varphi(x)\cdot\nu(x)}\nu(x)\cdot p - \frac{1}{2}(q - b(x))^T a(x)^{-1}(q - b(x)).$$

PROOF. If $\nu(x)\cdot\nabla f(x) \le x$, then the supremum in the definition of $\mathbf{H}$ will be achieved with $r = 0$, so (10.44) follows.

The first inequality in (10.45) follows by setting $r = 0$. The second inequality follows by observing that

$$\mathbf{H}f(x) \le \sup_q \left((q + \beta(x,q)\nu(x))\cdot\nabla f(x) - \frac{1}{2}(q - b(x))^T a(x)^{-1}(q - b(x))\right).$$

If this supremum is achieved for q satisfying $\beta(x,q) = 0$, then $\mathbf{H}f(x) \le \widetilde{H}(x,\nabla f(x))$. If not, the supremum is achieved for $q = \widehat{q}(x,\nabla f(x))$. In either case, the second inequality follows. □

Let $\mathbf{H}_\dagger$ and $\mathbf{H}_\ddagger$ be the same as $H_\dagger$ and $H_\ddagger$ but with $H_b(x,p)$ replaced by

$$H_b(x,p) \vee G_b(x,p) \vee \left|\frac{\nabla\varphi(x)\cdot(a(x)p + b(x))}{\nabla\varphi(x)\cdot\nu(x)} \wedge 0\right|.$$

Then $\mathbf{H}_\ddagger f(x) \le \mathbf{H}f(x) \le \mathbf{H}_\dagger f(x)$ and $\mathbf{H}_\ddagger f(x) \le H_\ddagger f(x) \le H_\dagger f(x) \le \mathbf{H}_\dagger f(x)$, and it follows that every subsolution of $(I - \alpha\mathbf{H})f = h$ and every subsolution of $(I - \alpha H_\dagger)f = h$ is a subsolution of $(I - \alpha\mathbf{H}_\dagger)f = h$ and every supersolution of $(I - \alpha\mathbf{H})f = h$ and every supersolution of $(I - \alpha H_\ddagger)f = h$ is a supersolution of $(I - \alpha\mathbf{H}_\ddagger)f = h$. Lemma 10.30 still holds for $\mathbf{H}_\dagger$ and $\mathbf{H}_\ddagger$, and hence $\mathbf{H}$ satisfies the comparison principle under the same conditions on the coefficients. Condition (a) of Corollary 8.28 is also satisfied.

Condition 8.9 is satisfied. Since $\mathbf{H}_\ddagger 1 = 0$, Condition 8.10 will follow immediately once we verify Condition 8.11.

Fixing $f \in \mathcal{D}(A)$, consider the equation

$$x(t) = x(0) + \int_0^t (a(x(s))\nabla f(x(s)) + b(x(s)) + r(s)\nu(x(s)))ds,$$

that is,

$$q(t) = a(x(t))\nabla f(x(t)) + b(x(t)).$$

Existence follows by approximation and compactness since the functions involved are continuous. By (10.43)

$$r(t) = -\left(\frac{\nabla\varphi(x(t)) \cdot (a(x(t))\nabla f(x(t)) + b(x(t)))}{\nabla\varphi(x(t)) \cdot \nu(x(t))}\right) \wedge 0. \tag{10.46}$$

Then, with reference to (8.15),

$$\begin{aligned}
&\int_{U\times[t_1,t_2]} (Af(x(s),u) - L(x(s),u))\lambda(ds \times du) \\
&= \int_{t_1}^{t_2} (\frac{1}{2}|\sigma(x(s))^T\nabla f(x(s))|^2 + b(x(s)) \cdot \nabla f(x(s)) + r(s)\nu(x(s)) \cdot \nabla f(x(s)))ds \\
&\geq \int_{t_1}^{t_2} \mathbf{H}_\ddagger f(x(s))ds,
\end{aligned}$$

where the inequality follows by (10.46) and the definition of $\mathbf{H}_\ddagger$.

Consequently, by Corollary 8.28,

$$I(x) = I_0(x(0)) + \inf_{\lambda:(x,\lambda)\in\mathcal{J}^\Gamma_{x(0)}} \int_{U\times[0,\infty)} L(x(s),u)\lambda(du \times ds).$$

Since any solution of the control problem will satisfy

$$x(t) = x(0) + \int_0^t (q(s) + r(s)\nu(x(s)))ds,$$

we must have $\dot{x}(t) = q(t) + r(t)\nu(x(t))$, with $r(t) \neq 0$ only when $\varphi(x(t)) = 0$, that is, $x(t) \in \partial E$. If $x(t) \in \partial E$, we can select $q \in R^d$ and $0 \leq r \leq \beta(x,q)$ in any way that satisfies $\dot{x}(t) = q(t) + r(t)\nu(x(t))$. Note that if x is absolutely continuous with values in E, then $\nabla\varphi(x(t)) \cdot \dot{x}(t) = 0$ for almost every t in $\{t : \varphi(x(t)) = 0\}$, and hence, for almost every such t,

$$r(t)\left(-\frac{\nabla\varphi(x(t)) \cdot (\dot{x}(t) - r(t)\nu(x(t)))}{\nabla\varphi(x(t)) \cdot \nu(x(t))} \wedge 0 - r(t)\right) = 0.$$

Defining

$$\widehat{L}(x,z) = \begin{cases} L(x,z,0) & x \in E^o \\ \inf\{L(x,z-r\nu(x),r) : 0 \leq r \leq \beta(x,z-r\nu(x))\} & x \in \partial E, \end{cases}$$

we have the following theorem.

THEOREM 10.33. *Assume that a and b are continuous and that (10.42) holds. Then the large deviation principle holds for $\{X_n\}$ in $D_E[0,\infty)$ with good rate function*

$$I(x) = \begin{cases} I_0(x(0)) + \int_0^\infty \widehat{L}(x(s),\dot{x}(s))ds & \text{if } x \text{ is absolutely continuous with values in } E \\ +\infty & \text{otherwise} \end{cases}$$

If σ and b are Lipschitz, we can take

$$Af(x,u) = q \cdot (\sigma^T(x)\nabla f(x)) + (b(x) + r\nu(x)) \cdot \nabla f(x),$$

and setting

$$\beta(x,q) = -\frac{\nabla\varphi(x)\cdot(\sigma(x)q+b(x))}{\nabla\varphi(x)\cdot\nu(x)} \wedge 0,$$

define

$$L(x,u) = \frac{1}{2}|q|^2 + r(\beta(x,q)-r).$$

Again, by essentially the same argument as before, we have

$$\mathbf{H}_\ddagger f(x) \leq \mathbf{H} f(x) \leq \mathbf{H}_\dagger f(x)$$

with $\mathbf{H}_\ddagger$ and $\mathbf{H}_\dagger$ defined as above.

Every solution of the control problem is of the form

$$x(t) = x(0) + \int_0^t \sigma(x(s))q(s)ds + \int_0^t (b(x(s)) + r(s)\nu(x(s)))ds.$$

Since $r(t) = 0$ unless $x(t) \in \partial E$, for almost every t with $r(t) > 0$,

$$\nabla\varphi(x(t))\cdot\dot{x}(t) = \nabla\varphi(x(t))\cdot(\sigma(x(s))q(t) + b(x(s)) + r(t)\nu(x(t)) = 0$$

and hence

$$r(t) = -\frac{\nabla\varphi(x(t))\cdot(\sigma(x(t))q(t)+b(x(t)))}{\nabla\varphi(x(t))\cdot\nu(x(t))} \wedge 0 = \beta(x(t),q(t)). \tag{10.47}$$

Consequently,

$$x(t) = x(0) + \int_0^t (b(x(s)) + \mathbf{1}_{\partial E}(x(s))r(s)\nu(x(s)))ds + \int_0^t \sigma(x(s))q(s)ds \tag{10.48}$$

with r given by (10.47).

THEOREM 10.34. *Assume that σ and b are Lipschitz. Then the large deviation principle holds for $\{X_n\}$ with good rate function*

$$I(x) = I_0(x(0)) + \frac{1}{2}\inf\{\int_0^\infty |q(s)|^2 ds : q \in L^2[0,\infty) \text{ satisfies (10.47) and (10.48)}\}.$$

CHAPTER 11

Random evolutions

In this chapter, we consider stochastic models with multiple time scales. The models include the random evolutions given in Examples 1.8 and 1.9, a model with small diffusion and periodic coefficients (Example 1.10), and a system with rapidly varying coefficients and small diffusion. The examples considered here are not exhaustive of the multiscale schemes having interesting large deviation behavior, and the results we give may not be the most general possible. Our main interest in this chapter is to give a relatively straightforward presentation for a variety of models exhibiting averaging behavior.

Let (E_0, r_0) be a complete, separable metric space, and let Y_n be a sequence of E_0-valued processes. Y_n models a randomly varying environment in which a second process evolves. We take the second process X_n to be an R^d-valued process satisfying

$$X_n(t) = x_n + \int_0^t F_n(X_n(s), Y_n(s))ds, \tag{11.1}$$

where $F_n : R^d \times E_0 \to R^d$ and $x_n \in R^d$. Different scalings of Y_n and F_n give different large deviation behavior for X_n. We consider two cases. In the first, $F_n(x, y) = F(x, y)$ and $Y_n(t) = Y(nt)$, where Y is an ergodic, E_0-valued Markov process with stationary distribution π. In this case, if $x_n \to x_0$ and F satisfies appropriate regularity and growth conditions, X_n converges to a solution of

$$X(t) = x_0 + \int_0^t \overline{F}(X(s))ds,$$

where $\overline{F}(x) = \int_{E_0} F(x, y)\pi(dy)$. Since $\{X_n\}$ essentially satisfies a law of large numbers, we will refer to this case as the law of large numbers scaling.

In the second case, $F_n(x, y) = \beta_n F(x, y)$, $\int F(x, y)\pi(dy) = 0$, and $Y_n(t) = Y(n\beta_n^2 t)$, for $\beta_n \to \infty$. We will see that the large deviation behavior of $\{X_n\}$ is analogous to that of a small diffusion problem and that, as in the examples of Section 10.1.6, the central limit theorem dominates the large deviation behavior. We will refer to this scaling as the central limit scaling.

We also consider discrete time analogs of these models. Let Y_n, $n = 1, 2, \ldots$, be E_0-valued Markov chains with transition operator $P_n \equiv P$ and time step $\epsilon_n > 0$, that is,

$$E[f(Y_n(t))|Y_n(0)] = P^{[t/\epsilon_n]} f(Y_n(0)),$$

and let X_n satisfy

$$X_n(t + \epsilon_n) = X_n(t) + \epsilon_n F_n(X_n(t), Y_n(t)), \qquad X_n(0) = x_n. \tag{11.2}$$

It follows that

$$X_n(t) = x_n + \int_0^{[t/\epsilon_n]\epsilon_n} F_n(X_n(s), Y_n(s))ds.$$

As in continuous time, we have two scalings. In the law of large numbers scaling, $F_n(x,y) = F(x,y)$ and $n\epsilon_n \to 1$. In the central limit scaling, $F_n(x,y) = (n\epsilon_n)^{-1/2}F(x,y)$, where $\int_{E_0} F(x,y)\pi(dy) = 0$, and $n\epsilon_n \to 0$.

Weak convergence results for models of this type date back to Khas'minskii [**66**] (see Pinsky [**94**] for a recent review). Large deviation results for the law of large numbers scaling have been studied by Freidlin and Wentzell [**52**], Chapter 7.

In the case of law of large number scaling as well as in the example of systems of diffusion with averaging, the main condition ((11.16), (11.56) and (11.104)) is formulated through an abstract inequality. This inequality is closely related to the principle eigenvalue problem for the driving Markov process as discussed in Donsker and Varadhan [**32**], Kurtz [**74**], among others. In Appendix B, we discuss the verification of this inequality in detail. In summary, this condition is more general than Condition F of Freidlin-Wentzell [**52**] (Theorem B.12) and Conditions U and $\widetilde{U}$ of Deuschel and Stroock [**30**] (Lemma B.9); this condition can be verified through the methods of Kontoyiannis and Meyn [**67**], [**68**], using geometric ergodicity of the Markov process (see Remarks 11.7 and 11.31). When the driving process is a nondegenerate multidimensional diffusion, the minimax theorems of Pinsky [**95**] which are a consequence of the Harnack inequality, can be applied. In the case of central limit scaling random evolutions, conditions in the main theorems are compatible with those appearing in the weak convergence context.

The methods developed for random evolutions apply to diffusions with periodic coefficients, and in Section 11.5 we consider an example of Baldi [**6**]. The methods also apply to models with small diffusion and averaging. Theorem 11.65, extends results of Freidlin and Wentzell [**52**] and Veretennikov [**126**].

Throughout this section, $K(E_0) \subset C_b(E_0)$ is the set of nonnegative, bounded continuous functions on E_0, $K_0(E_0) \subset K(E_0)$ is the set of strictly positive, bounded continuous functions, and $K_1(E_0) \subset K_0(E_0)$ is the set of bounded continuous functions satisfying $\inf_{y\in E_0} f(y) > 0$. For a generic operator $A \subset M(E_0) \times M(E_0)$, we denote $\mathcal{D}^+(A) = \mathcal{D}(A) \cap K_0(E_0)$ and $\mathcal{D}^{++}(A) = \mathcal{D}(A) \cap K_1(E_0)$. If $y \in E_0$ and $K \subset E_0$, then $r_0(y,K) = \inf_{z\in K} r_0(y,z)$.

11.1. Discrete time, law of large numbers scaling

Let $\epsilon_n > 0$ and $\{Y(k) : k = 0,1,2,\ldots\}$ be an E_0-valued Markov chain with transition operator P on $B(E_0)$,

$$Pf(y) = \int_{E_0} f(z)P(y,dz),$$

where $P(y,dz)$ is a transition probability on E_0. We consider the discrete time random evolution defined by (11.2) with $F_n(x,y) = F(x,y)$ and

$$Y_n(t) = Y([\frac{t}{\epsilon_n}]), \qquad n = 1,2,\ldots.$$

We assume the following:

CONDITION 11.1.

(1) $\lim_{n\to\infty} n\epsilon_n = 1$.
(2) *P is Feller in the sense that $P : C_b(E_0) \mapsto C_b(E_0)$.*
(3) *F satisfies*

$$\sup_{y\in E_0} |F(0,y)| < \infty \tag{11.3}$$

and

$$|F(x_1,y) - F(x_2,y)| \le K|x_1 - x_2|, \quad x_i \in R^d, y \in E_0, \tag{11.4}$$

for some $K > 0$.

(4) *There is an index set $\mathcal{Q}$ and a family of subsets of E_0, $\{\widetilde{K}_n^q \subset E_0 : q \in \mathcal{Q}\}$, such that for $q_1, q_2 \in \mathcal{Q}$, there exists $q_3 \in \mathcal{Q}$ with $\widetilde{K}_n^{q_1} \cup \widetilde{K}_n^{q_2} \subset \widetilde{K}_n^{q_3}$; and for each $y \in E_0$, there exists $q \in \mathcal{Q}$ such that $\lim_{n\to\infty} r_0(y, \widetilde{K}_n^q) = 0$. Moreover, for each $q \in \mathcal{Q}$, $T > 0$ and $a > 0$, there exists a $\widehat{q}(q,a,T) \in \mathcal{Q}$ satisfying*

$$\limsup_{n\to\infty} \sup_{y\in\widetilde{K}_n^q} \frac{1}{n} \log P(Y(t) \notin \widetilde{K}_n^{\widehat{q}(q,a,T)}, \text{ some } t \le \epsilon_n^{-1} T | Y(0) = y) \le -a. \tag{11.5}$$

(5) *There exist an upper semicontinuous function ψ on E_0, $\{\varphi_n\} \subset K_1(E_0)$, and $q_0 \in \mathcal{Q}$ such that ψ is bounded above, $\{y \in E_0 : \psi(y) \ge c\}$ is compact for each $c \in R$, $0 < \inf_{y\in K_n^{q_0}} \varphi_n(y) < 2\inf_{y\in E_0} \varphi_n(y)$, $\inf_{n,y\in E_0} \varphi_n(y) > 0$, $\lim_{n\to\infty} \frac{1}{n} \log \|\varphi_n\| = 0$, and for each $q \in \mathcal{Q}$,*

$$\limsup_{n\to\infty} \sup_{y\in\widetilde{K}_n^q} \left(\log \frac{P\varphi_n(y)}{\varphi_n(y)} - \psi(y)\right) \le 0. \tag{11.6}$$

If E_0 is compact, we can take $\mathcal{Q}$ to be an arbitrary set, $\widetilde{K}_n^q = E_0$, $\varphi_n = 1$ and $\psi = 0$. Then Conditions 11.1.4 and 11.1.5 are always satisfied. See Example 11.24 for the verification of these conditions for a noncompact E_0. Condition 11.1.5 is essentially a Lyapunov condition for the chain. One consequence of the Feller property and the Lyapunov condition is that the chain has a stationary distribution.

LEMMA 11.2. *Assume that Condition 11.1 holds. Then there exists a stationary distribution π for P satisfying $\int_{E_0} \psi d\pi \ge 0$.*

PROOF. Let $y_n \in K_n^{q_0}$ satisfy $\varphi_n(y_n) \le 2\inf_{y\in E_0} \varphi_n(y)$, and let Y_n be a Markov chain corresponding to P with time step 1 and $Y_n(0) = y_n$. By Condition 11.1.4, there is a $\widetilde{q} \in \mathcal{Q}$ such that $P\{Y_n(k) \in K_n^{\widetilde{q}}, 0 \le k < n\} \to 1$. Define

$$\nu_n(A) \equiv E[\frac{1}{n} \sum_{k=0}^{n-1} I_{\{Y_n(k)\in A\}}]$$

and

$$\widetilde{\nu}_n(A) \equiv E[\frac{1}{n} \sum_{k=0}^{n-1} I_{\{Y_n(k)\in A\}} | Y_n(k) \in K_n^{\widetilde{q}}, 0 \le k < n].$$

Then

$$\begin{aligned}
1 &= \lim_{n\to\infty} E[\frac{\varphi_n(Y_n(n))}{\varphi_n(y_n)} e^{-\sum_{k=0}^{n-1}\log\frac{P\varphi_n}{\varphi_n}(Y_n(k))}] \\
&\geq \limsup_{n\to\infty}\frac{1}{2}E[e^{-\sum_{k=0}^{n-1}\psi(Y_n(k))}\mathbf{1}_{\{Y_n(k)\in K_n^{\tilde{q}},0\leq k<n\}}] \\
&\geq \frac{1}{2}\limsup_{n\to\infty} e^{-E[\sum_{k=0}^{n-1}\psi(Y_n(k))|Y_n(k)\in K_n^{\tilde{q}},0\leq k<n]} \\
&= \frac{1}{2}\limsup_{n\to\infty} e^{-n\int\psi d\widetilde{\nu}_n},
\end{aligned}$$

where the last inequality follows from Jensen's inequality and the fact that $P\{Y_n(k)\in K_n^{\tilde{q}}, 0\leq k<n\}\to 1$. It follows that $\liminf_{n\to\infty}\int\psi d\widetilde{\nu}_n\geq 0$, so $\{\widetilde{\nu}_n\}$ is relatively compact. Any limit point π of $\{\widetilde{\nu}_n\}$ is also a limit point for $\{\nu_n\}$. Since P is Feller, π will be a stationary measure for P, and the upper semicontinuity of ψ implies $\int_{E_0}\psi d\pi\geq 0$. □

11.1.1. Convergence of H_n. Let $E = R^d\cup\{\infty\}$ be the one point compactification of R^d and $E_n = E\times E_0$. Following the Stone-Čech compactification (see Chapter 5 of Kelley [73]), there is a metric $\widehat{r}_0$ that is topologically equivalent to r_0 and is totally bounded. The completion $\widehat{E}_0$ of E_0 under the metric $\widehat{r}_0$ is compact. If E_0 is already compact, then $\widehat{E}_0 = E_0$.

Define $E' = E\times\widehat{E}_0$, and for $\{\widetilde{K}_n^q, q\in\mathcal{Q}\}$ as in Condition 11.1.4, set $K_n^q = E\times\widetilde{K}_n^q$. The mapping η_n' is just the natural injection of $E\times E_0$ into E', and η_n and γ are just the projections $\eta_n(x,y)=x$, $(x,y)\in E_n$ and $\gamma(x,y)=x$, $(x,y)\in E'$. Clearly, $\eta_n(x,y)=\gamma\circ\eta_n'(x,y)=x$ for $(x,y)\in E_n$. Condition 11.1.4 implies that Condition 2.8 holds. Furthermore, by the compactness of E', Condition 7.9 also holds. Finally, we define the LIM convergence according to Definition 2.5.

Let φ be a function on E_0 that is upper semicontinuous, bounded above, and such that $\{z\in E_0:\varphi(z)\geq c\}$ is compact for each $c\in R$. We extend φ to a function on $\widehat{E}_0$ by setting

$$\widehat{\varphi}(y) = \lim_{\delta\to 0}\sup_{z\in B_\delta(y)\cap E_0}\varphi(z), \quad y\in\widehat{E}_0.$$

LEMMA 11.3. *$\widehat{\varphi}$ is upper semicontinuous on $\widehat{E}$, $\widehat{\varphi}(y)=\varphi(y)$ for $y\in E_0$, and $\widehat{\varphi}(y)=-\infty$ for $y\in\widehat{E}_0-E_0$.*

PROOF. The assertion that $\widehat{\varphi}(y)=\varphi(y)$ for $y\in E_0$ follows from the upper semicontinuity of φ on E_0, and from the definition of $\widehat{\varphi}$. If E_0 is not compact and $y\in\widehat{E}_0-E_0$, then for $c\in R$, $\{z\in E_0:\varphi(z)\geq c\}^c$ is is an open set containing y. For each $c\in R$, there exists $\delta>0$ sufficiently small so that $B_\delta(y)\cap E_0\subset E_0-K_c=\{z\in E_0:\varphi(z)<c\}$, and hence $\widehat{\varphi}(y)\leq c$. Since c is arbitrary, $\widehat{\varphi}(y)=-\infty$. The upper semicontinuity of $\widehat{\varphi}$ follows from the upper semicontinuity of φ and the fact that if $y_m\to y\in\widehat{E}_0-E_0$ and $c\in R$, then for m sufficiently large $y_m\in\widehat{E}_0-K_c$ implying that $\widehat{\varphi}(y_m)\leq c$ and hence $\widehat{\varphi}(y_m)\to-\infty$. □

Define $T_n: C_b(E_n)\to C_b(E_n)$ by

$$T_nf(x,y) = \int_{E_0} f(x+\epsilon_nF(x,y),z)P(y,dz), \qquad f\in C_b(E_n),$$

and set

$$H_n f = \frac{1}{n\epsilon_n} \log e^{-nf} T_n e^{nf}.$$

Defining $f_n(x,y) = f(x) + n^{-1}\log g(y)$, for $f \in D_d$ and $g \in K_1(E_0)$, we have

$$\lim_{n\to\infty} H_n f_n(x,y) = G_{f,g}(x,y) \equiv F(x,y)\cdot\nabla f(x) + \log\frac{Pg(y)}{g(y)}, \tag{11.7}$$

so the limiting H is naturally multivalued. In fact, in order to handle noncompact E_0, we will work with an operator pair $H_\dagger$, $H_\ddagger$ where $H_\dagger \supset H$ and $H_\ddagger \supset H$.

Let ψ satisfy Condition 11.1.5, $f \in D_d$, and $g \in K_1(E_0)$. For $0 < \kappa < 1$, define

$$G_{1,f,g,\kappa}(x,y) = \begin{cases} F(x,y)\cdot\nabla f(x) + (1-\kappa)\log\frac{Pg(y)}{g(y)} + \kappa\psi(y) & y \in E_0 \\ -\infty & y \in \widehat{E}_0 - E_0 \end{cases}$$

and for $\kappa > 0$, define

$$G_{2,f,g,\kappa}(x,y) = \begin{cases} F(x,y)\cdot\nabla f(x) + (1+\kappa)\log\frac{Pg(y)}{g(y)} - \kappa\psi(y) & y \in E_0 \\ \infty & y \in \widehat{E}_0 - E_0. \end{cases}$$

By Lemma 11.3

$$G_{1,f,g,\kappa}(x,y) = \lim_{\delta\to 0+}\sup_{z\in B_\delta(y)\cap E_0}\left(F(x,z)\cdot\nabla f(x) + (1-\kappa)\log\frac{Pg(z)}{g(z)} + \kappa\psi(z)\right)$$

and

$$G_{2,f,g,\kappa}(x,y) = \lim_{\delta\to 0+}\inf_{z\in B_\delta(y)\cap E_0}\left(F(x,z)\cdot\nabla f(x) + (1+\kappa)\log\frac{Pg(z)}{g(z)} - \kappa\psi(z)\right).$$

For $\kappa = 0$, we define $G_{1,f,g,0} = G_{f,g}$, extended to be upper semicontinuous on E', and $G_{2,f,g} = G_{f,g}$ extended to by lower semicontinuous on E'. We define $H_\dagger \subset C_b(E) \times M^u(E',\overline{R})$ and $H_\ddagger \subset C_b(E)\times M^l(E',\overline{R})$ by

$$H_\dagger = \{(f, G_{1,f,g,\kappa}) : f \in D_d, g \in K_1(E_0), 0 \le \kappa < 1\}, \tag{11.8}$$

$$H_\ddagger = \{(f, G_{2,f,g,\kappa}) : f \in D_d, g \in K_1(E_0), \kappa \ge 0\}. \tag{11.9}$$

Next we verify the Convergence Condition 7.11. For $f \in D_d$, $g \in K_1(E_0)$, $\varphi_n \in K_1(E_0)$ as in Condition 11.1.5, and $0 \le \kappa < 1$, define

$$f_{n,\kappa}(x,y) = f(x) + (1-\kappa)\frac{1}{n}\log g(y) + \kappa\frac{1}{n}\log\varphi_n(y),$$

and

$$h_{n,\kappa}(x,y) = \frac{1}{\epsilon_n}(f(x+\epsilon_n F(x,y)) - f(x)) + \frac{1-\kappa}{n\epsilon_n}\log\frac{Pg(y)}{g(y)} + \frac{\kappa}{n\epsilon_n}\log\frac{P\varphi_n(y)}{\varphi_n(y)}.$$

Then by the Hölder inequality,

$$\begin{aligned} & e^{-nf_{n,\kappa}(x,y)} T_n e^{nf_{n,\kappa}}(x,y) \\ = \;& e^{n(f(x+\epsilon_n F(x,y)) - f(x))} \int_{E_0} \left(\frac{g(z)}{g(y)}\right)^{1-\kappa} \left(\frac{\varphi_n(z)}{\varphi_n(y)}\right)^{\kappa} P(y,dz) \\ \le \;& e^{n(f(x+\epsilon_n F(x,y))-f(x)} \left(\frac{Pg(y)}{g(y)}\right)^{1-\kappa} \left(\frac{P\varphi_n(y)}{\varphi_n(y)}\right)^{\kappa} \\ = \;& \exp\{n\epsilon_n h_{n,\kappa}(x,y)\}, \end{aligned}$$

so $H_n f_{n,\kappa} \le h_{n,\kappa}$. Let $v_n = n$. Then noting that by Condition 11.1.5, $\sup_n \|f_{n,\kappa}\| < \infty$, (7.18) and (7.19) follow. By (11.6), for $(x_n, y_n) \in K_n^q = E \times \widetilde{K}_n^q$ satisfying $(x_n, y_n) \to (x, y) \in E'$,

$$\begin{aligned}\limsup_{n\to\infty} h_{n,\kappa}(x_n, y_n) &\le \limsup_{n\to\infty}(F(x, y_n)\cdot\nabla f(x) + (1-\kappa)\log\frac{Pg(y_n)}{g(y_n)} + \kappa\psi(y_n)) \\ &\le G_{1,f,g,\kappa}(x,y),\end{aligned}$$

and (7.20) holds.

Similarly, define

$$\widetilde{f}_{n,\kappa}(x,y) = f(x) + (1+\kappa)\frac{1}{n}\log g(y) - \kappa\frac{1}{n}\log\varphi_n(y),$$

and

$$\widetilde{h}_{n,\kappa}(x,y) = \frac{1}{\epsilon_n}(f(x+\epsilon_n F(x,y)) - f(x)) + (1+\kappa)\frac{1}{n\epsilon_n}\log\frac{Pg(y)}{g(y)} - \kappa\frac{1}{n\epsilon_n}\log\frac{P\varphi_n(y)}{\varphi_n(y)}.$$

By the Hölder inequality,

$$Pg = P\Big((g^{1+\kappa}\varphi_n^{-\kappa})^{(1+\kappa)^{-1}}\varphi_n^{\kappa(1+\kappa)^{-1}}\Big) \le P^{(1+\kappa)^{-1}}(g^{1+\kappa}\varphi_n^{-\kappa})P^{\kappa(1+\kappa)^{-1}}\varphi_n,$$

or equivalently,

$$P(g^{1+\kappa}\varphi_n^{-\kappa}) \ge P^{1+\kappa}gP^{-\kappa}\varphi_n.$$

Then

$$T_n e^{n\widetilde{f}_{n,\kappa}}(x,y) \ge \exp\{n\widetilde{f}_{n,\kappa}(x,y) + n\epsilon_n\widetilde{h}_{n,\kappa}(x,y)\},$$

and (7.21)-(7.23) follow as above.

11.1.2. Exponential tightness. Since E is compact, the convergence of H_n to H given in (11.7) implies exponential tightness for $\{X_n\}$ by Corollary 4.17.

11.1.3. The comparison principle. Let $\alpha > 0$ and $h \in C_b(E)$. By a slight modification of Lemma 9.9, every subsolution $\overline{f}$ of $(I - \alpha H_\dagger)f = h$ is a strong subsolution (see Remark 7.2), so if $f_0 \in D_d$ and

$$\overline{f}(x_0) - f_0(x_0) = \sup_x(\overline{f}(x) - f_0(x)),$$

then for each $0 \le \kappa < 1$ and $g \in K_1(E_0)$, there exist $\{y_n\} \subset E_0$ and $\{x_n\} \subset E$ such that $x_n \to x_0$ and

$$\begin{aligned}&\alpha^{-1}(\overline{f}(x_0) - h(x_0)) \\ &\le \limsup_{n\to\infty}(F(x_n,y_n)\cdot\nabla f_0(x_n) + (1-\kappa)\log\frac{Pg(y_n)}{g(y_n)} + \kappa\psi(y_n)) \\ &\le \limsup_{n\to\infty}(F(x_0,y_n)\cdot\nabla f_0(x_0) + (1-\kappa)\log\frac{Pg(y_n)}{g(y_n)}) + \kappa\psi(y_n)),\end{aligned}$$

and hence

$$\begin{aligned}(11.10)\ &\alpha^{-1}(\overline{f}(x_0) - h(x_0)) \\ &\le \inf_{\kappa\in(0,1)}\inf_{g\in K_1(E_0)}\sup_{y\in E_0}\Big(F(x_0,y)\cdot\nabla f_0(x_0) + (1-\kappa)\log\frac{Pg(y)}{g(y)} + \kappa\psi(y)\Big).\end{aligned}$$

Similarly, if $\underline{f}$ is a supersolution of $(I - \alpha H_\ddagger)f = h$, $f_0 \in D_d$, and

$$f_0(x_0) - \underline{f}(x_0) = \sup_x(f_0(x) - \underline{f}(x)),$$

then

$$\alpha^{-1}(\underline{f}(x_0) - h(x_0))$$
$$\geq \sup_{\kappa>0} \sup_{g\in K_1(E_0)} \inf_{y\in E_0} \Big(F(x_0,y)\cdot\nabla f_0(x_0) + (1+\kappa)\log\frac{Pg(y)}{g(y)}) - \kappa\psi(y)\Big).$$

Define

$$H_1(x,p) = \inf_{0<\kappa\leq 1} \inf_{g\in K_1(E_0)} \sup_{y\in E_0} (F(x,y)\cdot p + (1-\kappa)\log\frac{Pg(y)}{g(y)} + \kappa\psi(y)), \tag{11.11}$$

$$H_2(x,p) = \sup_{\kappa>0} \sup_{g\in K_1(E_0)} \inf_{y\in E_0} (F(x,y)\cdot p + (1+\kappa)\log\frac{Pg(y)}{g(y)} - \kappa\psi(y)), \tag{11.12}$$

and

$$H_i f(x) = H_i(x,\nabla f(x)), \quad f\in D_d,\ i=1,2. \tag{11.13}$$

Every subsolution of $(I-\alpha H_\dagger)f = h$ is a subsolution of

$$(I-\alpha H_1)f = h, \tag{11.14}$$

and similarly, every supersolution of $(I-\alpha H_\ddagger)f = h$ is a supersolution of

$$(I-\alpha H_2)f = h. \tag{11.15}$$

The requirement that

$$H_1(x,p) \leq H_2(x,p), \quad x,p\in R^d. \tag{11.16}$$

will be the critical hypothesis in the large deviation theorem. This inequality is closely related to hypotheses used by a number of authors. These connections are discussed more fully in Appendix B. Here we indicate several situations in which (11.16) can be verified.

LEMMA 11.4. *If for each $x\in E$ and $p\in R^d$, there exists $g_{x,p}\in K_1(E_0)$ such that*

$$\lambda_{x,p} \equiv F(x,y)\cdot p + \log\frac{Pg_{x,p}(y)}{g_{x,p}(y)} \tag{11.17}$$

does not depend upon y, then (11.16) holds.

REMARK 11.5. If the $\{Y(k)\}$ are independent with distribution ν, then we can take $g_{x,p}(y) = e^{F(x,y)\cdot p}$ and (11.17) becomes $\log\int_{E_0} e^{F(x,z)\cdot p}\nu(dz)$. More generally, the desired $g_{x,p}$ is an eigenfunction of the operator $e^{F(x,\cdot)p}P$ satisfying

$$e^{F(x,y)\cdot p}Pg_{x,p}(y) = \lambda_{x,p}g_{x,p}(y).$$

Consequently, if E_0 is finite and P is the transition matrix for an irreducible Markov chain, then $e^{F(x,\cdot)\cdot p}P$ is an irreducible, nonnegative matrix, and the Perron-Frobenius theorem ensures the existence of $g_{x,p}$ and $\lambda_{x,p}$. Generalizations of the Perron-Frobenius theorem are discussed in Section B.1. In particular, Theorem B.5 gives conditions for existence.

PROOF. Note that for $0 < \kappa \leq 1$,

$$\begin{aligned} H_1(x,p) &\leq \sup_{y\in E_0}(F(x,y)\cdot p + (1-\kappa)\log\frac{Pg_{x,p}(y)}{g_{x,p}(y)} + \kappa\psi(y)) \\ &= \log\lambda_{x,p} + \kappa\sup_y(\psi(y) - \log\frac{Pg_{x,p}(y)}{g_{x,p}(y)}) \end{aligned}$$

and

$$\begin{aligned} H_2(x,p) &\geq \inf_{y\in E_0}(F(x,y)\cdot p + (1+\kappa)\log\frac{Pg_{x,p}(y)}{g_{x,p}(y)} - \kappa\psi(y)) \\ &= \log\lambda_{x,p} + \kappa\inf_y(\log\frac{Pg_{x,p}(y)}{g_{x,p}(y)} - \psi(y)). \end{aligned}$$

Recalling that ψ is bounded above, letting $\kappa \to 0$ gives the inequality. □

LEMMA 11.6. *Let $T_{x,p}f(y) = e^{F(x,y)\cdot p}Pf(y)$. Suppose that there exist $\lambda_{x,p} \in R$ and $g_{x,p} \in K_1(E_0)$ such that for each $c \in R$,*

$$\lim_{n\to\infty}\sup_{\{y:\psi(y)\geq c\}}\left|\frac{T^{n+1}_{x,p}g_{x,p}(y)}{T^n_{x,p}g_{x,p}(y)} - \lambda_{x,p}\right| = 0. \tag{11.18}$$

Then (11.16) holds.

REMARK 11.7. Under Harris recurrence conditions, Kontoyiannis and Meyn [**67**], Theorem 4.1, and [**68**], Theorem 3.1, give results that imply (11.18). Take ψ in (11.6) to be PV/V for the V in [**67**], and multiply the numerator and denominator in (11.18) by $e^{-n\Lambda}$, where Λ is the constant in the exponent in (50) of [**67**].

PROOF. Let $K_{x,p} = \sup_y|F(x,y)\cdot p|$, $c_{x,p} = \inf_y g_{x,p}(y)$, $C_{x,p} = \sup_y g_{x,p}(y)$, and note that

$$\frac{c_{x,p}}{C_{x,p}}e^{-K_{x,p}} \leq \frac{T^{n+1}_{x,p}g_{x,p}(y)}{T^n_{x,p}g_{x,p}(y)} \leq \frac{C_{x,p}}{c_{x,p}}e^{K_{x,p}}$$

and

$$\frac{c_{x,p}}{C_{x,p}}e^{-2K_{x,p}} \leq \frac{PT^n_{x,p}g_{x,p}(y)}{T^n_{x,p}g_{x,p}(y)} = e^{-F(x,y)\cdot p}\frac{T^{n+1}_{x,p}g_{x,p}(y)}{T^n_{x,p}g_{x,p}(y)} \leq \frac{C_{x,p}}{c_{x,p}}e^{2K_{x,p}}.$$

For $\kappa, \epsilon > 0$ and $c \in R$, there exists n such that

$$\begin{aligned} H_1(x,p) &\leq \sup_{y\in E_0}(F(x,y)\cdot p + (1-\kappa)\log\frac{PT^n_{x,p}g_{x,p}(y)}{T^n_{x,p}g_{x,p}(y)} + \kappa\psi(y)) \\ &= \sup_{y\in E_0}(\log\frac{e^{F(x,y)\cdot p}PT^n_{x,p}g_{x,p}(y)}{T^n_{x,p}g_{x,p}(y)} - \kappa\log\frac{PT^n_{x,p}g_{x,p}(y)}{T^n_{x,p}g_{x,p}(y)} + \kappa\psi(y)) \\ &= \sup_{y\in E_0}(\log\frac{T^{n+1}_{x,p}g_{x,p}(y)}{T^n_{x,p}g_{x,p}(y)} - \kappa\log\frac{PT^n_{x,p}g_{x,p}(y)}{T^n_{x,p}g_{x,p}(y)} + \kappa\psi(y)) \\ &= \left[\log(\lambda_{x,p}+\epsilon) + \kappa\sup_{\{y:\psi(y)\geq c\}}(\psi(y) - \log\frac{PT^n_{x,p}g_{x,p}(y)}{T^n_{x,p}g_{x,p}(y)})\right] \\ &\qquad \vee\left[(1+2\kappa)(K_{x,p} + \log\frac{C_{x,p}}{c_{x,p}}) + \kappa c\right] \end{aligned}$$

and

$$\begin{aligned} H_2(x,p) &\geq \inf_{y\in E_0}(F(x,y)\cdot p+(1+\kappa)\log\frac{PT^n_{x,p}g_{x,p}(y)}{T^n_{x,p}g_{x,p}(y)}-\kappa\psi(y)) \\ &= \left[\log(\lambda_{x,p}-\epsilon)+\kappa\inf_{\{y:\psi(y)\geq c\}}(\log\frac{Pg_{x,p}(y)}{g_{x,p}(y)}-\kappa\psi(y))\right] \\ &\qquad\qquad \wedge\left[-(1+2\kappa)(K_{x,p}+\log\frac{C_{x,p}}{c_{x,p}})-\kappa c\right]. \end{aligned}$$

By selecting ϵ and κ sufficiently small and c sufficiently negative (that is $c<0$ and $|c|$ sufficiently large), we see that the desired inequality must hold. □

LEMMA 11.8. *If E_0 is compact and for each $x,p\in R^d$, there exists a constant $c(x,p)$ such that*

$$\lim_{n\to\infty}\frac{1}{n}\log E[e^{\sum_{k=0}^{n-1}p\cdot F(x,Y(k))}|Y(0)=y]=c(x,p) \tag{11.19}$$

uniformly in y, then

$$c(x,p)=\sup_{\mu\in\mathcal{P}(E_0)}(\int_{E_0}F(x,y)\cdot p\mu(dy)+\inf_{g\in K_1(E_0)}\int_{E_0}\log\frac{Pg}{g}d\mu), \tag{11.20}$$

and (11.16) holds with

$$H_1(x,p)\leq c(x,p)\leq H_2(x,p). \tag{11.21}$$

REMARK 11.9. The convergence in (11.19) is essentially Condition F of [**52**]. For more general situations where E_0 may be noncompact, methods for verifying (11.21) are discussed in Appendix B.

PROOF. The lemma is a special case of Theorem B.20. □

LEMMA 11.10. *Assume that (11.3) and (11.4) hold, that H_1 and H_2 are given by (11.11) and (11.12), and that (11.16) is satisfied. Then for $h\in C_b(E)$ and $\alpha>0$, the comparison principle holds for subsolutions of $(I-\alpha H_1)f=h$ and supersolutions of $(I-\alpha H_2)f=h$.*

PROOF. We verify Condition 9.10. We have

$$\begin{aligned} &H_1(x_1,p)-H_1(x_2,p) \\ &=\inf_{\kappa\in(0,1)}\inf_{g\in K_1(E_0)}\sup_{y\in E_0}(F(x_1,y)\cdot p+(1-\kappa)\log\frac{Pg(y)}{g(y)})+\kappa\psi(y) \\ &\quad-\inf_{\kappa\in(0,1)}\inf_{g\in K_1(E_0)}\sup_y(F(x_2,y)\cdot p+(1-\kappa)\log\frac{Pg(y)}{g(y)}+\kappa\psi(y)) \\ &\leq\inf_{\kappa\in(0,1)}\inf_{g\in K_1(E_0)}\sup_y(F(x_2,y)\cdot p+K|x_1-x_2||p|+(1-\kappa)\log\frac{Pg(y)}{g(y)}+\kappa\psi(y)) \\ &\quad-\inf_{\kappa\in(0,1)}\inf_{g\in K_1(E_0)}\sup_y(F(x_2,y)\cdot p+(1-\kappa)\log\frac{Pg(y)}{g(y)}+\kappa\psi(y)) \\ &=K|x_1-x_2||p|. \end{aligned}$$

By (11.16),

$$H_1(x_1,p)-H_2(x_2,p)\leq K|x_1-x_2||p|,$$

verifying Condition 9.10.1, by Lemma 9.17.

To verify Condition 9.10.2, let $x_m \to \infty$ and $|x_m||p_m| \to 0$. Then

$$\begin{aligned}
\limsup_{m\to\infty} H_1(x_m,p_m) &= \inf_{\kappa\in(0,1)} \inf_{g\in K_1(E_0)} \sup_{y\in E_0} ((1-\kappa)\log\frac{Pg(y)}{g(y)} + \kappa\psi(y)) \\
&\le \inf_{\kappa\in(0,1)} \sup_{y\in E_0} ((1-\kappa)\log\frac{P1(y)}{1(y)} + \kappa\psi(y)) \\
&= \inf_{\kappa\in(0,1)} (\kappa \sup_{y\in E_0} \psi(y)) = 0,
\end{aligned}$$

and similarly,

$$\begin{aligned}
\liminf_{m\to\infty} H_2(x_m,p_m) &= \sup_{\kappa\in(0,1)} \sup_{g\in K_1(E_0)} \inf_{y\in E_0} ((1+\kappa)\log\frac{Pg(y)}{g(y)} - \kappa\psi(y)) \\
&\ge \sup_{\kappa\in(0,1)} \inf_{y\in E_0} (-\kappa\psi(y)) \\
&\ge -\inf_{\kappa\in(0,1)} \kappa \sup_{y\in E_0} \psi(y) = 0.
\end{aligned}$$

The comparison principle follows by Lemma 9.15. □

Since every subsolution of $(I-\alpha H_\dagger)f = h$ is a subsolution of $(I-\alpha H_1)f = h$ and every supersolution of $(I-\alpha H_\ddagger)f = h$ is a supersolution of $(I-\alpha H_2)f = h$, we have the following immediate consequence of Lemma 11.10.

LEMMA 11.11. *Assume that (11.3), (11.4) and (11.16) hold. Then for $h \in C_b(E)$ and $\alpha > 0$, the comparison principle holds for subsolutions of $(I-\alpha H_\dagger)f = h$ and supersolutions of $(I-\alpha H_\ddagger)f = h$.*

11.1.4. Variational representation of H. The following lemma gives the desired control representation.

LEMMA 11.12. *Let Condition 11.1 hold, and let H_1 be given by (11.11). Then*

$$H_1(x,p) = \mathbf{H}(x,p) \equiv \sup_{\nu\in\mathcal{P}(E_0)} (\int_{E_0} F(x,y)\cdot p d\nu - I_P(\nu)), \tag{11.22}$$

where

$$I_P(\nu) = -\inf_{g\in K_1(E_0)} \int_{E_0} \log\frac{Pg}{g} d\nu \wedge \int \psi d\nu.$$

REMARK 11.13. Under Condition 11.1.5, typically

$$\int \psi d\nu \ge \limsup_{n\to\infty} \int_{E_0} \log\frac{P\varphi_n}{\varphi_n} d\nu$$

so that

$$I_P(\nu) = -\inf_{g\in K_1(E_0)} \int_{E_0} \log\frac{Pg}{g} d\nu.$$

PROOF. Note that

$$\begin{aligned}
& H_1(x,p) \\
= & \inf_{0<\kappa\leq 1} \inf_{g\in K_1(E_0)} \sup_{\nu\in\mathcal{P}(E_0)} \int_{E_0} (F(x,y)\cdot p + (1-\kappa)\log\frac{Pg(y)}{g(y)} + \kappa\psi(y))\nu(dy) \\
\geq & \inf_{0<\kappa\leq 1} \sup_{\nu\in\mathcal{P}(E_0)} \inf_{g\in K_1(E_0)} (\int_{E_0} F(x,\cdot)\cdot p d\nu + (1-\kappa)\int_{E_0}\log\frac{Pg}{g}d\nu \wedge \int_{E_0}\psi d\nu \\
& \qquad + \kappa\int_{E_0}\psi d\nu) \\
= & \inf_{0<\kappa\leq 1} \sup_{\nu\in\mathcal{P}(E_0)} (\int_{E_0} F(x,\cdot)\cdot p d\nu + (1-\kappa)\inf_{g\in K_1(E_0)}\int_{E_0}\log\frac{Pg}{g}d\nu \wedge \int_{E_0}\psi d\nu \\
& \qquad + \kappa\int_{E_0}\psi d\nu) \\
= & \sup_{\nu\in\mathcal{P}(E_0)} (\int_{E_0} F(x,\cdot)\cdot p d\nu + \inf_{g\in K_1(E_0)}\int_{E_0}\log\frac{Pg}{g}d\nu \wedge \int_{E_0}\psi d\nu) \\
= & \mathbf{H}(x,p).
\end{aligned}$$

Let $g_1,\ldots,g_m \in K_1(E_0)$, $\alpha_i \geq 0$, and $\sum_{i=1}^m \alpha_i = 1$. Observing that $P\prod g_i^{\alpha_i} \leq \prod(Pg_i)^{\alpha_i}$, we have

$$\log\frac{P\prod g_i^{\alpha_i}}{\prod g_i^{\alpha_i}} \leq \sum_{i=1}^m \alpha_i \log\frac{Pg_i}{g_i},$$

and hence

$$H_1(x,p) = \inf_{0<\kappa\leq 1} \inf_{\sum_{i=1}^m \alpha_i=1} \inf_{g_i\in K_1(E_0)} \sup_{y\in E_0} \Big(F(x,y)\cdot p + (1-\kappa)\sum_{i=1}^m \alpha_i\log\frac{Pg_i(y)}{g_i(y)} + \kappa\psi(y)\Big).$$

There exists a sequence $\{g_i\} \subset K_1(E_0)$ such $K_1(E_0\}$ is the bounded, pointwise closure of $\{g_i\}$, and $\{g_i\}$ can be selected so that for each $\nu \in \mathcal{P}(E_0)$,

$$\inf_i \int_{E_0} \log\frac{Pg_i}{g_i}d\nu = \inf_{g\in K_1(E_0)} \int_{E_0}\log\frac{Pg}{g}d\nu.$$

Consequently,

$$
\begin{aligned}
&H_1(x,p)\\
\le\ & \lim_{m\to\infty}\inf_{0<\kappa\le 1}\inf_{\sum_{i=0}^m\alpha_i=1}\sup_{y\in E_0}\Big(F(x,y)\cdot p+(1-\kappa)\sum_{i=1}^m\alpha_i\log\frac{Pg_i(y)}{g_i(y)}+\kappa\psi(y)\Big)\\
=\ & \lim_{m\to\infty}\inf_{0<\kappa\le 1}\inf_{\sum_{i=1}^m\alpha_i=1}\sup_{\nu\in\mathcal{P}(E_0)}\int_{E_0}\Big(F(x,y)\cdot p+(1-\kappa)\sum_{i=1}^m\alpha_i\log\frac{Pg_i(y)}{g_i(y)}\\
&\qquad +\kappa\psi(y)\Big)\nu(dy)\\
=\ & \lim_{m\to\infty}\inf_{0<\kappa\le 1}\sup_{\nu\in\mathcal{P}(E_0)}\inf_{\sum_{i=1}^m\alpha_i=1}\int_{E_0}\Big(F(x,y)\cdot p+(1-\kappa)\sum_{i=1}^m\alpha_i\log\frac{Pg_i(y)}{g_i(y)}\\
&\qquad +\kappa\psi(y)\Big)\nu(dy)\\
=\ & \lim_{m\to\infty}\inf_{0<\epsilon\le 1}\sup_{\nu\in\mathcal{P}(E_0)}\inf_{\epsilon\le\kappa\le 1}\Big[\int_{E_0}F(x,\cdot)\cdot p d\nu+(1-\kappa)\min_{1\le i\le m}\int_{E_0}\log\frac{Pg_i}{g_i}d\nu\\
&\qquad +\kappa\int\psi d\nu\Big]\\
=\ & \lim_{m\to\infty}\inf_{0<\epsilon\le 1}\sup_{\nu\in\mathcal{P}(E_0)}\Big[\int_{E_0}F(x,\cdot)\cdot p d\nu+(1-\epsilon)\min_{1\le i\le m}\int\log\frac{Pg_i}{g_i}d\nu\wedge\int\psi d\nu\\
&\qquad +\epsilon\int\psi d\nu\Big]\\
=\ & \lim_{m\to\infty}\sup_{\nu\in\mathcal{P}(E_0)}\Big[\int_{E_0}F(x,\cdot)\cdot p d\nu+\min_{1\le i\le m}\int\log\frac{Pg_i}{g_i}d\nu\wedge\int\psi d\nu\Big]\\
=\ & \sup_{\nu\in\mathcal{P}(E_0)}\Big(\int_{E_0}F(x,\cdot)\cdot p d\nu-I_P(\nu)\Big)=\mathbf{H}(x,p).
\end{aligned}
$$

The second and third equalities follow by Sion's theorem, [**111**], Corollary 3.3. The final equality follows by observing that if the limit is greater than $-\infty$, then there exist ν_m such that $\inf\int\psi d\nu_m>-\infty$ and hence $\{\nu_m\}$ is relatively compact (assume it converges to ν_0) and

$$
\begin{aligned}
&\lim_{m\to\infty}\sup_{\nu\in\mathcal{P}(E_0)}\Big[\int_{E_0}F(x,y)\cdot p d\nu+\min_{1\le i\le m}\int\log\frac{Pg_i}{g_i}d\nu\wedge\int\psi d\nu\Big]\\
&\quad=\lim_{m\to\infty}\Big[\int_{E_0}F(x,y)\cdot p d\nu_m+\min_{1\le i\le m}\int\log\frac{Pg_i}{g_i}d\nu_m\wedge\int\psi d\nu_m\Big]\\
&\quad\le\lim_{m_0\to\infty}\limsup_{m\to\infty}\Big[\int_{E_0}F(x,y)\cdot p d\nu_m+\min_{1\le i\le m_0}\int\log\frac{Pg_i}{g_i}d\nu_m\wedge\int\psi d\nu_m\Big]\\
&\quad\le\lim_{m_0\to\infty}\Big[\int_{E_0}F(x,y)\cdot p d\nu_0+\min_{1\le i\le m_0}\int\log\frac{Pg_i}{g_i}d\nu_0\wedge\int\psi d\nu_0\Big]\\
&\quad=\int_{E_0}F(x,y)\cdot p d\nu_0-I_P(\nu_0)\\
&\quad\le\mathbf{H}(x,p),
\end{aligned}
$$

so equality holds throughout. $\square$

Next, we discuss some properties of $\mathbf{H}$ and I_P.

LEMMA 11.14. *Let π be a stationary distribution for P. Then*

$$\int_{E_0} \log \frac{Pg}{g} d\pi \geq 0, \text{ for every } g \in K_1(E_0),$$

and

$$\mathbf{H}(x,p) \geq \int_{E_0} F(x,y) \cdot p\pi(dy).$$

PROOF. By Jensen's inequality, $\log Pg(x) \geq P(\log g)(x)$, so

$$\int_{E_0} \log \frac{Pg}{g} d\pi \geq \int_{E_0} (P(\log g) - \log g) d\pi = 0.$$

Consequently,

$$\mathbf{H}(x,p) \geq \int_{E_0} F(x,y) \cdot p\pi(dy) + \inf_{g \in K_1(E_0)} \int_{E_0} \log \frac{Pg}{g} d\pi \geq \int_{E_0} F(x,y) \cdot p\pi(dy).$$

□

One consequence of Lemma 11.14 is that $\mathbf{H}(x,0) = 0$. (Take $g = 1$ in the definition of I_P.) It also follows that $I_P(\pi) = 0$ for any stationary π satisfying $\int_{E_0} \psi d\pi \geq 0$. Such a π exists by Lemma 11.2.

From (11.22), we see that I_P should be the running cost in a control representation of the type considered in Chapter 8 (take $\mathbf{H}_\dagger = \mathbf{H}_\ddagger = \mathbf{H}$). Note that

$$\Gamma_P^M = \{\mu \in \mathcal{P}(E_0) : I_P(\mu) \leq M\} \subset \{\mu \in \mathcal{P}(E_0) : \int_{E_0} \psi d\mu \geq M\} \tag{11.23}$$

is compact for each $M \in R$.

LEMMA 11.15. *Let F be continuous and Conditions 11.1 hold. Assume that (11.16) holds, and set*

$$Af(x,\mu) = \int_{E_0} F(x,y) \cdot \nabla f(x) \mu(dy), \quad f \in D_d, \mu \in \mathcal{P}(E_0).$$

Then

$$H_1 f(x) = \mathbf{H} f(x) = \sup_{\mu \in \mathcal{P}(E_0)} (Af(x,\mu) - I_P(\mu)) \leq H_2 f(x) \tag{11.24}$$

and taking $U = \mathcal{P}(E_0)$, Conditions 8.30 and 8.10 are satisfied.

REMARK 11.16. Suppose that λ is a relaxed control with $U = \mathcal{P}(E_0)$, and define $\mu_t \in \mathcal{P}(E_0)$ by

$$\mu_t(\Gamma) = \int_{\mathcal{P}(E_0)} \mu(\Gamma) \lambda_t(d\mu), \quad \Gamma \in \mathcal{B}(E_0).$$

Since $Af(x,\mu)$ is linear in μ, for $x \in D_E[0,\infty)$,

$$\int_0^t \int_{\mathcal{P}(E_0)} Af(x(s),\mu) \lambda_s(d\mu) ds = \int_0^t \int_{E_0} F(x(s),y) \cdot \nabla f(x(s)) \mu_s(dy) ds,$$

and

$$\begin{aligned}\int_{\mathcal{P}(E_0)} I_P(\mu)\lambda_t(d\mu) &= -\int_{\mathcal{P}(E_0)} \inf_{g\in K_1(E_0)} \int \log \frac{Pg}{g} d\mu \wedge \int \psi d\mu \, \lambda_t(d\mu) \\ &\geq -\inf_{g\in K_1(E_0)} \int_{\mathcal{P}(E_0)} \int \log \frac{Pg}{g} d\mu \wedge \int \psi d\mu \, \lambda_t(d\mu) \\ &\geq I_P(\mu_t),\end{aligned}$$

so relaxed controls are not needed. For any $x_0 \in R^d$, there is a unique solution x of

$$x(t) = x_0 + \int_0^t \int_{E_0} F(x(s), y)\mu_s(dy)ds. \tag{11.25}$$

All solutions of the control problem with $x_0 \in R^d$ are of this form. Note that under (11.3) and (11.4), no solution can reach ∞ in finite time. If $x_0 = \infty$, then $x \equiv \infty$ for all controls.

PROOF. (11.24) follows by (11.16) and (11.22). Since $E = R^d \cup \{\infty\}$, Condition 8.30.1 follows by the assumptions on F and the definition of D_d. $L(x, \mu) = I_P(\mu)$ is lower semicontinuous since it is the negative of the minimum of two upper semicontinuous functions, and Γ_P^M is compact, so Condition 8.30.2 holds with $\Phi(\mu) = I_P(\mu)$. The continuity of $\mathbf{H}f$ follows from the Lipschitz assumptions on F, so Condition 8.30.3 follows. $\|Af\| < \infty$, so Condition 8.30.4 is immediate.

Condition 8.10 follows by taking $\lambda(d\mu \times ds) = \delta_{\{\pi\}}(d\mu) \times ds$ in (8.12), where π is a stationary distribution for P. □

For some purposes, a different form for the rate function may be useful. Note that by definition, $\mathbf{H}(x,p)$ is a convex function of p.

LEMMA 11.17. *Assume that Conditions 11.1 holds. Let $\rho(x) = 1 + |x|$, and define*

$$L_\rho(x, q) = \sup_{p\in R^d} (p \cdot q - \mathbf{H}(x, p/\rho(x))), \quad x, q \in R^d, \tag{11.26}$$

and

$$L_\rho(\infty, q) = \liminf_{|x|\to\infty, q'\to q} L_\rho(x, q').$$

Take $U = R^d$, and define

$$Af(x, q) = \rho(x) q \cdot \nabla f(x), \quad f \in D_d, q \in U, x \in R^d,$$

and $Af(\infty, q) = 0$. Then $L_\rho \geq 0$ is lower semicontinuous,

$$\mathbf{H}f(x) = \sup_q (Af(x, q) - L_\rho(x, q)), \tag{11.27}$$

and Conditions 8.30 and 8.10 are satisfied.

REMARK 11.18. Note that under (11.3) and (11.4), the introduction of ρ ensures that $\mathbf{H}(x, p/\rho(x))$ is a bounded function of x.

PROOF. The properties of L_ρ and (11.27) follow as in the proof of Lemma 10.21 using the fact that $\mathbf{H}(x, 0) = 0$, the convexity of $\mathbf{H}(x,p)$, and the finiteness of $\sup_{|p|\leq c} \sup_{x\in R^d} \mathbf{H}(x, p/\rho(x)) < \infty$, for each $c > 0$.

Conditions 8.30.1 and 8.30.3 follow from the definition of A and the continuity assumptions; Condition 8.30.2 and 8.30.4 follow by (10.31).

To see that Condition 8.10 holds, let $\overline{F}(x) = \int_{E_0} F(x,y)\pi(dy)$ and recall that $\mathbf{H}(x,p) \geq \overline{F}(x) \cdot p$ and $L_\rho(x,q) \geq 0$. Taking $q = \overline{F}(x)/\rho(x)$, we have

$$L_\rho(x, \overline{F}(x)/\rho(x)) = \sup_{p \in R^d} (p \cdot \overline{F}(x)/\rho(x) - \mathbf{H}(x, p/\rho(x))) = 0.$$

The desired solution is given by

$$x(t) = x_0 + \int_0^t \overline{F}(x(s))ds,$$

if $x_0 \in R^d$, and $x \equiv \infty$, if $x_0 = \infty$. □

Under Condition 11.1, Lemma 11.12 gives $\mathbf{H} = H_1$, and the proof of Lemma 11.10, gives the following lemma.

LEMMA 11.19. *Assume Condition 11.1. Then for $\alpha > 0$ and $h \in C_b(E)$, the comparison principle holds for*

$$(I - \alpha\mathbf{H})f = h.$$

11.1.5. The large deviation theorem.

For each $f \in B(E \times E_0)$, define

$$V_n(t)f(x,y) = \frac{1}{n} \log E[e^{nf(X_n(t),Y_n(t))}|X_n(0) = x, Y_n(0) = y].$$

$M_{\mathcal{P}(E_0)}[0,\infty)$ is the space of measurable, $\mathcal{P}(E_0)$-valued functions on $[0,\infty)$.

THEOREM 11.20. *Suppose that Condition 11.1 holds and that H_1 and H_2 are given by (11.11) and (11.12) and satisfy $H_1(x,p) \leq H_2(x,p)$, $x,p \in R^d$. Assume that the large deviation principle holds for $\{X_n(0)\}$ with good rate function I_0. Then the following hold:*

a) *$H_\dagger$ and $H_\ddagger$ defined in (11.8) and (11.9) together determine a semigroup $V(t)$ on $C_b(E)$ in the sense of Theorem 7.17. For each $f \in C_b(E)$ and $f_n \in B(E \times E_0)$ satisfying*

$$\lim_{n\to\infty} \sup_{(x,y)\in E\times \widetilde{K}_n^q} |f_n(x,y) - f(x)| = 0, \quad q \in \mathcal{Q},$$

and $t_n \to t$, we have

$$\lim_{n\to\infty} \sup_{(x,y)\in E\times \widetilde{K}_n^q} |V_n(t_n)f_n(x,y) - V(t)f(x)| = 0. \tag{11.28}$$

b) *$\{X_n\}$ satisfies a large deviation principle in $C_{R^d}[0,\infty)$ with good rate function I.*

Suppose in addition that F is continuous. Then

c)

$$I(x) = \inf\{I_0(x(0)) + \int_0^\infty I_P(\mu_s)ds :$$

$$x(t) = x(0) + \int_0^t \int_{E_0} F(x(s),y)\mu_s(dy)ds, t \geq 0, \mu \in M_{\mathcal{P}(E_0)}[0,\infty)\}, \tag{11.29}$$

where $\inf \emptyset = \infty$.

d) *For* $L(x,q)$ *given by*

$$L(x,q) = \sup_{p\in R^d}(p\cdot q - \mathbf{H}(x,p)),$$

$$(11.30) \qquad I(x) = \begin{cases} I_0(x(0)) + \int_0^\infty L(x(s),\dot{x}(s))ds & x \text{ absolutely continuous} \\ \infty & \text{otherwise.} \end{cases}$$

PROOF. Apply Theorem 7.18 with $K_n^q \equiv E \times \widetilde{K}_n^q$. The Convergence Condition 7.11 was verified in Section 11.1.1. Part (a) follows and the large deviation principle holds for $\{X_n\}$ in $D_E[0,\infty)$ (and, hence, in $C_E[0,\infty)$) with a good rate function I. Since, $H_1 = \mathbf{H} \le H_2$, each subsolution of $f - \alpha H_1 f = h$ (supersolution of $f - \alpha H_2 f = h$)is a subsolution (supersolution) of $f - \alpha\mathbf{H}f = h$, and since the comparison principle for $f - \alpha\mathbf{H}f = h$ holds for $h \in C(E)$, Part (c) follows from Lemmas 11.15 and 8.32 and Corollary 8.28. Part (d) follows by Lemma 11.17 and 8.32 and Corollary 8.28. □

11.2. Continuous time, law of large numbers scaling

Let Y be an E_0-valued Markov process with semigroup

$$S(t)f(y) \equiv E[f(Y(t))|Y(0) = y], \quad f \in C_b(E_0).$$

We assume that for each initial $Y(0) = y \in E_0$, there is a cadlag version of Y, and that $\{S(t)\}$ is *buc*-continuous, that is, (11.33) holds. Let $B \subset C_b(E_0) \times C_b(E_0)$ be the weak infinitesimal generator for $\{S(t)\}$, that is, for $f \in \mathcal{D}(B)$,

$$(11.31) \quad \sup_t t^{-1}\|S(t)f - f\| < \infty, \qquad Bf(y) = \lim_{t\to 0+}\frac{S(t)f(y) - f(y)}{t}, \quad y \in E_0.$$

If $f \in \mathcal{D}(B)$, then

$$S(t)f(y) - f(y) = \int_0^t S(r)Bf(y)ds,$$

and if (11.33) holds, the convergence in (11.31) is *buc*-convergence. We also consider the strong infinitesimal generator $B_0 \subset B$ for which the convergence in (11.31) is uniform. If $f \in \mathcal{D}(B)$, then $f_h = h^{-1}\int_0^t S(s)fds \in \mathcal{D}(B_0)$, $buc - \lim_{h\to 0} f_h = f$, and

$$buc - \lim_{h\to 0} B_0 f_h = buc - \lim_{h\to 0} h^{-1}\int_0^h S(s)Bfds = Bf.$$

It follows that B is the *buc*-closure of B_0. Note that

$$\inf_{y\in E_0} f_h(y) \ge \inf_{y\in E_0} f(y).$$

We consider (11.1) with $Y_n(s) = Y(ns)$, $F_n(x,y) = F(x,y)$, and $\lim_{n\to\infty} x_n = x_0 \in R^d$.

We assume the following:

CONDITION 11.21.

(1) *The martingale problem for* B *is well-posed. The semigroup is Feller in the sense that*

$$(11.32) \qquad S(t) : C_b(E_0) \to C_b(E_0),$$

and buc-continuous in the sense that

$$(11.33) \qquad buc - \lim_{t\to 0} S(t)f = f, \quad f \in C_b(E_0).$$

(2) *(11.3) and (11.4) hold for F.*

(3) *There is an index set $\mathcal{Q}$ and a family of subsets of E_0, $\{\widetilde{K}_n^q \subset E_0 : q \in \mathcal{Q}\}$, such that for $q_1, q_2 \in \mathcal{Q}$, there exists $q_3 \in \mathcal{Q}$ with $\widetilde{K}_n^{q_1} \cup \widetilde{K}_n^{q_2} \subset \widetilde{K}_n^{q_3}$; and for each $y \in E_0$, there exists $q \in \mathcal{Q}$ such that $\lim_{n\to\infty} r_0(y, \widetilde{K}_n^q) = 0$. Moreover, for each $q \in \mathcal{Q}$, $T > 0$ and $a > 0$, there exists a $\widehat{q}(q,a,T) \in \mathcal{Q}$ satisfying*

$$\limsup_{n\to\infty} \sup_{y\in\widetilde{K}_n^q} \frac{1}{n} \log P(Y(t) \notin \widetilde{K}_n^{\widehat{q}(q,a,T)}, \text{ some } t \le nT | Y(0) = y) \le -a. \tag{11.34}$$

(4) *There exist an upper semicontinuous function ψ on E_0, $\{\varphi_n\} \subset \mathcal{D}^{++}(B_0)$, and $q_0 \in \mathcal{Q}$ such that ψ is bounded above, $\{y \in E_0 : \psi(y) \ge c\}$ is compact for each $c \in R$, $0 < \inf_{y\in K_n^{q_0}} \varphi_n(y) < 2\inf_{y\in E_0} \varphi_n(y)$, $\inf_{n,y\in E_0} \varphi_n(y) > 0$,*

$$\lim_{n\to\infty} \frac{1}{n} \log \|\varphi_n\| = 0, \quad \sup_{n,y} B\varphi_n(y)/\varphi_n(y) < \infty, \tag{11.35}$$

and for each $q \in \mathcal{Q}$,

$$\lim_{n\to\infty} \sup_{y\in\widetilde{K}_n^q} (\frac{B\varphi_n(y)}{\varphi_n(y)} - \psi(y)) \le 0, \quad q \in \mathcal{Q}. \tag{11.36}$$

In addition, for each n and $\beta \in (-\infty, 1]$,

$$\lim_{t\to 0} \|S(t)\varphi_n^\beta - \varphi_n^\beta\| = 0. \tag{11.37}$$

REMARK 11.22. Let $P_y \in \mathcal{P}(D_{E_0}[0,\infty))$ be the distribution of Y with $Y(0) = y$. If $y \to P_y$ is continuous in the weak topology on $\mathcal{P}(D_E[0,\infty))$, then (11.33) holds. (Consider $S(t_n)f(y_n)$ for $y_n \to y$ and $t_n \to 0$.) This continuity is not, in general, implied by the Feller condition (11.32). It is equivalent to (11.32) plus the assumption that $\{P_y : y \in K\}$ is tight for each compact $K \subset E_0$.

Note the (11.37) holds if $\varphi_n^\beta \in \mathcal{D}(B)$ for each $\beta \in (-\infty, 1]$.

LEMMA 11.23. *Assume that Condition 11.21 holds. Then there exists a stationary distribution π for B satisfying $\int_{E_0} \psi d\pi \ge 0$.*

PROOF. The proof is essentially the same as that of Lemma 11.2 □

Again, we let $E = R^d \cup \{\infty\}$ be the one point compactification of R^d, $E_n \equiv E \times E_0$ and

$$D_d = \{f : f|_{R^d} - f(\infty) \in C_c^2(R^d)\} \subset C(E).$$

Let $\mathcal{D}(A_n)$ be the linear span of

$$\{f_1 f_2 : f_1 \in D_d, f_2 \in \mathcal{D}(B)\},$$

and define A_n by

$$A_n g(x,y) = F(x,y) \cdot \nabla_x g(x,y) + nBg(x,\cdot)(y),$$

for $g \in \mathcal{D}(A_n)$. Since B is the generator of the Markov process Y, for each $g \in \mathcal{D}(A_n)$ and $x \in R^d$,

$$g(x, Y_n(t)) - g(x, Y_n(0)) - \int_0^t nBg(x, Y_n(s))ds$$

is a martingale. If X_n satisfies (11.1), then by Lemma 3.4, Chapter 4 of Ethier and Kurtz [**36**],

$$g(X_n(t), Y_n(t)) - g(X_n(0), Y_n(0)) - \int_0^t A_n g(X_n(s), Y_n(s))ds \tag{11.38}$$

is a martingale, that is, (X_n, Y_n) is a solution of the martingale problem for A_n.

Since for each initial value $Y_n(0)$, the distribution of Y_n is uniquely determined by B and any solution of the martingale problem for A_n must satisfy (11.1), it follows that for each initial value $(X_n(0), Y_n(0))$, there is a distributionally unique solution to the martingale problem in (11.38). Since the distribution of Y_n is a measurable function of the initial value $Y_n(0)$, the distribution of (X_n, Y_n) is a measurable function of its initial value. In particular, there exists an extension of A_n that is the full generator for a well-posed martingale problem.

As in the discrete time case, if E_0 is compact, we can take $\mathcal{Q}$ to be an arbitrary set with $\widetilde{K}_n^q = E_0$, $\varphi_n = 1$, and $\psi = 0$, and Conditions 11.21.3 and 11.21.4 will be satisfied. Below, we show that a large class of R^d-valued diffusion processes satisfies these two conditions as well.

EXAMPLE 11.24. Let Y be a diffusion process such that $B_0 g = Lg$, $g \in D_d$, where L is the differential operator

$$Lg(y) = \frac{1}{2}\sum_{i,j=1}^m a_{ij}(y)\frac{\partial^2}{\partial y_i \partial y_j}g(y) + \sum_{i=1}^m b_i(y)\frac{\partial}{\partial y_i}g(y).$$

Suppose that

CONDITION 11.25.
There exists $\varphi_1, \varphi_2 \in C^2(R^d)$ *satisfying*

(1) $\varphi_i \geq 1$, $i = 1, 2$
(2) $\lim_{|y|\to\infty} \varphi_i(y) = \infty$, $i = 1, 2$
(3)

$$\lim_{|y|\to\infty} \frac{L\varphi_i(y)}{\varphi_i(y)} = -\infty, \quad i = 1, 2 \tag{11.39}$$

(4)

$$\lim_{|y|\to\infty} \frac{\log \varphi_1(y)}{\log \varphi_2(y)} = 0 \tag{11.40}$$

Setting $\widetilde{K}_n^q = \{y : \varphi_2(y) \leq e^{qn}\}$, Condition 11.21.3 follows by Lemma 4.20.

Let $\rho \in C^2(R)$ be nondecreasing and concave and satisfy $\rho(r) = r$, $0 \leq r \leq 1$, and $\rho(r) = 2$, $r \geq 3$. For $n \geq 2$, define

$$r_n = \quad \sup\{\log \varphi_1(y) : y \text{ satisfying } \sqrt{\log \varphi_1(y) \log \varphi_2(y)} \leq n\}$$

$$\varphi_n(y) = e^{r_n}\rho(e^{-r_n}\varphi_1(y)). \tag{11.41}$$

Note that (11.40) implies $\lim_{n\to\infty} n^{-1} r_n = 0$ and hence $\lim_{n\to\infty} n^{-1}\log\|\varphi_n\| = 0$. Then

$$\frac{B_0\varphi_n(y)}{\varphi_n(y)} = \frac{\rho'(e^{-r_n}\varphi_1(y))L\varphi_1(y) + 2^{-1}e^{-r_n}\rho''(e^{-r_n}\varphi_1(y))\nabla\varphi_1^T(y)a(y)\nabla\varphi_1(y)}{e^{r_n}\rho(e^{-r_n}\varphi_1(y))}.$$

Since $\rho'' \leq 0$, $\rho(r) \geq r\rho'(r)$ and

$$\frac{B_0\varphi_n(y)}{\varphi_n(y)} \leq \frac{L\varphi_1(y) \vee 0}{\varphi_1(y)},$$

and $\sup_{n,y} B_0\varphi_n(y)/\varphi_n(y) < \infty$. For $y \in \{y : \log\varphi_1(y) \leq r_n\}$,

$$\frac{B_0\varphi_n(y)}{\varphi_n(y)} = \frac{L\varphi_1(y)}{\varphi_1(y)}.$$

Since

$$\begin{aligned}\{y : \log\varphi_1(y) \leq r_n\} \quad &\supset \quad \{y : \sqrt{\log\varphi_1(y)\log\varphi_2(y)} \leq n\} \\ &\supset \{y : \log\varphi_2(y) \leq \sqrt{\frac{\log\varphi_2(y)}{\log\varphi_1(y)}}n\},\end{aligned}$$

for each q, there exists n_q such that $n > n_q$ implies

$$\inf_{y:\log\varphi_1(y)>r_n} \sqrt{\frac{\log\varphi_2(y)}{\log\varphi_1(y)}} > q,$$

and hence $\{y : \log\varphi_1(y) \leq r_n\} \supset \widetilde{K}_n^q$. Consequently, taking $\psi = L\varphi_1/\varphi_1$, we have

$$\frac{B_0\varphi_n(y)}{\varphi_n(y)} = \psi(y), \quad y \in \widetilde{K}_n^q,$$

for $n \geq n_q$. Note that $\varphi_n^\beta \in \mathcal{D}(B_0)$ for all $\beta \in R$. Consequently, (11.37) holds completing the verification of Condition 11.21.4.

If Y is the Ornstein-Uhlenbeck process with

$$Bg = B_0g(y) = \frac{1}{2}\Delta g(y) - \alpha y \cdot \nabla g(y), \quad g \in D_d,$$

for some $\alpha > 0$, then we can take $\varphi_1(y) = \exp\{\sqrt{1+|y|^2}\}$ and $\varphi_2(y) = \exp\{\delta|y|^2\}$, for $\delta < \alpha$.

11.2.1. Convergence of H_n. Let $H_nf = \frac{1}{n}e^{-nf}A_ne^{nf}$. For $f_n(x,y) = f(x) + \frac{1}{n}\log g(y)$, $f \in D_d$ and $g \in \mathcal{D}^{++}(B)$,

$$\begin{aligned}(11.42) \qquad & H_nf_n(x,y) \\ = \quad & \frac{1}{n}e^{-nf(x)}g^{-1}(y)\Big(nF(x,y)\cdot(\nabla_x f(x))e^{nf(x)}g(y) + ne^{nf(x)}Bg(y)\Big) \\ = \quad & F(x,y)\cdot\nabla_x f(x) + \frac{Bg(y)}{g(y)} \\ \equiv \quad & G_{f,g}(x,y),\end{aligned}$$

so $(f, G_{f,g}) \in H$.

LEMMA 11.26. *Let $g_1, g_2 \in \mathcal{D}^{++}(B)$ and $0 < \kappa < 1$, and define*

$$g_h = h^{-1}\int_0^h S(s)g_1^\kappa g_2^{1-\kappa}ds.$$

Then

$$\frac{Bg_h}{g_h} \leq \frac{g_1^\kappa g_2^{1-\kappa}}{g_h}h^{-1}\left(\left(\frac{S(h)g_1}{g_1}\right)^\kappa\left(\frac{S(h)g_2}{g_2}\right)^{1-\kappa} - 1\right) \tag{11.43}$$

and as $h \to 0$, the buc-limit of the right side is

$$\kappa \frac{Bg_1}{g_1} + (1-\kappa)\frac{Bg_2}{g_2}. \tag{11.44}$$

For $\kappa > 0$, let

$$\widetilde{g}_h = h^{-1}\int_0^h S(s) g_1^{1+\kappa} g_2^{-\kappa} ds.$$

Then

$$\frac{B\widetilde{g}_h}{\widetilde{g}_h} \geq \frac{g_1^{1+\kappa} g_2^{-\kappa}}{\widetilde{g}_h}\left(\left(\frac{S(h)g_1}{g_1}\right)^{(1+\kappa)}\left(\frac{S(h)g_2}{g_2}\right)^{-\kappa} - 1\right) \tag{11.45}$$

and the buc-limit of the right side is

$$(1+\kappa)\frac{Bg_1}{g_1} - \kappa\frac{Bg_2}{g_2}$$

PROOF. Following the Hölder inequality arguments of the previous section,

$$\begin{aligned}\frac{Bg_h}{g_h} &= \frac{h^{-1}(S(h)g_1^\kappa g_2^{1-\kappa} - g_1^\kappa g_2^{1-\kappa})}{g_h} \\ &\leq \frac{g_1^\kappa g_2^{1-\kappa}}{g_h}\left(\left(\frac{S(h)g_1}{g_1}\right)^\kappa \left(\frac{S(h)g_2}{g_2}\right)^{1-\kappa} - 1\right)\end{aligned}$$

giving (11.43).

For the second part, observe that

$$\frac{S(h)g_1}{g_1} \leq \left(\frac{S(h)g_1^{1+\kappa}g_2^{-\kappa}}{g_1^{1+\kappa}g_2^{-\kappa}}\right)^{(1+\kappa)^{-1}}\left(\frac{S(h)g_2}{g_2}\right)^{\kappa(1+\kappa)^{-1}}$$

so that

$$\left(\frac{S(h)g_1}{g_1}\right)^{(1+\kappa)}\left(\frac{S(h)g_2}{g_2}\right)^{-\kappa} \leq \left(\frac{S(h)g_1^{1+\kappa}g_2^{-\kappa}}{g_1^{1+\kappa}g_2^{-\kappa}}\right)$$

giving (11.45) □

We apply Lemma 11.26 to obtain sufficiently rich limit operators $H_\dagger$ and $H_\ddagger$.

Let $(\widehat{E}_0, \widehat{r}_0)$ be the compactification of (E_0, r_0) as obtained in the discrete time case (Section 11.1.1), and let $E' = E \times \widehat{E}_0$ and $K_n^q = E \times \widetilde{K}_n^q$, for $q \in \mathcal{Q}$. Define $\eta_n' : E_n \mapsto E'$ and $\gamma : E' \mapsto E$ as in Section 11.1.1. Condition 11.21.3 implies Condition 2.8 holds. In addition, by the compactness of E', Condition 7.9 is satisfied. Again define the LIM convergence according to Definition 2.5.

Let ψ satisfy Condition 11.21.4, $f \in D_d$, and $g \in \mathcal{D}^{++}(B)$. For $0 < \kappa < 1$, define

$$G_{1,f,g,\kappa}(x,y) = \begin{cases} F(x,y)\cdot\nabla f(x) + (1-\kappa)\frac{Bg(y)}{g(y)} + \kappa\psi(y) & y \in E_0 \\ -\infty & y \in \widehat{E}_0 - E_0 \end{cases}$$

and for $\kappa > 0$, define

$$G_{2,f,g,\kappa}(x,y) = \begin{cases} F(x,y)\cdot\nabla f(x) + (1+\kappa)\frac{Bg(y)}{g(y)} - \kappa\psi(y) & y \in E_0 \\ \infty & y \in \widehat{E}_0 - E_0. \end{cases}$$

As in the discrete time case, by Lemma 11.3,

$$G_{1,f,g,\kappa}(x,y) = \lim_{\delta\to 0+} \sup_{z\in B_\delta(y)\cap E_0} (F(x,z)\cdot\nabla f(x) + (1-\kappa)\frac{Bg(z)}{g(z)} + \kappa\psi(z))$$

and

$$G_{2,f,g,\kappa}(x,y) = \lim_{\delta\to 0+} \inf_{z\in B_\delta(y)\cap E_0} (F(x,z)\cdot\nabla f(x) + (1+\kappa)\frac{Bg(z)}{g(z)} - \kappa\psi(z)).$$

For $\kappa = 0$, define $G_{1,f,g,0} = G_{2,f,g,0} = G_{f,g}$, and set

$$H_\dagger = \{(f, G_{1,f,g,\kappa}) : f \in D_d, g \in \mathcal{D}^{++}(B_0), \kappa \in [0,1)\}, \tag{11.46}$$

$$H_\ddagger = \{(f, G_{2,f,g,\kappa}) : f \in D_d, g \in \mathcal{D}^{++}(B_0), \kappa \geq 0\}. \tag{11.47}$$

We verify the Convergence Condition 7.11 with $v_n = n(\|\varphi_n\| \vee 1)$. For $\kappa = 0$, the convergence follows from (11.42). For $g \in \mathcal{D}^{++}(B_0)$ and $\kappa > 0$, let

$$g_{n,\kappa} = h_n^{-1}\int_0^{h_n} S(s)g^{1-\kappa}\varphi_n^\kappa ds$$

and for $f \in D_d$, let

$$f_{n,\kappa}(x,y) = f(x) + \frac{1}{n}\log g_{n,\kappa}(y). \tag{11.48}$$

Then by (11.43)

$$\begin{aligned} H_n f_{n,\kappa}(x,y) &= F(x,y)\cdot\nabla_x f(x) + \frac{Bg_{n,\kappa}(y)}{g_{n,\kappa}(y)} \\ &\leq F(x,y)\cdot\nabla_x f(x) \qquad (11.49)\\ &\quad + \frac{g^{1-\kappa}\varphi_n^\kappa}{g_{n,\kappa}} h_n^{-1}\left(\left(\frac{S(h_n)g(y)}{g(y)}\right)^{1-\kappa}\left(\frac{S(h_n)\varphi_n(y)}{\varphi_n(y)}\right)^\kappa - 1\right) \end{aligned}$$

Since g and φ_n are in $\mathcal{D}(B_0)$ and (11.37) holds, h_n can be selected so that (11.50)

$$\sup_y \left| h_n^{-1}\left(\left(\frac{S(h_n)g}{g}\right)^{1-\kappa}\left(\frac{S(h_n)\varphi_n}{\varphi_n}\right)^\kappa - 1\right) - \left((1-\kappa)\frac{Bg(y)}{g(y)} + \kappa\frac{B\varphi_n(y)}{\varphi_n(y)}\right)\right| \to 0,$$

$$\lim_{n\to\infty} \sup_{\{y:\psi(y)\geq c\}} \left|\frac{g^{1-\kappa}\varphi_n^\kappa}{g_{n,\kappa}} - 1\right| = 0, \tag{11.51}$$

and

$$\lim_{n\to\infty} \sup_{0\leq t\leq h_n} \|S(t)\varphi_n^\kappa - \varphi_n^\kappa\| = 0. \tag{11.52}$$

By Condition 11.21.4, $\frac{1}{n}\log\|\varphi_n\| \to 0$, and hence

$$\lim_{n\to\infty} \sup_{(x,y)\in E_n} |f_{n,\kappa}(x,y) - f(x)| = 0,$$

verifying (7.18). The bounds in (7.19) follow by the definition of v_n, (11.51), (11.35), and the fact that there exist c_1 and c_2 depending on g such that

$$c_1\frac{\varphi_n^\kappa}{h_n^{-1}\int_0^{h_n} S(s)\varphi_n^\kappa ds} \leq \frac{g^{1-\kappa}\varphi_n^\kappa}{g_{n,\kappa}} \leq c_2\frac{\varphi_n^\kappa}{h_n^{-1}\int_0^{h_n} S(s)\varphi_n^\kappa ds}$$

and the ratio on the left and right converges uniformly to 1.

For $q \in \mathcal{Q}$, let $(x_n, y_n) \in K_n^q = E_0 \times \widetilde{K}_n^q$ satisfy $(x_n, y_n) \to (x, y) \in E' = E \times \widehat{E}_0$. Then by (11.36) and the properties of ψ, for any subsequence satisfying $\psi(y_n) \to -\infty$, we have $H_n f_{n,\kappa}(x_n, y_n) \to -\infty$, and for any subsequence satisfying $\psi(y_n) \geq c > -\infty$, (11.50) and (11.51) imply

$$\limsup_{n\to\infty} H_n f_{n,\kappa}(x_n, y_n) \leq \limsup_{n\to\infty} G_{1,f,g,\kappa}(x, y_n) \leq G_{1,f,g,\kappa}(x, y),$$

verifying (7.20).

Similarly, we can verify the convergence to $H_\ddagger$.

11.2.2. Exponential tightness. Exponential tightness for $\{X_n\}$ follows by the convergence of H_n and Corollary 4.17.

11.2.3. The comparison principle. Let $\alpha > 0$, $h \in C_b(E)$. We define

$$H_1(x,p) = \inf_{\kappa\in(0,1)} \inf_{g\in\mathcal{D}^{++}(B)} \sup_{y\in E_0} \Big(F(x,y)\cdot p + (1-\kappa)\frac{Bg(y)}{g(y)} + \kappa\psi(y)\Big), \tag{11.53}$$

$$H_2(x,p) = \sup_{\kappa>0} \sup_{g\in\mathcal{D}^{++}(B)} \inf_{y\in E_0} \Big(F(x,y)\cdot p + (1+\kappa)\frac{Bg(y)}{g(y)} - \kappa\psi(y)\Big), \tag{11.54}$$

and

$$H_i f(x) = H_i(x, \nabla f(x)), \quad f \in D_d. \tag{11.55}$$

If $g \in \mathcal{D}^{++}(B)$, then for $\epsilon > 0$,

$$g_\epsilon = \epsilon^{-1}\int_0^\epsilon S(t)g\,dt \in \mathcal{D}^{++}(B_0), \quad B_0 g_\epsilon = \epsilon^{-1}\int_0^\epsilon S(t)Bg\,dt,$$

and assuming (11.33),

$$buc-\lim_{\epsilon\to 0}\frac{B_0 g_\epsilon}{g_\epsilon} = \frac{Bg}{g}.$$

The properties of ψ then assure that taking the inf over $g \in \mathcal{D}^{++}(B)$ in the definition of H_1 gives the same result as taking the inf over $g \in \mathcal{D}^{++}(B_0)$ and similarly with the sup in the definition of H_2.

LEMMA 11.27. *Assume that (11.3) and (11.4) hold and that H_1 and H_2 are given by (11.53) and (11.54). Then the comparison principle holds for H_1 and H_2.*

PROOF. The proof is similar to that of Lemma 11.10. □

By the same argument as in Section 11.1.3, if

$$H_1(x,p) \leq H_2(x,p), \quad x, p \in R^d, \tag{11.56}$$

and the comparison principle holds for $(I - \alpha H_1)f = h$ or $(I - \alpha H_2)f = h$, then the comparison principle holds for subsolutions of $(I - \alpha H_\dagger)f = h$ and supersolutions of $(I - \alpha H_\ddagger)f = h$.

We have conditions that imply (11.56) similar to the conditions giving (11.16) in Section 11.1.3.

LEMMA 11.28. *If for each $x \in E$ and $p \in R^d$, there exists $g_{x,p} \in \mathcal{D}^{++}(B)$ such that*

$$\lambda_{x,p} \equiv F(x,y)\cdot p + \frac{Bg_{x,p}(y)}{g_{x,p}(y)} \tag{11.57}$$

does not depend upon y, then (11.56) holds.

REMARK 11.29. The desired $g_{x,p}$ is an eigenfunction of the operator $B+F(x,\cdot)\cdot p$ satisfying

$$Bg_{x,p}(y) + F(x,\cdot)\cdot pg_{x,p}(y) = \lambda_{x,p}g_{x,p}(y).$$

Consequently, if E_0 is finite and B is the intensity matrix for an irreducible Markov chain, then the Perron-Frobenius theorem ensures the existence of $g_{x,p}$ and $\lambda_{x,p}$. Generalizations of the Perron-Frobenius theorem are discussed in Section B.1. In particular, Theorem B.3 gives conditions for existence.

PROOF. The proof is similar to that of Lemma 11.4. □

LEMMA 11.30. *Let*

$$T_{x,p}(t)f(y) = E[e^{\int_0^t F(x,Y(s))\cdot pds}f(Y(t))|Y(0) = y].$$

Suppose that there exist $\delta > 0$, $\lambda_{x,p}(s) \in R$, $0 \le s \le \delta$, *and* $g_{x,p} \in K_1(E_0)$ *such that for each* $c \in R$,

$$\lim_{t\to\infty} \sup_{\{y:\psi(y)\ge c\}} \left| \frac{T_{x,p}(t+s)g_{x,p}(y)}{T_{x,p}(t)g_{x,p}(y)} - \lambda_{x,p}(s)\right| = 0. \tag{11.58}$$

Then (11.56) holds.

REMARK 11.31. Again, under Harris recurrence conditions, Kontoyiannis and Meyn [**67**], Theorem 4.1, and [**68**], Theorem 1.2, give results that imply (11.58).

PROOF. Let

$$g_{x,p}^{\delta,t} = \frac{1}{\delta}\int_t^{t+\delta} T_{x,p}(s)g_{x,p}.$$

Then

$$Bg_{x,p}^{\delta,t} + F(x,y)\cdot pg_{x,p}^{\delta,t} = \delta^{-1}(T_{x,p}(t+\delta)g_{x,p} - T_{x,p}(t)g_{x,p}),$$

and the remainder of the proof is similar to the proof of Lemma 11.6. □

LEMMA 11.32. *Suppose that for each* $x, p \in R^d$, *there exist* $\{g_{x,p}^n\} \subset \mathcal{D}^{++}(B)$ *and* $\{\lambda_{x,p}^n\}$ *such that*

$$\sup_{y,n}\left|\frac{Bg_{x,p}^n(y)}{g_{x,p}^n(y)}\right| < \infty, \tag{11.59}$$

$$\sup_n |\lambda_{x,p}^n| < \infty, \tag{11.60}$$

and for each $c \in R$,

$$\lim_{n\to\infty} \sup_{\{y:\psi(y)\ge c\}} \left|F(x,y)\cdot p + \frac{Bg_{x,p}^n(y)}{g_{x,p}^n(y)} - \lambda_{x,p}^n\right| = 0. \tag{11.61}$$

Then (11.56) holds.

PROOF. Since we can select a subsequence along which $\{\lambda_{x,p}^n\}$ converges, without loss of generality we can assume that $\lambda_{x,p}^n = \lambda_{x,p}$ is independent of n. Select $a_{x,p}, b_{x,p} \in R$ so that

$$a_{x,p} \le \frac{Bg_{x,p}^n}{g_{x,p}^n} \le b_{x,p}.$$

For $\epsilon > 0$, $c \in R$, there exists n_c such that $n \geq n_c$ implies

$$\begin{aligned} H_1(x,p) &\leq \sup_{y\in E_0} (F(x,y)\cdot p + (1-\kappa)\frac{Bg^n_{x,p}(y)}{g^n_{x,p}(y)} + \kappa\psi(y)) \\ &\leq \left[\lambda_{x,p} + \epsilon + \kappa \sup_{\{y:\psi(y)\geq c\}} (\psi(y) - a_{x,p})\right] \\ &\qquad \vee \left[\sup_{y\in E_0} F(x,y)\cdot p + (1-\kappa)b_{x,p} + \kappa c\right] \end{aligned}$$

and

$$\begin{aligned} H_2(x,p) &\geq \inf_{y\in E_0} \Big(F(x,y)\cdot p + (1+\kappa)\frac{Bg^n_{x,p}(y)}{g^n_{x,p}(y)} - \kappa\psi(y)\Big) \\ &\geq \left[\lambda_{x,p} - \epsilon - \kappa \sup_{\{y:\psi(y)\geq c\}} (\psi(y) - a_{x,p})\right] \\ &\qquad \wedge \left[\inf_{y\in E_0} F(x,y)\cdot p + (1+\kappa)a_{x,p} - \kappa c\right]. \end{aligned}$$

Selecting κ sufficiently small, $c < 0$ with $|c|$ sufficiently large, and ϵ sufficiently small, we see that $H_1(x,p) \leq \lambda_{x,p} \leq H_2(x,p)$. □

The conditions of Lemma 11.32 hold for a large class of nondegenerate diffusion processes.

EXAMPLE 11.33. For a domain $\Gamma \subset R^m$ and $0 < \alpha < 1$, define the Hölder norm

$$\|h\|_{\alpha,\Gamma} = \sup_{y\in\Gamma} |h(y)| + \sup_{x,y\in\Gamma, x\neq y} \frac{|h(y)-h(x)|}{|y-x|^\alpha}.$$

Let

$$Lg(y) = \frac{1}{2}\sum_{i,j=1}^m a_{ij}(y)\frac{\partial^2}{\partial y_i\partial y_j}g(y) + \sum_{i=1}^m b_i(y)\frac{\partial}{\partial y_i}g(y), \quad g \in C^2(R^m).$$

Suppose that for each bounded domain Γ and each $x \in R^d$, there exist $\epsilon_\Gamma > 0$ and $\Lambda_{\Gamma,x}$ such that

$$\sum_{i,j} a_{ij}(y)\xi_i\xi_j \geq \epsilon_\Gamma|\xi|^2, \quad y\in\Gamma, \xi\in R^m,$$

and

$$\|a_{i,j}\|_{\alpha,\Gamma}, \|b_i\|_{\alpha,\Gamma}, \|F_i(x,\cdot)\|_{\alpha,\Gamma} \leq \Lambda_{\Gamma,x}.$$

(cf. Assumption H, page 85, Pinsky [**95**]). Let Y be the diffusion corresponding to L. Y exists and is unique up to the first time Y leaves a bounded domain.

These conditions ensure that for each smooth, bounded, open domain $\Gamma \subset R^m$ and $x,p \in R^d$, there exist $\lambda_{\Gamma,x,p} \in R$ and $\varphi_{\Gamma,x,p} \in C^2(\overline{\Gamma})$ such that $\varphi_{\Gamma,x,p} > 0$ on Γ, $\varphi_{\Gamma,x,p} = 0$ on $\partial\Gamma$, and

$$F(x,y)\cdot p\varphi_{\Gamma,x,p}(y) + L\varphi_{\Gamma,x,p}(y) = \lambda_{\Gamma,x,p}\varphi_{\Gamma,x,p}(y), \quad y\in\Gamma.$$

(See Theorem 5.5, page 94, Pinsky [**95**].) For $y \notin \overline{\Gamma}$, set $\varphi_{\Gamma,x,p}(y) = 0$. Setting $V_{x,p}(y) = F(x,y)\cdot p$ and $\tau_\Gamma = \inf\{t : Y(t) \notin \Gamma\}$,

$$E[e^{\int_0^t V_{x,p}(Y(s))ds}\varphi_{\Gamma,x,p}(Y(t))\mathbf{1}_{\{\tau_\Gamma > t\}}|Y(0)=y] = e^{\lambda_{\Gamma,x,p}t}\varphi_{\Gamma,x,p}(y). \tag{11.62}$$

In particular,

$$\inf_y V_{x,p}(y) \leq \lambda_{\Gamma,x,p} \leq \sup_y V_{x,p}(y). \tag{11.63}$$

Assume that $0 \in \Gamma$, and we can normalize $\varphi_{\Gamma,x,p}$ so that $\varphi_{\Gamma,x,p}(0) = 1$. Then the Harnack inequality (Theorem 0.1, page 124, Pinsky [**95**]) implies that for each $r > 0$, there exist $0 < d_{1,x,p,r} < d_{2,x,p,r}$, such that for all $\Gamma \supset B_r(0)$

$$d_{1,x,p,r} \leq \varphi_{\Gamma,x,p}(y) \leq d_{2,x,p,r}, \quad |y| \leq r.$$

Setting $Bf = Lf$ for $f \in D_l$, we assume that the martingale problem for B is well-posed and that the corresponding semigroup $\{S(t)\}$ satisfies

$$S(t) : \widehat{C}(R^l) \mapsto \widehat{C}(R^l), \tag{11.64}$$

where $\widehat{C}(R^l)$ is the space of continuous functions vanishing at infinity. Note that (11.64) implies

$$\lim_{|y|\to\infty} P\{Y(t) \in K | Y(0) = y\} = 0, \tag{11.65}$$

for each $t \geq 0$ and each compact $K \subset R^l$. Conversely, if $S(t) : C_b(R^l) \mapsto C_b(R^l)$ and (11.65) holds, then (11.64) holds. To verify (11.65), it is enough to show the existence of $f \in \widehat{C}(R^l) \cap C^2(R^l)$ such that $f(y) > 0$, $y \in R^l$, and

$$c \equiv \sup_y \frac{Lf(y)}{f(y)} < \infty. \tag{11.66}$$

Then

$$f(Y(t))e^{-\int_0^t \frac{Lf(Y(s))}{f(Y(s))}ds}$$

is a local martingale and

$$P\{Y(t) \in K | Y(0) = y\} \leq \frac{1}{\inf_{z\in K} f(z)} e^{ct} f(y).$$

In particular, if $|b(y)| \leq c_1 + c_2|y|$ and $|a(y)| \leq c_1 + c_2|y|^2$, then (11.66) holds for $f(y) = (1 + |y|^2)^{-1}$.

We assume that there there exist φ_1 and φ_2 satisfying Condition 11.25, and define $\widetilde{K}_n^q = \{y : \varphi_2(y) \leq e^{qn}\}$. Define $\varphi_0 = \sqrt{\varphi_1}$, and note that

$$\frac{L\varphi_0}{\varphi_0} \leq \frac{L\varphi_1}{2\varphi_1}$$

so

$$d_j = \sup_y \frac{L\varphi_j(y)}{\varphi_j(y)}, \quad j = 0, 1$$

are finite. Let $\mu_{x,p}$ satisfy

$$\mu_{x,p} > \|V_{x,p}\| + \sup_\Gamma |\lambda_{\Gamma,x,p}| + d_0 + \frac{1}{2}d_1.$$

For simplicity, let $E_y[Z]$ denote $E[Z|Y(0) = y]$.

LEMMA 11.34. *There exists a constant $C_{x,p} > 0$ independent of Γ such that*

$$\varphi_{\Gamma,x,p}(y) \leq C_{x,p}\varphi_0(y).$$

PROOF. Let $r > 0$ satisfy

$$\sup_{|y|\geq r} \frac{L\varphi_0(y)}{\varphi_0(y)} < -\mu_{x,p}$$

and let $\gamma_r = \inf\{t > 0 : Y(t) \leq r\}$. Let $\Gamma \supset B_r(0)$. Since

$$\varphi_{\Gamma,x,p}(Y(t \wedge \tau_\Gamma))e^{\int_0^{t\wedge\tau_\Gamma}(V_{x,p}(Y(s))-\lambda_{\Gamma,x,p})ds} \tag{11.67}$$

is a martingale, for $y \in \Gamma$, we have

$$\begin{aligned}\varphi_{\Gamma,x,p}(y) &= E_y[\varphi_{\Gamma,x,p}(Y(\gamma_r \wedge \tau_\Gamma))e^{\int_0^{\gamma_r\wedge\tau_\Gamma}(V_{x,p}(Y(s))-\lambda_{\Gamma,x,p})ds}] \\ &\leq \frac{d_{2,x,p,r}}{\inf_{|y|\leq r}\varphi_0(y)} E_y[\varphi_0(Y(\gamma_r \wedge \tau_\Gamma))e^{\gamma_r\wedge\tau_\Gamma(\|V_{x,p}\|-\lambda_{\Gamma,x,p})}].\end{aligned}$$

Since

$$\begin{aligned}\varphi_0(y) &= E_y[\varphi_0(Y(\gamma_r \wedge \tau_\Gamma)e^{-\int_0^{\gamma_r\wedge\tau_\Gamma}\frac{L\varphi_0(Y(s))}{\varphi_0(Y(s))}ds}] \\ &\geq E_y[\varphi_0(Y(\gamma_r \wedge \tau_\Gamma)e^{\gamma_r\wedge\tau_\Gamma\mu_{x,p}}],\end{aligned}$$

the inequality follows with

$$C_{x,p} = \frac{d_{2,x,p,r}}{\inf_{|y|\leq r}\varphi_0(y)}.$$

□

For $n = 1, 2, \ldots$, let $\Gamma_n = \{y : |y| < n\}$, and let $\rho_n \in C_b(R^m)$ satisfy $0 \leq \rho_n(y) \leq 1$, $\rho_n(y) = 1$ on Γ_{n+1}^c and $\rho_n(y) = 0$ on Γ_n. Define $\lambda_{x,p}^n = \lambda_{\Gamma_n,x,p}$ and

$$g_{x,p}^n(y) = E_y[\int_0^\infty e^{\int_0^t(V_{x,p}(Y(s))-\mu_{x,p})ds}(\varphi_{\Gamma_n,x,p}(Y(t)) + \epsilon_n\rho_n(Y(t)))dt],$$

where $\epsilon_n > 0$ is to be determined. Note that

$$V_{x,p}(y) + \frac{Bg_{x,p}^n(y)}{g_{x,p}^n(y)} = \mu_{x,p} - \frac{\varphi_{\Gamma_n,x,p}(y) + \epsilon_n\rho_n(y)}{g_{x,p}^n(y)} \tag{11.68}$$

and by (11.62) and the strong Markov property

$$\begin{aligned}&g_{x,p}^n(y) \\ &= E_y[\int_0^\infty e^{\int_0^t(V_{x,p}(Y(s))-\mu_{x,p})ds}\varphi_{\Gamma_n,x,p}(Y(t))\mathbf{1}_{\{\tau_{\Gamma_n}>t\}}dt] \\ &\qquad + E_y[\int_0^\infty e^{\int_0^t(V_{x,p}(Y(s))-\mu_{x,p})ds}(\varphi_{\Gamma_n,x,p}(Y(t)) + \epsilon_n\rho_n(Y(t)))\mathbf{1}_{\{\tau_{\Gamma_n}\leq t\}}dt] \\ &= \frac{1}{\mu_{x,p}-\lambda_{x,p}^n}\varphi_{\Gamma_n,x,p}(y) + E_y[e^{\int_0^{\tau_{\Gamma_n}}(V_{x,p}(Y(s))-\mu_{x,p})ds}g_{x,p}^n(Y(\tau_{\Gamma_n}))] \\ &= \frac{1}{\mu_{x,p}-\lambda_{x,p}^n}\varphi_{\Gamma_n,x,p}(y) + E_y[\int_0^\infty e^{\int_0^t(V_{x,p}(Y(s))-\mu_{x,p})ds}\varphi_{\Gamma_n,x,p}(Y(t))\mathbf{1}_{\{\tau_{\Gamma_n}\leq t\}}dt] \\ &\qquad + \epsilon_n E_y[\int_0^\infty e^{\int_0^t(V_{x,p}(Y(s))-\mu_{x,p})ds}\rho_n(Y(t)))dt].\end{aligned}$$

For $c_{x,p} > 0$ satisfying $-c_{x,p} \le \inf_y (V_{x,p}(y) - \mu_{x,p})$

$$E_y[\int_0^\infty e^{\int_0^t (V_{x,p}(Y(s))-\mu_{x,p})ds} \rho_n(Y(t)))dt] \\ \ge \int_0^\infty e^{-c_{x,p}t} S(t)\rho_n(y)dt,$$

and since $\lim_{|y|\to\infty} S(t)\rho_n(y) = 1$,

$$K_n = \{y : \int_0^\infty e^{-c_{x,p}t} S(t)\rho_n(y)dt \le \frac{1}{2c_{x,p}}\}$$

is compact. Let

$$\epsilon_n = n^{-1} \wedge \inf_{y\in K_n} E_y[\int_0^\infty e^{\int_0^t (V_{x,p}(Y(s))-\mu_{x,p})ds} \varphi_{\Gamma_n,x,p}(Y(t))\mathbf{1}_{\{\tau_{\Gamma_n}\le t\}}dt],$$

and note that $\epsilon_n > 0$.

Since $\rho_n > 0$ only when $\varphi_{\Gamma_n,x,p} = 0$, it follows from (11.68) and the definition of ϵ_n and $c_{x,p}$ that

$$\begin{aligned} \left|V_{x,p}(y) + \frac{Bg^n_{x,p}(y)}{g^n_{x,p}(y)}\right| &\le |\mu_{x,p}| + \frac{\varphi_{\Gamma_n,x,p}(y) + \epsilon_n\rho_n(y)}{g^n_{x,p}(y)} \\ &\le |\mu_{x,p}| + (\mu_{x,p} - \lambda^n_{x,p}) \vee 1 \vee (2c_{x,p}), \end{aligned}$$

which implies (11.59).

The bound (11.60) follows from (11.63). To see that (11.61) is satisfied, it is enough to show that for each compact K

$$\lim_{n\to\infty} \sup_{y\in K} E_y[\int_0^\infty e^{\int_0^t (V_{x,p}(Y(s))-\mu_{x,p})ds} \varphi_{\Gamma_n,x,p}(Y(t))\mathbf{1}_{\{\tau_{\Gamma_n}\le t\}}dt] = 0. \tag{11.69}$$

By Lemma 11.34 and the fact that

$$\varphi_j(Y(t))e^{-\int_0^t \frac{L\varphi_j(Y(s))}{\varphi_j(Y(s))}ds}$$

is a supermartingale,

$$\begin{aligned} &E_y[\int_0^\infty e^{\int_0^t (V_{x,p}(Y(s))-\mu_{x,p})ds} \varphi_{\Gamma_n,x,p}(Y(t))\mathbf{1}_{\{\tau_{\Gamma_n}\le t\}}dt] \\ &\le C_{x,p}E_y[e^{-\tau_{\Gamma_n}(\mu_{x,p}-\|V_{x,p}\|)} \int_0^\infty e^{-t(\mu_{x,p}-\|V_{x,p}\|)}\varphi_0(Y(\tau_{\Gamma_n}+t))dt] \\ &\le C_{x,p}E_y[e^{-\tau_{\Gamma_n}(\mu_{x,p}-\|V_{x,p}\|)}\varphi_0(Y(\tau_{\Gamma_n})) \int_0^\infty e^{-t(\mu_{x,p}-\|V_{x,p}\|-d_0)}dt] \\ &\le \frac{C_{x,p}}{\mu_{x,p}-\|V_{x,p}\|-d_0} E_y[e^{-2\tau_{\Gamma_n}(\mu_{x,p}-\|V_{x,p}\|)+\int_0^{\tau_{\Gamma_n}} \frac{L\varphi_1(Y(s))}{\varphi_1(Y(s))}ds}]^{1/2} \\ &\qquad \times E_y[\varphi_1(Y(\tau_{\Gamma_n}))e^{-\int_0^{\tau_{\Gamma_n}} \frac{L\varphi_1(Y(s))}{\varphi_1(Y(s))}ds}]^{1/2} \\ &\le \frac{C_{x,p}}{\mu_{x,p}-\|V_{x,p}\|-d_0} E[e^{-2\tau_{\Gamma_n}(\mu_{x,p}-\|V_{x,p}\|-\frac{1}{2}d_1)}]^{1/2}. \end{aligned}$$

Since $\tau_{\Gamma_n} \to \infty$, (11.69) holds.

11.2.4. Variational representation of H. As in Section 11.1.4, the following lemma gives the desired control representation.

LEMMA 11.35. *Let Condition 11.21 hold, and let H_1 be given by (11.53). Then*

$$H_1(x,p) = \mathbf{H}(x,p) \equiv \sup_{\nu\in\mathcal{P}(E_0)} \Big(\int_{E_0} F(x,y)\cdot p\nu(dy) - I_B(\nu)\Big), \tag{11.70}$$

where

$$I_B(\nu) = -\inf_{g\in\mathcal{D}^{++}(B)} \int_{E_0} \frac{Bg}{g} d\nu \wedge \int_{E_0} \psi d\nu.$$

REMARK 11.36. Under Condition 11.21.4, typically

$$\int \psi d\nu \geq \limsup_{n\to\infty} \int_{E_0} \frac{B\varphi_n}{\varphi_n} d\nu$$

so that

$$I_B(\nu) = -\inf_{g\in\mathcal{D}^{++}(B)} \int_{E_0} \frac{Bg}{g} d\nu.$$

PROOF. Note that

$$\begin{aligned}
&H_1(x,p)\\
&= \inf_{0<\kappa\leq 1} \inf_{g\in\mathcal{D}^{++}(B)} \sup_{\nu\in\mathcal{P}(E_0)} \int_{E_0} (F(x,y)\cdot p + (1-\kappa)\frac{Bg(y)}{g(y)} + \kappa\psi(y))\nu(dy)\\
&\geq \inf_{0<\kappa\leq 1} \sup_{\nu\in\mathcal{P}(E_0)} \inf_{g\in\mathcal{D}^{++}(B)} \Big(\int_{E_0} F(x,\cdot)\cdot pd\nu + (1-\kappa)\int_{E_0}\frac{Bg}{g}d\nu \wedge \int_{E_0}\psi d\nu\\
&\qquad + \kappa\int_{E_0}\psi d\nu\Big)\\
&= \inf_{0<\kappa\leq 1} \sup_{\nu\in\mathcal{P}(E_0)} \Big(\int_{E_0} F(x,\cdot)\cdot pd\nu + (1-\kappa)\inf_{g\in\mathcal{D}^{++}(B)}\int_{E_0}\frac{Bg}{g}d\nu \wedge \int_{E_0}\psi d\nu\\
&\qquad + \kappa\int_{E_0}\psi d\nu\Big)\\
&= \sup_{\nu\in\mathcal{P}(E_0)} \Big(\int_{E_0} F(x,\cdot)\cdot pd\nu + \inf_{g\in\mathcal{D}^{++}(B)}\int_{E_0}\frac{Bg}{g}d\nu \wedge \int_{E_0}\psi d\nu\Big)\\
&= \mathbf{H}(x,p).
\end{aligned}$$

Let $g_1,\ldots,g_m \in \mathcal{D}^{++}(B_0)$, $\alpha_i \geq 0$, and $\sum_{i=1}^m \alpha_i = 1$, and set

$$g_h = h^{-1}\int_0^h S(s)\prod g_i^{\alpha_i} ds.$$

Observing that $S(t)\prod g_i^{\alpha_i} \leq \prod (S(t)g_i)^{\alpha_i}$, as $h\to 0$, we have

$$Bg_h = h^{-1}(S(h)g_h - g_h) \leq h^{-1}(\prod (S(h)g_i)^{\alpha_i} - \prod g_i^{\alpha_i}) \to \prod g_i^{\alpha_i}\sum_{i=1}^m \alpha_i\frac{Bg_i}{g_i},$$

where the convergence is uniform because we have required the g_i to be in the domain of the strong generator. It follows that

$$H_1(x,p) \leq \inf_{0<\kappa\leq 1}\inf_{\sum_{i=1}^m\alpha_i=1}\inf_{g_i\in\mathcal{D}^{++}(B_0)}\sup_{y\in E_0}\Big(F(x,y)\cdot p + (1-\kappa)\sum_{i=1}^m \alpha_i\frac{Bg_i(y)}{g_i(y)} + \kappa\psi(y)\Big).$$

LEMMA 11.37. *There exists a sequence $\{g_i\} \subset \mathcal{D}^{++}(B_0)$ such that B is the buc-closure of $\{(g_i, Bg_i)\}$ and*

$$\inf_i \int_{E_0} \frac{Bg_i}{g_i} d\nu = \inf_{g\in\mathcal{D}^{++}(B)} \int_{E_0} \frac{Bg}{g} d\nu, \tag{11.71}$$

for each $\nu \in \mathcal{P}(E_0)$.

PROOF. For $0 < a < b$, let $C_{a,b} = \{f \in C_b(E) : a \leq f \leq b\}$. It follows from Corollary A.9 (note Remark A.10) that there exists a countable subset $D_0 \subset C_b(E)$ such that $C_{a,b}$ is the *buc*-closure of $D_0 \cap C_{a,b}$. Let Λ be a countable dense subset of $(0,\infty)$. Setting $R_\lambda f = \int_0^\infty e^{-\lambda t}S(t)f$, note that if $f \in C_{a,b} \cap D_0$, then $\lambda^2 R_\lambda^2 f \in C_{a,b} \cap \mathcal{D}(B_0)$ with $B_0\lambda^2 R_\lambda^2 f = \lambda^3 R_\lambda^2 f - \lambda^2 R_\lambda f$. The *buc*-closure of $\{(\lambda^2 R_\lambda^2 f, B_0\lambda^2 R_\lambda^2 f) : f \in C_{a,b} \cap D_0, \lambda \in \Lambda\}$ contains

$$\{(\lambda^2 R_\lambda^2 f, B_0\lambda^2 R_\lambda^2 f) : f \in C_{a,b}, \lambda > 0\} \supset \{(\lambda^2 R_\lambda^2 f, \lambda^2 R_\lambda^2 Bf) : f \in C_{a,b} \cap \mathcal{D}(B)\}.$$

Letting $\lambda \to \infty$, we see that the *buc*-closure contains $\{(f, Bf) : f \in C_{a,b} \cap \mathcal{D}(B)$.

Note also that the *buc*-closure of $\{\frac{B_0\lambda^2 R_\lambda^2 f}{\lambda^2 R_\lambda^2 f} : f \in C_{a,b} \cap D_0, \lambda \in \Lambda\}$ contains $\{\frac{Bf}{f} : f \in C_{a,b} \cap \mathcal{D}(B)\}$, and (11.71) follows. □

Consequently,

$$\begin{aligned}
&H_1(x,p) \\
\leq\ & \lim_{m\to\infty} \inf_{0<\kappa\leq 1} \inf_{\sum_{i=0}^m \alpha_i = 1} \sup_{y\in E_0} \Big(F(x,y)\cdot p + (1-\kappa)\sum_{i=1}^m \alpha_i \frac{Bg_i(y)}{g_i(y)} + \kappa\psi(y)\Big) \\
=\ & \lim_{m\to\infty} \inf_{0<\kappa\leq 1} \inf_{\sum_{i=1}^m \alpha_i = 1} \sup_{\nu\in\mathcal{P}(E_0)} \int_{E_0} \Big(F(x,y)\cdot p + (1-\kappa)\sum_{i=1}^m \alpha_i \frac{Bg_i(y)}{g_i(y)} \\
&\qquad + \kappa\psi(y)\Big)\nu(dy) \\
=\ & \lim_{m\to\infty} \inf_{0<\kappa\leq 1} \sup_{\nu\in\mathcal{P}(E_0)} \inf_{\sum_{i=1}^m \alpha_i = 1} \int_{E_0} \Big(F(x,y)\cdot p + (1-\kappa)\sum_{i=1}^m \alpha_i \frac{Bg_i(y)}{g_i(y)} \\
&\qquad + \kappa\psi(y)\Big)\nu(dy) \\
=\ & \lim_{m\to\infty} \inf_{0<\epsilon\leq 1} \sup_{\nu\in\mathcal{P}(E_0)} \inf_{\epsilon\leq\kappa\leq 1} \Big[\int_{E_0} F(x,\cdot)\cdot p d\nu + (1-\kappa)\min_{1\leq i\leq m} \int_{E_0} \frac{Bg_i}{g_i} d\nu \\
&\qquad + \kappa\int \psi d\nu\Big] \\
=\ & \lim_{m\to\infty} \inf_{0<\epsilon\leq 1} \sup_{\nu\in\mathcal{P}(E_0)} \Big[\int_{E_0} F(x,\cdot)\cdot p d\nu + (1-\epsilon)\min_{1\leq i\leq m} \int \frac{Bg_i}{g_i} d\nu \wedge \int \psi d\nu \\
&\qquad + \epsilon\int \psi d\nu\Big] \\
=\ & \lim_{m\to\infty} \sup_{\nu\in\mathcal{P}(E_0)} \Big[\int_{E_0} F(x,\cdot)\cdot p d\nu + \min_{1\leq i\leq m} \int \frac{Bg_i}{g_i} d\nu \wedge \int \psi d\nu\Big] \\
=\ & \sup_{\nu\in\mathcal{P}(E_0)} \Big(\int_{E_0} F(x,\cdot)\cdot p d\nu - I_B(\nu)\Big) = \mathbf{H}(x,p).
\end{aligned}$$

As in the discrete time case, the second and third equalities follow by Sion's theorem, [**111**], Corollary 3.3, and the final equality follows as in the proof of Lemma 11.12. □

Next, we discuss some properties of $\mathbf{H}$ and I_B.

LEMMA 11.38. *Let π be a stationary distribution for B. Then if $g \in \mathcal{D}^+(B) = \{g \in \mathcal{D}(B) : g > 0\}$ and $\int_{E_0} \left|\frac{Bg}{g}\right| d\pi < \infty$,*

$$\int_{E_0} \frac{Bg}{g} d\pi \geq 0$$

and

$$\mathbf{H}(x,p) \geq \int_{E_0} F(x,y) \cdot p\pi(dy).$$

PROOF. By considering $g + c$ with $c > 0$ and taking the limit as $c \to 0$, we see that it is enough to consider g bounded away from zero. Then

$$\int_{E_0} \frac{Bg}{g} d\pi = \lim_{t\to 0} \frac{1}{t}\left(E_\pi[\frac{g(Y(t))}{g(Y(0))}] - 1\right) = \lim_{t\to 0} \frac{1}{t}\left(E_\pi[e^{\log g(Y(t)) - \log g(Y(0))}] - 1\right) \geq 0,$$

where the last inequality follows by Jensen's inequality and stationarity.

From the definition of H_1 in (11.53), we have

$$\begin{aligned} H_1(x,p) &= \inf_{g\in\mathcal{D}^{++}(B)} \sup_{y\in E_0} (F(x,y)\cdot\nabla p + \frac{Bg(y)}{g(y)}) \\ &= \inf_{g\in\mathcal{D}^{++}(B)} \sup_{\mu\in\mathcal{P}(E_0)} \int_{E_0} (F(x,y)\cdot\nabla p + \frac{Bg(y)}{g(y)})\mu(dy) \\ &\geq \int_{E_0} F(x,y)\cdot p\pi(dy) + \inf_{g\in\mathcal{D}^{++}(B)} \int_{E_0} \frac{Bg}{g} d\pi \geq \int_{E_0} F(x,y)\cdot p\pi(dy). \end{aligned}$$

□

One consequence of Lemma 11.38 is that $\mathbf{H}(x,0) = 0$. (Take $g = 1$ in (11.53).) It also follows that $I_B(\pi) = 0$ for any stationary π satisfying $\int_{E_0} \psi d\pi \geq 0$. Such a π exists by Lemma 11.23.

From (11.70), we see that I_B should be the running cost in a control representation of the type considered in Chapter 8 (take $\mathbf{H}_\dagger = \mathbf{H}_\ddagger = \mathbf{H}$). Note that

$$\Gamma_B^M = \{\mu \in \mathcal{P}(E_0) : I_B(\mu) \leq M\}$$

is compact.

The next two lemmas are derived using arguments identical to those used for Lemmas 11.15 and 11.17.

LEMMA 11.39. *Let F be continuous and satisfy (11.3) and (11.4). Set*

$$Af(x,\mu) = \int_{E_0} F(x,y)\cdot\nabla f(x)\mu(dy), \quad f \in D_d, \mu \in \mathcal{P}(E_0).$$

Define $\mathbf{H}$ as in (11.70). Then

$$\mathbf{H}f(x) = \sup_{\mu\in\mathcal{P}(E_0)} (Af(x,\mu) - I_B(\mu)), \tag{11.72}$$

and, taking $U = \mathcal{P}(E_0)$, Conditions 8.30 and 8.10 are satisfied.

LEMMA 11.40. *Suppose that there exists a stationary distribution π for B. Let $\rho(x) = 1 + |x|$, and define*

$$L_\rho(x,q) = \sup_{p\in R^d}(p\cdot q - \mathbf{H}(x,p/\rho(x))), \quad x,q\in R^d, \tag{11.73}$$

and

$$L_\rho(\infty,q) = \liminf_{|x|\to\infty,q'\to q} L_\rho(x,q').$$

Take $U = R^d$, and define

$$Af(x,q) = \rho(x)q\cdot\nabla f(x), \quad f\in D_d, q\in U, x\in R^d,$$

and $Af(\infty,q) = 0$. Then $L_\rho \geq 0$ is lower semicontinuous,

$$\mathbf{H}f(x) \equiv \sup_q(Af(x,q) - L_\rho(x,q)), \tag{11.74}$$

and Conditions 8.30 and 8.10 are satisfied.

11.2.5. The large deviation theorem. For each $f\in B(E\times E_0)$, define

$$V_n(t)f(x,y) = \frac{1}{n}\log E[e^{nf(X_n(t),Y_n(t))}|X_n(0)=x, Y_n(0)=y].$$

THEOREM 11.41. *Suppose that Condition 11.21 holds and that H_1 and H_2 are given by (11.53) and (11.54) and satisfy $H_1(x,p)\leq H_2(x,p)$ for $x,p\in R^d$. Assume that the large deviation principle holds for $\{X_n(0)\}$ with good rate function I_0. Then the following hold:*

a) *$H_\dagger, H_\ddagger$ defined by (11.46) and (11.47) determine a semigroup $V(t)$ on $C_b(E)$ in the sense of Theorem 7.17. For each $f\in C_b(E)$ and $f_n\in B(E\times E_0)$ satisfying*

$$\lim_{n\to\infty}\sup_{(x,y)\in E\times\widetilde{K}_n^q}|f_n(x,y) - f(x)| = 0, \quad \forall q\in\mathcal{Q},$$

we have

$$\lim_{n\to\infty}\sup_{(x,y)\in E\times\widetilde{K}_n^q}|V_n(t)f_n(x,y) - V(t)f(x)| = 0. \tag{11.75}$$

b) *$\{X_n\}$ satisfies a large deviation principle in $C_{R^d}[0,\infty)$ with good rate function I.*

Suppose, in addition, that F is continuous. Then

c)

$$\begin{aligned} I(x) = \inf\{I_0(x(0)) + \int_0^\infty I_B(\mu_s)ds : \\ x(t) = x(0) + \int_0^t\int_{E_0}F(x(s),y)\mu_s(dy)ds\} \end{aligned} \tag{11.76}$$

where $\inf\emptyset = \infty$.

d) *For $L(x,q)$ given by*

$$L(x,q) = \sup_{p\in R^d}(p\cdot q - \mathbf{H}(x,p)),$$

$$I(x) = \begin{cases} I_0(x(0)) + \int_0^\infty L(x(s),\dot{x}(s))ds & \text{for } x \text{ absolutely continuous} \\ \infty & \text{otherwise.} \end{cases} \tag{11.77}$$

PROOF. The proof is similar to that of Theorem 11.20. □

11.3. Continuous time, central limit scaling

As in Section 11.2, let Y be an E_0-valued Markov process whose semigroup

$$S(t)f(y) \equiv E[f(Y(t))|Y(0) = y], \quad f \in C_b(E_0),$$

satisfies $S(t) : C_b(E_0) \mapsto C_b(E_0)$. We assume that for each initial $Y(0) = y \in E_0$, there is a cadlag version of Y, and let B denote the weak infinitesimal generator for $\{S(t)\}$.

For $\beta_n \to \infty$, consider (11.1) with $F_n(x,y) = \beta_n F(x,y)$, $Y_n(t) = Y(n\beta_n^2 t)$, and $\lim_{n\to\infty} x_n = x_0 \in R^d$.

Let $E = R^d \cup \{\infty\}$, and $E_n \equiv E \times E_0$. We assume the following:

CONDITION 11.42.

(1) *There is an index set $\mathcal{Q}$ and a family of subsets of E_0, $\{\widetilde{K}_n^q \subset E_0 : q \in \mathcal{Q}\}$, such that for $q_1, q_2 \in \mathcal{Q}$, there exists $q_3 \in \mathcal{Q}$ with $\widetilde{K}_n^{q_1} \cup \widetilde{K}_n^{q_2} \subset \widetilde{K}_n^{q_3}$; and for each $y \in E_0$, there exists $q \in \mathcal{Q}$ such that $\lim_{n\to\infty} r_0(y, \widetilde{K}_n^q) = 0$. Moreover, for each $q \in \mathcal{Q}$, $T > 0$ and $a > 0$, there exists a $\widehat{q}(q,a,T) \in \mathcal{Q}$ satisfying*

$$\limsup_{n\to\infty} \sup_{y\in\widetilde{K}_n^q} \frac{1}{n} \log P\{Y(t) \notin \widetilde{K}_n^{\widehat{q}(q,a,T)}, \text{ some } t \le n\beta_n^2 T | Y(0) = y\} \le -a. \tag{11.78}$$

(2) *There exists $\psi \in M^u(E_0, \overline{R})$ such that ψ is upper semicontinuous and bounded from above, and $\{y \in E_0 : \psi(y) \ge c\}$ is compact for each $c \in R$. There exist $\varphi_n \in \mathcal{D}(B)$ such that $\lim_{n\to\infty} \frac{1}{\beta_n}\|\varphi_n\| = 0$,*

$$\sup_{y\in E_0} \sup_n B\varphi_n(y) < \infty,$$

and

$$\limsup_{n\to\infty} \sup_{y\in\widetilde{K}_n^q} \Big(B\varphi_n(y) \vee c - \psi(y) \vee c\Big) = 0, \; c \in R, q \in \mathcal{Q}, \tag{11.79}$$

and Y is ergodic with stationary distribution π, satisfying

$$\lim_{\lambda\to 0} \sup_{y\in\{\psi\ge c\}} |\lambda \int_0^\infty e^{-\lambda t} S(t)f(y)dt - \int_{E_0} f d\pi| = 0, \quad f \in C_b(E_0), c \in R. \tag{11.80}$$

(3) *F satisfies*

$$\sup_y |F(0,y)| < \infty, \quad |F(x_1,y) - F(x_2,y)| \le K|x_1 - x_2|,$$

for some constant K, and

$$\int_{E_0} F(x,y)\pi(dy) = 0, \quad x \in R^d.$$

(4) *There exists $h : R^d \times E_0 \to R^d$ such that for each $x \in R^d$, $h(x,\cdot) \in \mathcal{D}(B)$ and*

$$Bh(x,y) = -F(x,y),$$

$h(\cdot,\cdot)$ is continuous, and continuously differentiable in x with

$$\sup_{x\in\Gamma, y\in E_0} |\nabla_x h(x,y)| < \infty$$

for each compact $\Gamma \subset R^d$. *In particular,*

$$\int |h(x_1, y) - h(x_2, y)|\pi(dy) < C_\Gamma |x_1 - x_2|, \quad x_1, x_2 \in \Gamma.$$

Note that under Condition 11.42.3, Condition 11.42.4 is reasonable since, under strong enough ergodicity hypotheses, h can be obtained as

$$h(x, y) = \lim_{t\to\infty} \int_0^t S(r)F(x, y)dr.$$

If E_0 is compact, we can always take $\widetilde{K}_n^q = E_0$ and $\varphi_n = 1, \psi = 0$. Then Condition 11.42.1) is always satisfied and Condition 11.42.2 reduces to

$$\lim_{\lambda\to 0+} \sup_{y\in E_0} |\lambda \int_0^\infty e^{-\lambda t} S(t)f(y)dt - \int_{E_0} f d\pi| = 0, \quad f \in C_b(E_0).$$

If E_0 is not compact, then Conditions 11.42.1 and 11.42.2 are the critical assumptions. Below, we show that an R^d-valued Ornstein-Uhlenbeck satisfies these assumptions.

EXAMPLE 11.43. Let $E_0 = R^d$ and Y be the Ornstein-Uhlenbeck process with generator B satisfying

$$Bg(y) = \frac{1}{2}\Delta g(y) - \alpha y \nabla g(y), \quad g \in C_c^2(R^d).$$

We take $\mathcal{Q} = [0, \infty)$ and look for $\widetilde{K}_n^q$ of the form $\{y \in R^d : |y| \le k_n^q\}$ to establish (11.78). We apply Lemma 4.20 with

$$f(y) = \kappa_n |y|^2, \quad \mathcal{H}^s f(y) = \frac{1}{2}\Delta f(y) + \frac{1}{2}|\nabla f(y)|^2 - \alpha y \nabla f(y) = d\kappa_n + 2\kappa_n^2 |y|^2 - 2\kappa_n \alpha |y|^2.$$

Then for $\kappa_n < \alpha$, $\tau_q = \inf\{t : |Y(t)| \ge k_n^q\}$ satisfies

$$P\{\tau_q \le Tn\beta_n^2 | Y(0) = y\} \le e^{-\kappa_n((k_n^q)^2 - |y|^2) + d\kappa_n Tn\beta_n^2}.$$

Setting $\kappa_n = 1/(n\beta_n^2)$ and $k_n^q = n\sqrt{q}\beta_n$, we have

$$P\{\tau_q \le Tn\beta_n^2 | Y(0) = y\} \le e^{-nq + \kappa_n |y|^2 + dT}.$$

Hence

$$\sup_{|y|\le k_n^q} P\{\tau_{\widehat{q}} \le Tn\beta_n^2 | Y(0) = y\} \le e^{-n(\widehat{q} - q) + dT},$$

which gives (11.78) with $\widehat{q}(q, a, T) = q + a$. Therefore, Condition 11.42.1 holds.

Now take $\varphi(y) = (1 + |y|^2)^r$ for $0 < r < 1$. Define φ_n so that $\varphi_n \in C^2(R^d)$, $\varphi_n(y) = \varphi(y)$, for $|y| \le n\beta_n^2$, $0 \le \varphi_n \le 2\varphi(n\beta_n^2)$, and φ_n constant for $|y| \ge 2n\beta_n^2$.

$$B\varphi_n(y) = 2r(r-1)(1+|y|^2)^{r-2}|y|^2 + rd(1+|y|^2)^{r-1} - 2\alpha r(1+|y|^2)^{r-1}|y|^2, \quad |y| < n\beta_n^2.$$

Let $q \in [0, \infty)$. Since when $\beta_n > \sqrt{q}$, $\widetilde{K}_n^q = \{y : |y| \le n\sqrt{q}\beta_n\} \subset \{y : |y| < n\beta_n^2\}$, (11.79) holds with

$$\psi(y) = 2r(r-1)(1 + |y|^2)^{r-2}|y|^2 + rd(1 + |y|^2)^{r-1} - 2\alpha r(1 + |y|^2)^{r-1}|y|^2.$$

Since $\{y \in R^d : \psi \ge c\}$ is compact, (11.80) becomes

$$\lim_{\lambda\to 0+} \sup_{|y|\le c} |\lambda \int_0^\infty e^{-\lambda t} S(t)f(y)dt - \int_{E_0} f d\pi| = 0, \quad f \in C_b(E_0), c \in [0, \infty).$$

The above ergodic identity can be directly verified, since we can explicitly identify the transition probability for the Ornstein-Uhlenbeck process Y. The resulting stationary measure is $\pi(dy) = \sqrt{\frac{2}{\pi}}e^{-\alpha|y|^2}dy$. Finally, if $\beta_n \geq cn^\delta$, for some $c > 0$ and $0 < \delta < 1$, and $r < \delta/4$, then $\beta_n^{-1}\|\varphi_n\| \to 0$ and $\sup_{n,y\in E_0} B\varphi_n(y) < \infty$. Hence Condition 11.42.2 holds.

11.3.1. Convergence of H_n. Define

$$A_n f(x,y) = \beta_n F(x,y)\cdot\nabla_x f(x,y) + n\beta_n^2 Bf(x,\cdot)(y)$$

on

$$\mathcal{D}(A_n) = \{f \in C_b(E_n) : f(x,\cdot) \in \mathcal{D}(B); Bf, F\cdot\nabla_x f \in C(E_n)\}.$$

As before, assuming the Lipschitz condition on F, for each initial value $(X_n(0), Y_n(0))$ there is a distributionally unique solution of the martingale problem for A_n, and A_n extends to a full generator for a well-posed martingale problem.

Again, let $D_d = \{f : f|_{R^d} - f(\infty) \in C_c^2(R^d)\}$ and $E' = E \times \widehat{E}_0$, where $\widehat{E}_0$ is the compactification of E_0 introduced in Section 11.1.1, and take $K_n^q = E \times \widetilde{K}_n^q, q \in \mathcal{Q}$, where the $\mathcal{Q}$ and $\widetilde{K}_n^q$ are given by Condition 11.42.1. As before, this condition assures that Condition 2.8 holds; by the compactness of E', Condition 7.9 is satisfied. LIM is defined by Definition 2.5.

The operator $H_n f = n^{-1}e^{-nf}A_n e^{nf}$ is given by

$$H_n f(x,y) = \beta_n F(x,y)\cdot\nabla_x f(x,y) + \beta_n^2 H_B(nf)(x,y),$$

where $H_B f = e^{-f}Be^f$. Let $f \in D_d$, $g \in \mathcal{D}(B)$, and h satisfy Condition 11.42.4. Define

$$f_n(x,y) = f(x) + \frac{1}{n}\log(1 + \beta_n^{-1}h(x,y)\cdot\nabla f(x) + \beta_n^{-2}g(y)),$$

for n large enough so that the expression inside the logarithm is positive. Then for sufficiently large n,

$$\begin{aligned} H_n f_n(x,y) = \beta_n F(x,y)\cdot\nabla f(x) &+ \frac{1}{n}\frac{F(x,y)^T(\nabla_x h(x,y)\nabla f(x) + \nabla^2 f(x)h(x,y))}{1 + \beta_n^{-1}h(x,y)\cdot\nabla f(x) + \beta_n^{-2}g(y)} \\ &+ \frac{Bg(y) - \beta_n F(x,y)\cdot\nabla f(x)}{1 + \beta_n^{-1}h(x,y)\cdot\nabla f(x) + \beta_n^{-2}g(y)}, \end{aligned}$$

which for $g \in \mathcal{D}(B)$, converges uniformly on $E \times E_0$ to

$$G_{f,g}(x,y) \equiv (F(x,y)\cdot\nabla f(x))(h(x,y)\cdot\nabla f(x)) + Bg(y).$$

Let φ_n, ψ be given by Condition 11.42,2. For $f \in D_d$, $g \in \mathcal{D}(B)$, and $\epsilon \geq 0$, define

$$G_{1,f,g,\epsilon}(x,y) = \begin{cases} G_{f,g}(x,y) + \epsilon\psi(y) & y \in E_0 \\ -\infty & y \in \widehat{E}_0 - E_0 \end{cases}$$

and

$$G_{2,f,g,\epsilon}(x,y) = \begin{cases} G_{f,g}(x,y) - \epsilon\psi(y) & y \in E_0 \\ \infty & y \in \widehat{E}_0 - E_0. \end{cases}$$

By Lemma 11.3

$$G_{1,f,g,\epsilon}(x,y) = \lim_{\delta\to 0}\sup_{z\in B_\delta(y)\cap E_0}(G_{f,g}(x,z) + \epsilon\psi(z)), \quad x \in E, y \in \widehat{E}_0,$$

and

$$G_{2,f,g,\epsilon}(x,y) = \lim_{\delta\to 0}\inf_{z\in B_\delta(y)\cap E_0}(G_{f,g}(x,z) - \epsilon\psi(z)), \quad x \in E, y \in \widehat{E}_0,$$

where $B_\delta(y) = \{z \in \widehat{E}_0 : \widehat{r}(y,z) < \delta\}$.

Replacing g by $g + \epsilon\varphi_n$, $\epsilon > 0$, in the definition of f_n and recalling that $\beta_n^{-1}\|\varphi_n\| \to 0$, we have for every $q \in \mathcal{Q}$

$$\lim_{n\to\infty} \sup_{(x,y)\in K_n^q = E\times\widetilde{K}_n^q} |f_n(x,y) - f(x)| \le \lim_{n\to\infty} \|f_n - f\|_{E\times E_0} = 0$$

and for $(x_n, y_n) \to (x,y) \in E \times E_0$, $y_n \in K_n^q$, for some $q > 0$,

$$H_n f_n(x_n, y_n) \to G_{f,g}(x,y) + \epsilon\psi(y) = G_{1,f,g,\epsilon}(x,y).$$

If E_0 is noncompact, then for $y \in \widehat{E}_0 - E_0$ and $E_n = E \times E_0 \ni (x_n, y_n) \to (x,y) \in E \times \widehat{E}_0$, by (11.79),

$$\limsup_{n\to\infty} H_n f_n(x_n, y_n) \le \limsup_{n\to\infty}(G_{f,g}(x, y_n) + \epsilon\psi(y_n)) \le G_{1,f,g,\epsilon}(x,y).$$

Consequently, we set

$$H_\dagger = \{(f, G_{1,f,g,\epsilon}) : f \in D_d, g \in \mathcal{D}(B), \epsilon \ge 0\}. \tag{11.81}$$

Similarly,

$$H_\ddagger = \{(f, G_{2,f,g,\epsilon}) : f \in D_d, g \in \mathcal{D}(B), \epsilon \ge 0\}. \tag{11.82}$$

As in Section 11.2, every subsolution $\overline{f}$ of $(I - \alpha H_\dagger)f = v$ is a strong subsolution, so if $f_0 \in D_d$ and

$$\overline{f}(x_0) - f_0(x_0) = \sup_x (\overline{f}(x) - f_0(x)),$$

then

$$\begin{aligned} &\alpha^{-1}(\overline{f}(x_0) - v(x_0)) \\ &\quad \le \inf_{g\in\mathcal{D}(B)} \sup_{y\in E_0} ((F(x_0,y)\cdot\nabla f_0(x_0))(h(x_0,y)\cdot\nabla f_0(x_0)) + Bg(y) + \epsilon\psi(y)). \end{aligned} \tag{11.83}$$

Define $h_0(y) = (F(x_0,y)\cdot\nabla f_0(x_0))(h(x_0,y)\cdot\nabla f_0(x_0))$ for $y \in E_0$, and let $g_\lambda = \int_0^\infty e^{-\lambda t}S(t)h_0 dt$. Then $Bg_\lambda = \lambda g_\lambda - h_0$, so

$$\alpha^{-1}(\overline{f}(x_0) - v(x_0)) \le \sup_{y\in E_0} (\lambda g_\lambda(y) + \epsilon\psi(y)).$$

Since $\|\lambda g_\lambda\| \le \|h_0\|$, for $\epsilon > 0$, the supremum will be realized on the set $\{\psi > -2\epsilon^{-1}\|h_0\|\}$. Letting $\lambda \to 0$, by Condition 11.42.2, we have

$$\alpha^{-1}(\overline{f}(x_0) - v(x_0)) \le \sup_{y\in E_0} (H_0(x_0, \nabla f_0(x_0)) + \epsilon\psi(y)),$$

where

$$H_0(x,p) = \int_{E_0} (F(x,y)\cdot p)(h(x,y)\cdot p)\pi(dy),$$

and letting $\epsilon \to 0$

$$\alpha^{-1}(\overline{f}(x_0) - v(x_0)) \le H_0(x_0, \nabla f_0(x_0)).$$

Similarly, if $\underline{f}$ is a supersolution of $(I - \alpha H_\ddagger)f = v$, $f_0 \in D_d$, and

$$f_0(x_0) - \underline{f}(x_0) = \sup_x (f_0(x) - \underline{f}(x)),$$

then

$$\alpha^{-1}(\underline{f}(x_0) - v(x_0)) \ge H_0(x_0, \nabla f_0(x_0)).$$

Consequently, every subsolution of $(I - \alpha H_\dagger)f = v$ is a subsolution of $(I - \alpha H_0)f = v$ and every supersolution of $(I - \alpha H_\ddagger)f = v$ is a supersolution of $(I -$

$\alpha H_0)f = v$,and if the comparison principle holds for $(I - \alpha H_0)f = v$, then the comparison principle holds for subsolutions of $(I - \alpha H_\dagger)f = v$ and supersolutions of $(I - \alpha H_\ddagger)f = v$.

11.3.2. Exponential tightness. Exponential tightness for $\{X_n\}$ follows by the convergence of H_n and Corollary 4.17.

11.3.3. The comparison principle. Define

$$a_{ij}(x) = \int_{E_0} \frac{1}{2}(h_i(x,y)F^{(j)}(x,y) + h_j(x,y)F^{(i)}(x,y))\pi(dy), \tag{11.84}$$

where $F(x,y) = (F^{(1)}(x,y), \cdots, F^{(d)}(x,y))$ and $h(x,y) = (h_1(x,y), \cdots, h_d(x,y))$. Then $a(x)$ is nonnegative definite and continuous, and $H_0(x,p) = p^T a(x)p$, which is a special case of the H treated in Section 10.3. Under Condition 11.42, a is continuous and satisfies $|a(x)| \leq K_1 + K_2|x|^2$ for some K_1 and K_2, so H_0 satisfies Condition 9.10.2.

The operator H_0 is closely related to a (pre-)Dirichlet form. For each $f, g \in \mathcal{D}(B)$, define

$$\mathcal{E}(f,g) = -\int_{E_0} (fBg + gBf)d\pi,$$

and write

$$\mathcal{E}(f) \equiv \mathcal{E}(f,f).$$

Then, under Condition 11.42,

$$H_0(x,p) = \frac{1}{2}\mathcal{E}(p \cdot h(x,\cdot)).$$

We derive some properties of $\mathcal{E}(f,g)$. Let Y be a stationary solution of the martingale problem for B, that is, the solution with initial distribution π. For each $f \in \mathcal{D}(B)$, let

$$M^f(t) = f(Y(t)) - f(Y(0)) - \int_0^t Bf(Y(s))ds.$$

LEMMA 11.44. *Suppose $f, g \in \mathcal{D}(B)$, then*

a) $E[M_f^2(t)] = \mathcal{E}(f)t$;
b) $E[M_f(t)M_g(t)] = \mathcal{E}(f,g)t$;
c) $\mathcal{E}(f,g) \leq \sqrt{\mathcal{E}(f)\mathcal{E}(g)}$.

PROOF. Applying a small amount of algebra and Itô's formula gives

$$\begin{aligned}
M_f^2(t) &= (f(Y(t)) - f(Y(0)))^2 - 2M_f(t)\int_0^t Bf(Y(s))ds - (\int_0^t Bf(Y(s))ds)^2 \\
&= (f(Y(t)) - f(Y(0)))^2 - 2\int_0^t M_f(s)Bf(Y(s))ds \\
&\quad -2\int_0^t\int_0^r Bf(Y(s))dsdM_f(r) - (\int_0^t Bf(Y(s))ds)^2 \\
&= (f(Y(t)) - f(Y(0)))^2 - 2\int_0^t f(Y(s))Bf(Y(s))ds \\
&\quad +2\int_0^t (f(Y(0)) + \int_0^s Bf(Y(r))dr)Bf(Y(s))ds \\
&\quad -2\int_0^t Bf(Y(s))\int_0^s Bf(Y(r))drds - 2\int_0^t\int_0^r Bf(Y(s))dsdM_f(r) \\
&= f^2(Y(t)) - f^2(Y(0)) - 2f(Y(0))(f(Y(t)) - f(Y(0)) - \int_0^t Bf(Y(s))ds) \\
&\quad -2\int_0^t f(Y(s))Bf(Y(s))ds - 2\int_0^t\int_0^r Bf(Y(s))dsdM_f(r).
\end{aligned}$$

Therefore

$$\begin{aligned}
E[M^2(t)] &= -2\int_0^t E[f(Y(s)Bf(Y(s))]ds \\
&= \mathcal{E}(f)t.
\end{aligned}$$

It follows that

$$4E[M_f(t)M_g(t)] = E[M_{f+g}^2(t)] - E[M_{f-g}^2(t)] = (\mathcal{E}(f+g) - \mathcal{E}(f-g))t = 4\mathcal{E}(f,g)t$$

and

$$\mathcal{E}(f,g) = E[M_f(1)M_g(1)] \le \sqrt{E[M_f^2(1)]E[M_g^2(1)]} = \sqrt{\mathcal{E}(f)\mathcal{E}(g)}.$$

□

LEMMA 11.45. *Suppose π is the stationary distribution of B. Then $\mathcal{E}(\cdot)$ is convex in the sense that for each $f, g \in \mathcal{D}(B)$ and $0 \le \alpha \le 1$,*

$$\mathcal{E}(\alpha f + (1-\alpha)g) \le \alpha\mathcal{E}(f) + (1-\alpha)\mathcal{E}(g).$$

PROOF. By Lemma 11.44,

$$\begin{aligned}
\mathcal{E}(\alpha f + (1-\alpha)g) &= \alpha^2\mathcal{E}(f) + (1-\alpha)^2\mathcal{E}(g) + 2\mathcal{E}(\alpha f, (1-\alpha)g) \\
&\le \alpha^2\mathcal{E}(f) + (1-\alpha)^2\mathcal{E}(g) + 2\sqrt{\mathcal{E}(\alpha f)\mathcal{E}((1-\alpha)g)} \\
&= (\alpha\sqrt{\mathcal{E}(f)} + (1-\alpha)\sqrt{\mathcal{E}(g)})^2 \\
&\le \alpha\mathcal{E}(f) + (1-\alpha)\mathcal{E}(g),
\end{aligned}$$

where the last inequality follows by Jensen's inequality. □

LEMMA 11.46. *Under Condition 11.42, the comparison principle holds for H_0.*

PROOF. The proof is a slight variation of Lemma 9.19.

Let $\Gamma \subset R^d$ be compact and convex. By convexity, for $\lambda > 1$ and $x, y \in \Gamma$,

$$\begin{aligned}\mathcal{E}(\frac{p\cdot h(x,\cdot)}{\lambda}) &= \mathcal{E}(\frac{\lambda-1}{\lambda}\frac{p\cdot(h(x,\cdot)-h(y,\cdot))}{\lambda-1}+\frac{1}{\lambda}p\cdot h(y,\cdot)) \\ &\leq (1-\frac{1}{\lambda})\mathcal{E}(\frac{p\cdot(h(x,\cdot)-h(y,\cdot))}{\lambda-1})+\frac{1}{\lambda}\mathcal{E}(p\cdot h(y,\cdot)).\end{aligned}$$

Therefore,

$$\begin{aligned}&\lambda H_0(x,\frac{p}{\lambda})-H_0(y,p) \\ =\ & \frac{\lambda}{2}\mathcal{E}(\frac{p\cdot h(x,\cdot)}{\lambda})-\frac{1}{2}\mathcal{E}(p\cdot h(y,\cdot)) \\ \leq\ & \frac{\lambda-1}{2}\mathcal{E}(\frac{p\cdot(h(x,\cdot)-h(y,\cdot))}{\lambda-1}) \\ =\ & \frac{1}{2(\lambda-1)}\int p\cdot(h(x,z)-h(y,z))(p\cdot(F(x,z)-F(y,z)))\pi(dz) \\ \leq\ & \frac{1}{2(\lambda-1)}C_\Gamma K|p|^2|x-y|^2,\end{aligned}$$

where Condition 11.42 was used to derive the last inequality.

Therefore, Condition 9.11 is satisfied. By Lemma 9.12 and Lemma 9.15, the comparison principle holds for H_0. □

11.3.4. The large deviation theorem. Take $U = R^d$, and define $L : E \times U \mapsto [0, \infty)$ by

$$L(x,q)=\sup_{p\in R^d}\{p\cdot q-p^Ta(x)p\},\quad x\in R^d,$$

and $L(\infty, q) = 0$.

THEOREM 11.47. *Suppose that Condition 11.42 is satisfied and that the large deviation principle holds for $\{X_n(0)\}$ with good rate function I_0. Then $\{X_n\}$ satisfies the large deviation principle in $C_{R^d}[0,\infty)$ with a good rate function*

$$I(x)=\begin{cases} I_0(x(0))+\int_0^\infty L(x(s),\dot{x}(s))ds & x \text{ absolutely continuous} \\ \infty & \text{otherwise.}\end{cases}$$

PROOF. The large deviation principle holds by Theorem 7.24. The rate function representation of I is the same as in Theorem 10.22. □

11.4. Discrete time, central limit scaling

As in Section 11.1, for $n = 1, 2, \ldots$, let Y be an E_0-valued, discrete-time Markov chain with transition operator P and $Y_n(t) = Y([\frac{t}{\epsilon_n}])$. Let $E = R^d \cup \{\infty\}$. We take $F_n(x, y) = (n\epsilon_n)^{-1/2}F(x, y)$ in (11.2), with $n\epsilon_n \to 0$, and assume

CONDITION 11.48.

(1) *There is an index set $\mathcal{Q}$ and a family of subsets of E_0, $\{\widetilde{K}_n^q \subset E_0 : q \in \mathcal{Q}\}$, such that for $q_1, q_2 \in \mathcal{Q}$, there exists $q_3 \in \mathcal{Q}$ with $\widetilde{K}_n^{q_1} \cup \widetilde{K}_n^{q_2} \subset \widetilde{K}_n^{q_3}$; and for each $y \in E_0$, there exists $q \in \mathcal{Q}$ such that $\lim_{n\to\infty} r_0(y, \widetilde{K}_n^q) = 0$.*

Moreover, for each $q \in \mathcal{Q}$, $T > 0$ and $a > 0$, there exists a $\widehat{q}(q, a, T) \in \mathcal{Q}$ satisfying

$$\limsup_{n\to\infty} \sup_{y\in\widetilde{K}_n^q} \frac{1}{n} \log P\{Y(t) \notin \widetilde{K}_n^{\widehat{q}(q,a,T)}, \text{ some } t \leq \epsilon_n^{-1}T | Y(0) = y\} \leq -a. \tag{11.85}$$

(2) *There exists $\psi \in M^u(E_0, \overline{R})$ which is upper semicontinuous and $\{y \in E_0 : \psi(y) \geq c\}$ is compact for each $c \in R$. There exist $\varphi_n \in B(E_0)$ such that $\lim_{n\to\infty} \sqrt{n\epsilon_n}\|\varphi_n\| = 0$, $\sup_n \sup_{y\in E_0}(P - I)\varphi_n(y) < \infty$, and*

$$\limsup_{n\to\infty} \sup_{y\in\widetilde{K}_n^q} \Big((P - I)\varphi_n(y) \vee c - \psi(y) \vee c\Big) \leq 0, \quad c \in R, q > 0,$$

and Y is ergodic with stationary distribution π, satisfying

$$\lim_{\rho\to 1} \sup_{y\in\{\psi\geq c\}} |(1-\rho)\sum_{k=0}^{\infty} \rho^k P^k f(y) - \int_{E_0} f d\pi| = 0, \quad f \in C_b(E_0), c \in R.$$

(3) *F satisfies*

$$\sup_y |F(0, y)| < \infty, \quad |F(x_1, y) - F(x_2, y)| \leq K|x_1 - x_2|,$$

for some constant K, and

$$\int_{E_0} F(x, y)\pi(dy) = 0, \quad x \in R^d.$$

(4) *There exists $h : R^d \times E_0 \to R^d$ such that*

$$(P - I)h(x, y) = -F(x, y),$$

$h(\cdot, \cdot)$ is continuous, and for each compact $\Gamma \subset R^d$,

$$\int |h(x_1, y) - h(x_2, y)|\pi(dy) < C_\Gamma|x - y|, \quad x_1, x_2 \in \Gamma.$$

Let $E_n \equiv E \times E_0$. The transition operator for (X_n, Y_n) is

$$T_n f(x, y) = \int_{E_0} f(x + \sqrt{\epsilon_n/n}F(x, y), z))P(y, dz), \qquad f \in B(E_n),$$

and for $f_n(x,y) = f(x) + \frac{1}{n}\log(1+\sqrt{n\epsilon_n}h(x,y)\cdot\nabla f(x) + n\epsilon_n g(y))$, $f \in D_d$ and $g \in B(E_0)$, we have

$$\begin{aligned}
H_n f_n(x,y) &= \frac{1}{n\epsilon_n}\log e^{-nf_n}T_n e^{nf}(x,y) \\
&= \frac{1}{\epsilon_n}(f(x+\sqrt{\epsilon_n/n}F(x,y)) - f(x)) \\
&\qquad + \frac{1}{n\epsilon_n}\log\frac{1+\sqrt{n\epsilon_n}T_n(h\cdot\nabla f)(x,y) + n\epsilon_n Pg(y)}{1+\sqrt{n\epsilon_n}h(x,y)\cdot\nabla f(x) + n\epsilon_n g(y)} \\
&\approx \frac{1}{\sqrt{n\epsilon_n}}F(x,y))\cdot\nabla f(x) \\
&\qquad + \frac{(n\epsilon_n)^{-1/2}(T_n(h\cdot\nabla f)(x,y) - h(x,y)\cdot\nabla f(x)) + (P-I)g(y)}{1+\sqrt{n\epsilon_n}h(x,y)\cdot\nabla f(x) + n\epsilon_n g(y)} \\
&\qquad - \frac{1}{2}(T_n(h\cdot\nabla f)(x,y) - h(x,y)\cdot\nabla f(x))^2 \\
&\approx F(x,y)\cdot\nabla f(x)h(x,y)\cdot\nabla f(x) - \frac{1}{2}(F(x,y)\cdot\nabla f(x))^2 + (P-I)g(y) \\
&= \frac{1}{2}\left((h(x,y)\cdot\nabla f(x))^2 - (Ph(x,y))\cdot\nabla f(x))^2\right) + (P-I)g(y),
\end{aligned}$$

where the last equality follows from the fact that $(P-I)h = -F$.

Following the continuous time case in Section 11.3, we take $K_n^q = E \times \widetilde{K}_n^q$ and $E' = E \times \widehat{E}_0$, where $(\widehat{E}_0, \widehat{r}_0)$ is the compactification of (E_0, r_0) in Section 11.1.1. Let φ_n, ψ be given by Condition 11.48.2. For each $f \in D_d$, $g \in B(E_0)$, Let φ_n, ψ be given by Condition 11.42,2. For $f \in D_d$, $g \in \mathcal{D}(B)$, and $\epsilon \geq 0$, setting

$$G_{f,g}(x,y) = \frac{1}{2}\left((h(x,y)\cdot\nabla f(x))^2 - (Ph(x,y)\cdot\nabla f(x))^2\right) + (P-I)g(y),$$

define

$$G_{1,f,g,\epsilon}(x,y) = \begin{cases} G_{f,g}(x,y) + \epsilon\psi(y) & y \in E_0 \\ -\infty & y \in \widehat{E}_0 - E_0 \end{cases}$$

and

$$G_{2,f,g,\epsilon}(x,y) = \begin{cases} G_{f,g}(x,y) - \epsilon\psi(y) & y \in E_0 \\ \infty & y \in \widehat{E}_0 - E_0. \end{cases}$$

Again, by Lemma 11.3,

$$G_{1,f,g,\epsilon}(x,y) = \lim_{\delta\to 0}\sup_{z\in B_\delta(y)\cap E_0} G_{1,f,g,\epsilon}(x,z)$$

and

$$G_{2,f,g,\epsilon}(x,y) = \lim_{\delta\to 0}\inf_{z\in B_\delta(y)\cap E_0} G_{2,f,g,\epsilon}(x,z).$$

Then we can verify the Convergence Condition 7.11 for convergence to limit operators

$$\begin{aligned}
H_\dagger &= \{(f, G_{1,f,g,\epsilon}) : f \in D_d, g \in \mathcal{D}(B), \epsilon \geq 0\} \\
H_\ddagger &= \{(f, G_{2,f,g,\epsilon}) : f \in D_d, g \in \mathcal{D}(B), \epsilon \geq 0\}.
\end{aligned}$$

Setting

$$H_0 f(x) = \int_{E_0}\frac{1}{2}\left((h(x,y)\cdot\nabla f(x))^2 - (Ph(x,y))\cdot\nabla f(x))^2\right)\pi(dy)$$

and

$$\mathcal{E}(f,g) = \int_{E_0} (fg - PfPg)d\pi,$$

the remainder of the analysis is similar to the continuous time case. In particular, observe that if $\{Y(k)\}$ is a stationary Markov chain with transition function P and initial distribution π, then

$$E[(f(Y(k)) - Pf(Y(k-1)))(g(Y(k)) - Pg(Y(k-1)))] = \mathcal{E}(f,g).$$

Define a by (11.84), and let

$$L(x,q) = \sup_{p \in R^d} \{p \cdot q - p^T a(x) p\}, \quad x \in R^d.$$

THEOREM 11.49. *Suppose that Condition 11.48 is satisfied and that the large deviation principle holds for $\{X_n(0)\}$ with good rate function I_0. Then $\{X_n\}$ satisfies the large deviation principle in $C_{R^d}[0,\infty)$ with a good rate function*

$$I(x) = \begin{cases} I_0(x(0)) + \int_0^\infty L(x(s), \dot{x}(s))ds & x \text{ absolutely continuous} \\ \infty & \text{otherwise.} \end{cases}$$

11.5. Diffusions with periodic coefficients

In this section we consider a large deviation result of Baldi [**6**] on diffusions with periodic coefficients. The technicalities needed to obtain the limiting H are essentially the same as in Section 11.3. The comparison principle is immediate since H is spatially homogeneous. Our conditions are slightly weaker than Baldi's, requiring only nondegeneracy and continuity of the diffusion coefficients. We also include a drift term. Freidlin and Sowers [**51**] consider a similar problem. The scaling treated here corresponds to $\delta/\epsilon \to 0$ in their notation.

Let σ be a $d \times d$ matrix-valued function and b be a R^d-valued function that are continuous and periodic in the sense that for each $1 \le i \le d$, there is a period $p_i > 0$ such that $\sigma(y) = \sigma(y + p_i e_i)$ and $b(y) = b(y + p_i e_i)$ for all $y \in R^d$. Define $a = \sigma\sigma^T$, and assume that $a(y)$ is positive definite for all y. Let

$$dX_n(t) = \frac{1}{\sqrt{n}} \sigma(\alpha_n X_n(t))dW(t) + b(\alpha_n X_n(t))dt, \tag{11.86}$$

where $\alpha_n > 0$ and $\lim_{n\to\infty} n^{-1}\alpha_n = \infty$.

11.5.1. Convergence of H_n.

Define

$$E_0 \equiv [0, p_1) \times \cdots \times [0, p_d),$$

topologized as a torus so that it is compact, and let $E_n = R^d$, $E = R^d \cup \{\infty\}$, $E' = E \times E_0$, $\eta'_n(x) = (x, y)$ where $y_i = \alpha_n x_i \bmod p_i$, and $\eta_n(x) = x$. As before, $D_d = \{f : f|_{R^d} - f(\infty) \in C_c^2(R^d)\}$. X_n solves the martingale problem given by

$$A_n f(x) = \frac{1}{2n} \sum_{ij} a_{ij}(\alpha_n x) \frac{\partial^2}{\partial x_i \partial x_j} f(x) + \sum_i b_i(\alpha_n x) \frac{\partial}{\partial x_i} f(x), \quad f \in D_d,$$

and

$$H_n f(x) = \frac{1}{2n} \sum_{ij} a_{ij}(\alpha_n x) \partial_i \partial_j f(x) + \sum_i b_i(\alpha_n x) \partial_i f(x) + \frac{1}{2} \sum_{ij} a_{ij}(\alpha_n x) \partial_i f(x) \partial_j f(x)$$

for $f \in D_d$.

Define

$$Bg(y) = \frac{1}{2}\sum_{ij} a_{ij}(y)\frac{\partial^2}{\partial y_i \partial y_j} g(y),$$

for all $g \in \mathcal{D}(B) \subset C(E_0)$ that extend periodically to a function $g \in C^2(R^d)$. The continuity and positive definiteness of a imply that the martingale problem for B is well posed, and the compactness of E_0 and nondegeneracy imply that the corresponding process is ergodic. In particular, if $\{S(t)\}$ is the corresponding semigroup, then Condition 11.42.2 holds with $\varphi \equiv 1$ and $\psi \equiv 0$. That is, there exists a $\pi \in \mathcal{P}(E_0)$,

$$\lim_{\lambda \to 0+} \sup_{y \in E_0} |\lambda \int_0^\infty e^{-\lambda t} S(t) f(y) - \int_{E_0} f d\pi| = 0, \quad f \in C_b(E_0).$$

For each $f \in D_d$ and $g \in \mathcal{D}(B)$, let $f_n(x) = f(x) + \epsilon_n g(\alpha_n x)$, where $\epsilon_n = n\alpha_n^{-2}$. Then

$$\begin{aligned} & H_n f_n(x) \\ = \; & \frac{1}{2n}\sum_{ij} a_{ij}(\alpha_n x)\Big(\frac{\partial^2}{\partial x_i \partial x_j} f(x) + +\epsilon_n \alpha_n^2 \frac{\partial^2}{\partial y_i \partial y_j} g(\alpha_n x)\Big) \\ & + \frac{1}{2}\sum_{ij} a_{ij}(\alpha_n x)\Big(\frac{\partial}{\partial x_i} f(x) + \epsilon_n \alpha_n \frac{\partial}{\partial y_i} g(\alpha_n x)\Big)\Big(\frac{\partial}{\partial x_j} f(x) + \epsilon_n \alpha_n \frac{\partial}{\partial y_j} g(\alpha_n x)\Big) \\ & + \sum_i b_i(\alpha_n x)\Big(\frac{\partial}{\partial x_i} f(x) + \epsilon_n \alpha_n \frac{\partial}{\partial y_i} g(\alpha_n x)\Big). \end{aligned}$$

Define

$$G_{f,g}(x,y) = \frac{1}{2}\nabla f(x)^T a(y) \nabla f(x) + b(y)\cdot \nabla f(x) + Bg(y).$$

Then

$$\lim_{n\to\infty} \sup_x |H_n f_n(x) - G_{f,g}(x, \alpha_n x)| = 0,$$

and we define

$$H = \{(f, G_{f,g}) : f \in D_d, g \in \mathcal{D}(B)\}.$$

Then, as before, if $\overline{f}$ is a subsolution of $(I - \alpha H)f = h$, it is a strong subsolution, and if $f_0 \in D_d$ and

$$\overline{f}(x_0) - f_0(x_0) = \sup_x(\overline{f}(x) - f_0(x)),$$

then

$$\alpha^{-1}(\overline{f}(x_0) - v(x_0)) \leq \inf_{g\in\mathcal{D}(B)} \sup_y (\frac{1}{2}\nabla f(x)^T a(y)\nabla f(x) + b(y)\cdot \nabla f(x) + Bg(y)).$$

Define $h_0(y) = \frac{1}{2}\nabla f(x)^T a(y) \nabla f(x) + b(y)\cdot\nabla f(x)$ and $g_\lambda = \int_0^\infty e^{-\lambda t} S(t) h_0 dt$. Then $Bg_\lambda = \lambda g_\lambda - h_0$, so

$$\alpha^{-1}(\overline{f}(x_0) - v(x_0)) \leq \lim_{\lambda\to\infty} \sup_y \lambda g_\lambda(y) = H_0(\nabla f(x_0)),$$

where

$$H_0(p) = \frac{1}{2}p^T \overline{a} p + \overline{b}\cdot p$$

for $\overline{a}_{ij} = \int_{E_0} a_{ij} d\pi$, $\overline{b}_i = \int_{E_0} b_i d\pi$. The corresponding inequality holds for supersolutions and since H_0 does not depend on x, the comparison principle follows.

As in Section 10.1.5, the next result follows from Theorems 7.18 and 8.28.

THEOREM 11.50. *Let X_n satisfy (11.86) where b and σ are continuous and periodic and $\sigma(y)$ is nonsingular for all y. Suppose $\{X_n(0)\}$ satisfies a large deviation principle on R^d with good rate function I_0. Then $\{X_n\}$ satisfies a large deviation principle with good rate function*

$$I(x) = \begin{cases} I_0(x(0)) + \frac{1}{2}\int_0^\infty \frac{|\dot{x}(s) - \bar{b}|^2}{\bar{a}} ds & x \text{ absolutely continuous} \\ \infty & \text{otherwise} \end{cases} . \tag{11.87}$$

11.6. Systems with small diffusion and averaging

Veretennikov [**126**] considers a system of the following form:

$$\begin{aligned} X_n(t) &= X_n(0) + \int_0^t F(X_n(s), Y_n(s))ds + \frac{1}{\sqrt{n}}\int_0^t \sigma^X(X_n(s), Y_n(s))dW(s) \\ Y_n(t) &= Y_n(0) + \int_0^t nb(X_n(s), Y_n(s))ds + \sqrt{n}\int_0^t \sigma^Y(X_n(s), Y_n(s))dW(s), \end{aligned}$$

where W is a m-dimensional standard Brownian motion. In [**126**], σ^Y is independent of x. Theorem 7.9.2 of Freidlin and Wentzell [**52**] considers the case $\sigma^X = 0$, σ^Y independent of x, and Y_n confined to a compact manifold. Guillin [**56**] considers related models.

We assume that $F : R^d \times R^l \mapsto R^d$, $b : R^d \times R^l \mapsto R^l$, $\sigma^X : R^d \times R^l \mapsto M^{d\times m}$, and $\sigma^Y : R^d \times R^l \mapsto M^{l\times m}$ are locally bounded and that the solution of the system exists and is unique so that (X_n, Y_n) is a R^{d+l}-valued Markov process with full generator A_n.

For f sufficiently smooth and satisfying appropriate growth conditions,

$$\begin{aligned} A_nf(x,y) &= \frac{1}{2n}\sum a^X_{ij}(x,y)\partial_{x_i}\partial_{x_j}f(x,y) + F(x,y)\cdot\nabla_x f(x,y) \\ &\quad + \sum a^{XY}_{ij}(x,y)\partial_{x_i}\partial_{y_j}f(x,y) \\ &\quad + \frac{n}{2}\sum a^Y_{ij}(x,y)\partial_{y_i}\partial_{y_j}f(x,y) + nb(x,y)\cdot\nabla_y f(x,y), \end{aligned}$$

where $a^X = \sigma^X(\sigma^X)^T$, $a^{XY} = \sigma^X(\sigma^Y)^T$, and $a^Y = \sigma^Y(\sigma^Y)^T$, and

$$\begin{aligned} H_nf(x,y) &= \frac{1}{n}e^{-nf(x,y)}A_ne^{nf}(x,y) \\ &= \frac{1}{2n}\sum a^X_{ij}(x,y)\partial_{x_i}\partial_{x_j}f(x,y) + F(x,y)\cdot\nabla_x f(x,y) \\ &\quad + \sum a^{XY}_{ij}(x,y)\partial_{x_i}\partial_{y_j}f(x,y) \\ &\quad + n\frac{1}{2}\sum a^Y_{ij}(x,y)\partial_{y_i}\partial_{y_j}f(x,y) + nb(x,y)\cdot\nabla_y f(x,y) \\ &\quad + \frac{1}{2}\sum a^X_{ij}(x,y)\partial_{x_i}f(x,y)\partial_{x_j}f(x,y) \\ &\quad + n\sum a^{XY}_{ij}(x,y)\partial_{x_i}f(x,y)\partial_{y_j}f(x,y) \\ &\quad + \frac{n^2}{2}\sum a^Y_{ij}(x,y)\partial_{y_i}f(x,y)\partial_{y_j}f(x,y). \end{aligned}$$

As before,

$$D_d = \{f \in C_b(R^d \cup \{\infty\}) : f|_{R^d} - f(\infty) \in C_c^2(R^d)\},$$

and we define $D_d^{++} = \{f \in D_d : \inf_{y \in R^d} f(y) > 0\}$. For each $g \in C^2(R^l)$, $x \in R^d$, and $p \in R^d$, define

(11.88)

$$B_{x,p}g(y) = \frac{1}{2}\sum_{i,j} a_{ij}^Y(x,y)\partial_{y_i}\partial_{y_j}g(y) + b(x,y)\cdot\nabla_y g(y) + \sum_{i,j} a_{ij}^{XY}(x,y)p_i\partial_{y_j}g(y), \quad y \in R^l.$$

For $f \in C_c^2(R^l)$ and $g = e^f$,

$$\begin{aligned} \frac{B_{x,p}g(y)}{g(y)} &= e^{-f(y)}B_{x,p}e^f(y) \\ &= \frac{1}{2}\sum_{i,j=1}^{l} a_{ij}^Y(x,y)\partial_{y_i}\partial_{y_j}f(y) + \frac{1}{2}|(\sigma^Y(x,y))^T\cdot\nabla_y f(y)|^2 \\ &\quad + b(x,y)\cdot\nabla_y f(y) + p^T\sigma^X(x,y)(\sigma^Y(x,y))^T\nabla_y f(y). \end{aligned} \tag{11.89}$$

We assume that

CONDITION 11.51.

(1) *F, b, σ^X, and σ^Y are continuous on $R^d \times R^l$, and there exist c_1, c_2 such that*

$$\sup_{y\in R^l}(|F(x,y)| + |\sigma^X(x,y)|) \le c_1 + c_2|x|.$$

(2) *There exists $\varphi_1, \varphi_2 \in C^2(R^d)$ satisfying*
  (a) $\varphi_i \ge 1$, $i = 1,2$
  (b) $\lim_{|y|\to\infty}\varphi_i(y) = \infty$, $i = 1,2$
  (c) *For each $k > 0$*

$$\lim_{|y|\to\infty}\sup_{|x|,|p|\le k}\frac{B_{x,p}\varphi_i(y)}{\varphi_i(y)} = -\infty, \quad i = 1,2 \tag{11.90}$$

  (d)

$$\lim_{|y|\to\infty}\frac{\log\varphi_1(y)}{\log\varphi_2(y)} = 0 \tag{11.91}$$

  (e) *There exists $b_0 > 0$ such that for $i = 1,2$,*

$$\sup_{y\in R^l}\frac{B_{x,p}\varphi_i(y)}{\varphi_i(y)} < b_0(1 + |p|(1+|x|)) \tag{11.92}$$

LEMMA 11.52. *Suppose that Condition 11.51 holds. For $q > 0$, let $\widetilde{K}_n^q = \{y : \varphi_2(y) \le e^{qn}\}$. Then for $q > 0$, $k > 0$, $T > 0$, and $a > 0$, there exist $\widehat{q}(q,k,a,T) > 0$ and $\widehat{k}(q,k,a,T) > 0$ for which*

$$\begin{aligned} \limsup_{n\to\infty}\sup_{|x|\le k, y\in\widetilde{K}_n^q}\frac{1}{n}\log P(Y_n(t) \notin \widetilde{K}_n^{\widehat{q}(q,k,a,T)} \text{ or } |X_n(t)| > \widehat{k}(q,k,a,T), \\ \text{some } t \le T | X_n(0) = x, Y_n(0) = y) \quad \le -a. \end{aligned} \tag{11.93}$$

PROOF. The lemma follows by Lemma 4.20 taking $f(x,y) = \frac{1}{n}\log\varphi_2(y) + \log(1+|x|^2)$. □

LEMMA 11.53. *Let Condition 11.51.2 hold, and define* $\widetilde{K}_n^q = \{y : \varphi_2(y) \leq e^{qn}\}$. *Then*

$$\{y \in R^l : \sup_{|x|,|p|\leq k} B_{x,p}\varphi_1(y)/\varphi_1(y) \geq c\} \tag{11.94}$$

is compact for each $c \in R$, *and for each* $q > 0$,

$$\lim_{n\to\infty} \frac{1}{n} \log \sup_{y\in \widetilde{K}_n^q} \varphi_1(y) = 0. \tag{11.95}$$

PROOF. The compactness of (11.94) follows from from (11.90), and (11.95) follows from (11.91) as in Example 11.24. □

The following lemma gives a simple criterion for Condition 11.51.2.

LEMMA 11.54. *Suppose there exist* $\delta_2 > \delta_1 > 0$ *and* b_1 *such that*

$$\frac{\sum_{i=1}^l a_{ii}^Y(x,y) + 2b(x,y)\cdot y + 2p^T a^{XY}(x,y)y}{(1+|y|^2)^{1-\delta_2}} \leq b_1(1+|p|(1+|x|))$$

$$\frac{y^T a^Y(x,y)y}{(1+|y|^2)^{1-\delta_2}} \leq b_1(1+|p|(1+|x|))$$

and for each k,

$$\lim_{|y|\to\infty} \sup_{|x|\leq k} \frac{\sum_{i=1}^l a_{ii}^Y(x,y) + 2b(x,y)\cdot y + 2k|a^{XY}(x,y)y|}{(1+|y|^2)^{1-\delta_1}} = -\infty \tag{11.96}$$

Then Condition 11.51.2 is satisfied with $\varphi_1(y) = \exp\{(1+|y|^2)^{\delta_1}\}$ *and* $\varphi_2(y) = \exp\{(1+|y|^2)^{\delta_2}\}$.

PROOF. For $\delta = \delta_1, \delta_2$,

$$\begin{aligned}\frac{B_{x,p}\varphi_i(y)}{\varphi_i(y)} &= \frac{\delta(\sum_j a_{jj}^Y(x,y) + 2b(x,y)\cdot y + 2p^T a^{XY}(x,y)y)}{(1+|y|^2)^{1-\delta}} \\ &\qquad + \frac{2(\delta - (1-\delta)(1+|y|^2)^{-\delta})y^T a^Y(x,y)y}{(1+|y|^2)^{2-2\delta}}\end{aligned}$$

and noting that $(1+|y|^2)^{1-\delta_2} \leq (1+|y|^2)^{1-\delta_1}$, the lemma follows. □

REMARK 11.55. Let $|a|$ denote the matrix norm $\sqrt{\sum_{ij} a_{ij}^2}$. Assume that $\inf_{x,y} |a^Y(x,y)| > 0$ and that for each $k > 0$,

$$\lim_{|y|\to\infty} \sup_{|x|,|p|\leq k,x} \frac{(b^T(x,y) + p^T a^{XY}(x,y))\frac{y}{|y|}}{|a^Y(x,y)|} = -\infty. \tag{11.97}$$

Then (11.96) is satisfied with $\delta_1 = 1/2$. This condition is a slightly generalized version of (2) in [**126**].

11.6.1. Convergence of H_n. For $f \in D_d$ and $h \in D_{d+l}$, define $f_n(x,y) = f(x) + n^{-1}h(x,y)$. Then the pointwise limit of $H_n f_n$ is

$$\begin{aligned} Hf(x,y) &= F(x,y)\cdot\nabla_x f(x) + \frac{1}{2}\sum a_{ij}^X(x,y)\partial_{x_i}f(x)\partial_{x_j}f(x) \\ &\quad + \frac{1}{2}\sum a_{ij}^Y(x,y)\partial_{y_i}\partial_{y_j}h(x,y) + b(x,y)\cdot\nabla_y h(x,y) \\ &\quad + \sum a_{ij}^{XY}(x,y)\partial_{x_i}f(x)\partial_{y_j}h(x,y) \\ &\quad + \frac{1}{2}\sum a_{ij}^Y(x,y)\partial_{y_i}h(x,y)\partial_{y_j}h(x,y). \end{aligned}$$

We must, however, verify the Convergence Condition 7.11. Let $E_n = R^d \times R^l$ and $E_0 = R^l \cup \{\infty\}$, the one point compactification of R^l, and take $E' = R^d \times E_0$.

Assume Condition 11.51. For $q > 0$, define $\widetilde{K}_n^q = \{y : \varphi_2(y) \le e^{qn}\}$ and $K_n^q = \{(x,y) : |x| \le q, y \in \widetilde{K}_n^q\}$. Let η' be the natural injection of E_n into E', and let $\gamma(x,y) = x$. As in Section 11.1.1, Conditions 2.8 and 7.9 are satisfied, and we can define LIM convergence according to Definition 2.5.

For $n > 2$, define $\varphi_n \in D_l^{++}$ as in (11.41) so that $\varphi_n(y) = \varphi_1(y)$, $y \in \widetilde{K}_n^q$, for all n sufficiently large and

$$\lim_{n\to\infty} \frac{1}{n}\log\|\varphi_n\| = 0.$$

Let $f \in D_d$ and $g \in D_l^{++}$, and for $\kappa \in (0,1)$, define $g_{n,\kappa}(y) = g(y)^{1-\kappa}\varphi_n(y)^\kappa$ and

$$f_n(x,y) = f(x) + \frac{1}{n}\log g_{n,\kappa}(y).$$

$H_n f_n(x,y)$ becomes

$$\begin{aligned} H_n f_n(x,y) &= \frac{1}{2n}\sum_{i,j} a_{ij}^X(x,y)\partial_{x_i}\partial_{x_j}f(x) + F(x,y)\cdot\nabla_x f(x) \\ &\quad + \frac{1}{2}\sum_{i,j} a_{ij}^Y(x,y)\partial_{y_i}\partial_{y_j}\log g_{n,\kappa}(y) \\ &\quad + b(x,y)\cdot\nabla_y \log g_{n,\kappa}(y) \\ &\quad + \frac{1}{2}\nabla_x f(x)^T a^X(x,y)\nabla_x f(x) \\ &\quad + \nabla_x f(x)^T a^{XY}(x,y)\nabla_y \log g_{n,\kappa}(y) \\ &\quad + \frac{1}{2}\nabla_y \log g_{n,\kappa}(y)^T a^Y(x,y)\nabla_y \log g_{n,\kappa}(y) \\ &= \frac{1}{2n}\sum_{i,j} a_{ij}^X(x,y)\partial_{x_i}\partial_{x_j}f(x) + F(x,y)\cdot\nabla_x f(x) \\ &\quad + \frac{1}{2}\nabla_x f(x)^T a^X(x,y)\nabla_x f(x) + \frac{B_{x,\nabla f(x)}g_{n,\kappa}(y)}{g_{n,\kappa}(y)}. \end{aligned}$$

For fixed x and p, $B_{x,p}$ is the generator for a Markov process, and applying Lemma 11.26, we have

$$\begin{aligned}&H_nf_n(x,y)\\ &\quad\leq F(x,y)\cdot\nabla_x f(x)+\frac{1}{2}\nabla_x f(x)^T a^X(x,y)\nabla_x f(x)+(1-\kappa)\frac{B_{x,\nabla f(x)}g(y)}{g(y)}\\ &\qquad+\kappa\frac{B_{x,\nabla f(x)}\varphi_n(y)}{\varphi_n(y)}+\frac{1}{2n}\sum a_{ij}^X(x,y)\partial_{x_i}\partial_{x_j}f(x).\end{aligned}$$

Similarly, if we take $\widetilde{g}_{n,\kappa}(x,y)=g(y)^{1+\kappa}\varphi_n(y)^{-\kappa}$, $\kappa>0$, and

$$\widetilde{f}_n(x,y)=f(x)+\frac{1}{n}\log\widetilde{g}_{n,\kappa}(y),$$

then

$$\begin{aligned}&H_n\widetilde{f}_n(x,y)\\ &\quad\geq F(x,y)\cdot\nabla_x f(x)+\frac{1}{2}\nabla_x f(x)^T a^X(x,y)\nabla_x f(x)+(1+\kappa)\frac{B_{x,\nabla f(x)}g(y)}{g(y)}\\ &\qquad-\kappa\frac{B_{x,\nabla f(x)}\varphi_n(y)}{\varphi_n(y)}+\frac{1}{2n}\sum a_{ij}^X(x,y)\partial_{x_i}\partial_{x_j}f(x).\end{aligned}$$

Let $\widetilde{H}(x,y,p)=F(x,y)\cdot p+\frac{1}{2}p^Ta^X(x,y)p$ and $\psi_{x,p}(y)=\frac{B_{x,p}\varphi_1(y)}{\varphi_1(y)}$. For $0<\kappa<1$, define

$$G_{1,f,g,\kappa}(x,y)=\begin{cases}\widetilde{H}(x,y,\nabla_x f(x))+(1-\kappa)\frac{B_{x,\nabla_x f(x)}g(y)}{g(y)}+\kappa\psi_{x,\nabla_x f(x)}(y), & y\in R^l\\ -\infty, & y=\infty,\end{cases}$$

and for $\kappa>0$, define

$$G_{2,f,g,\kappa}(x,y)=\begin{cases}\widetilde{H}(x,y,\nabla_x f(x))+(1+\kappa)\frac{B_{x,\nabla_x f(x)}g(y)}{g(y)}-\kappa\psi_{x,\nabla_x f(x)}(y), & y\in R^l\\ \infty, & y=\infty.\end{cases}$$

Set

$$H_\dagger=\{(f,G_{1,f,g,\kappa}):f\in D_d,g\in D_l^{++},\kappa\in[0,1)\}$$

and

$$H_\ddagger=\{(f,G_{2,f,g,\kappa}):f\in D_d,g\in D_l^{++},\kappa\in[0,1)\}.$$

As in Section 11.1.1, the Convergence Condition 7.11 holds.

11.6.2. Exponential tightness. The exponential compact containment condition follows by Condition 11.51.2. Exponential tightness for $\{X_n\}$ follows by the convergence of H_n and Corollary 4.17.

11.6.3. The comparison principle. Let

(11.98)
$$H_1(x,p)=\inf_{\kappa\in(0,1)}\inf_{g\in D_l^{++}}\sup_{y\in R^l}(F(x,y)\cdot p+\frac{1}{2}p^Ta^X(x,y)p+(1-\kappa)\frac{B_{x,p}g(y)}{g(y)}+\kappa\psi_{x,p}(y)),$$

(11.99)
$$H_2(x,p)=\sup_{\kappa>0}\sup_{g\in D_l^{++}}\inf_{y\in R^l}(F(x,y)\cdot p+\frac{1}{2}p^Ta^X(x,y)p+(1+\kappa)\frac{B_{x,p}g(y)}{g(y)}-\kappa\psi_{x,p}(y)),$$

and

(11.100)
$$H_if(x)=H_i(x,\nabla f(x)),\quad f\in D_d.$$

As in the previous examples, the critical hypothesis will be that

$$H_1(x,p) \le H_2(x,p). \tag{11.101}$$

Note that the results of Section 11.2 and Appendix B can be applied to the operator $B_{x,p}$ for fixed x and p to verify (11.101). In particular, we have the following:

LEMMA 11.56. *Assume that for each $x, p \in R^d$, $B_{x,p}$ satisfies the Hölder conditions of Example 11.33 and that Condition 11.51 holds. Then (11.101) holds.*

PROOF. The computations in Example 11.33 verify the conditions of the Lemma 11.32. □

For nondegenerate a^X, the comparison principle is an immediate consequence of Lemma 9.16.

LEMMA 11.57. *Let $h \in C_b(R^d)$ and $\alpha > 0$. Suppose that Condition 11.51 and (11.101) hold. If for each $k > 0$, there exists $\epsilon_k > 0$ such that*

$$c_k \equiv \sup_{|x|\le k, y\in R^l} |F(x,y)| < \infty$$

and

$$\inf_{|x|\le k, y\in R^l} p^T a^X(x,y)p \ge \epsilon_k |p|^2, \quad p \in R^d,$$

then the comparison principle holds for subsolutions of $(I - \alpha H_1)f = h$ and supersolutions of $(I - \alpha H_2)f = h$.

PROOF. For $g \in D_l^{++}$ and $0 < \kappa < 1$, the continuity of the coefficients and the compactness of (11.94) implies

$$h_{g,\kappa}(x,p) \equiv \sup_{y\in R^l} (F(x,y)\cdot p + \frac{1}{2}p^T a^X(x,y)p + (1-\kappa)\frac{B_{x,p}g(y)}{g(y)} + \kappa\psi_{x,p}(y))$$

is upper semicontinuous. Since the infimum of a collection of upper semicontinuous functions is upper semicontinuous, H_1 is upper semicontinuous. Similarly, H_2 is lower semicontinuous. Let $\kappa > 0$ and $g = 1$. Then

$$\begin{aligned} H_2(x,p) &\ge \inf_{y\in R^l} (F(x,y)\cdot p + \frac{1}{2}p^T a^X(x,y)p - \kappa\psi_{x,p}(y)) \\ &\ge -c_k|p| + \frac{1}{2}\epsilon_k|p|^2 - \kappa \sup_y \psi_{x,p}(y), \end{aligned}$$

for $|x| \le k$, and hence

$$\lim_{|p|\to\infty} \inf_{|x|\le k} H_2(x,p) = \infty.$$

Consequently, Lemma 9.16 implies that Condition 9.10.1 holds.

Suppose $|x_m| \to \infty$ and $|x_m||p_m| \to 0$. Then

$$\begin{aligned} &H_1(x_m,p_m) \\ &\quad \le \inf_{0<\kappa<1} \sup_{y\in R^l} (F(x_m,y)\cdot p_m + \frac{1}{2}p_m^T a^X(x_m,y)p_m + \kappa\psi_{x_m,p_m}(y)) \\ &\quad \le \inf_{0<\kappa<1} (c_1 + c_2|x_m|)|p_m| + \frac{1}{2}(c_1+c_2|x_m|)^2|p_m|^2 + \kappa c_3(1+|p_m|)(1+|x_m|)) \\ &\quad \to 0, \end{aligned}$$

and similarly $\liminf H_2(x_m, p_m) \geq 0$, so Condition 9.10.2 holds, and the lemma follows by Lemma 9.15. □

In Section 10.3, we proved the comparison principle under nondegeneracy conditions, as in Lemma 11.57, and under Lipschitz assumptions on the coefficients of the generator. Proving the comparison principle under Lipschitz assumptions in the present context appears to be much more difficult. Here we give some partial results.

As in Lemma 11.35,

$$\begin{aligned} (11.102)\quad H_1(x,p) &= \mathbf{H}(x,p) \\ &\equiv \sup_{\mu\in\mathcal{P}(R^l)} \Big(\int_{R^l} F(x,y)\cdot p\mu(dy) + \frac{1}{2}p^T \big(\int_{R^l} a^X(x,y)\mu(dy)\big)p - I_B(\mu|x,p)\Big) \end{aligned}$$

where

$$I_B(\mu|x,p) = -\inf_{g\in D_l^{++}} \int_{R^l} \frac{B_{x,p}g}{g} d\mu \wedge \int_{R^l} \psi_{x,p} d\mu, \quad x,p\in R^d, \tag{11.103}$$

and we define $\mathbf{H}f(x) = \mathbf{H}(x, \nabla f(x))$. Noting that $D_l^{++} = \{e^g : g \in C_c^2(R^l)\}$, we have

$$I_B(\mu|x,p) = -\inf_{g\in C_c^2(R^l)} \int_{R^l} e^{-g} B_{x,p} e^g d\mu \wedge \int_{R^l} \psi_{x,p} d\mu.$$

For each $g \in D_l^{++}$, $-\int \frac{B_{x,p}g}{g} d\mu$ is continuous in $(x,p,\mu) \in R^d \times R^d \times \mathcal{P}(R^l)$. Hence $I_B(\mu|x,p)$ is lower semicontinuous in (x,p,μ).

Let $\alpha > 0, h \in C_b(E)$. As before, if

$$\mathbf{H}(x,p) = H_1(x,p) \leq H_2(x,p), \quad x,p \in R^d, \tag{11.104}$$

then any subsolution of

$$(I - \alpha H_\dagger)f = h \tag{11.105}$$

is a subsolution of

$$(I - \alpha\mathbf{H})f = h, \tag{11.106}$$

and any supersolution of

$$(I - \alpha H_\ddagger)f = h \tag{11.107}$$

is a supersolution for (11.106). Hence the comparison principle between subsolutions of (11.105) and supersolutions of (11.107) is implied by the comparison principle for the single equation (11.106). In order to prove the comparison principle for (11.106), we need a regularity condition on $I(\mu|x,p)$.

CONDITION 11.58. *For each compact $\Gamma \subset R^d$ and $\lambda > 1$, there exists $\omega_{\Gamma,\lambda}(s,t)$: $[0,\infty)^2 \to [0,\infty]$ with the following properties:*

a)

$$\liminf_{\lambda\to 1+} \lim_{s\to 0+, t\to 0+} \omega_{\Gamma,\lambda}(s,t) = 0$$

b) *For each $\mu \in \mathcal{P}(R^l)$, there exists $\nu \in \mathcal{P}(R^l)$ such that*

$$I(\nu|z,p) - \lambda I(\mu|x,\frac{p}{\lambda}) \leq \omega_{\Gamma,\lambda}(|z-x|, |z-x||p|), \quad x,z\in\Gamma, \tag{11.108}$$

and

$$\int F(x,y)p\mu(dy) - \int F(z,y)p\nu(dy) + \frac{1}{2\lambda}\int |\sigma^X(x,y)^T p|^2\mu(dy) - \frac{1}{2}\int |\sigma^X(z,y)^T p|^2\nu(dy) \le \omega_{\Gamma,\lambda}(|z-x|,|z-x||p|), \quad x,z\in\Gamma. \tag{11.109}$$

Usually, we can choose $\nu = \mu$. In that case (11.109) is implied by local Lipschitz conditions on F and σ^X.

LEMMA 11.59. *If there exists $L_\Gamma > 0$ satisfying*

$$\sup_{y\in R^l} |F(x,y) - F(z,y)| + \sup_{y\in R^l} |\sigma^X(x,y) - \sigma^X(z,y)| \le L_\Gamma |x-z|, \quad x,z\in\Gamma,$$

then (11.109) is satisfied with $\omega_{\Gamma,\lambda}$ of the form

$$K_\Gamma(|x-y||p| + \frac{4\lambda^2+\lambda-1}{2\lambda(\lambda-1)}|x-z|^2|p|^2)$$

PROOF. The necessary estimate on the first term on the left of (11.109) is immediate. To estimate the second, note that $-au^2 + 2bu \le b^2/a$ for $a > 0$. Then for $x, z \in \Gamma$ and $\lambda > 1$,

$$\begin{aligned}
&\frac{1}{2\lambda}\int |\sigma^X(x,y)^T p|^2\mu(dy) - \frac{1}{2}\int |\sigma^X(z,y)^T p|^2\nu(dy)\\
&= \int \Big(\frac{1}{2\lambda}|(\sigma^X(x,y)^T - \sigma^X(z,y)^T + \sigma^X(z,y))p|^2 - \frac{1}{2}|\sigma^X(z,y)^T p|^2\Big)\mu(dy)\\
&\le \int \Big(\frac{1}{2\lambda}((L_\Gamma|x-z||p|)^2 + 2L_\Gamma|x-z||p||\sigma^X(x,y)^T p|)\\
&\qquad\qquad -\frac{\lambda-1}{2\lambda}|\sigma^X(z,y)^T p|^2\Big)\mu(dy)\\
&\le \frac{4\lambda^2+\lambda-1}{2\lambda(\lambda-1)}L_\Gamma^2|x-z|^2|p|^2.
\end{aligned}$$

□

Verification of (11.108) is more difficult. If $B_{x,p}$ generates a reversible process with respect to a stationary distribution, then usually $I(\mu|x,p)$ can be computed through its associated Dirichlet form (e.g. Theorem 7.44 of Stroock [**115**]). Such an explicitly parameterized representation (in x, p) of I may be helpful in verifying (11.108). Sometimes, (11.108) can be verified directly. One set of conditions is given in the following lemma.

LEMMA 11.60. *Assume that*

(1) *$\sigma^Y(x,y) = \sigma^Y(y)$ is independent of x.*

(2) *a^Y is uniformly nondegenerate*

$$a^Y(y) = \sigma^Y(y)\sigma^Y(y)^T \ge \epsilon_0 I, \qquad \text{some } \epsilon_0 > 0. \tag{11.110}$$

(3) *$\sigma^X(x,y)$ is locally Lipschitz in x uniformly in y*

$$\sup_{|x|+|z|\le N, x\ne z; y\in R^l} \frac{|\sigma^X(x,y) - \sigma^X(z,y)|}{|x-z|} < \infty, \quad \forall N > 0;$$

(4) *$b(x,y)$ is locally Lipschitz in x, uniformly in y*

$$\sup_{|x|+|z|\leq N, x\neq z; y\in R^l} \frac{|(b(x,y)-b(z,y))|}{|x-z|} < \infty, \qquad N>0;$$

(5) *$p^T\sigma^X(\sigma^Y)^T = 0$ for all $p \in R^d$.*

Then Condition 11.58 is satisfied with

$$\omega_{\Gamma,\lambda}(|x-z|, |x-z||p|) = L_\Gamma \frac{\lambda}{\lambda-1}|x-z|^2.$$

REMARK 11.61. If $p^T\sigma^X(\sigma^Y)^T = 0$ for all $p \in R^d$, then $B_{x,p} = B_x$ and $I(\mu|x,p) = I(\mu|x)$ do not depend on p.

PROOF. Let $\Gamma \subset R^d$ be compact, $\lambda > 1$, $x,z \in \Gamma$, and $p \in R^d$. By definition, $I(\mu|z,p) = \sup_{g\in C_c^2(R^l)} -\int e^{-g}B_{z,p}e^g d\mu$. For each $\mu \in \mathcal{P}(R^l)$, $\epsilon > 0$, there exists a $g_0 \in C_c^2(R^d)$ depending on μ, z, p, ϵ such that

$$I(\mu|z) \leq \epsilon - \int e^{-g_0(y)}B_z e^{g_0}(y)\mu(dy).$$

Hence

$$\begin{aligned}
&I(\mu|z) - \lambda I(\mu|x)\\
&\leq \epsilon - \int e^{-g_0(y)}B_z e^{g_0}(y)\mu(dy) + \lambda \inf_{g\in C_c^2(R^l)} \int e^{-g(y)}B_x e^g(y)\mu(dy)\\
&\leq \epsilon + \lambda \int \Big(\frac{1}{2}\sum_{i,j=1}^{l} a_{ij}^Y(y)\partial_{y_i}\partial_{y_j}(\lambda^{-1}g_0)(y) + \frac{1}{2}|(\sigma^Y(y))^T\cdot\nabla_y(\lambda^{-1}g_0)(y)|^2\\
&\qquad + b(x,y)\cdot\nabla_y(\lambda^{-1}g_0)(y)\Big)\mu(dy)\\
&\qquad - \int \Big(\frac{1}{2}\sum_{i,j=1}^{l} a_{ij}^Y(y)\partial_{y_i}\partial_{y_j}g_0(y) + \frac{1}{2}|(\sigma^Y(y))^T\cdot\nabla_y g_0(y)|^2\\
&\qquad + b(z,y)\cdot\nabla_y g_0(y)\Big)\mu(dy)\\
&= \epsilon + \int \Big((b(x,y)-b(z,y))\nabla g_0(y) + \frac{1}{2}\frac{1-\lambda}{\lambda}|\sigma^Y(y)^T\nabla g_0(y)|^2\Big)\mu(dy)\\
&\leq \epsilon + \int \Big(|b(x,y)-b(z,y)||\nabla g_0(y)| - \epsilon_0\frac{\lambda-1}{2\lambda}|\nabla g_0(y)|^2\Big)\mu(dy)\\
&\leq \epsilon + \int \frac{\lambda|b(x,y)-b(z,y)|^2}{2\epsilon_0(\lambda-1)}\mu(dy),
\end{aligned}$$

where $\epsilon_0 > 0$ is given by (11.110). Note that the integral in the last expression does not depend on ϵ, so by the Lipschitz continuity of b, there exists $L_\Gamma \geq 0$ such that for $x,z \in \Gamma$ and $\lambda > 1$,

$$I(\mu|z) - \lambda I(\mu|x) \leq L_\Gamma \frac{\lambda}{\lambda-1}|x-z|^2 \tag{11.111}$$

and (11.108) holds. □

CONDITION 11.62. *For each $x,p \in R^d$, there exists $\pi_{x,p} \in \mathcal{P}(R^l)$ such that $I(\pi_{x,p}|x,p) = 0$.*

Under condition Conditions 11.51.2, $B_{x,p}$ has a stationary distribution $\pi_{x,p}$ and $I(\pi_{x,p}|x,p) = 0$. (See Lemmas 11.23, 11.2, and 11.38.)

LEMMA 11.63. *Assume that Conditions 11.51.1, 11.58 and 11.62 hold, and (11.104) is satisfied. Let $\alpha > 0$ and $h \in C_b(E)$. Then the comparison principle holds for $(I - \alpha\mathbf{H})f = h$, and hence, the comparison principle holds between subsolutions of $(I - \alpha H_\dagger)f = h$ and supersolutions of $(I - \alpha H_\ddagger)f = h$.*

PROOF. Let $x, p \in R^d$, $\mu \in \mathcal{P}(R^l)$, and denote

$$F(x,\mu) = \int F(x,y) \cdot p\mu(dy), \qquad a^X(x,\mu) = \int a^X(x,y)\mu(dy).$$

Let $\lambda > 1$. Then

$$\begin{aligned}
&\lambda\mathbf{H}(x, \frac{p}{\lambda}) - \mathbf{H}(z,p) \\
&= \sup_{\mu\in\mathcal{P}(R^l)} \inf_{\nu\in\mathcal{P}(R^l)} \Big(F(x,\mu)\cdot p + \frac{1}{2\lambda}p^T a^X(x,\mu)p - \lambda I(\mu|x,\frac{p}{\lambda}) \\
&\qquad\qquad -F(z,\nu)\cdot p - \frac{1}{2}p^T a^X(z,\nu)p + I(\nu|z,p)\Big) \\
&\le \sup_{\mu\in\mathcal{P}(R^l)} \Big((F(x,\mu) - F(z,\nu))\cdot p + \frac{1}{2\lambda}p^T a^X(x,\mu)p - \frac{1}{2}p^T a^X(z,\nu)p \\
&\qquad\qquad -\lambda I(\mu|x,\frac{p}{\lambda}) + I(\nu|z,p)\Big)
\end{aligned}$$

for any $\nu \in \mathcal{P}(R^l)$. For compact $\Gamma \subset R^d$ and $x, z \in \Gamma$, Condition 11.58 gives

$$\lambda\mathbf{H}(x,\frac{p}{\lambda}) - \mathbf{H}(z,p) \le 2\omega_{\Gamma,\lambda}(|z-x|, |z-x||p|). \tag{11.112}$$

Hence Condition 9.10.1 is satisfied.

Noting that $I(\mu|x,p) \ge 0$, Conditions 11.51.1 and 11.62 imply that Condition 9.10.2 is satisfied.

By Lemma 9.15, the comparison principle holds for $\mathbf{H}$. □

11.6.4. Variational representation. If $I(\mu|x,p)$ does not depend on p, then

$$\mathbf{H}(x,p) = \sup_{\mu\in\mathcal{P}(R^l)} \Big(\int_{R^l} F(x,y)\cdot p\mu(dy) + \frac{1}{2}p^T(\int_{R^l} a^X(x,y)\mu(dy))p - I_B(\mu|x)\Big),$$

and $\mathbf{H}(x,p)$ is convex in p. This is the case when $a^{XY} = 0$, or slightly more generally, when $a^{XY}\cdot q = 0$ for every $q \in R^d$. Frequently, $\mathbf{H}(x,p)$ is also continuous in (x,p). This is the case, for example, when Condition 11.58 is satisfied (hence (11.112) holds) and when $I(\mu|x,p)$ is p-independent. If $\mathbf{H}(x,p)$ is both continuous in $(x,p) \in R^d \times R^d$ and is convex in p, the large deviation rate function admits a simple variational representations.

Let $\varphi(x) = 1 + |x|$. We define

$$L_\varphi(x,q) = \sup_{p\in R^d} (p\cdot q - \mathbf{H}(x, \frac{p}{\varphi(x)})), \quad x \in R^d.$$

If $\mathbf{H}(x,p)$ is convex in p, then

$$\mathbf{H}(x,p) = \sup_{q\in R^d} (\varphi(x)p\cdot q - L_\varphi(x,q)), \quad x \in R^d.$$

As in Lemma 10.21, we have the following.

LEMMA 11.64. *Suppose that $a^{XY} \cdot q = 0$ for every $q \in R^d$ (hence $I(\mu|x,p) = I(\mu|x)$ and $B_{x,p} = B_x$ do not depend on p) and that $\mathbf{H}(x,p)$ is continuous in $(x,p) \in R^d \times R^d$. Assume that for each $x \in R^d$, there exists a stationary distribution π_x for B_x and that $\int F(x,y)\pi_x(dy)$ is Lipschitz continuous in x. Let $U = R^d$ and*

$$Af(x,q) = \varphi(x)q\nabla f(x), \quad q \in U, f \in D_d,$$

where $\nabla f(\infty) = 0$, and extend the definition of L_φ by

$$L_\varphi(\infty, q) = \liminf_{|x|\to\infty, q'\to q} L_\varphi(x, q').$$

Then $L_\varphi(x,q) \geq 0$ is lower semicontinuous on $E \times U$,

$$\mathbf{H}f(x) = \sup_{q\in R^d}(Af(x,u) - L_\varphi(x,u)), \quad x \in E, f \in D_d,$$

and Conditions 8.9 and 8.10 are satisfied. Furthermore, Condition 8.11 is satisfied with $\mathbf{H}_\dagger = \mathbf{H}_\ddagger = \mathbf{H}$.

PROOF. As in Lemma 10.21, Condition 8.9 is satisfied. In particular,

$$\lim_{N\to\infty} \inf_{x\in R^d} \inf_{|q|=N} \frac{L_\varphi(x,q)}{|q|} = \infty,$$

Condition 8.30 holds.

We note that

$$\mathbf{H}(x,p) \geq \int F(x,y) \cdot p\pi_x(dy)$$

and that $L_\varphi(x,q) \geq 0$. For $q(x) = \int F(x,y)\pi_x(dy)$, we must have

$$L_\varphi(x, q(x)) = \sup_{p\in R^d}(p \cdot \int F(x,y)\pi_x(dy) - \mathbf{H}(x, \frac{p}{\varphi(x)}) = 0.$$

Hence Condition 8.10 holds with

$$x(t) = x_0 + \int_0^t F(x(s),y)\pi_{x(s)}(dy)$$

if $x_0 \in R^d$, and $x(t) = \infty$ if $x_0 = \infty$.

By Lemma 8.32, Conditions 8.9 and 8.11 hold. □

If $\sigma^X = 0$, then (11.102) simplifies to

$$\mathbf{H}(x,p) = \sup_{\mu\in\mathcal{P}(R^l)} \{\int F(x,y) \cdot p\mu(dy) - I(\mu|x)\}, \tag{11.113}$$

and a modification of Lemma 11.15 gives a representation of the form

$$I(x) = \inf\{I_0(x(0)) + \int_0^\infty I(\mu_s|x(s))ds :$$

$$x(t) = x(0) + \int_0^t \int_{E_0} F(x(s),y)\mu_s(dy)ds\}; \tag{11.114}$$

however, because of the x-dependency in $I(\mu|x)$, the conditions are more complicated, and we do not pursue this representation further.

11.6.5. The large deviation theorem. We have the following large deviation theorem.

THEOREM 11.65. *Suppose that Conditions 11.51, 11.58, and 11.62 hold, and inequality (11.104) is satisfied. Then*

(1) *$\{X_n\}$ satisfies the large deviation principle in $C_{R^d}[0,\infty)$ with a good rate function.*
(2) *Suppose that the conditions of Lemma 11.64 are satisfied. Then for L given by*

$$L(x,p) = \sup_{p\in R^d} (p\cdot q - \mathbf{H}(x,p)),$$

we have

$$I(x) = \begin{cases} I_0(x(0)) + \int_0^\infty L(x(s),\dot{x}(s))ds & \text{for } x \text{ absolutely continuous} \\ \infty & \text{otherwise.} \end{cases} \tag{11.115}$$

CHAPTER 12

Occupation measures

Large deviations for occupation measures, such as Example 1.11, have been studied by many authors beginning with the fundamental work of Donsker and Varadhan [**33**]. The books [**52, 57, 29, 30, 35**] discuss these results at length. In this chapter, we derive large deviation results for occupation measures (both continuous- and discrete-time) as corollaries of the random evolution results in Chapter 11. Theorem 12.4 gives the main discrete time result and Theorem 12.7 the main continuous time result. The critical assumptions are again the inequalities (12.10) and (12.19). Conditions implying these inequalities and their relationship to earlier work are discussed in Chapter 11 and Appendix B.

Let E_0 be a complete, separable metric space, and let $\mathcal{M}_f(E_0)$ be the space of finite, Borel measures on E_0. Define $E = \mathcal{M}_f(E_0)$. We take the weak topology on E, so that $\mu_n \to \mu$ if and only if $\int_{E_0} f d\mu_n \to \int_{E_0} f d\mu$ for every $f \in C_b(E_0)$, and set $E_n = E_0 \times E$.

In the continuous time case, let Y, $\{S(t)\}$, and B be as in Section 11.2. Define $Y_n(t) = Y(nt)$ and

$$Z_n(C,t) = \frac{1}{n}\int_0^{nt} I_C(Y(s))ds = \int_0^t I_C(Y_n(s))ds. \tag{12.1}$$

Then X_n defined by $X_n(t) = (Y_n(t), Z_n(\cdot,t))$ is a Markov process with state space E_n.

We can consider Z_n as a random variable with values in

$$\mathcal{L}(E_0) = \{z \in \mathcal{M}(E_0 \times [0,\infty)) : z(E_0 \times [0,t]) = t, t \geq 0\}.$$

(Topologize $\mathcal{L}(E_0)$ by weak convergence on bounded time intervals, that is, $z_n \to z$ if and only if

$$\int_{E_0 \times [0,t]} f(u,s) z_n(du \times ds) \to \int_{E_0 \times [0,t]} f(u,s) z(du \times ds)$$

for all $f \in C_b(E_0 \times [0,\infty)$ and $t \geq 0$.) Let $M_{\mathcal{P}(E_0)}[0,\infty)$ be the space of measurable, $\mathcal{P}(E_0)$-valued functions on $[0,\infty)$. If $z \in \mathcal{L}(E_0)$, then there exists $\mu \in M_{\mathcal{P}(E_0)}[0,\infty)$ such that

$$z(C \times [0,t]) = \int_0^t \mu(C,s)ds, \quad C \in \mathcal{B}(E_0), \tag{12.2}$$

and we will write $\dot{z}(t) = \mu(t)$. Conversely, if $\mu \in M_{\mathcal{P}(E_0)}[0,\infty)$, then (12.2) defines an element of $\mathcal{L}(E_0)$. We will write $z(t) = \int_0^t \mu(s)ds$.

In the discrete time case, let $\widetilde{Y} = \{\widetilde{Y}(k), k \geq 0\}$ be an E_0-valued Markov chain with one step transition probability $P(x, dy)$, and define

$$Z_n(C,t) = \frac{1}{n}\sum_{k=0}^{[nt]-1} I_C(\widetilde{Y}(k)). \tag{12.3}$$

Setting $Y_n(t) = \widetilde{Y}([nt])$, we have

$$Z_n(C,t) = \int_0^{\frac{[nt]}{n}} I_C(Y_n(s))ds.$$

It follows that $X_n(t) = (Y_n(t), Z_n(\cdot, t))$ is a Markov chain with time-step $\frac{1}{n}$. As in the continuous time case, we can consider Z_n as a random variable with values in

$$\mathcal{L}_n(E_0) = \{z \in \mathcal{M}(E_0 \times [0,\infty)) : z(E_0 \times [0,t]) = [nt]/n, t \geq 0\}.$$

In both cases, we assume $\widetilde{Y}$ and Y are Feller processes, and we are interested in large deviations for $\{Z_n\}$.

12.1. Occupation measures of a Markov process - Discrete time

Define

$$Pf(x) = \int_{E_0} f(y)P(x,dy),$$

for each $f \in B(E_0)$. We assume the Feller property $P : C_b(E_0) \mapsto C_b(E_0)$.

12.1.1. Convergence of H_n. We modify the definition of Z_n to allow for a nonzero initial value, so

$$Z_n(C,t) = Z_n(C,0) + \frac{1}{n}\sum_{k=0}^{[nt]-1} I_C(\widetilde{Y}(k)).$$

Let T_n be the transition operator for the chain $X_n = (Y_n, Z_n)$ with step-size $\frac{1}{n}$. Then

$$E[f(Y_n(t), Z_n(t))|Y_n(0) = y, Z_n(0) = z] = T_n^{[nt]}f(y,z), \qquad f \in B(E_n).$$

Let $\beta_i \in B(E_0)$, $i = 1, \ldots, d$. To simplify notation, for $\beta = (\beta_1, \ldots, \beta_d)$, we define

$$\langle \beta, z\rangle = (\langle \beta_1, z\rangle, \ldots, \langle \beta_d, z\rangle), \quad z \in \mathcal{M}_f(E_0).$$

If $f(y,z) = f_0(y)f_1(\langle\beta, z\rangle)$ for $f_0 \in B(E_0)$ and $f_1 \in C(R^d \cup \{\infty\})$, then

$$T_nf(y,z) = f_1(\langle\beta,z\rangle + \frac{1}{n}\beta(y))Pf_0(y),$$

and if

$$f_n(y,z) = f_1(\langle\beta,z\rangle) + \frac{1}{n}f_0(y),$$

then

$$\begin{aligned} H_nf_n(y,z) &\equiv \log(e^{-nf_n}T_ne^{nf_n})(y,z) \\ &= \log(e^{nf_1(\langle\beta,z\rangle+\frac{1}{n}\beta(y))-nf_1(\langle\beta,z\rangle)}e^{-f_0(y)}Pe^{f_0}(y)). \end{aligned} \tag{12.4}$$

Assume that $f_1 \in D_d$, and let $g = e^{f_0}$. Then $f_n \to f$ given by

$$f(z) = f_1(\langle\beta,z\rangle) \tag{12.5}$$

and $H_n f_n \to h$ given by

$$h(y,z) = \nabla f_1(\langle \beta, z\rangle)\cdot\beta(y) + \log\frac{Pg(y)}{g(y)}, \quad (y,z)\in E_0\times E. \tag{12.6}$$

In particular,

$$\sup_{y\in E_0, z\in E} |H_n f_n(y,z)| < \infty. \tag{12.7}$$

12.1.2. Exponential tightness.

LEMMA 12.1. *Let φ be a Borel measurable function on E_0 satisfying*

$$\rho \equiv \inf_{y\in E_0}\varphi(y) > 0, \quad \alpha_K \equiv \sup_{y\in K}\varphi(y) < \infty$$

for each compact $K\subset E_0$, and

$$K_a = \{y\in E_0 : \frac{\varphi}{P\varphi}(y) \le a\}$$

compact for each $0<a<\infty$. If $\{Y_n(0)\}$ is exponentially tight, then $\{Z_n\}$ satisfies the exponential compact containment condition.

Similarly, if Conditions 11.1.4 and 11.1.5 hold and for each $a>0$, there exists $q\in Q$ such that

$$\limsup_{n\to\infty}\frac{1}{n}\log P\{Y_n(0)\notin K\cap\widetilde{K}_n^q\} \le -a, \tag{12.8}$$

then $\{Z_n\}$ satisfies the exponential compact containment condition.

PROOF. Let $\varphi_m = \varphi\wedge m$ and $\psi_m = \log\frac{P\varphi_m}{\varphi_m}$. Then

$$\exp\{-n\langle\psi_m, Z_n(t)\rangle + \log\varphi_m(Y_n(t)) - \log\varphi_m(Y_n(0))\}$$

is a martingale. Let $m\to\infty$. Then by Fatou's lemma,

$$U_n(t) = \exp\{-n\langle\log\frac{P\varphi}{\varphi}, Z_n(t)\rangle + \log(\varphi(Y_n(t)) - \log\varphi(Y_n(0))\}$$

is a cadlag super-martingale. Consequently, for each $T>0$ and $L>0$,

$$\begin{aligned} P\{\inf_{0\le t\le T}\langle\log\frac{P\varphi}{\varphi}, Z_n(t)\rangle \le -L\} &\\ \le P\{\sup_{0\le t\le T} U_n(t) \ge e^{nL+\log\rho-\log\alpha_K}\} + P\{Y_n(0)\notin K\} &\\ \le e^{-nL-\log\rho+\log\alpha_K} + P\{Y_n(0)\notin K\}. & \end{aligned}$$

Since $\log\frac{P\varphi}{\varphi}$ is bounded above and $\{y : \log\frac{P\varphi(y)}{\varphi(y)} \ge -L\}$ is compact, for each $0<L<\infty$,

$$K_L \equiv \{z : \langle\log\frac{P\varphi}{\varphi}, z\rangle \ge -L\}$$

is compact in $\mathcal{M}_f(E_0)$, and we have

$$\limsup_{n\to\infty}\frac{1}{n}\log P\{\exists t\le T \ni Z_n(t)\notin K_L\} \le \max\{-L, \limsup_{n\to\infty}\frac{1}{n}\log P\{Y_n(0)\notin K\}\}.$$

The exponential compact containment condition follows by the fact that L is arbitrary and $\{Y_n(0)\}$ is exponentially tight.

The second statement follows by a similar argument. (See Lemma 12.5.) □

$\mathcal{D}(H) \subset C_b(E)$ is closed under addition and separates points. By (12.7), the exponential tightness of $\{Z_n\}$ follows by Corollary 4.17.

12.1.3. Convergence of nonlinear semigroups. In this section, we suppose that the Y satisfies Conditions 11.1.2, 11.1.4 and 11.1.5 in Section 11.1. Let $X_n^\beta(t) = \langle \beta, Z_n(t)\rangle$. Then (X_n^β, Y_n) is a special case of the random evolution models considered in that section.

Since (X_n^β, Y_n) is a Markov process,

$$U_n^\beta(t)h(x,y) = \frac{1}{n}\log E[e^{nh(X_n^\beta(t),Y_n(t))}|X_n^\beta(0) = x, Y_n(0) = y],$$

defines a semigroup on $B(\left(R^d \cup \{\infty\}\right) \times E_0)$, and letting

$$V_n(t)f(y,z) = \frac{1}{n}\log E[e^{nf(Y_n(t),Z_n(t))}|Y_n(0) = y, Z_n(0) = z],$$

for $f_\beta(y,z) = h(\langle\beta,z\rangle, y)$,

$$V_n(t)f_\beta(y,z) = U_n^\beta(t)h(\langle\beta,z\rangle,y). \tag{12.9}$$

Let ψ be given by Condition 11.1.5. Define

$$H_1^\beta(p) = \inf_{\kappa\in[0,1)}\inf_{g\in K_1(E_0)}\sup_{y\in E_0}(\beta(y)\cdot p + (1-\kappa)\log\frac{Pg(y)}{g(y)} + \kappa\psi(y)), \quad p \in R^d,$$

and

$$H_2^\beta(p) = \sup_{\kappa\in[0,1)}\sup_{g\in K_1(E_0)}\inf_{y\in E_0}(\beta(y)\cdot p + \frac{1}{1-\kappa}\log\frac{Pg(y)}{g(y)} - \frac{\kappa}{1-\kappa}\psi(y)), \quad p \in R^d.$$

Suppose, as in (11.16),

$$H_1^\beta \le H_2^\beta \tag{12.10}$$

Then by Theorem 11.20, there exists a semigroup $U^\beta(t)$ on $C(R^d \cup \{\infty\})$ such that

$$\begin{aligned}\lim_{n\to\infty}&\sup_{(z,y)\in E\times\widetilde{K}_n^q}|U_n^\beta(t)f_n(\langle\beta,z\rangle,y) - U^\beta(t)f(\langle\beta,z\rangle)|\\ &+\sup_{(z,y)\in E\times\widetilde{K}_n^q}|U_n^\beta(t+n^{-1})f_n(\langle\beta,z\rangle,y) - U^\beta(t)f(\langle\beta,z\rangle)| = 0, \quad q\in\mathcal{Q},\end{aligned} \tag{12.11}$$

for $\{\widetilde{K}_n^q, q \in \mathcal{Q}\}$ in Condition 11.1.4, where $f_n \in B(\left(R^d \cup \{\infty\}\right) \times E_0)$ and $f \in C(R^d \cup \{\infty\})$ satisfy

$$\lim_{n\to\infty}\sup_{x\in R^d\cup\{\infty\}, y\in K_n^q}|f_n(x,y) - f(x)| = 0, \quad \forall q \in \mathcal{Q}.$$

By (12.9), it follows that for $f_\beta(z) = h(\langle\beta,z\rangle)$, $h \in C(R^d \cup \{\infty\})$,

$$V(t)f_\beta(z) = \lim_{n\to\infty}V_n(t)f_\beta(z) = \lim_{n\to\infty}U_n^\beta(t)h(\langle\beta,z\rangle) = U^\beta(t)h(\langle\beta,z\rangle).$$

Let

$$D(E) = \{h(\langle\beta,\cdot\rangle) : h \in D_d, \beta \in C(E_0)^d, d = 1,2,\ldots\}.$$

Clearly, V extends to a contraction semigroup on $\overline{D}(E)$. By the contraction property of $V_n(t)$ and the convergence in (12.11), for $g \in \overline{D}(E)$ and $g_n \in B(E_0 \times E)$ satisfying $g = \mathrm{LIM}g_n$,

$$V(t)g = \mathrm{LIM}V_n(t)g_n = \mathrm{LIM}V_n(t+\frac{1}{n})g_n,$$

where LIM is given by Definition 2.5 with $K_n^q = (R^d \cup \{\infty\}) \times \widetilde{K}_n^q$.

12.1.4. Variational representation. As in Section 11.1, define

$$\mathbf{H}^\beta(p) = \sup_{\mu \in \mathcal{P}(E_0)} (\langle \beta, \mu \rangle \cdot p - I_P(\mu)) \tag{12.12}$$

where

$$I_P(\mu) = -\inf_{K_1(E_0)} \int_{E_0} \log \frac{Pg}{g} d\mu \wedge \int_{E_0} \psi d\mu.$$

Then under Conditions 11.1.2, 11.1.4 and 11.1.5

$$H_1^\beta = \mathbf{H}^\beta, . \tag{12.13}$$

LEMMA 12.2. *If Conditions 11.1.2, 11.1.4 and 11.1.5 hold and (12.10) holds for each $\beta \in C_b(E_0)^d$, $d = 1, 2, \ldots$, then Γ_P^M given by (11.23) is compact for each $M > 0$ and for $h \in C(R^d \cup \{\infty\})$,*

$$U^\beta(t)h(x) = \sup\{h(x(t)) - \int_0^t I_P(\mu(s))ds : x(t) = x + \int_0^t \langle \beta, \mu(s) \rangle ds\},$$

where the supremum is over $\mu \in M_{\mathcal{P}(E_0)}[0, \infty)$.

PROOF. Since $\mathbf{H}^\beta$depends only on p, the comparison principle follows by Lemma 9.15. The representation follows from Lemma 11.15 and Theorem 8.27, taking

$$\mathbf{H}_\dagger = \mathbf{H}_\ddagger = \mathbf{H}^\beta.$$

□

LEMMA 12.3. *Let $h \in C(R^d \cup \{\infty\})$, $\beta \in C_b(E_0)^d$ and $f_\beta(z) = h(\langle \beta, z \rangle)$. Then $f_\beta \in \overline{D}(E)$, and under the conditions of Lemma 12.2,*

$$\begin{aligned} V(t)f_\beta(z_0) &= U^\beta(t)h(\langle \beta, z_0 \rangle) \\ &= \sup\{f_\beta(z(t)) - \int_0^t I_P(\mu(s))ds : z(t) = z_0 + \int_0^t \mu(s)ds\}, \end{aligned}$$

where the supremum is over $\mu \in M_{\mathcal{P}(E_0)}[0, \infty)$.

PROOF. By the variational representation of $U^\beta(t)$ in Lemma 12.2 and the definition of $V(t)$,

$$\begin{aligned} V(t)f_\beta(z) &= U^\beta(t)h(\langle \beta, z \rangle) \\ &= \sup\{h(x(t)) - \int_0^t I_P(\mu(s))ds : x(t) = \langle \beta, z \rangle + \int_0^t \langle \beta, \mu(s) \rangle ds\} \\ &= \sup\{f_\beta(z(t)) - \int_0^t I_P(\mu(s))ds : z(t) = z + \int_0^t \mu(s)ds\}. \end{aligned}$$

□

12.1.5. The large deviation theorem.

THEOREM 12.4. *Suppose that Conditions 11.1.2, 11.1.4 and 11.1.5 hold and that $\{Y_n(0)\}$ satisfies (12.8). Let Z_n be defined by (12.3). Let $\mathcal{C} \subset C_b(E_0)$ be separating, and suppose that $H_1^\beta \le H_2^\beta$, for $\beta \in \mathcal{C}^d$ and $d = 1, 2, \ldots$. Then $\{Z_n\}$ satisfies a large deviation principle in $C_E[0,\infty)$ with good rate function I given by*

$$I(z) = \begin{cases} \int_0^\infty I_P(\dot{z}(s))ds & \text{for } z \in \mathcal{L}(E_0) \\ \infty & \text{otherwise.} \end{cases} \tag{12.14}$$

PROOF. Theorem 5.15 and Remark 5.16 give the large deviation principle for $\{Z_n\}$ in $D_E[0,\infty)$ with a good rate function I. By Lemma 12.3 and Theorem 8.14, I can be represented as (12.14). □

12.2. Occupation measures of a Markov process - Continuous time

Let Y, $\{S(t)\}$, and B be as in Section 11.2. Define $Y_n(t) = Y(nt)$ and

$$Z_n(C,t) = Z_n(C,0) + \frac{1}{n}\int_0^{nt} I_C(Y(s))ds = \int_0^t I_C(Y_n(s))ds. \tag{12.15}$$

Then X_n defined by $X_n(t) = (Y_n(t), Z_n(\cdot,t))$ is a Markov process with state space $E_n = E_0 \times \mathcal{M}_f(E_0)$.

Let $f_0 \in \mathcal{D}(B)$, $f_1 \in D_d$, and $\beta_i \in C_b(E_0)$, $i = 1, \ldots, d$. Define

$$f(y,z) = f_0(y) f_1(\langle \beta, z\rangle), \quad (y,z) \in E_n,$$

and

$$A_n f(y,z) = n f_1(\langle \beta, z\rangle) B f_0(y) + f_0(y)\beta(y)\cdot\nabla f_1(\langle\beta,z\rangle).$$

Then A_n has a linear extension that is the full generator of X_n. Note that for fixed β, $(Y_n, \langle \beta, Z_n\rangle)$ is a special case of the model considered in Section 11.2 with $F(x,y) = \beta(y)$.

12.2.1. Convergence of H_n.

Convergence follows by essentially the same computation as in Section 11.2. If $f_1 \in D_d$, $\beta_1, \ldots, \beta_d \in C_b(E_0)$, $e^{nf_0} \in \mathcal{D}(B)$, and

$$f(y,z) = f_0(y) + f_1(\langle\beta,z\rangle),$$

then $H_n f = \frac{1}{n} e^{-nf} A_n e^{nf}$ is given by

$$H_n f(y,z) = \beta(y)\cdot\nabla f_1(\langle\beta,z\rangle) + e^{-nf_0(y)} B e^{nf_0}(y).$$

In particular, if $f_1 \in D_d$, $e^{f_0} \in \mathcal{D}(B)$, and

$$f_n(y,z) = f_1(\langle\beta,z\rangle) + \frac{1}{n} f_0(y),$$

then

$$H_n f_n(y,z) = \beta(y)\cdot\nabla f_1(\langle\beta,z\rangle) + e^{-f_0} B e^{f_0}(y). \tag{12.16}$$

Let $g = e^{f_0}$. Then $f_n \to f$ given by

$$f(z) = f_1(\langle\beta,z\rangle), \tag{12.17}$$

and $H_n f_n \to h$ given by

$$h(z,y) = \nabla f_1(\langle\beta,z\rangle)\cdot\beta(y) + \frac{Bg(y)}{g(y)}. \tag{12.18}$$

Let H consist of all pairs (f, h) given by (12.17) and (12.18). Then, by definition,

$$H \subset ex\text{-}\lim_n H_n,$$

where $ex\text{-}\lim_n H_n$ is defined in Definition A.12.

12.2.2. Exponential tightness.

LEMMA 12.5. *Suppose Conditions 11.34.3 and 11.34.4 hold. If for each $a > 0$, there exists compact $q \in \mathcal{Q}$ such that*

$$\limsup_{n\to\infty} \frac{1}{n} \log P\{Y_n(0) \notin K \cap \widetilde{K}_n^q\} \leq -a,$$

then $\{Z_n\}$ satisfies the exponential compact containment condition.

PROOF. For each n,

$$U_n(t) = \frac{\varphi_n(Y_n(t))}{\varphi_n(Y_n(0))} \exp\{-n \int_0^t \frac{B\varphi_n(Y_n(s))}{\varphi_n(Y_n(s))} ds\}$$

is a martingale. Let $\rho = \inf_{n, y\in E_0} \varphi_n(y)$ Then, for each $T > 0$ and $L > 0$,

$$\begin{aligned} P\{\inf_{0\leq t\leq T} \langle \frac{B\varphi_n}{\varphi_n}, Z_n(t)\rangle \leq -L\} &\leq P\{\sup_{0\leq t\leq T} U_n(t) \geq e^{nL+\log\rho-\log\|\varphi_n\|}\} \\ &\leq e^{-nL-\log\rho+\log\|\varphi_n\|}. \end{aligned}$$

Since ψ is bounded above and $\{y : \psi(y) \geq c\}$ is compact for each $c \in R$, for each $L > 0$,

$$K_L \equiv \{z : \langle \psi, z\rangle \geq -L\}$$

is compact in $\mathcal{M}_f(E_0)$. By Condition 11.34.3, there exists $\widehat{q}$ such that

$$\begin{aligned} &\limsup_{n\to\infty} \frac{1}{n} \log P\{\exists t \leq T \ni Z_n(t) \notin K_L\} \\ &\quad \leq \max\{-L, \limsup_{n\to\infty} \frac{1}{n} \log P\{\inf_{0\leq t\leq T} \langle \frac{B\varphi_n}{\varphi_n}, Z_n(t)\rangle \\ &\qquad \leq -L + T \sup_{y\in\widetilde{K}_n^{\widehat{q}}} (\frac{B\varphi_n(y)}{\varphi_n(y)} - \psi(y))\}\} \\ &\quad\ \leq -L \end{aligned}$$

The exponential compact containment condition follows by the fact that L is arbitrary. □

The exponential tightness of $\{Z_n\}$ follows from the convergence of H_n and Corollary 4.17.

12.2.3. Convergence of nonlinear semigroups. For $\beta \in C_b(E_0)^d$, let $X_n^\beta(t) = \langle \beta, Z_n(t)\rangle$. Since (X_n^β, Y_n) is a Markov process,

$$U_n^\beta(t)h(x, y) = \frac{1}{n} \log E[e^{nh(X_n^\beta(t), Y_n(t))} | X_n^\beta(0) = x, Y_n(0) = y],$$

defines a semigroup on $B(\left(R^d \cup \{\infty\}\right) \times E_0)$, and letting

$$V_n(t)f(y, z) = \frac{1}{n} \log E[e^{nf(Y_n(t), Z_n(t))} | Y_n(0) = y, Z_n(0) = z],$$

for $f_\beta(y,z) = h(\langle\beta,z\rangle,y)$,

$$V_n(t)f_\beta(y,z) = U_n^\beta(t)h(\langle\beta,z\rangle,y).$$

We suppose throughout that Y satisfies Conditions 11.21.1, 11.21.3 and 11.21.4. In particular, let ψ be given by Condition 11.21.4. Define

$$H_1^\beta(p) = \inf_{\kappa\in[0,1)} \inf_{g\in\mathcal{D}^{++}(B)} \sup_y (\beta(y)\cdot p + (1-\kappa)\frac{Bg(y)}{g(y)} + \kappa\psi(y)), \quad p \in R^d,$$

and

$$H_2^\beta(p) = \sup_{\kappa\in[0,1)} \sup_{g\in\mathcal{D}^{++}(B)} \inf_y (\beta(y)\cdot p + \frac{1}{1-\kappa}\frac{Bg(y)}{g(y)} - \frac{\kappa}{1-\kappa}\psi(y)), \quad p \in R^d.$$

Suppose

$$H_1^\beta(p) \le H_2^\beta(p) \tag{12.19}$$

for all $p \in R^d$, then by Theorem 11.41, there exists a semigroup $U^\beta(t)$ on $C(R^d \cup \{\infty\})$ such that

$$\lim_{n\to\infty} \sup_{(z,y)\in E\times\widetilde{K}_n^q} |U_n^\beta(t)f_n(\langle\beta,z\rangle,y) - U^\beta(t)f(\langle\beta,z\rangle)| = 0, \tag{12.20}$$

whenever $f_n \in B(\Big(R^d \cup \{\infty\}\Big) \times E_0)$ and $f \in C(R^d \cup \{\infty\})$ satisfy

$$\lim_{n\to\infty} \sup_{x\in R^d\cup\{\infty\}, y\in\widetilde{K}_n^q} |f_n(x,y) - f(x)| = 0, \quad \forall q \in \mathcal{Q}.$$

For $f_\beta(z) = h(\langle\beta,z\rangle)$, $h \in C(R^d \cup \{\infty\})$, define

$$V(t)f_\beta(z) = U^\beta(t)h(\langle\beta,z\rangle),$$

and as before, let

$$D(E) = \{h(\langle\beta,\cdot\rangle) : h \in D_d, \beta \in C(E_0)^d, d = 1,2,\ldots\}.$$

Clearly, V extends to a contraction semigroup on $\overline{D}(E)$. By the contraction property of $V_n(t)$ and the convergence in (12.20), for $g \in \overline{D}(E)$ and $g_n \in B(E_0 \times E)$ satisfying $g = \text{LIM} g_n$, we have

$$V(t)g = \text{LIM}V_n(t)g_n,$$

where the LIM is defined in Definition 2.5.

12.2.4. Variational representation. As in Section 11.2, define

$$\mathbf{H}^\beta(p) = \sup_{\mu\in\mathcal{P}(E_0)} (\langle\beta,\mu\rangle\cdot p - I_B(\mu)), \tag{12.21}$$

where

$$I_B(\mu) = -\inf_{g\in\mathcal{D}^{++}(B)} \int_{E_0} \frac{Bg}{g}d\mu \wedge \int_{E_0} \psi d\mu.$$

Then

$$H_1^\beta = \mathbf{H}^\beta. \tag{12.22}$$

The same arguments as in Section 12.1.4 lead to the following conclusion.

LEMMA 12.6. *Let $h \in D_d$, $\beta \in C_b(E_0)^d$ and $f_\beta(z) = h(\langle \beta, z \rangle)$. If (12.19) holds for all $\beta \in C_b(E_0)^d$, $d = 1, 2, \ldots$, Γ_B^M is compact for all $M > 0$, and there is a stationary distribution for B, then*

$$\begin{aligned} V(t)f_\beta(z_0) &= U^\beta(t)h(\langle \beta, z_0 \rangle) \\ &= \sup\{f_\beta(z(t)) - \int_0^t I_B(\mu(s))ds : z(t) = z_0 + \int_0^t \mu(s)ds\}, \end{aligned}$$

where the supremum is over $\mu \in M_{\mathcal{P}(E_0)}[0, \infty)$.

12.2.5. The large deviation theorem. As in Theorem 12.4, we have the following results.

THEOREM 12.7. *Suppose that Conditions 11.21.1, 11.21.3 and 11.21.4 hold and that $\{Y_n(0)\}$ satisfies the conditions of Lemma 12.5. Let Z_n be defined by (12.15) with $Z_n(0) = 0$. Let $C \subset C_b(E_0)$ be separating, and suppose that $H_1^\beta \leq H_2^\beta$, for $\beta \in C^d$ and $d = 1, 2, \ldots$. Then $\{Z_n\}$ satisfies a large deviation principle in $C_E[0, \infty)$ with good rate function given by*

$$I(z) = \begin{cases} \int_0^\infty I_B(\dot{z}(s))ds & z \in \mathcal{L}(E_0) \\ \infty & \text{otherwise.} \end{cases} \tag{12.23}$$

CHAPTER 13

Stochastic equations in infinite dimensions

In this chapter, we prove large deviation theorems for Examples 1.12, 1.13 and 1.14. Since the state space E is either a function space or the space of probability measures, we use ρ, γ or τ to denote a typical element in E and let x denote an element in the space on which the functions or measures are defined. As discussed in the beginning of Chapter 9, in addition to providing a proof of the large deviation principle, the results in this chapter also imply existence and uniqueness for a number of first order Hamilton-Jacobi equations in infinite dimensions. The methods we use here are new both for large deviations and for partial differential equations. In particular, the probability-measure-valued example in Section 13.3 cannot be treated using existing results in partial differential equations. Our method for this example relies on careful estimates obtained using mass transport techniques (Appendix D).

13.1. Stochastic reaction-diffusion equations on a rescaled lattice

Let $\Lambda_m \subset \mathcal{O} \equiv [0,1)^d$ be the periodic lattice with mesh size m^{-1} in each direction; $m = m(n) \to \infty$ as $n \to \infty$. Setting $E_n = R^{|\Lambda_m|}$ and $E = L^2(\mathcal{O})$, the map $\eta_n \equiv \widehat{\eta}_m : E_n \mapsto E$ is the piecewise constant interpolation map defined by (C.2). We define

$$\langle p, q \rangle_m = \frac{1}{m^d} \sum_{x \in \Lambda_m} p(x) q(x), \quad \|p\|_m^2 \equiv \|p\|_{L^2(\Lambda_m)}^2 = \langle p, p \rangle_m$$

and similarly

$$\langle p, q \rangle = \int_{\mathcal{O}} p(x) q(x) dx, \quad \|p\|^2 \equiv \|p\|_{L^2(\mathcal{O})}^2 \equiv \langle p, p \rangle.$$

Let $\pi_m : L^2(\mathcal{O}) \mapsto R^{|\Lambda_m|}$ be the projection defined by (C.1). To simplify notation, for $p, q \in L^2(\mathcal{O})$, we write

$$\langle p, q \rangle_m = \langle \pi_m p, \pi_m q \rangle_m = \langle \widehat{\eta}_m \pi_m p, \widehat{\eta}_m \pi_m q \rangle.$$

Let $\{B(\cdot, x) : x \in \Lambda_m\}$ be independent, standard Brownian motions, and let $Y_n(t, x)$ satisfy

$$dY_n(t,x) = (\Delta_m Y_n(t,x) - F'(Y_n(t,x)))dt + \frac{m^{d/2}}{\sqrt{n}} dB(t,x), \quad x \in \Lambda_m, \tag{13.1}$$

with Δ_m the discrete Laplacian defined in Appendix C. We study the large deviation behavior of $\{\eta_n(Y_n)\}$ as a sequence of $L^2(\mathcal{O})$-valued Markov processes. We assume that F'' is bounded so that F' is Lipschitz, that there exist $c_1 \in R$, $c_2 > 0$ such that $F(r) \geq c_1 + c_2 r^2$, and that $m^{2+d}/n \to c_0 \geq 0$.

We take $\mathcal{Q}$ to be the collection of compact subsets of $L^2(\mathcal{O})$ and define $K_n^q = \pi_m(q)$, $q \in \mathcal{Q}$.

13.1.1. Convergence of H_n. The convergence in (1.32) does not, in general, satisfy Condition 7.11, so care needs to be taken in the selection of the domains $\mathcal{D}(H_\dagger)$ and $\mathcal{D}(H_\ddagger)$ and the approximating sequences $\{f_n\}$ for each element in the domains.

To compute A_n and H_n, define $\mathbb{D}_m(\rho) = \Delta_m\rho - F'(\rho)$, $\rho \in L^2(\Lambda_m)$ and note that

$$\begin{aligned} Z_{n,p}(t) &= \langle Y_n(t), p\rangle_m \\ &= \langle Y_n(0), p\rangle_m + \int_0^t \langle \mathbb{D}_m(Y_n(s)), p\rangle_m ds + \frac{m^{d/2}}{\sqrt{n}}\langle B(t), p\rangle_m \end{aligned}$$

and that the covariation for two such processes is

$$[Z_{n,p_1}, Z_{n,p_2}]_t = \frac{1}{n}\langle p_1, p_2\rangle_m t.$$

Similarly,

$$\begin{aligned} R_n(t) &= \langle Y_n(t), Y_n(t)\rangle_m \\ &= \langle Y_n(0), Y_n(0)\rangle_m + \int_0^t \left(2\langle \mathbb{D}_m(Y_n(s)), Y_n(s)\rangle_m + \frac{m^d}{n}\right) ds \\ &\quad + \frac{2}{\sqrt{m^d n}} \sum_{x\in\Lambda_m} \int_0^t Y_n(s,x) dB(s,x) \end{aligned}$$

More generally, let $\{e_{k,m} : k = (k_1, \dots, k_m)\}$ be the orthonormal basis of eigenfunctions for $-\Delta_m$ (see Appendix C), $\gamma \in R^{|\Lambda_m|}$, and let $\{a_k\} \subset R$. Then

$$\begin{aligned} R_{n,\gamma}^{\{a_k\}}(t) &= \sum_k a_k \langle Y_n(t) - \gamma, e_{k,m}\rangle^2 \\ &= \sum_k a_k \langle Y_n(0) - \gamma, e_{k,m}\rangle^2 \\ &\quad + \frac{m^{d/2}}{\sqrt{n}} \sum_k a_k \int_0^t 2\langle Y_n(s) - \gamma, e_{k,m}\rangle d\langle B(s), e_{k,m}\rangle_m \\ &\quad + \int_0^t \sum_k a_k \left(2\langle Y_n(s) - \gamma, e_{k,m}\rangle \langle \mathbb{D}_m(Y_n(s), e_{k,m}\rangle_m + \frac{1}{n}\right) ds. \end{aligned}$$

Consequently,

$$[R_{n,\gamma}^{\{a_k\}}, R_{n,\gamma}^{\{a_k\}}]_t = 4\frac{1}{n}\int_0^t \sum_k a_k^2 \langle Y_n(s) - \gamma, e_{k,m}\rangle_m^2 ds$$

and

$$[R_{n,\gamma_1}^{\{a_k\}}, R_{n,\gamma_2}^{\{b_k\}}]_t = 4\frac{1}{n}\int_0^t \sum_k a_k b_k \langle Y_n(s) - \gamma_1, e_{k,m}\rangle \langle Y_n(s) - \gamma_2, e_{k,m}\rangle ds.$$

Let $\{\lambda_{k,m}\}$ be the eigenvalues for $-\Delta_m$ such that $-\Delta e_{k,m} = \lambda_{k,m} e_{k,m}$. Define $B_m = (I - \Delta_m)^{-1}$ and $\|\rho\|_{-1,m}^2 = \langle B_m\rho, \rho\rangle_m$. Then $B_m e_{k,m} = (1 + \lambda_{k,m})^{-1} e_{k,m}$ and for every $\rho, \gamma \in L^2(\Lambda_m) = R^{|\Lambda_m|}$,

$$\|\rho - \gamma\|_{-1,m}^2 = \sum_k (1 + \lambda_{k,m})^{-1} \langle \rho - \gamma, e_{k,m}\rangle_m^2,$$

and

$$\sum_k (1+\lambda_{k,m})^{-1}\langle \rho-\gamma, e_{k,m}\rangle\langle \mathbb{D}_m(\rho), e_{k,m}\rangle_m = \langle B_m(\rho-\gamma), \mathbb{D}_m(\rho)\rangle_m.$$

Let $\varphi \in C_b^2(R^{l+1})$ and $\gamma_0, \gamma_1, \ldots, \gamma_l \in L^2(\mathcal{O})$. We consider

$$f_n(\rho) = \varphi(\|\rho-\pi_m\gamma_0\|_m^2, \|\rho-\pi_m\gamma_1\|_{-1,m}^2, \ldots, \|\rho-\pi_m\gamma_l\|_{-1,m}^2). \tag{13.2}$$

For simplicity of notation, we write

$$\vec{\gamma} = (\gamma_1, \ldots, \gamma_l), \quad \|\rho-\pi_m\vec{\gamma}\|_{-1,m}^2 = (\|\rho-\pi_m\gamma_1\|_{-1,m}^2, \ldots, \|\rho-\pi_m\gamma_l\|_{-1,m}^2). \tag{13.3}$$

Then

$$f_n(\rho) = \varphi(\|\rho-\pi_m\gamma_0\|_m^2, \|\rho-\pi_m\vec{\gamma}\|_{-1,m}^2).$$

Define $\kappa_m = \sum_k (1+\lambda_{k,m})^{-1}$, and index the $l+1$ coordinates of R^{l+1} by $0, 1, \ldots, l$. By Itô's formula, the generator of Y_n applied to f_n is

$$\begin{aligned}
&A_n f_n(\rho) \\
&= \sum_{i=1}^{l} \partial_i\varphi(\|\rho-\gamma_0\|_m^2, \|\rho-\vec{\gamma}\|_{-1,m}^2)\Big(\langle 2B_m(\rho-\gamma_i), \mathbb{D}_m(\rho)\rangle_m + \frac{\kappa_m}{n}\Big) \\
&\quad + \partial_0\varphi(\|\rho-\gamma_0\|_m^2, \|\rho-\vec{\gamma}\|_{-1,m}^2)\Big(\langle 2(\rho-\gamma_0), \mathbb{D}_m(\rho)\rangle_m + \frac{m^d}{n}\Big) \\
&\quad + \frac{2}{n}\sum_{i,j=1}^{l} \partial_i\partial_j\varphi(\|\rho-\gamma_0\|_m^2, \|\rho-\vec{\gamma}\|_{-1,m}^2)\langle B_m(\rho-\gamma_i), B_m(\rho-\gamma_j)\rangle_m \\
&\quad + \frac{2}{n}\sum_{i=1}^{l} \partial_0\partial_i\varphi(\|\rho-\gamma_0\|_m^2, \|\rho-\vec{\gamma}\|_{-1,m}^2)\langle(\rho-\gamma_0), B_m(\rho-\gamma_i)\rangle_m \\
&\quad + \frac{2}{n}\partial_0^2\varphi(\|\rho-\gamma_0\|_m^2, \|\rho-\vec{\gamma}\|_{-1,m}^2)\|\rho-\gamma_0\|_m^2
\end{aligned}$$

and

$$\begin{aligned}
&H_n f_n(\rho)\\
&= \sum_{i=1}^{l} \partial_i\varphi(\|\rho-\gamma_0\|_m^2, \|\rho-\vec{\gamma}\|_{-1,m}^2)\Big(\langle 2B_m(\rho-\gamma_i), \mathbb{D}_m(\rho)\rangle_m + \frac{\kappa_m}{n}\Big)\\
&+\partial_0\varphi(\|\rho-\gamma_0\|_m^2, \|\rho-\vec{\gamma}\|_{-1,m}^2)\Big(\frac{m^d}{n} + 2\langle \rho-\gamma_0, \mathbb{D}_m(\rho)\rangle_m\Big)\\
&+2\sum_{i,j=1}^{l} \partial_i\varphi\partial_j\varphi(\|\rho-\gamma_0\|_m^2, \|\rho-\vec{\gamma}\|_{-1,m}^2)\langle B_m(\rho-\gamma_i), B_m(\rho-\gamma_j)\rangle_m\\
&+2\sum_{i=1}^{l} \partial_0\varphi\partial_i\varphi(\|\rho-\gamma_0\|_m^2, \|\rho-\vec{\gamma}\|_{-1,m}^2)\langle \rho-\gamma_0, B_m(\rho-\gamma_i)\rangle_m\\
&+2(\partial_0\varphi)^2(\|\rho-\gamma_0\|_m^2, \|\rho-\vec{\gamma}\|_{-1,m}^2)\|\rho-\gamma_0\|_m^2\\
&+\frac{2}{n}\sum_{i,j=1}^{l} \partial_i\partial_j\varphi(\|\rho-\gamma_0\|_m^2, \|\rho-\vec{\gamma}\|_{-1,m}^2)\langle B_m(\rho-\gamma_i), B_m(\rho-\gamma_j)\rangle_m\\
&+\frac{2}{n}\sum_{i=1}^{l} \partial_0\partial_i\varphi(\|\rho-\gamma_0\|_m^2, \|\rho-\vec{\gamma}\|_{-1,m}^2)\langle \rho-\gamma_0, B_m(\rho-\gamma_i)\rangle_m\\
&+\frac{2}{n}\partial_0^2\varphi(\|\rho-\gamma_0\|_m^2, \|\rho-\vec{\gamma}\|_{-1,m}^2)\|\rho-\gamma_0\|_m^2.
\end{aligned}$$

Let $B=(I-\Delta)^{-1}$. Following the notation of Example 9.29, for $\beta=1,2,\ldots$, let $H_\beta(\mathcal{O})$ be the completion of $C^\infty(\mathcal{O})$ under the norm $\|\rho\|_\beta^2 = \langle (I-\Delta)^\beta\rho, \rho\rangle$ and $H_{-\beta}(\mathcal{O})$ be the completion of $H_0(\mathcal{O})$ under the norm $\|\rho\|_{-\beta}^2 = \langle (I-\Delta)^{-\beta}\rho,\rho\rangle$. The inner product $\langle\rho,\gamma\rangle$ extends consistently to $\rho\in H_{-\beta}(\mathcal{O})$ and $\gamma\in H_\beta(\mathcal{O})$, and

$$|\langle\rho,\gamma\rangle| \le \|\rho\|_{-\beta}\|\gamma\|_\beta.$$

The following inequality is useful in later calculations throughout this section: for each $\rho,\gamma\in L^2(\mathcal{O})$, $\Delta\rho\in H_{-2}(\mathcal{O})$ and $B(\rho-\gamma)\in H_2(\mathcal{O})$. Noting that $\Delta B = B-I$,

$$\begin{aligned}
\langle B(\rho-\gamma), \Delta\rho - F'(\rho)\rangle &= \|\rho\|_{-1}^2 - \|\rho\|^2 - \langle (B-I)\gamma,\rho\rangle - \langle B(\rho-\gamma), F'(\rho)\rangle)\\
&\le \|\rho\|_{-1}^2 - \|\rho\|^2 + c_{\gamma,F}(1+\|\rho\|_{-1})\|\rho\|, \qquad (13.4)
\end{aligned}$$

where $c_{\gamma,F}$ is a constant depending only on γ and F. Furthermore, if $\rho\in H_1(\mathcal{O})$, then $\Delta\rho\in H_{-1}(\mathcal{O})$ and

$$\langle \rho-\gamma, \Delta\rho - F'(\rho)\rangle = -\|\nabla\rho\|^2 - \langle\Delta\gamma,\rho\rangle - \langle\rho-\gamma, F'(\rho)\rangle.$$

Similarly, for $\rho,\gamma\in L^2(\Lambda_m)$,

$$\begin{aligned}
(13.5)\qquad &\langle B_m(\rho-\gamma), \Delta_m\rho - F'(\rho)\rangle_m\\
&= \|\rho\|_{-1,m}^2 - \|\rho\|_m^2 - \langle (B_m-I)\gamma,\rho\rangle_m - \langle B_m(\rho-\gamma), F'(\rho)\rangle_m)\\
&\le \|\rho\|_{-1,m}^2 - \|\rho\|_m^2 + c_{\gamma,F,m}(1+\|\rho\|_{-1,m})\|\rho\|.
\end{aligned}$$

LEMMA 13.1. *Let $\mathcal{D}(H_\dagger)$ be the collection of functions of the form*

$$(13.6)\quad f(\rho) = \varphi(\|\rho-\gamma_0\|^2, \|\rho-\gamma_1\|_{-1}^2, \ldots, \|\rho-\gamma_l\|_{-1}^2) = \varphi(\|\rho-\gamma_0\|^2, \|\rho-\vec{\gamma}\|_{-1}^2),$$

where $l = 0,1,2,\dots$, $\gamma_0 \in C_b^2(\mathcal{O})$, $\gamma_1,\dots,\gamma_l \in L^2(\mathcal{O})$, $\varphi \in C_b^2(R^{l+1})$ *satisfies* $\inf_{r\in R^{l+1}}\varphi(r) > -\infty$, *and for* $i = 0,1,\dots,l$, $\partial_i\varphi \geq 0$ *and there exists* d_i *such that* $\partial_i\varphi(r_0,r_1,\dots,r_l) = 0$ *if* $r_i \geq d_i$. *Define* $H_\dagger f$ *given by*

$$\begin{aligned}
&H_\dagger f(\rho)\\
&= 2\sum_{i=1}^{l}\partial_i\varphi(\|\rho-\gamma_0\|^2, \|\rho-\vec{\gamma}\|_{-1}^2)\langle B(\rho-\gamma_i), \mathbb{D}(\rho)\rangle\\
&\quad +2\partial_0\varphi(\|\rho-\gamma_0\|^2, \|\rho-\vec{\gamma}\|_{-1}^2)\Big(-\|\nabla\rho\|^2 - \langle\rho, F'(\rho)\rangle - \langle\gamma_0, \mathbb{D}(\rho)\rangle\Big)\\
&\quad +2\sum_{i,j=1}^{l}\partial_i\varphi\partial_j\varphi(\|\rho-\gamma_0\|^2, \|\rho-\vec{\gamma}\|_{-1}^2)\langle B(\rho-\gamma_i), B(\rho-\gamma_j)\rangle\\
&\quad +2\sum_{i=1}^{l}\partial_0\varphi\partial_i\varphi(\|\rho-\gamma_0\|^2, \|\rho-\vec{\gamma}\|_{-1}^2)\langle \rho-\gamma_0, B(\rho-\gamma_i)\rangle\\
&\quad +2(\partial_0\varphi)^2(\|\rho-\gamma_0\|^2, \|\rho-\vec{\gamma}\|_{-1}^2)\|\rho-\gamma_0\|^2\\
&= 2\sum_{i=1}^{l}\partial_i\varphi(\|\rho-\gamma_0\|^2, \|\rho-\vec{\gamma}\|_{-1}^2)\langle B(\rho-\gamma_i), \mathbb{D}(\rho)\rangle\\
&\quad +2\partial_0\varphi(\|\rho-\gamma_0\|^2, \|\rho-\vec{\gamma}\|_{-1}^2)\Big(-\|\nabla\rho\|^2 - \langle\rho, F'(\rho)\rangle - \langle\gamma_0, \mathbb{D}(\rho)\rangle\Big)\\
&\quad +\frac{1}{2}\|2\partial_0\varphi_0(\|\rho-\gamma_0\|^2, \|\rho-\vec{\gamma}\|_{-1}^2)(\rho-\gamma_0)\\
&\quad +\sum_{i=1}^{l}\partial_i\varphi(\|\rho-\gamma_0\|^2, \|\rho-\vec{\gamma}\|_{-1}^2)B(\rho-\gamma_i)\|^2,
\end{aligned}$$

where $\mathbb{D}(\rho) = \Delta\rho - F'(\rho)$ *and, noting that* $\|\nabla\rho\|^2$ *may be infinite, we take* $0\cdot\infty = 0$.

Let $\mathcal{D}(H_\ddagger) = \{f : -f \in \mathcal{D}(H_\dagger)\}$, *and define* $H_\ddagger$ *by the same formula as* $H_\dagger$. *Then Condition 7.11 is satisfied with* $v_n = n$.

REMARK 13.2. By the assumptions on φ and (13.4), $\sup_{\rho\in E} H_\dagger f(\rho) < \infty$ and $H_\dagger f \in M^u(E,\overline{R})$ and is upper semicontinuous. Similarly, $H_\ddagger f \in M^l(E,\overline{R})$ and is lower semicontinuous.

PROOF. Note that $\langle\rho, \pi_m p\rangle_m = \langle\widehat{\eta}_m\rho, p\rangle$ for $p \in L^2(\mathcal{O})$. Let

$$f_n(\rho) = \varphi(\|\rho - \pi_m\gamma_0\|_m^2, \|\rho - \pi_m\vec{\gamma}\|_{-1,m}^2), \tag{13.7}$$

where φ satisfies the conditions in the definition of $\mathcal{D}(H_\dagger)$. If $\|\rho_n - \pi_m\rho\| \to 0$ as $n\to\infty$, then

$$f_n(\rho_n) \to f(\rho) = \varphi(\|\rho-\gamma_0\|^2, \|\rho-\gamma_1\|_{-1}^2, \dots, \|\rho-\gamma_l\|_{-1}^2).$$

Since $K_n^q = \pi_m(q)$, where q is a compact subset of $L^2(\mathcal{O})$, (7.18) follows.

Since $H_nf_n(\rho) = 0$ if $\|\rho\|_{-1,m}^2$ is large enough and (13.5) holds,

$$\sup_{n,\rho\in E_n} H_nf_n(\rho) < \infty, \tag{13.8}$$

giving the second part of (7.19). To obtain the first part, note that $\langle -\Delta_m\rho,\rho\rangle_m \le dm^2\|\rho\|_m^2$, so that

$$\begin{aligned}&\partial_0\varphi(\|\rho-\pi_m\gamma_0\|_m^2,\|\rho-\pi_m\vec{\gamma}\|_{-1,m}^2)\Big|\langle\rho-\gamma_0,\Delta_m\rho-F'(\rho)\rangle_m\Big|\\&\le\partial_0\varphi(\|\rho-\pi_m\gamma_0\|_m^2,\|\rho-\pi_m\vec{\gamma}\|_{-1,m}^2)\Big(c_0+c_1(1+dm^2\|\rho\|_m^2)\Big)\\&\le c_2(1+dm^2),\end{aligned}$$

for constants $c_0,c_1,c_2>0$, and

$$\begin{aligned}\|\rho\|_m^2 &= \langle(I-\Delta_m)B_m\rho,\rho\rangle_m\\&\le \langle B_m\rho,\rho\rangle_m+\langle(-\Delta_m)B_m^{1/2}\rho,B_m^{1/2}\rho\rangle_m\\&\le (1+dm^2)\|\rho\|_{-1,m}^2,\end{aligned}$$

so that

$$\begin{aligned}&\partial_i\varphi(\|\rho-\pi_m\gamma_0\|_m^2,\|\rho-\pi_m\vec{\gamma}\|_{-1,m}^2)\Big|\langle B_m(\rho-\gamma_i),\Delta_m\rho-F'(\rho)\rangle_m\Big|\\&\le\partial_i\varphi(\|\rho-\pi_m\gamma_0\|_m^2,\|\rho-\pi_m\vec{\gamma}\|_{-1,m}^2)c_3(1+\|\rho\|_m^2)\\&\le\partial_i\varphi(\|\rho-\gamma_0\|_m^2,\|\rho-\vec{\gamma}\|_{-1,m}^2)c_3(2+dm^2\|\rho\|_{-1,m}^2)\\&\le c_4(1+dm^2)\end{aligned}$$

for some constants $c_3,c_4>0$. It follows that

$$\sup_{\rho\in E_n}|H_nf_n(\rho)|\le c_5(1+dm^2)$$

for some $c_5>0$, and since $m^{2+d}/n\to c_0\ge 0$, the first part of (7.19) holds with $v_n=n$.

For $\rho_n\in E_n=R^{|\Lambda_m|}$ satisfying $\eta_n(\rho_n)\to\rho$, by Lemma C.1,

$$\liminf_{n\to\infty}\langle\rho_n,-\Delta_m\rho_n\rangle_m\ge\|\nabla\rho\|^2.$$

The nonnegativity of $\partial_0\varphi$ and Lemma C.2 then imply

$$\limsup_{n\to\infty}H_nf_n(\rho_n)\le H_\dagger^*f(\rho), \tag{13.9}$$

which gives (7.20). □

Let $f\in M(E,\overline{R})$. Define $\mathrm{grad}f(\rho)$ according to Definition 9.30, that is, we identify $\mathrm{grad}f(\rho)$ as the unique element in $\cup_{\beta=1,2,\ldots}H_{-\beta}(\mathcal{O})$ satisfying

$$\lim_{t\to0}\frac{f(\rho+tq)-f(\rho)}{t}=\langle\mathrm{grad}f(\rho),q\rangle,\qquad\forall q\in C^\infty(\mathcal{O}),$$

if it exists. If $f(\rho)=\langle\rho,p\rangle$, then $\mathrm{grad}f(\rho)=p$, and if $f(\rho)=\|C\rho\|^2$ for some bounded operator C, $\mathrm{grad}f(\rho)=2C\rho$. More generally, for the f in (13.6),

$$\begin{aligned}\text{grad } f(\rho) &= 2\partial_0\varphi_0(\|\rho-\gamma_0\|^2,\|\rho-\vec{\gamma}\|_{-1}^2)(\rho-\gamma_0)\\&\quad+2\sum_{i=1}^{l}\partial_i\varphi(\|\rho-\gamma_0\|^2,\|\rho-\vec{\gamma}\|_{-1}^2)B(\rho-\gamma).\end{aligned}\tag{13.10}$$

Since $\langle\Delta\rho,\rho\rangle=-\langle\nabla\rho,\nabla\rho\rangle=-\|\nabla\rho\|^2$, we have

$$H_\dagger f(\rho)=\langle\Delta\rho-F'(\rho),\mathrm{grad}f(\rho)\rangle+\frac{1}{2}\|\mathrm{grad}f(\rho)\|^2. \tag{13.11}$$

In particular, if $\partial_0\varphi\equiv0$, $\mathrm{grad}f(\rho)\in H_2(\mathcal{O})$, for each $\rho\in L^2(\mathcal{O})$, and $H_\dagger f\in C(E)$.

Let $\mathcal{E} : E \to \overline{R}$ be defined according to (9.64):

$$\mathcal{E}(\rho) = \int_{\mathcal{O}} (\frac{1}{2}|\nabla \rho(x)|^2 + F(\rho(x)))dx,$$

Then

$$\text{grad}\mathcal{E}(\rho) = -\Delta\rho + F'(\rho), \quad \forall \rho \in H_1(\mathcal{O}),$$

and $H_\dagger$ is of the form (9.62):

$$H_\dagger f(\rho) = \langle -\text{grad}\mathcal{E}(\rho), \text{grad} f(\rho)\rangle + \frac{1}{2}\|\text{grad} f(\rho)\|^2.$$

Similarly, $H_\ddagger$ admits the same representations.

In the proof of the comparison principle, a special class of test functions

$$\widehat{\mathcal{D}} = \{f(\rho) = \alpha\|\rho - \gamma\|_{-1}^2 + d : \alpha, d \in R, \gamma \in L^2(\mathcal{O})\} \tag{13.12}$$

will play a major role. $\widehat{\mathcal{D}}_+ \subset \widehat{\mathcal{D}}$ will denote the subclass with $\alpha > 0$ and $\widehat{\mathcal{D}}_-$ the subclass with $\alpha < 0$. Since they are unbounded, these functions do not belong to the domains of $H_\dagger$ or $H_\ddagger$; however, we can apply Lemma 7.6 to obtain appropriate extensions.

To extend $H_\dagger$, we approximate $f \in \widehat{\mathcal{D}}$ by

$$f_n(\rho) = \alpha\varphi_n(\|\rho - \gamma\|_{-1}^2) + d \tag{13.13}$$

where $\varphi_n \in C_b^2(R)$ is nondecreasing,

$$\varphi_n(r) = r, \quad 0 \le r \le n, \quad \varphi_n(r) = n+1, \quad r \ge n+2.$$

By definition, $f_n \in \mathcal{D}(H_\dagger)$ and

$$\begin{aligned} H_\dagger f_n(\rho) &= 2\varphi_n'(\|\rho-\gamma\|_{-1}^2)(\langle \rho - \gamma, B\rho - \rho\rangle - \langle B(\rho-\gamma), F'(\rho)\rangle) \\ &\quad + \frac{1}{2}\left\|2\varphi_n'(\|\rho-\gamma\|_{-1}^2)B(\rho-\gamma)\right\|^2. \end{aligned}$$

LEMMA 13.3. *Let*

$$\mathcal{D}(\widetilde{H}_\dagger) = \widehat{\mathcal{D}}_+ \cup \mathcal{D}(H_\dagger), \quad \mathcal{D}(\widetilde{H}_\ddagger) = \widehat{\mathcal{D}}_- \cup \mathcal{D}(H_\ddagger).$$

Define

$$\widetilde{H}_\dagger f(\rho) = \langle \Delta\rho - F'(\rho), \text{grad} f(\rho)\rangle + \frac{1}{2}\|\text{grad} f(\rho)\|^2, \quad f \in \mathcal{D}(\widetilde{H}_\dagger),$$

and define $\widetilde{H}_\ddagger$ similarly.

Then for $h \in C_b(E)$, any subsolution of $f - \alpha H_\dagger f = h$ is a subsolution of $f - \alpha\widetilde{H}_\dagger f = h$ and any supersolution of $f - \alpha H_\ddagger f = h$ is a supersolution of $f - \alpha\widetilde{H}_\ddagger f = h$.

PROOF. Let $f(\rho) = \alpha\|\rho-\gamma\|_{-1}^2 + d$ and let f_n be given by (13.13). If $\|\rho-\gamma\|_{-1}^2 < n$, $f_n(\rho) = f(\rho)$ and $H_\dagger f_n(\rho) = \widetilde{H}_\dagger f(\rho)$, and hence for each $c \in R$,

$$\lim_{n\to\infty} \sup_{\rho\in E} |f_n(\rho) \wedge c - f(\rho) \wedge c| = 0$$

and (7.10) holds. Lemma 7.6 then gives the conclusion for subsolutions. The supersolution case is similar. □

13.1.2. Exponential tightness. We apply Corollary 4.17. The exponential compact containment condition is verified in Example 2.9 using the following lemma.

LEMMA 13.4. *Suppose that Y_n and Z_n both satisfy (13.1), but with different initial values:*

$$\begin{aligned} dY_n(t,x) &= (\Delta_m Y_n(t,x) - F'(Y_n(t,x)))dt + \frac{m^{d/2}}{\sqrt{n}} dB(t,x) \\ dZ_n(t,x) &= (\Delta_m Z_n(t,x) - F'(Z_n(t,x)))dt + \frac{m^{d/2}}{\sqrt{n}} dB(t,x). \end{aligned}$$

Then for each $T > 0$, there exists a constant $C_0 = C_0(T,F) > 0$ such that

$$\sup_{0 \le t \le T} \|Y_n(t) - Z_n(t)\|_{L^2(\Lambda_m)} \le C_0 \|Y_n(0) - Z_n(0)\|_{L^2(\Lambda_m)} \quad a.s.$$

PROOF. Observing that

$$d(Y_n(t,x) - Z_n(t,x)) = \Delta_m(Y_n(t,x) - Z_n(t,x))dt - \Big(F'(Y_n(t,x)) - F'(Z_n(t,x))\Big)dt,$$

we have

$$\begin{aligned} &\frac{1}{2}\|Y_n(t) - Z_n(t)\|^2_{L^2(\Lambda_m)} - \frac{1}{2}\|Y_n(0) - Z_n(0)\|^2_{L^2(\Lambda_m)} \\ = &\int_0^t \Big(\langle \Delta_m (Y_n(s) - Z_n(s)), Y_n(s) - Z_n(s)\rangle_m \\ &\qquad - \langle F'(Y_n(s)) - F'(Z_n(s)), Y_n(s) - Z_n(s)\rangle_m \Big) ds \\ \le &\int_0^t L_{F'} \|Y_n(s) - Z(s)\|^2_{L^2(\Lambda_m)} ds, \end{aligned}$$

where $L_{F'}$ is the Lipschitz constant for F'. The conclusion follows by Gronwall's inequality. □

With reference to Corollary 4.17, functions of the form (13.6) are in $\mathcal{D}(H_\dagger)$, and, consequently, $\mathcal{D}(H_\dagger)$ approximates the metric $q(\rho_1,\rho_2) = \|\rho_1 - \rho_2\|$ in the sense of Definition 4.16. Since $\lambda > 0$ and $f \in \mathcal{D}(H_\dagger)$ implies $\lambda f \in \mathcal{D}(H_\dagger)$, Conditions (b) and (c) of Corollary 4.17 follow from the convergence of H_n verified above. The corollary gives the desired exponential tightness.

13.1.3. The comparison principle. Let $D_0 \subset C_b(E)$ consist of functions of the form

$$h(\rho) = \varphi(\langle \rho, q_1\rangle, \dots, \langle \rho, q_m\rangle), \tag{13.14}$$

where $\varphi \in C^\infty(R^m) \cap C_b(R^m)$ is Lipschitz, $q_k \in C^\infty(\mathcal{O})$, and $m = 1, 2, \dots$. Then D_0 is an algebra which separates points in E and vanishes nowhere. Therefore, by Theorem A.8, D_0 is buc-dense in $C_b(E)$. Let $\alpha > 0$ and $h \in D_0$. Suppose that $\overline{f}$ is a viscosity subsolution to $(I - \alpha H_\dagger)f = h$ and $\underline{f}$ is a viscosity supersolution to $(I - \alpha H_\ddagger)f = h$.

Our strategy is to show that slightly perturbed versions of $\overline{f}$ and $\underline{f}$ satisfy Condition 9.27 and that the perturbed solutions solve the nonlinear equations given by $\widehat{H}_\dagger$ and $\widehat{H}_\ddagger$ in Example 9.29 in an approximate sense. We then apply Theorem 9.31.

Each $h \in D_0$ is uniformly continuous with respect to $\|\cdot\|_{-1}$. Let ω_h denote the modulus of continuity, that is, $h(\rho) - h(\gamma) \leq \omega_h(\|\rho - \gamma\|_{-1})$. Let $d(\rho,\gamma) = \|\rho - \gamma\|$. For $\delta > 0$, define

$$\begin{aligned} \overline{f}_\delta(\rho) &= \lambda_{0,\delta} \sup_{\gamma \in E}\{\overline{f}(\gamma) - \frac{\|\rho-\gamma\|_{-1}^2}{2\delta}\}, \\ \underline{f}_\delta(\gamma) &= \lambda_{1,\delta} \inf_{\rho \in E}\{\underline{f}(\rho) + \frac{\|\rho-\gamma\|_{-1}^2}{2\delta}\}; \end{aligned}$$

where

$$\lambda_{0,\delta} = (1 + \frac{(1+L_F)^2\delta}{2}), \text{ and } \lambda_{1,\delta} = (1 - \frac{(1+L_F)^2\delta}{2})^{-1}.$$

By Theorem 2.64 on page 228 of Attouch [5], $\overline{f}_\delta$ and $\underline{f}_\delta$ are continuous with respect to $\|\cdot\|_{-1}$, hence also with respect to $\|\cdot\|$. Let

$$h_{0,\delta} = \lambda_{0,\delta}\Big(h + \omega_h(2\sqrt{\delta\|\|\overline{f}\|\|})\Big),$$

and

$$h_{1,\delta} = \lambda_{1,\delta}\Big(h - \omega_h(2\sqrt{\delta\|\|\underline{f}\|\|})\Big);$$

and let $\widehat{H}_\dagger, \widehat{H}_\ddagger$ be defined according to (9.67) and (9.68).

LEMMA 13.5. *$\overline{f}_\delta \in C_b(E)$ is a viscosity subsolution of*

$$(I - \alpha \widehat{H}_\dagger) f = h_{0,\delta}$$

and $\underline{f}_\delta \in C_b(E)$ is a viscosity supersolution of

$$(I - \alpha \widehat{H}_\ddagger) f = h_{1,\delta}.$$

PROOF. As before, we only prove the subsolution case. The supersolution case is similar.

Let f_0 be defined according to (9.66) with $\mathcal{E}$ defined by (9.64). Since $\overline{f}_\delta$ is bounded above and f_0 has compact level sets under the L^2-norm, there exists $\rho_0 \in H_1(\mathcal{O})$ such that

$$(\overline{f}_\delta - f_0)(\rho_0) = \sup_{\rho,\gamma \in E}\Big(\lambda_{0,\delta}(\overline{f}(\gamma) - \frac{\|\rho-\gamma\|_{-1}^2}{2\delta}) - f_0(\rho)\Big).$$

By the viscosity subsolution property of $\overline{f}$ and by Lemma 13.3, there exist $\epsilon_k \to 0$ and $\gamma_k \in E$ such that

(13.15)

$$\Big(\lambda_{0,\delta}(\overline{f}(\gamma_k) - \frac{\|\rho_0-\gamma_k\|_{-1}^2}{2\delta}) - f_0(\rho_0)\Big) + \epsilon_k \geq \sup_{\rho,\gamma \in E}\Big(\lambda_{0,\delta}(\overline{f}(\gamma) - \frac{\|\rho-\gamma\|_{-1}^2}{2\delta}) - f_0(\rho)\Big),$$

and

$$\begin{aligned} \alpha^{-1}(\overline{f} - h)(\gamma_k) &\leq \epsilon_k + (\widetilde{H}_\dagger \frac{\|\rho_0 - \cdot\|_{-1}^2}{2\delta})(\gamma_k) \\ &= \epsilon_k + \langle \Delta\gamma_k - F'(\gamma_k), \frac{B(\gamma_k - \rho_0)}{\delta}\rangle + \frac{1}{2}\|\frac{B(\gamma_k-\rho_0)}{\delta}\|^2. \end{aligned} \quad (13.16)$$

It follows from (13.15) that $\sup_k \|\rho_0 - \gamma_k\|_{-1} < \infty$. Since

$$\begin{aligned}&\langle \Delta\gamma_k - F'(\gamma_k), B(\gamma_k - \rho_0)\rangle - \langle \Delta\rho_0 - F'(\rho_0), B(\gamma_k - \rho_0)\rangle \\ &\quad = -\|\gamma_k - \rho_0\|^2 + \langle \gamma_k - \rho_0, B(\gamma_k - \rho_0)\rangle - \langle F'(\gamma_k) - F'(\rho_0), B(\gamma_k - \rho_0)\rangle \\ &\quad \leq -\|\gamma_k - \rho_0\|^2 + \|\gamma_k - \rho_0\|\|B(\gamma_k - \rho_0)\|(1 + L_F),\end{aligned}$$

where $L_F = \sup_r |F'(r)|$. (13.16) implies

$$\begin{aligned}\alpha^{-1}(\overline{f} - h)(\gamma_k) \quad \leq \quad & \epsilon_k + \langle \Delta\rho_0 - F'(\rho_0), \frac{B(\gamma_k - \rho_0)}{\delta}\rangle + \frac{1}{2}\|\frac{B(\gamma_k - \rho_0)}{\delta}\|^2 \\ & - \frac{\|\gamma_k - \rho_0\|^2}{\delta} + \frac{\|\gamma_k - \rho_0\|}{\delta^{1/2}}\frac{\|B(\gamma_k - \rho_0)\|(1 + L_F)}{\delta^{1/2}}.\end{aligned}$$

By the boundedness of $\overline{f}$ and h and by the estimate

$$\sup_k |\langle \Delta\rho_0, B\gamma_k\rangle| = \sup_k |\langle \rho_0, (B - I)\gamma_k\rangle| \leq \sup_k(\|\rho_0\| + \|\rho_0\|_1)\|\gamma_k\|_{-1} < \infty,$$

we have $\sup_k \|\rho_0 - \gamma_k\| < \infty$, implying the weak sequential relative compactness of γ_k in $L^2(\mathcal{O})$. Let assume γ_0 be a limit point.

By (13.15), $\lambda_{0,\delta}\overline{f}(\gamma_k) + \epsilon_k \geq \overline{f}_\delta(\rho_0)$, and since $-x^2 + ax \leq a^2/4$,

$$\begin{aligned}(13.17)\quad & \alpha^{-1}(\lambda_{0,\delta}^{-1}\overline{f}_\delta(\rho_0) - h(\gamma_0)) \leq \liminf_{k\to\infty} \alpha^{-1}(\overline{f} - h)(\gamma_k) \\ & \leq \lim_{k\to\infty}\Big(\langle \Delta\rho_0 - F'(\rho_0), \frac{B(\gamma_k - \rho_0)}{\delta}\rangle + \frac{1}{2}(1 + \frac{(1 + L_F)^2\delta}{2})\|\frac{B(\gamma_k - \rho_0)}{\delta}\|^2\Big) \\ & = \langle \Delta\rho_0 - F'(\rho_0), \frac{B(\gamma_0 - \rho_0)}{\delta}\rangle + \frac{1}{2}(1 + \frac{(1 + L_F)^2\delta}{2})\|\frac{B(\gamma_0 - \rho_0)}{\delta}\|^2.\end{aligned}$$

Since B is a self-adjoint compact operator on $L^2(\mathcal{O})$, $\lim_{k\to\infty}\|B\gamma_k - B\gamma_0\| = 0$, and

$$\lim_{k\to\infty}\|\gamma_k - \gamma_0\|_{-1}^2 = \lim_{k\to\infty}\langle B(\gamma_k - \gamma_0), \gamma_k - \gamma_0\rangle = 0.$$

From (13.15),

$$-\lambda_{0,\delta}\frac{\|\gamma_k - \rho_0\|_{-1}^2}{2\delta} - f_0(\rho_0) + \epsilon_k \geq -\lambda_{0,\delta}\frac{\|\rho - \gamma_k\|_{-1}^2}{2\delta} - f_0(\rho), \quad \rho \in L^2(\mathcal{O}).$$

In the limit as $k \to \infty$,

$$-\lambda_{0,\delta}\frac{\|\gamma_0 - \rho_0\|_{-1}^2}{2\delta} - f_0(\rho_0) \geq -\lambda_{0,\delta}\frac{\|\rho - \gamma_0\|_{-1}^2}{2\delta} - f_0(\rho), \quad \rho \in L^2(\mathcal{O}).$$

Let $\rho = \rho_0 + tp$, where $p \in C^\infty(\mathcal{O})$, and apply Definition 9.30. Since p is arbitrary,

$$\text{grad} f_0(\rho_0) = -\lambda_{0,\delta}\frac{B(\rho_0 - \gamma_0)}{\delta} \in H_2(\mathcal{O}).$$

By (13.15), we also have

$$\|\rho_0 - \gamma_0\|_{-1}^2 \leq 4\delta\|\overline{f}\|,$$

and (13.17) becomes

$$\begin{aligned}&\alpha^{-1}\Big(\lambda_{0,\delta}^{-1}\overline{f}_\delta(\rho_0) - h(\rho_0) - \omega_h(2\sqrt{\|\overline{f}\|\delta})\Big) \\ &\quad \leq \langle \Delta\rho_0 - F'(\rho_0), \frac{\text{grad} f_0(\rho_0)}{\lambda_{0,\delta}}\rangle + \frac{1}{2}(1 + \frac{(1 + L_F)^2\delta}{2})\|\frac{\text{grad} f_0(\rho_0)}{\lambda_{0,\delta}}\|_{L^2(\mathcal{O})}^2,\end{aligned}$$

and hence

$$\begin{aligned}\alpha^{-1}(\overline{f}_\delta(\rho_0) - h_{0,\delta}(\rho_0)) &\leq \langle \Delta\rho_0 - F'(\rho_0), \mathrm{grad} f_0(\rho_0)\rangle + \frac{1}{2}\|\mathrm{grad} f_0(\rho_0)\|^2_{L^2(\mathcal{O})} \\ &= \widehat{H}_\dagger f_0(\rho_0),\end{aligned}$$

giving the result. □

THEOREM 13.6. *Let h be given by (13.14 and $\alpha > 0$. Suppose $\overline{f}$ is a subsolution of $(I - \alpha H_\dagger)f = h$ and $\underline{f}$ is a supersolution of $(I - \alpha H_\ddagger)f = h$. Then*

$$\overline{f} \leq \underline{f}.$$

PROOF. By Lemma 13.5 and Theorem 9.31,

$$\lambda_{0,\delta}\overline{f}(\rho) - \lambda_{1,\delta}\underline{f}(\rho) \leq \overline{f}_\delta(\rho) - \underline{f}_\delta(\rho) \leq \sup_{\gamma\in E}(h_{0,\delta}(\gamma) - h_{1,\delta}(\gamma)),$$

and letting $\delta \to 0+$ gives the desired inequality. □

13.1.4. Variational representation. Since

$$\frac{1}{2}\|\mathrm{grad}\ f(\rho)\|^2 = \sup_{u\in L^2(\mathcal{O})}(\langle u, \mathrm{grad}\ f(\rho)\rangle - \frac{1}{2}\|u\|^2),$$

we have

$$H_\dagger f(\rho) = \sup_{u\in L^2(\mathcal{O})}\{\langle \Delta\rho - F'(\rho) + u, \mathrm{grad}\ f(\rho)\rangle - \frac{1}{2}\|u\|^2\}.$$

An identical representation holds for $H_\ddagger$ as well. These representations suggest that the results of Chapter 8 apply; however, to verify the conditions, we need to choose the operators and the control space carefully.

Let

$$D_\dagger = \{f(\rho) = \varphi(\|\rho - \gamma\|^2_{-1}) : \gamma \in E, \varphi \in C_b^2(R), \varphi' \geq 0, \exists d_\varphi, \varphi'(r) = 0, r \geq d_\varphi\},$$

and

$$D_\ddagger = \{f(\rho) = \varphi(\|\rho - \gamma\|^2_{-1}) : \gamma \in E, \varphi \in C_b^2(R), \varphi' \leq 0, \exists d_\varphi, \varphi'(r) = 0, r \geq d_\varphi\}$$

Then $D_\dagger$ and $D_\ddagger$ separate points in E. Note that for $\rho \in E = L^2(\mathcal{O})$ and $f \in D_\dagger \cup D_\ddagger$, $\mathrm{grad} f(\rho) \in H_2(\mathcal{O})$.

Let $\mathcal{D}(A)$ be the linear span of $D_\dagger \cup D_\ddagger$, and for $\rho \in E$ and $u \in U \equiv H_{-1}(\mathcal{O})$, define

$$Af(\rho, u) = \langle \Delta\rho - F'(\rho) + u, \mathrm{grad}\ f(\rho)\rangle$$

and

$$L(\rho, u) = \frac{1}{2}\|u\|^2.$$

Then

$$\mathbf{H}f(\rho) = \sup_{u\in U}\{Af(\rho, u) - L(\rho, u)\}. \tag{13.18}$$

Define $\mathbf{H}_\dagger$ and $\mathbf{H}_\ddagger$ by letting $\mathcal{D}(\mathbf{H}_\dagger) = D_\dagger$, $\mathbf{H}_\dagger f = H_\dagger f$, $\mathcal{D}(\mathbf{H}_\ddagger) = D_\ddagger$, and $\mathbf{H}_\ddagger f = H_\ddagger f$. Then, (8.33) is satisfied.

We now verify the other conditions required in Corollary 8.28.

Let $(\rho, \lambda) \in \mathcal{J}$ and

$$\int_{U\times[0,\infty)} L(\rho(s), u)\lambda(du \times ds) < \infty. \tag{13.19}$$

Write $\lambda(du\times ds) = \lambda_s(du)ds$, and define $u(s) = \int_U u\lambda_s(du) \in L^2(\mathcal{O})$ in the Bochner sense. By Example 8.3, the control equation (8.1) is just the weak form of the controlled partial differential equation

$$\frac{\partial}{\partial t}\rho = \Delta\rho - F'(\rho) + u. \tag{13.20}$$

In particular, for λ satisfying (13.19), the equation uniquely determines the solution ρ.

Except for Condition 8.9.4, Conditions 8.9 and 8.10 can be directly verified. For $\rho_0 \in L^2(\mathcal{O})$ and $f \in \mathcal{D}(\mathbf{H})$, let ρ be the solution to

$$\frac{\partial}{\partial t}\rho = \Delta\rho - F'(\rho) + \text{grad } f(\rho).$$

Since $\text{grad} f(\rho)$ is Lipschitz in ρ in the $L^2(\mathcal{O})$ norm, existence and uniqueness of the solution follows from classical semilinear equation theory.

Taking $u(t) = \text{grad} f(\rho(t))$, (ρ, u) satisfies (13.20) and Condition 8.11 is verified:

$$\int_0^t \mathbf{H}f(\rho(s))ds = \int_0^t \{Af(\rho(s)u(s)) - L(\rho(s), u(s))\}ds, \quad t \geq 0.$$

Verification of Condition 8.9.4 consists of three steps. Let u in (13.20) satisfy

$$C_0 \equiv \int_0^\infty \int_{\mathcal{O}} u^2(t,x)dxdt < \infty.$$

If $\mathcal{E}(\rho(0)) < \infty$, standard a priori estimates give

$$\begin{aligned}\mathcal{E}(\rho(t)) - \mathcal{E}(\rho(0)) &\leq \int_0^t \Big(- \|\Delta\rho - F'(\rho)\|^2 + \|u\|\|\Delta\rho - F'(\rho)\|\Big)ds \\ &\leq \frac{1}{4}\int_0^\infty \int_{\mathcal{O}} u^2(t,x)dxds,\end{aligned}$$

implying that $\rho(t)$ is in the compact set

$$K(\rho(0), C_0) = \{\gamma : \mathcal{E}(\gamma) \leq \mathcal{E}(\rho(0)) + \frac{1}{4}C_0\}.$$

If ρ, γ are solutions of (13.20) with identical control u,

$$\|\rho(t) - \gamma(t)\| \leq C_2\|\rho(0) - \gamma(0)\|, \qquad 0 \leq t \leq T,$$

where $C_2 = C_2(T, F)$ depends only on $T > 0$ and F and is independent of u. This inequality follows Gronwall's inequality from the estimate

$$\begin{aligned}&\frac{1}{2}\|\rho(t) - \gamma(t)\|^2 - \frac{1}{2}\|\rho(0) - \gamma(0)\|^2 \\ \leq\ & \int_0^t \Big(- \|\nabla(\rho(s) - \gamma(s))\|^2 + \|F'(\rho(s)) - F'(\gamma(s))\|\|\rho - \gamma\|ds \\ \leq\ & \int_0^t L_F\|\rho(s) - \gamma(s)\|^2 ds,\end{aligned}$$

where L_F is the Lipschitz constant for F'.

Finally, for any compact $K_0 \subset L^2(\mathcal{O})$ and $\epsilon > 0$, there exists $N = N(\epsilon, K_0)$ and $\gamma_1, \ldots, \gamma_N \in L^2(\mathcal{O})$ with $\mathcal{E}(\gamma_k) < \infty$ such that

$$K_0 \subset \cup_{k=1}^N B(\gamma_k; \frac{\epsilon}{C_2}).$$

Let $\gamma_k(t)$ be the solution of (13.20) with $\gamma_k(0) = \gamma_k$. For any solution $\rho(t)$ of (13.20) with $\rho(0) \in K_0$, there exists γ_i satisfying $\|\rho(0) - \gamma_i(0)\| < C_2^{-1}\epsilon$ and hence

$$\sup_{0\le t\le T} \|\rho(t) - \gamma_i(t)\| < \epsilon.$$

Consequently,

$$\rho(t) \in \left(\cup_{i=1}^{N(\epsilon)} K(\gamma_i(0), C_0)\right)^\epsilon, \qquad 0 \le t \le T,$$

where A^ϵ is the ϵ-fattening of a set A. Since $\epsilon > 0$ is arbitrary, there exists compact $K(K_0, C_0)$ such that $\rho(0) \in K_0$ implies $\rho(t) \in K(K_0, C_0)$, $0 \le t \le T$.

Finally, for ρ satisfying (13.20), by the convexity in u of $L(\rho, u) = \frac{1}{2}\|u\|^2$,

$$\begin{aligned}
&\inf_\lambda\{\int_{U\times[0,\infty)} L(\rho(s), u)\lambda(du \times ds) : (\rho, \lambda) \in \mathcal{J}\} \\
&= \frac{1}{2}\int_0^\infty \|u(s)\|^2 ds \\
&= \frac{1}{2}\int_0^\infty \int_{\mathcal{O}} |\frac{\partial}{\partial t}\rho - \Delta\rho + F'(\rho)|^2 dxdt.
\end{aligned}$$

13.1.5. The large deviation theorem.

THEOREM 13.7. *Let $Y_n, n = 1, 2, \ldots$ satisfy (13.1). Assume that $n^{-1}m^{2+d} \to c_0 \ge 0$, $F \in C^2(R)$ with F'' bounded, and $F(r) \ge c_1 + c_2 r^2$ with $c_2 > 0$. Suppose $\eta_n(Y_n(0))$ satisfies a large deviation principle with good rate function I_0. Then $\{\eta_n(Y_n)\}$ satisfies a large deviation principle in $C_{L^2(\mathcal{O})}[0, \infty)$ with good rate function*

$$I(\rho) = I_0(\rho(0)) + \frac{1}{2}\int_0^\infty \int_{\mathcal{O}} |\frac{\partial}{\partial t}\rho - \Delta\rho + F'(\rho)|^2 dxdt.$$

13.2. Stochastic Cahn-Hilliard equations on rescaled lattice

We now consider the large deviation problem in Example 1.13. As before, let $\Lambda_m \subset \mathcal{O} = [0, 1)^d$ be the periodic lattice with mesh size m^{-1} in each direction. We assume that m is odd so that Δ_m has only one zero eigenvalue. Let E, E_n be defined according to (1.38):

$$E_n = \{\rho \in R^{|\Lambda_m|} : \sum_{x\in\Lambda_m} \rho(x) = 0\}, \qquad E = \{\rho \in L^2(\mathcal{O}) : \int \rho(x)dx = 0\}.$$

The metric on E will be the metric given by the L^2 norm, $r(\rho, \gamma) = \|\rho - \gamma\|$. Define $\langle p, q\rangle_m$, $\|\rho\|_m$, $\eta_n = \widehat{\eta}_m : E_n \mapsto E$, and $\pi_m : E \mapsto E_n$ as in Section 13.1. Let $\mathcal{Q}$ in Definition 2.5 be the collection of compact subsets of E, and define $K_n^q = \pi_m(q), q \in \mathcal{Q}$. Let $B = (-\Delta)^{-1}$. Since $\overline{\rho} \equiv \int_{\mathcal{O}} \rho(x)dx = 0$ for every $\rho \in E$,

$$B\rho = \int_0^\infty S(t)\rho dt = \int_0^\infty S(t)(\rho - \overline{\rho})dt,$$

where $\{S(t)\}$ is the semigroup corresponding to Δ with periodic boundary conditions. The continuity properties of $\{S(t)\}$ imply that B is a bounded, compact operator on E. The Poincaré inequality gives

$$\|\rho\|^2 = \int_{\mathcal{O}} |\rho(x) - \overline{\rho}|^2 dx \le C\|\nabla\rho\|^2, \tag{13.21}$$

Define $H^{\beta}(\mathcal{O}) : \beta = \ldots, -1, 0, 1, \ldots$ as in Example 9.32, and recall that the L^2 inner product can be extended so that

$$|\langle \rho, \gamma \rangle| \leq \|\rho\|_{-\beta} \|\gamma\|_{\beta}, \quad \rho \in H^{-\beta}(\mathcal{O}), \gamma \in H^{\beta}(\mathcal{O}).$$

Note that B is not the same as in Section 13.1, and to emphasize the difference, we use $H^{\beta}(\mathcal{O})$ instead of the previous notation $H_{\beta}(\mathcal{O})$, $\beta = \ldots, -1, 0, 1, \ldots$.

Define the discrete partial derivative in the i-th direction ∇_n^i and the discrete Laplacian Δ_m as in Appendix C. Let $\{B_i(t,x) : x \in \Lambda_m, i = 1, \ldots, d\}$ be a collection of independent standard Brownian motions, and let $Y_n(t,x)$ satisfy (13.22)

$$dY_n(t,x) = \Delta_m(-\Delta_m Y_n(t,x) + F'(Y_n(t,x))dt + \frac{m^{d/2}}{\sqrt{n}} \sum_{i=1}^{d} \nabla_m^i dB_i(t,x), \quad x \in \Lambda_m.$$

We study the large deviation behavior of $\{\eta_n(Y_n)\}$ as a sequence of E-valued Markov processes (Theorem 13.13). As in the previous example, we assume that F'' is bounded so that F' is Lipschitz, and denote the Lipschitz constant by L_F. We also assume the existence of $c_1 \in R$, $c_2 > 0$ such that $F(r) \geq c_1 + c_2 r^2$, and that $m^{4+d}/n \to c_0$.

13.2.1. Convergence of H_n.

As in the stochastic reaction-diffusion example in Section 13.1, the test functions defining (1.40) is not good enough for verifying the convergence Condition 7.11. Therefore, we first identify a suitable class of test functions f_n which will verify that condition.

Let $\{(\lambda_{k,m}, e_{k,m}) : k = (k_1, \ldots, k_d), 0 \leq k_j \leq m-1\}$ be the eigenvalue-eigenfunction system for $-\Delta_m$ on E, that is, $-\Delta_m e_{k,m} = \lambda_{k,m} e_{k,m}$, where $\{e_{k,m}\}$ forms an orthonormal basis for E. (See Appendix C.) For $p \in R^{|\Lambda_m|}$, we note that

$$\begin{aligned} Z_{n,p}(t) &= \langle Y_n(t), p \rangle_m = \langle Y_n(0), p \rangle_m + \int_0^t \langle \Delta_m(-\Delta_m Y_n(s) + F'(Y_n(s))), p \rangle_m ds \\ &\quad + \frac{m^{d/2}}{\sqrt{n}} \sum_{i=1}^{d} \langle \nabla_m^i B_i(t), p \rangle_m. \end{aligned}$$

As before, let $\mathbb{D}_m(\rho) = \Delta_m \rho - F'(\rho)$. For $\gamma \in R^{|\Lambda_m|}$ and $\{a_k\} \subset R$, let

$$\begin{aligned} &R_{n,\gamma}^{\{a_k\}}(t) \\ &\equiv \sum_k a_k \langle Y_n(t) - \gamma, e_{k,m} \rangle_m^2 \\ &= \sum_k a_k \langle Y_n(0) - \gamma, e_{k,m} \rangle_m^2 \\ &\quad + \frac{m^{d/2}}{\sqrt{n}} \sum_k a_k \int_0^t 2\langle Y_n(s) - \gamma, e_{k,m} \rangle_m d \sum_{i=1}^{d} \langle \nabla_m^i B_i(s), e_{k,m} \rangle_m \\ &\quad + \int_0^t \sum_k a_k \Big(2\langle Y_n(s) - \gamma, e_{k,m} \rangle_m \langle -\Delta_m \mathbb{D}_m(\rho), e_{k,m} \rangle_m + \frac{1}{n} \lambda_{k,m} \Big) ds. \end{aligned}$$

The last equality follows from

$$\begin{aligned}
&[\langle Y_n - \gamma, e_{k,m}\rangle_m, \langle Y_n - \gamma, e_{l,m}\rangle_m]_t \\
= \; & \frac{m^d}{n}\sum_{i=1}^{d}[\langle B_i, (-\nabla_m^i)e_{k,m}\rangle_m, \langle B_i, (-\nabla_m^i)e_{l,m}\rangle_m]_t \\
= \; & \frac{1}{n}\sum_{i=1}^{d}\langle \nabla_m^i e_{k,m}, \nabla_m^i e_{l.m}\rangle_m t \\
= \; & \frac{1}{n}\langle -\Delta_m e_{k,m}, e_{l,m}\rangle_m t = \frac{1}{n}\delta_{kl}\lambda_{k,m} t.
\end{aligned}$$

Consequently, we also have

$$[R_{n,\gamma_1}^{\{a_k\}}, R_{n,\gamma_2}^{\{b_k\}}]_t = 4\frac{1}{n}\int_0^t \sum_k a_k b_k \lambda_k \langle Y_n(s) - \gamma_1, e_{k,m}\rangle_m \langle Y_n(s) - \gamma_2, e_{k,m}\rangle_m ds.$$

Let $B_m = (-\Delta_m)^{-1}$, so $B_m e_{k,m} = \lambda_{k,m}^{-1} e_{k,m}$, and define

$$\|\rho\|_{-2,m}^2 = \|B_m\rho\|_m^2 = \sum_k \lambda_{k,m}^{-1}\langle \rho, e_{k,m}\rangle_m^2, \quad \forall \rho \in E_m.$$

As in Section 13.1, let $\varphi \in C_b^2(R^{l+1}$ and $\gamma_i \in E_n$, $i = 0, 1, \ldots, l$, and define

$$f_n(\rho) = \varphi(\|\rho - \gamma_0\|_m^2, \|\rho - \vec{\gamma}\|_{-2,m}^2).$$

Then by Itô's formula, the generator of Y_n applied to f_n is

$$\begin{aligned}
&A_n f_n(\rho) \\
&= 2\sum_{i=1}^{l}\partial_i\varphi(\|\rho - \gamma_0\|_m^2, \|\rho - \vec{\gamma}\|_{-2,m}^2)\Big(\langle B_m^2(\rho - \gamma_i), -\Delta_m \mathbb{D}_m(\rho)\rangle_m + \frac{1}{n}\sum_k \lambda_{k,m}^{-1}\Big) \\
&\quad +2\partial_0\varphi(\|\rho - \gamma_0\|_m^2, \|\rho - \vec{\gamma}\|_{-2,m}^2)(\langle \rho - \gamma_0, -\Delta_m \mathbb{D}_m(\rho)\rangle_m + \frac{1}{n}\sum_k \lambda_{k,m}) \\
&\quad +\frac{2}{n}\sum_{i,j=1}^{l}\partial_i\partial_j\varphi(\|\rho - \gamma_0\|_m^2, \|\rho - \vec{\gamma}\|_{-2,m}^2)\langle -\Delta_m B_m^2(\rho - \gamma_i), B_m^2(\rho - \gamma_j)\rangle_m \\
&\quad +\frac{2}{n}\sum_{i=1}^{l}\partial_0\partial_i\varphi(\|\rho - \gamma_0\|_m^2, \|\rho - \vec{\gamma}\|_{-2,m}^2)\langle \rho - \gamma_0, B_m(\rho - \gamma_i)\rangle_m \\
&\quad +\frac{2}{n}\partial_0^2\varphi(\|\rho - \gamma_0\|_m^2, \|\rho - \vec{\gamma}\|_{-2,m}^2)\langle -\Delta_m(\rho - \gamma_0), \rho - \gamma_0\rangle_m.
\end{aligned}$$

and

$$\begin{aligned}
&H_n f_n(\rho)\\
&= 2\sum_{i=1}^{l}\partial_i\varphi(\|\rho-\gamma_0\|_m^2, \|\rho-\vec{\gamma}\|_{-2,m}^2)\Big(\langle B_m^2(\rho-\gamma_i), -\Delta_m\mathbb{D}_m(\rho)\rangle_m + \frac{1}{n}\sum_k \lambda_{k,m}^{-1}\Big)\\
&\quad +2\partial_0\varphi(\|\rho-\gamma_0\|_m^2, \|\rho-\vec{\gamma}\|_{-2,m}^2)(\langle\rho-\gamma_0, -\Delta_m\mathbb{D}_m(\rho)\rangle_m + \frac{1}{n}\sum_k \lambda_{k,m})\\
&\quad +2\sum_{i,j=1}^{l}\partial_i\varphi\partial_j\varphi(\|\rho-\gamma_0\|_m^2, \|\rho-\vec{\gamma}\|_{-2,m}^2)\langle -\Delta_m B_m^2(\rho-\gamma_i), B_m^2(\rho-\gamma_j)\rangle_m\\
&\quad +2\sum_{i=1}^{l}\partial_0\varphi\partial_i\varphi(\|\rho-\gamma_0\|_m^2, \|\rho-\vec{\gamma}\|_{-2,m}^2)\langle\rho-\gamma_0, -\Delta_m B_m^2(\rho-\gamma_i)\rangle_m\\
&\quad +2(\partial_0\varphi)^2(\|\rho-\gamma_0\|_m^2, \|\rho-\vec{\gamma}\|_{-2,m}^2)\langle(-\Delta_m)(\rho-\gamma_0), \rho-\gamma_0\rangle_m\\
&\quad +\frac{2}{n}\sum_{i,j=1}^{l}\partial_i\partial_j\varphi(\|\rho-\gamma_0\|_m^2, \|\rho-\vec{\gamma}\|_{-2,m}^2)\langle -\Delta_m B_m^2(\rho-\gamma_i), B_m^2(\rho-\gamma_j)\rangle_m\\
&\quad +\frac{2}{n}\sum_{i=1}^{l}\partial_0\partial_i\varphi(\|\rho-\gamma_0\|_m^2, \|\rho-\vec{\gamma}\|_{-2,m}^2)\langle\rho-\gamma_0, -\Delta_m B_m^2(\rho-\gamma_i)\rangle_m\\
&\quad +\frac{2}{n}\partial_0^2\varphi(\|\rho-\gamma_0\|_m^2, \|\rho-\vec{\gamma}\|_{-2,m}^2)\langle -\Delta_m(\rho-\gamma_0), \rho-\gamma_0\rangle_m.
\end{aligned}$$

Let $\gamma_0 \in C^4(\mathcal{O})$. Then for $\rho \in H^2(\mathcal{O})$,

$$\langle\rho-\gamma_0, \Delta(-\Delta\rho+F'(\rho))\rangle = -\|\Delta\rho\|^2 + \langle\Delta\rho, F'(\rho)\rangle + \langle\Delta^2\gamma_0, \rho\rangle - \langle\Delta\gamma_0, F'(\rho)\rangle.$$

Extend this expression to all $\rho \in E$ by noting that if $\rho_n \in H^2(\mathcal{O})$ and $\|\rho_n - \rho\| \to 0$ for $\rho \notin H^2(\mathcal{O})$, then

$$\limsup_{n\to\infty}\langle\rho_n-\gamma_0, \Delta(-\Delta\rho_n+F'(\rho_n))\rangle = -\infty.$$

It follows that $\langle\rho-\gamma_0, \Delta(-\Delta\rho+F'(\rho))\rangle$ is upper semicontinuous in $\rho \in E$ and is bounded above. Similarly, since $\|\nabla\rho\|$ is dominated by $\|\Delta\rho\|$,

$$\limsup_{n\to\infty}\langle\rho_n-\gamma_0, \Delta(-\Delta\rho_n+F'(\rho_n))\rangle + c\|\nabla(\rho_n-\gamma_0)\|^2 = -\infty \tag{13.23}$$

for $\rho_n \to \rho \notin H^2(\mathcal{O})$ and $c \in R$.

Following an argument similar to the proof of Lemma 13.1, we have

LEMMA 13.8. *Let $\mathcal{D}(H_\dagger)$ be the collection of functions of the form*

$$f(\rho) = \varphi(\|\rho-\gamma_0\|^2, \|\rho-\vec{\gamma}\|_{-2}^2), \tag{13.24}$$

where $\gamma_0 \in E\cap C_b^4(\mathcal{O})$, $\gamma_1,\ldots,\gamma_l \in E$, $l = 1,2,\ldots$, $\varphi \in C_b^2(R^{l+1})$, $\inf_{r\in R^{l+1}}\varphi(r) > -\infty$, and for $i = 0,1,\ldots,l$, $\partial_i\varphi \geq 0$ and there exists d_i such that $\partial_i\varphi(r_0,r_1,\ldots,r_l) =$

0 if $r_i \geq d_i$. Define

$$\begin{aligned}
&H_\dagger f(\rho)\\
&\quad = 2\sum_{i=1}^{l} \partial_i\varphi(\|\rho-\gamma_0\|^2, \|\rho-\vec{\gamma}\|_{-2}^2)\langle B^2(\rho-\gamma_i), \Delta(-\Delta\rho+F'(\rho))\rangle\\
&\quad +2\partial_0\varphi(\|\rho-\gamma_0\|^2, \|\rho-\vec{\gamma}\|_{-2}^2)\langle \rho-\gamma_0, \Delta(-\Delta\rho+F'(\rho))\rangle\\
&\quad +\frac{1}{2}\|\nabla\Big(2\sum_{i=1}^{l} \partial_i\varphi(\|\rho-\gamma_0\|^2, \|\rho-\vec{\gamma}\|_{-2}^2)B^2(\rho-\gamma_i)\\
&\qquad\qquad +2\partial_0\varphi(\|\rho-\gamma_0\|^2, \|\rho-\vec{\gamma}\|_{-2}^2)(\rho-\gamma_0)\Big)\|^2\\
&\quad = 2\sum_{i=1}^{l} \partial_i\varphi(\|\rho-\gamma_0\|^2, \|\rho-\vec{\gamma}\|_{-2}^2)\langle B(\rho-\gamma_i), \Delta\rho-F'(\rho)\rangle\\
&\quad +2\partial_0\varphi(\|\rho-\gamma_0\|^2, \|\rho-\vec{\gamma}\|_{-2}^2)\langle \rho-\gamma_0, \Delta(-\Delta\rho+F'(\rho))\rangle\\
&\quad +\frac{1}{2}\|\nabla\Big(2\sum_{i=1}^{l} \partial_i\varphi(\|\rho-\gamma_0\|^2, \|\rho-\vec{\gamma}\|_{-2}^2)B^2(\rho-\gamma_i)\\
&\qquad\qquad +2\partial_0\varphi(\|\rho-\gamma_0\|^2, \|\rho-\vec{\gamma}\|_{-2}^2)(\rho-\gamma_0)\Big)\|^2,
\end{aligned}$$

for $\rho \in H^2(\mathcal{O})$ and extend the definition to all $\rho \in E$ using (13.23).

Let $\mathcal{D}(H_\ddagger) = \{f : -f \in \mathcal{D}(H_\dagger)\}$, and define $H_\ddagger$ by the same formula as $H_\dagger$. Then Condition 7.11 is satisfied with $v_n = n$.

REMARK 13.9. By the assumptions on φ and (13.23), $\sup_{\rho\in E} H_\dagger f(\rho) < \infty$ and $H_\dagger f \in M^u(E, \overline{R})$ and is upper semicontinuous. Similarly, $H_\ddagger f \in M^l(E, \overline{R})$ and is lower semicontinuous.

PROOF. The proof follows that of Lemma 13.1; however, we need some additional estimates. Note that $\lim_{n\to\infty} n^{-1}\sum_k \lambda_{k,m}^{-1} = 0$, and since we are assuming $m^{4+d}/n \to c_0$, $m^{2+d}n^{-1} \to 0$ and

$$\lim_{n\to\infty} n^{-1}\sum_k \lambda_{k,m} = 0.$$

In addition, to control the growth rate of $\sup_{\rho\in E_n} |H_n f_n(\rho)|$, note that

$$\|\Delta_m\rho\|_m^2 = \|\nabla_m\cdot(\nabla_m\rho)\|_m^2 \leq dm^2\|\nabla_m\rho\|_m^2 \leq d^2m^4\|\rho\|_m^2, \quad \forall\rho\in E_n.$$

□

For $f : E \to \overline{R}$, we use the notion of gradient given in Definition 9.33. For f of the form (13.24),

$$\begin{aligned}
\operatorname{grad} f(\rho) &= -2\Delta\Big(\partial_0\varphi(\|\rho-\gamma_0\|^2, \|\rho-\vec{\gamma}\|_{-2}^2)(\rho-\gamma_0)\\
&\qquad +\sum_{i=1}^{l} \partial_i\varphi(\|\rho-\gamma_0\|^2, \|\rho-\vec{\gamma}\|_{-2}^2)B^2(\rho-\gamma_i)\Big)\\
&= -2\partial_0\varphi(\|\rho-\gamma_0\|^2, \|\rho-\vec{\gamma}\|_{-2}^2)\Delta(\rho-\gamma_0)\\
&\qquad +2\sum_{i=1}^{l} \partial_i\varphi(\|\rho-\gamma_0\|^2, \|\rho-\vec{\gamma}\|_{-2}^2)B(\rho-\gamma_i) \in H^{-2}(\mathcal{O}).
\end{aligned}$$

Define $\mathcal{E} : E \to \overline{R}$ as in (9.64). Then by (9.73), for ρ satisfying $\mathcal{E}(\rho) < \infty$,

$$\text{grad}\mathcal{E}(\rho) = \Delta(\Delta\rho - F'(\rho)).$$

Let

$$\langle p, q\rangle_{-1} = \langle Bp, q\rangle, \quad \forall q, p \in H^{-1}(\mathcal{O}).$$

Noting that

$$\|\nabla p\|^2 = \langle \Delta p, p\rangle = \|\Delta p\|_{-1}^2$$

and

$$\langle \Delta(-\Delta\rho + F'(\rho)), \text{grad}f(\rho)\rangle_{-1} = \langle \Delta\rho - F'(\rho), \text{grad}f(\rho)\rangle,$$

for $\rho \in H^2(\mathcal{O})$,

$$\begin{aligned} H_\dagger f(\rho) &= \langle \Delta(-\Delta\rho + F'(\rho)), \text{grad}f(\rho)\rangle_{-1} + \frac{1}{2}\|\text{grad}f(\rho)\|_{-1}^2 \\ &= -\langle \text{grad}\mathcal{E}(\rho), \text{grad}f(\rho)\rangle_{-1} + \frac{1}{2}\|\text{grad}f(\rho)\|_{-1}^2. \end{aligned}$$

A similar representation holds for $H_\ddagger$.

We next extend the definition of $H_\dagger, H_\ddagger$ to include test functions of the form

$$f(\rho) = \alpha\|\rho - \gamma\|_{-2}^2 + c = \alpha\|B(\rho - \gamma)\|^2 + c, \quad \forall \alpha, c \in R, \gamma \in E.$$

For $\alpha > 0$, define

$$\begin{aligned} (13.25)\widetilde{H}_\dagger f(\rho) &= \langle \Delta(-\Delta\rho + F'(\rho)), \text{grad}f(\rho)\rangle_{-1} + \frac{1}{2}\|\text{grad}f(\rho)\|_{-1}^2 \\ &= \langle \Delta(-\Delta\rho + F'(\rho)), 2\alpha B^2(\rho - \gamma)\rangle + \frac{1}{2}\|\nabla 2\alpha B^2(\rho - \gamma)\|^2. \end{aligned}$$

Then $\widetilde{H}_\dagger f$ is bounded above and continuous. Similarly, for $\alpha < 0$, define

$$\widetilde{H}_\ddagger f = \langle \Delta(-\Delta\rho + F'(\rho)), 2\alpha B^2(\rho - \gamma)\rangle + \frac{1}{2}\|\nabla 2\alpha B^2(\rho - \gamma)\|^2.$$

Then $\widetilde{H}_\ddagger f$ is bounded below and continuous.

We have the following analogue of Lemma 13.3, The proof is the same.

LEMMA 13.10. *For $h \in C_b(E)$, any subsolution of $f - \alpha H_\dagger f = h$ is a subsolution of $f - \alpha\widetilde{H}_\dagger f = h$ and any supersolution of $f - \alpha H_\ddagger f = h$ is a supersolution of $f - \alpha\widetilde{H}_\ddagger f = h$.*

13.2.2. Exponential tightness. Exponential tightness follows by Corollary 4.19 provided we can verify (4.25). The verification is similar to Example 2.9. We only highlight the differences.

Let $\mathcal{E}_m$ be defined by (2.14). We first establish an analogue of (2.15). By Itô's formula,

$$\begin{aligned} H_n\mathcal{E}_n(\rho) &= \langle \Delta_m(-\Delta_m\rho + F'(\rho)), -\Delta_m\rho + F'(\rho)\rangle_m + \frac{1}{2}\|\nabla_n(-\Delta_m\rho + F'(\rho))\|_m^2 \\ &\quad + \frac{1}{2n}\sum_{k=1}^d \sum_{z\in\Lambda_m} m^{-d}\left(\frac{m^{d+4}}{2} + \frac{F''(\rho(z))}{2}m^{d+2}\right) \\ &\le -\frac{1}{2}\|\nabla_m(-\Delta_m\rho + F'(\rho))\|_m^2 \\ &\quad + \frac{1}{2n}\sum_{k=1}^d \sum_{z\in\Lambda_m} m^{-d}\left(\frac{m^{d+4}}{2} + \frac{F''(\rho(z))}{2}m^{d+2}\right). \end{aligned}$$

Since $m^{4+d}/n \to c_0$,

$$\sup_n \sup_{\rho \in E_n} H_n \mathcal{E}_n(\rho) < \infty. \tag{13.26}$$

To obtain the analog of Lemma 13.4 for solutions Y_n, Z_n of (13.22), observe that

$$\begin{aligned}
&\frac{d}{ds}\frac{1}{2}\|Y_n(s) - Z_n(s)\|_m^2 \\
&= \langle \Delta_m(-\Delta_m Y_n(s) + F'(Y_n)), Y_n(s) - Z_n(s)\rangle_m \\
&\quad -\langle \Delta_m(-\Delta_m Z_n(s) + F'(Z_n(s))), Y_n(s) - Z_n(s)\rangle_m \\
&= -\|\Delta_m(Y_n(s) - Z_n(s))\|_m^2 - \langle F'(Y_n(s)) - F'(Z_n(s)), Y_n(s) - Z_n(s)\rangle_m \\
&\le \frac{1}{4} L_F \|Y_n(s) - Z_n(s)\|_m^2.
\end{aligned}$$

and Gronwall's inequality gives

$$\|Y_n(t) - Z_n(t)\|_m^2 \le \|Y_n(0) - Z_n(0)\|_m^2 e^{\frac{1}{2}L_F t}.$$

The rest of the argument follows as in Example 2.9. Note that we do not need (2.17) here, as the Poincaré inequality (13.21) and its discrete version hold for elements in E and on E_n. Consequently, if $\sup_n \mathcal{E}_m(\rho_n) < \infty$, $\sup_n \|\nabla_m \rho_n\|_m < \infty$ and $\{\eta_n(\rho_n)\}$ is relatively compact in E.

13.2.3. The comparison principle. Let D_0 be defined according to (13.14). Suppose $\alpha > 0$ and $h \in D_0$. Let $\overline{f}$ be a viscosity subsolution of $(I - \alpha H_\dagger)f = h$ and $\underline{f}$ a viscosity supersolution. As in Section 13.1.3, D_0 is buc-dense in $C_b(E)$, and each $h \in D$ is uniformly continuous with respect to $\|\cdot\|_{-2}$. Let ω_h be a modulus of continuity. The proof of the comparison principle is similar to the previous example.

Let

$$\begin{aligned}
\overline{f}_\delta(\rho) &= \lambda_{0,\delta} \sup_{\gamma \in E}\{\overline{f}(\gamma) - \frac{\|B(\rho - \gamma)\|^2}{2\delta}\}, \\
\underline{f}_\delta(\gamma) &= \lambda_{1,\delta} \inf_{\rho \in E}\{\underline{f}(\rho) + \frac{\|B(\rho - \gamma)\|^2}{2\delta}\},
\end{aligned}$$

where

$$\lambda_{0,\delta} = 1 + \delta C L_F^4/4, \quad \lambda_{1,\delta} = (1 - \delta C L_F^4/4)^{-1}$$

and C (depending only on the dimension d) is the constant in Poincaré inequality

$$\int_O |\rho(x) - \int_O \rho(z)dz|^2 dx \le C\|\nabla \rho\|^2.$$

Let $\widehat{H}_\dagger, \widehat{H}_\ddagger$ be given by (9.77) and (9.78).

LEMMA 13.11. *$\overline{f}_\delta$ is a continuous viscosity subsolution of*

$$(I - \alpha \widehat{H}_\dagger)f = h_{0,\delta}$$

and $\underline{f}_\delta$ is a continuous viscosity supersolution of

$$(I - \alpha \widehat{H}_\ddagger)f = h_{1,\delta},$$

where

$$h_{0,\delta} = \lambda_{0,\delta}\Big(h + \omega_h(2\sqrt{\delta \sup_{\rho \in E} |\overline{f}(\rho)|})\Big)$$

and

$$h_{1,\delta} = \lambda_{1,\delta}\Big(h - \omega_h(2\sqrt{\delta \sup_{\rho\in E}|\underline{f}(\rho)|})\Big).$$

PROOF. The general method is similar to the proof of Lemma 13.5. The major difference lies in technical estimates in which the second order differential operators are replaced by forth order ones.

Let $f_0 \in \mathcal{D}(\widehat{H}_\dagger)$ be of the form (9.75). Then $\overline{f}_\delta - f_0$ attains its supremum at some point $\rho_0 \in H^1(\mathcal{O})$. By Lemma 13.10 and the definition of viscosity subsolution, there exists $\epsilon_k \to 0$ and $\gamma_k \in E$ such that

$$\begin{aligned}
&\Big(\lambda_{0,\delta}(\overline{f}(\gamma_k) - \frac{\|B(\rho_0-\gamma_k)\|^2}{2\delta}) - f_0(\rho_0)\Big) + \epsilon_k \\
&\geq \sup_{\rho,\gamma\in E}\Big(\lambda_{0,\delta}(\overline{f}(\gamma) - \frac{\|B(\rho-\gamma)\|^2}{2\delta}) - f_0(\rho)\Big) \qquad (13.27)\\
&= \sup_{\rho\in E}\Big(\overline{f}_\delta(\rho) - f_0(\rho)\Big)
\end{aligned}$$

and

$$\begin{aligned}
(13.28)\quad \alpha^{-1}(\overline{f} - h)(\gamma_k) &\leq \epsilon_k + (\widetilde{H}_\dagger \frac{\|B(\rho_0 - \cdot)\|^2}{2\delta})(\gamma_k) \\
&= \epsilon_k + \frac{1}{\delta}\langle \Delta(-\Delta\gamma_k + F'(\gamma_k)), B^2(\gamma_k - \rho_0)\rangle + \frac{1}{2}\|\frac{\nabla B^2(\gamma_k-\rho_0)}{\delta}\|^2.
\end{aligned}$$

From (13.27), $\sup_k \|B(\rho_0 - \gamma_k)\| < \infty$. Noting that $\overline{f}$ and h are bounded, as in the proof of Lemma 13.5, (13.28) implies that $\sup_k \|\gamma_k\| < \infty$. Therefore, there exists $\gamma_0 \in L^2(\mathcal{O})$ that is a weak sequential limit point of $\{\gamma_k\}$. Without loss of generality, we can assume that $\{\gamma_k\}$ converges weakly to γ_0, and in particular, $\lim_{k\to\infty} B\gamma_k = B\gamma_0$.

Note that

$$\begin{aligned}
&\frac{1}{\delta}\langle \Delta(-\Delta\gamma_k + F'(\gamma_k)), B^2(\gamma_k-\rho_0)\rangle - \frac{1}{\delta}\langle \Delta(-\Delta\rho_0 + F'(\rho_0)), B^2(\gamma_k-\rho_0)\rangle \\
&= -\frac{1}{\delta}\|\gamma_k - \rho_0\|^2 + \frac{1}{\delta}\langle F'(\gamma_k) - F'(\rho_0), B(\gamma_k - \rho_0)\rangle) \\
&\leq -\frac{1}{2\delta}\|\gamma_k-\rho_0\|^2 + \frac{L_F^2}{2\delta}\|B(\gamma_k-\rho_0)\|^2 \\
&\leq -\frac{1}{2\delta}\|\gamma_k-\rho_0\|^2 + \frac{L_F^2}{2}\|\gamma_k-\rho_0\|\|\frac{B^2(\gamma_k-\rho_0)}{\delta}\| \\
&\leq \frac{\delta}{2}\frac{L_F^4}{4}\|\frac{B^2(\gamma_k-\rho_0)}{\delta}\|^2 \leq \frac{CL_F^4}{8}\delta\|\frac{\nabla B^2(\gamma_k-\rho_0)}{\delta}\|^2,
\end{aligned}$$

where the last step follows from the Poincaré inequality (13.21). Therefore from (13.27) and (13.28), we conclude

$$\begin{aligned}
(13.29)\quad \alpha^{-1}\Big(\lambda_{0,\delta}^{-1}\overline{f}_\delta(\rho_0) - h(\gamma_0)\Big) &\leq \liminf_{k\to\infty}\alpha^{-1}(\overline{f} - h)(\gamma_k) \\
&\leq \lim_{k\to\infty}\Big(\frac{1}{\delta}\langle \Delta(-\Delta\rho_0 + F'(\rho_0)), B^2(\gamma_k-\rho_0)\rangle \\
&\qquad + \frac{1}{2}(1+\delta\frac{CL_F^4}{4})\|\frac{\nabla B^2(\gamma_k-\rho_0)}{\delta}\|^2\Big) \\
&= \langle \Delta(-\Delta\rho_0 + F'(\rho_0)), \frac{B^2(\gamma_0-\rho_0)}{\delta}\rangle + \frac{\lambda_{0,\delta}}{2}\|\frac{\nabla B^2(\gamma_0-\rho_0)}{\delta}\|^2.
\end{aligned}$$

From (13.27),

$$-\lambda_{0,\delta}\frac{\|B(\rho_0-\gamma_k)\|^2}{2\delta}-f_0(\rho_0)+\epsilon_k\geq-\lambda_{0,\delta}\frac{\|B(\rho-\gamma_k)\|^2}{2\delta}-f_0(\rho),\quad \rho\in E,$$

and letting $k\to\infty$,

$$(13.30) f_0(\rho)-f_0(\rho_0)\geq-\lambda_{0,\delta}\frac{\|B(\rho-\gamma_0)\|^2}{2\delta}-\Big(-\lambda_{0,\delta}\frac{\|B(\rho_0-\gamma_0)\|^2}{2\delta}\Big).$$

Let $p\in C^\infty(\mathcal{O})$ and $\rho(t)=\rho_0-t\Delta p$. Replacing ρ in (13.30) by $\rho(t)$ and differentiation at $t=0$, we have

$$\int_{\mathcal{O}}\mathrm{grad} f_0(\rho_0)p dx\geq\int_{\mathcal{O}}\lambda_{0,\delta}\frac{B(\gamma_0-\rho_0)}{\delta}p dx.$$

Since p is arbitrary,

$$\mathrm{grad} f_0(\rho_0)=\lambda_{0,\delta}\frac{B(\rho_0-\gamma_0)}{\delta}\in H^2(\mathcal{O}). \tag{13.31}$$

Taking $\rho=\rho_0$ and $\gamma=\gamma_0$ on the right side of (13.27), we have

$$\liminf_{k\to\infty}(\overline{f}(\gamma_k)-\overline{f}(\rho_0))\geq\lim_{k\to\infty}\frac{\|B(\rho_0-\gamma_k)\|^2}{2\delta},$$

so

$$4\delta\sup_{\rho\in E}|\overline{f}(\rho)|\geq\|B(\rho_0-\gamma_0)\|^2. \tag{13.32}$$

By (13.31), (13.29) becomes

$$\alpha^{-1}\Big(\overline{f}_\delta(\rho_0)-\lambda_{0,\delta}h(\gamma_0)\Big)\leq\langle\Delta(-\Delta\rho_0+F'(\rho_0)),\mathrm{grad} f_0(\rho_0)\rangle_{-1}+\frac{1}{2}\|\mathrm{grad} f_0(\rho_0)\|_{-1}^2,$$

and since

$$\lambda_{0,\delta}h(\gamma_0)\leq\lambda_{0,\delta}h(\rho_0)+\lambda_{0,\delta}\omega_h(\|B(\rho_0-\gamma_0)\|)\leq h_{0,\delta}(\rho_0),$$

so the conclusion follows. □

THEOREM 13.12. *Let h be given by (13.14 and $\alpha>0$. Suppose $\overline{f}$ is a subsolution of $(I-\alpha H_\dagger)f=h$ and $\underline{f}$ is a supersolution of $(I-\alpha H_\ddagger)f=h$. Then*

$$\overline{f}\leq\underline{f}.$$

PROOF. By Lemma 13.11 and Theorem 9.34,

$$\sup_{\rho\in E}\Big(\overline{f}_\delta(\rho)-\underline{f}_\delta(\rho)\Big)\leq\sup_{\rho\in E}\Big(h_{0,\delta}(\rho)-h_{1,\delta}(\rho)\Big).$$

Sending $\delta\to0+$, we conclude. □

13.2.4. Variational representation. Recall that $\langle p,q\rangle_{-1}=\langle Bp,q\rangle$ for $q,p\in H^{-1}(\mathcal{O})$. If $\mathrm{grad} f(\rho)\in H^{-1}(\mathcal{O})$, then

$$\frac{1}{2}\|\mathrm{grad} f(\rho)\|_{-1}^2=\sup_{u\in H^{-1}(\mathcal{O})}(\langle\mathrm{grad} f(\rho),u\rangle_{-1}-\frac{1}{2}\|u\|_{-1}^2).$$

Therefore,

$$H_\dagger f(\rho)=\sup_{u\in H^{-1}(\mathcal{O})}(\langle\Delta(-\Delta\rho+F'(\rho))+u,\mathrm{grad} f(\rho)\rangle_{-1}-\frac{1}{2}\|u\|_{-1}^2),$$

and a representation of the same form holds for $H_\ddagger f$.

As in Section 13.1, we need to exercise care in defining $\mathbf{H},\mathbf{H}_\dagger$ and $\mathbf{H}_\ddagger$.

Let

$$D_\dagger = \{f(\rho) = \varphi(\|\rho-\gamma\|_{-2}^2) : \gamma \in E, \varphi \in C_b^2(R), \varphi' \geq 0, \exists d_\varphi, \varphi'(r) = 0, \text{ if } r \geq d_\varphi\},$$

and

$$D_\ddagger = \{f(\rho) = \varphi(\|\rho-\gamma\|_{-2}^2) : \gamma \in E, \varphi \in C_b^2(R), \varphi' \leq 0, \exists d_\varphi, \varphi'(r) = 0, \text{ if } r \geq d_\varphi\}$$

Then $D_\dagger, D_\ddagger \subset C_b(E)$ separate points, and for $f \in D_\dagger \cup D_\ddagger$,

$$\operatorname{grad} f(\rho) = 2\varphi'(\|\rho-\gamma\|_{-2}^2)(-\Delta)B^2(\rho-\gamma) = 2\varphi'(\|\rho-\gamma\|_{-2}^2)B(\rho-\gamma) \in H^2(\mathcal{O}).$$

Let $\mathcal{D}(A)$ be the linear span of $D_\dagger \cup D_\ddagger$, and for $\rho \in E$ and $u \in U \equiv H^{-2}(\mathcal{O})$, define

$$Af(\rho,u) = \langle \Delta(-\Delta\rho + F'(\rho)) + u, \operatorname{grad} f(\rho)\rangle_{-1} = \langle \Delta\rho - F'(\rho)) + u, \operatorname{grad} f(\rho)\rangle$$

and

$$L(\rho,u) = \frac{1}{2}\|u\|_{-1}^2.$$

Then Af is continuous on $E \times U$, $\{u \in U : L(\rho,u) \leq c\}$ is compact in U for each $c < \infty$, and

$$\mathbf{H}f(\rho) = \sup_{u \in U}\{Af(\rho,u) - L(\rho,u)\} = \sup_{u \in H^{-1}(\mathcal{O})}\{Af(\rho,u) - L(\rho,u)\}.$$

As in the previous example, define $\mathbf{H}_\dagger$ and $\mathbf{H}_\ddagger$ by letting $\mathcal{D}(\mathbf{H}_\dagger) = D_\dagger$, $\mathbf{H}_\dagger f = H_\dagger f$, $\mathcal{D}(\mathbf{H}_\ddagger) = D_\ddagger$ and $\mathbf{H}_\ddagger f = H_\ddagger f$. It follows that (8.33) is satisfied, $\mathbf{H}_\dagger \subset \mathcal{D}(A) \times M^u(E)$, $\mathbf{H}_\ddagger \subset \mathcal{D}(A) \times M^l(E)$.

Let $(\rho,\lambda) \in \mathcal{J}$ with

$$\int_{U\times[0,\infty)} L(\rho(s),u)\lambda(du \times ds) < \infty. \tag{13.33}$$

Writing $\lambda(du \times ds) = \lambda_s(du)ds$ we define $u(s) = \int_U u\lambda_s(du)$ in the Bochner sense. Then

$$f(\rho(t)) - f(\rho(0)) = \int_{U\times[0,t]} Af(\rho(s),u)\lambda(du\times ds) = \int_0^t Af(\rho(s),u(s))ds.$$

As in Example 8.3, it follows that any solution of the control equation is a weak solution of the controlled partial differential equation

$$\frac{\partial}{\partial t}\rho = \Delta(-\Delta\rho + F'(\rho)) + u, \tag{13.34}$$

which, by the Lipschitz condition on F', uniquely determines a solution for any control satisfying (13.33).

Verification of the regularity conditions in Chapter 8 follows as in Section 13.1.4. First, for $\rho_0 \in L^2(\mathcal{O})$ and $f \in \mathcal{D}(\mathbf{H})$, let ρ be the weak solution of

$$\frac{\partial}{\partial t}\rho = \Delta(-\Delta\rho + F'(\rho)) + \operatorname{grad} f(\rho),$$

which again exists and is uniquely determined since $\rho \mapsto F'(\rho) + \operatorname{grad}(\rho)$ is Lipschitz on $H^0(\mathcal{O})$. Taking $u(t) = \operatorname{grad} f(\rho(t))$, ρ, u satisfies (13.34) and Condition 8.11:

$$\int_0^t \mathbf{H}f(\rho(s))ds = \int_{U\times[0,t]}\{Af(\rho(s)u(s)) - L(\rho(s),u(s))\}ds, \quad t \geq 0.$$

To verify Condition 8.9.4, let the *free energy* functional $\mathcal{E}$ be defined by (9.64). If $\mathcal{E}(\rho(0)) < \infty$, standard estimates give

$$\begin{aligned}
&\mathcal{E}(\rho(t)) - \mathcal{E}(\rho(0)) \\
&\quad \le \int_0^t \Big(- \|\nabla(-\Delta\rho + F'(\rho))\|^2 + \|u\|_{-1}\|\nabla(-\Delta\rho + F'(\rho))\|\Big)ds \\
&\quad \le \frac{1}{4}\int_0^\infty \|u(s)\|_{-1}^2 ds < \infty,
\end{aligned}$$

so if

$$C_u \equiv \int_0^\infty \|u(t)\|_{-1}^2 dt < \infty,$$

then ρ takes values in the compact set $\{\rho : \mathcal{E}(\rho) \le C_u\}$. If ρ, γ are solutions of (13.34) with the same control process u, then

$$\begin{aligned}
&\frac{1}{2}\|\rho(t) - \gamma(t)\|^2 - \frac{1}{2}\|\rho(0) - \gamma(0)\|^2 \\
&\quad \le \int_0^t \Big(- \|\Delta(\rho - \gamma)\|_{L^2(\mathcal{O})}^2 + \|F'(\rho) - F'(\gamma)\|\|\Delta(\rho - \gamma)\|\Big)ds \\
&\quad \le \int_0^t L_F^2\|\rho(s) - \gamma(s)\|^2 ds,
\end{aligned}$$

where L_F is the Lipschitz constant for F', and Gronwall's inequality, gives

$$\|\rho(t) - \gamma(t)\|^2 \le e^{2L_F^2 t}\|\rho(0) - \gamma(0)\|^2.$$

Then Condition 8.9.4 follows as in Section 13.1.4.

Finally, for ρ satisfying (13.34), the convexity of $L(\rho, u)$ in u implies

$$\begin{aligned}
&\inf_\lambda\{\int_{U\times[0,\infty)} L(\rho(s), u)\lambda(du \times ds) : (\rho, \lambda) \in \mathcal{J}\} \\
&\quad = \frac{1}{2}\int_0^\infty \|u(s)\|_{-1}^2 ds \\
&\quad = \frac{1}{2}\int_0^\infty \|\frac{\partial}{\partial t}\rho - \Delta\Big(- \Delta\rho + F'(\rho)\Big)\|_{-1}^2 dt.
\end{aligned}$$

13.2.5. The large deviation theorem.

THEOREM 13.13. *Let $Y_n, n = 1, 2, \ldots$ satisfy (13.22). Assume that $F \in C^2(R)$ with F'' bounded and that $m^{4+d}/n \to c_0$. Suppose that $\{\eta_n(Y_n(0))\}$ satisfies a large deviation principle with good rate function I_0 in $(E, \|\cdot\|)$. Then $\{\eta_n(Y_n) : n = 1, 2\ldots\}$ satisfies a large deviation principle in $C_E[0, \infty)$ with good rate function*

$$I(\rho) = I_0(\rho(0)) + \frac{1}{2}\int_0^\infty \|\frac{\partial}{\partial t}\rho - \Delta(-\Delta\rho + F'(\rho))\|_{-1}^2 dt.$$

13.3. Weakly interacting stochastic particles

We now consider large deviations for the weakly interacting diffusions in Example 1.14. Our goal here is not only to prove the large deviation principle which was first established by Dawson and Gärtner [**22**] using a different argument, but also to show that the comparison principle for equation (1.49) holds under reasonable assumptions. In this context, the probabilistic large deviation problem is naturally linked with variational problems originating from controlled partial differential equations in the space of probability measures.

Throughout, we assume the following on Φ and Ψ.

CONDITION 13.14.

(1) $\Phi, \Psi \in C^2(R^d)$.
(2) $\inf_z \Phi(z) > -\infty$.
(3) Ψ, Φ *are semi-convex in the sense that, there exist* $\lambda_\Psi, \lambda_\Phi \in R$ *such that*

$$(\nabla\Psi(x) - \nabla\Psi(y)) \cdot (x-y) \geq \lambda_\Psi |x-y|^2 \tag{13.35}$$

and similarly for Φ.
(4) Φ *is even (i.e.* $\Phi(x) = \Phi(-x)$*).*
(5)

$$|\nabla\Phi(x)| \leq C_0, \text{ some } C_0 > 0.$$

(6)

$$\lim_{|x|\to\infty} \frac{\Psi(x)}{|x|^2} = \infty \tag{13.36}$$

REMARK 13.15. The requirement that Φ is even is common in physical applications where Φ is a pairwise interaction potential depending only on particle distance. It is possible to relax Condition 13.14 to allow sub-quadratic growth of $|\nabla\Phi|$ by using a super-exponential approximation argument (Lemma 3.14).

We will need the following estimate on $\nabla\Psi$.

LEMMA 13.16. *If Condition 13.14 holds, then*

$$\lim_{|x|\to\infty} \frac{\nabla\Psi(x) \cdot x}{|x|^2} = \infty. \tag{13.37}$$

PROOF. By (13.35),

$$\nabla\Psi(x) \cdot x = \Psi(x) - \Psi(0) + \int_0^1 (\nabla\Psi(x) - \nabla\Psi(sx)) \cdot x ds \geq \Psi(x) - \Psi(0) + \frac{1}{2}\lambda_\Psi |x|^2.$$

Dividing by $|x|^2$, the lemma follows by (13.36). □

We consider the finite system of stochastic equations

$$dX_{n,i}(t) = -\nabla\Psi(X_{n,i}(t)) - \frac{1}{n}\sum_{j=1}^n \nabla\Phi(X_{n,i}(t) - X_{n,j}(t))dt + dW_i(t) \tag{13.38}$$

$i = 1, \ldots, n$, where the $\{W_i : i = 1, \ldots, n\}$ are independent R^d-valued Brownian motions. We define empirical-measure-valued processes

$$\rho_n(t, dx) = \frac{1}{n}\sum_{k=1}^n \delta_{X_{n,k}(t)}(dx).$$

Note that for $\varphi \in C^2(R^d)$,

$$
\begin{aligned}
(13.39)\quad \langle \varphi, \rho_n(t)\rangle &= \langle \varphi, \rho_n(0)\rangle + \int_0^t \langle \frac{1}{2}\Delta\varphi - \nabla\Psi\cdot\nabla\varphi, \rho_n(s)\rangle ds \\
&\quad - \int_0^t \langle \int \nabla\Phi(\cdot - x)\cdot\nabla\varphi(\cdot)\rho_n(s,dx), \rho_n(s)\rangle ds \\
&\quad + \frac{1}{n}\sum_{i=1}^n \int_0^t \nabla\varphi(X_{n,i}(s))dW_i(s) \\
&= \langle \varphi, \rho_n(0)\rangle + \int_0^t \langle B(\rho_n(s))\varphi, \rho_n(s)\rangle ds \\
&\quad + \frac{1}{n}\sum_{i=1}^n \int_0^t \nabla\varphi(X_{n,i}(s))dW_i(s),
\end{aligned}
$$

where

$$
(13.40)\qquad B(\rho)p = \frac{1}{2}\Delta p - \nabla(\Psi + \Phi * \rho)\cdot\nabla p, \quad \forall \rho \in E, p \in C^2(R^d).
$$

As in the other examples, we first derive a limit operator H and then prove the comparison principle for a nonlinear equation given by H (Theorem 13.32). Finally, we show that the result implies the large deviation principle.

Let

$$
E_n = \{\rho(dx) \equiv \frac{1}{n}\sum_{k=1}^n \delta_{x_k}(dx), x_k \in R^d, k = 1, \ldots, n\}
$$

and

$$
E = \mathcal{P}_2(R^d) \equiv \{\rho \in \mathcal{P}(R^d) : \int_{R^d} |x|^2 d\rho < \infty\},
$$

and let $\eta_n : E_n \to E$ be the identity map. Let r and r_n be given by the 2-Wasserstein metric (Definition D.13). Then (E, r) and (E_n, r_n) are complete, separable metric spaces.

We now specify $\{K_n^q : q \in \mathcal{Q}\}$ determining LIM convergence by Definition 2.5. Let

$$
\begin{aligned}
(13.41)\quad \mathcal{Q} = \{q = (\varphi, L) : L \in (0,\infty), \varphi : [0,\infty) \mapsto [0,\infty), \\
\varphi \text{ increasing}, \lim_{r\to\infty} \frac{\varphi(r)}{r} = \infty\},
\end{aligned}
$$

and define

$$
K_n^q = \{\rho \in E_n : \int_{R^d} \varphi(|x|^2)\rho(dx) \le L\}, \quad q = (\varphi, L) \in \mathcal{Q}.
$$

This definition is motivated by the characterization of relatively compact sets in E (Lemma D.17 in Appendix D): K is relatively compact if and only if there exists a $q = (\varphi, L) \in \mathcal{Q}$ such that

$$
\sup_{\rho\in K} \int_{R^d} \varphi(|x|^2)\rho(dx) \le L.
$$

In particular, K_n^q is compact, as is

$$
(13.42)\qquad K^q \equiv \{\rho \in E : \int_{R^d} \varphi d\rho \le L\},
$$

and $\lim_{n\to\infty} K_n^q = K^q$. The conditions of Definition 2.5 are satisfied with $K_q = \widehat{K}_q$.

Let $f_n \in B(E_n)$ and $f \in C_b(E)$. Since every convergent sequence is in some K^q, $f = \text{LIM} f_n$ if and only if

(1) $\sup_n \|f_n\| < \infty$;
(2) $\lim_{n\to\infty} f_n(\rho_n) = f(\rho)$ whenever $\rho_n \in E_n, \rho \in E$ satisfies $r(\rho_n, \rho) \to 0$.

13.3.1. Convergence of H_n to H. Define

$$\begin{aligned}\mathcal{D}(H) = D_0 \equiv \{f(\rho) = \varphi(\langle \mathbf{p}, \rho\rangle) : \varphi \in C^2(R^l), \\ \mathbf{p} = (p_1, \ldots, p_l), p_i - C_i \in C_c^\infty(R^d), \\ C_i \in R, i = 1, \ldots, l; l = 1, 2, \ldots\} \subset C_b(E),\end{aligned} \tag{13.43}$$

and for $f \in \mathcal{D}(H)$, define

$$\frac{\delta f}{\delta \rho} = \sum_{i=1}^{l} \partial_i \varphi(\langle \mathbf{p}, \rho\rangle) p_i \tag{13.44}$$

and

$$Hf(\rho) = \langle \rho, B(\rho) \frac{\delta f}{\delta \rho}\rangle + \frac{1}{2}\int_{R^d} |\nabla \frac{\delta f}{\delta \rho}|^2 d\rho. \tag{13.45}$$

Since $|\langle p, \rho\rangle| \leq \sup_x |p(x)|$, $Hf \in C_b(E)$.

The convergence arguments provided in Example 1.14 show that $\{H_n\}$ satisfies the requirements of Condition 7.11 with $H_\dagger = H_\ddagger = H$.

13.3.2. Exponential tightness. $\mathcal{D}(H)$ is an algebra and separates points in E. Furthermore, for each $\rho \in E$, there exists a $f \in \mathcal{D}(H)$ such that $f(\rho) \neq 0$. By Theorem A.8, the buc-closure for $\mathcal{D}(H)$ is $C_b(E)$. Consequently, Corollary 4.17 gives exponential tightness once we verify the exponential compact containment condition.

LEMMA 13.17. *Assume Condition 13.14. Let $K \subset E$ be compact. Then there exists $\gamma : [0, \infty) \mapsto [0, \infty)$ such that $\gamma', \gamma'' \geq 0$, γ'' nonincreasing, and $\lim_{r\to\infty} r^{-1}\gamma(r) = \infty$ such that $L_K \equiv \sup_{\rho\in K} \int \gamma(|x|^2)\rho(dx) < \infty$ and $\varphi_K(x) \equiv \gamma(|x|^2)$ satisfies*

$$C_K \equiv \sup_{x\in R^d, \rho\in E} B(\rho)\varphi_K(x) + \frac{1}{2}|\nabla \varphi_K(x)|^2 < \infty. \tag{13.46}$$

PROOF. For $\varphi_K(x) = \gamma(|x|^2)$,

$$B(\rho)\varphi_K(x) = 2\gamma''(|x|^2)|x|^2 + \gamma'(|x|^2)(d - 2\nabla\Psi(x)\cdot x - 2\int \nabla\Phi(x-y)\rho(dy)\cdot x)$$

and $|\nabla \varphi_K(x)|^2 = 4|x|^2\gamma'(|x|^2)^2$. By (13.37), γ' increases slowly enough

$$\lim_{|x|\to\infty} \frac{\gamma'(|x|^2)|x|^2}{\nabla\Psi(x)\cdot x} = 0,$$

and (13.46) follows. Modifying γ so that γ' increases still more slowly if necessary, we can select γ so that $L_K \equiv \sup_{\rho\in K} \int \gamma(|x|^2)\rho(dx) < \infty$. □

Let K be compact, and let φ_K and L_K be given by Lemma 13.17. Since φ_K has super-quadratic growth, by Lemma D.17, for each $M > 0$,

$$K_M = \{\rho \in E : \langle \varphi_K, \rho\rangle \leq M\}$$

is a compact set in (E,r). Suppose that $\rho_n(0) \in K$. Let

$$\tau_{n,M} = \inf\{t > 0 : \langle \varphi_K, \rho_n(t)\rangle > M\}$$

and

$$V(\rho) = \log(1 + \langle \varphi_K, \rho\rangle).$$

Then by Itô's formula, (13.39), and the strong Markov property,

$$\begin{aligned} Z_n(t) &= \exp\Big\{nV(\rho_n(t\wedge\tau_{n,M})) - nV(\rho_n(0)) \\ &\quad -n\int_0^{t\wedge\tau_{n,M}} \big(\frac{\langle B(\rho(s))\varphi_K, \rho_n(s)\rangle}{1+\langle\varphi_K,\rho_n(s)\rangle} + (1-\frac{1}{n})\frac{1}{2}\frac{\langle|\nabla\varphi_K|^2,\rho(s)\rangle}{(1+\langle\varphi_K,\rho_n(s)\rangle)^2}\big)ds\Big\} \end{aligned}$$

is a mean one martingale. By (13.46),

$$\int_0^{t\wedge\tau_{n,M}} \big(\frac{\langle B(\rho_n(s))\varphi_K, \rho_n(s)\rangle}{1+\langle\varphi_K,\rho_n(s)\rangle} + (1-\frac{1}{n})\frac{1}{2}\frac{\langle|\nabla\varphi_K|^2,\rho_n(s)\rangle}{(1+\langle\varphi_K,\rho_n(s)\rangle)^2}\big)ds \le C_K(t\wedge\tau_{n,M}).$$

Therefore, as in Lemmas 4.20 and 4.22, for each $T > 0$,

$$P(\exists t, t\in[0,T], \rho_n(t)\notin K_M|\rho_n(0)\in K)e^{n\log\frac{1+M}{1+L_K} - nC_KT} \le E[Z(T)] = 1.$$

Summarizing the above, we arrive at the following.

LEMMA 13.18. *For each $a, T > 0$, there exists $M > 0$ such that for all $n = 1,2,\ldots$*

$$\sup_{\rho_0\in K\cap E_n} P(\exists t, 0\le t\le T, \rho_n(t)\notin K_M|\rho_n(0)=\rho_0) \le e^{-na}. \tag{13.47}$$

13.3.3. Viscosity extension of H. We obtain the comparison principle by extending the operator H in a number of ways, so that Theorem 9.41 applies. The extensions are involved because we only identified the limiting H for a small class of test functions $\mathcal{D}(H)$. Starting with a small class of test functions gives a stronger result that has additional applications.

The *free energy* functional $\mathcal{E}$ is defined by (9.94). In addition, we define the Fisher information I and $I^{\#}$ by (9.102) and (9.105) respectively. The gradient for a function on E is defined in Definition 9.36. We will make extensive use of the weighted Sobolev space $H^{-1}_\rho(R^d)$, whose definition and properties are discussed in Appendix D.5. The argument parallels the previous two examples; however, mass transport inequalities (Appendix D.7) will be used in place of the Sobolev inequalities. The mass transport techniques used here are summarized in Appendix D.

For the purpose of proving the comparison principle, it is enough to work with subsolutions that are upper semicontinuous taking the weak topology on E and supersolutions that are lower semicontinuous taking the weak topology on E. To see that is the case, let $\widehat{r}$ be a metric on $\mathcal{P}(R^d)$ giving the topology of weak convergence. (Note that $(E,\widehat{r})$ is not complete.) Define

$$\widehat{\overline{f}}(\rho) = \lim_{\epsilon\to 0+}\sup\{\overline{f}(\gamma):\gamma\in E, \widehat{r}(\rho,\gamma)<\epsilon\}, \widehat{\underline{f}}(\rho) = \lim_{\epsilon\to 0+}\inf\{\underline{f}(\gamma):\gamma\in E,\widehat{r}(\rho,\gamma)<\epsilon\}.$$

Then for $\rho_n,\rho_0\in E$ and $\widehat{r}(\rho_n,\rho_0)\to 0$ (that is $\rho_n\Rightarrow\rho_0$),

$$\limsup_{n\to\infty}\widehat{\overline{f}}(\rho_n) \le \widehat{\overline{f}}(\rho_0), \quad \liminf_{n\to\infty}\widehat{\underline{f}}(\rho_n)\ge\widehat{\underline{f}}(\rho_0).$$

LEMMA 13.19. *Let $h \in D_0$ (as defined in (13.43)) and $\alpha > 0$, and consider the equation*

$$f - \alpha H f = h. \tag{13.48}$$

Let $\overline{f}$ be a viscosity subsolution of (13.48) and $\underline{f}$ a viscosity supersolution. Then $\widehat{\overline{f}}$ is a viscosity subsolution of (13.48) and $\widehat{\underline{f}}$ is a viscosity supersolution.

PROOF. Let $f_0 \in \mathcal{D}(H)$. Since $\overline{f}$ is a subsolution, there exists $\rho_n \in E$ such that

$$\lim_{n\to\infty}(\overline{f}(\rho_n) - f_0(\rho_n)) = \sup_{\rho\in E}(\overline{f}(\rho) - f_0(\rho))$$

and

$$\limsup_{n\to\infty}(\overline{f}(\rho_n) - h(\rho_n) - \alpha H f_0(\rho_n)) \le 0.$$

By definition, $\widehat{\overline{f}} \ge \overline{f}$, and clearly

$$\sup_{\rho\in E}(\overline{f}(\rho) - f_0(\rho)) = \sup_{\rho\in E}(\widehat{\overline{f}}(\rho) - f_0(\rho)).$$

Consequently

$$\begin{aligned}\lim_{n\to\infty}(\widehat{\overline{f}}(\rho_n) - f_0(\rho_n)) &\ge \lim_n(\overline{f}(\rho_n) - f_0(\rho_n)) = \sup_{\rho\in E}(\overline{f}(\rho) - f_0(\rho))\\ &= \sup_{\rho\in E}(\widehat{\overline{f}}(\rho) - f_0(\rho)) \ge \lim_{n\to\infty}(\widehat{\overline{f}}(\rho_n) - f_0(\rho_n))\end{aligned}$$

and $\lim_n \left(\widehat{\overline{f}}(\rho_n) - \overline{f}(\rho_n)\right) = 0$. Therefore

$$\liminf_{n\to\infty}(\widehat{\overline{f}}(\rho_n) - h(\rho_n) - \alpha H f_0(\rho_n)) \le 0.$$

The proof for $\underline{f}$ is similar. □

REMARK 13.20. Because $\overline{f} \le \widehat{\overline{f}}$ and $\underline{f} \ge \widehat{\underline{f}}$, to prove the comparison principle $\overline{f} \le \underline{f}$, it is enough to show $\widehat{\overline{f}} \le \widehat{\underline{f}}$.

With Lemma 13.19 in mind, we have the following variation of Lemma 7.7.

LEMMA 13.21. *Let $(E, r) = (\mathcal{P}_2(R^d), d_2)$, $H_\dagger \subset M^l(E, \overline{R}) \times M(E, \overline{R})$, and $h \in D_0$ (D_0 defined in (13.43)), and let $\widehat{r}$ be a metric on $\mathcal{P}(R^d)$ giving the topology of weak convergence. Let $\overline{f}$ be a subsolution of $f - \alpha H_\dagger f = h$ such that $\overline{f}$ is upper semicontinuous on $(E, \widehat{r})$.*

Let $(f_0, g_0) \in M^l(E, \overline{R}) \times M(E, \overline{R})$. Suppose that there exists $\{(f_n, g_n)\} \subset H_\dagger$ satisfying the following:

a) *f_n is lower semicontinuous on $(E, \widehat{r})$.*
b) *$f_1 \le f_2 \le \cdots$ and $f_n \to f_0$ pointwise.*
c) *If $\{\rho_n\} \subset E$ satisfies $\sup_n f_n(\rho_n) < \infty$ and $\inf_n g_n^*(\rho_n) > -\infty$, then $\{\rho_n\}$ is relatively compact in $(E, \widehat{r})$, and if a subsequence $\{\rho_{n_k}\}$ converges to $\rho_0 \in E$ and $\lim_{n\to\infty} f_{n_k}(\rho_{n_k}) = f_0(\rho_0)$, then*

$$\limsup_{n\to\infty} g_n^*(\rho_{n_k}) \le g_0^*(\rho_0). \tag{13.49}$$

Then, setting $\widehat{H}_\dagger = H_\dagger \cup \{(f_0, g_0)\}$, $\overline{f}$ is a subsolution of $f - \alpha\widehat{H}_\dagger f = h$. In addition, there exists $\rho_0 \in E$ such that

$$\overline{f}(\rho_0) - f_0(\rho_0) = \sup_{\rho\in E}(\overline{f}(\rho) - f_0(\rho)) \tag{13.50}$$

and

$$\alpha^{-1}(\overline{f}(\rho_0) - h(\rho_0)) \le g_0^*(\rho_0). \tag{13.51}$$

Let $\underline{f}$ be a supersolution of $f - \alpha H_\ddagger f = h$ such that $\underline{f}$ is lower semicontinuous on $(E, \widehat{r})$. Let $(f_0, g_0) \in M^u(E, \overline{R}) \times M(E, \overline{R})$. Suppose that there exists $\{(f_n, g_n)\} \subset H_\ddagger$ satisfying the following:

a) *f_n is upper semicontinuous on $(E, \widehat{r})$.*
b) *$f_1 \ge f_2 \ge \cdots$ and $f_n \to f_0$ pointwise.*
c) *If $\{\rho_n\} \subset E$ satisfies $\inf_n f_n(\rho_n) > -\infty$ and $\sup_n (g_n)_*(\rho_n) < \infty$, then $\{\rho_n\}$ is relatively compact in $(E, \widehat{r})$, and if a subsequence $\{\rho_{n_k}\}$ converges to $\rho_0 \in E$ and $\lim_{n\to\infty} f_{n_k}(\rho_{n_k}) = f_0(\rho_0)$, then*

$$\liminf_{n\to\infty} (g_n)_*(\rho_{n_k}) \ge (g_0)_*(\rho_0). \tag{13.52}$$

Then, setting $\widehat{H}_\ddagger = H_\ddagger \cup \{(f_0, g_0)\}$, $\underline{f}$ is a subsolution of $f - \alpha\widehat{H}_\ddagger f = h$. In addition, there exists $\rho_0 \in E$ such that

$$f_0(\rho_0) - \underline{f}(\rho_0) = \sup_{\rho\in E}(f_0(\rho) - \underline{f}(\rho_0)) \tag{13.53}$$

and

$$\alpha^{-1}(\overline{f}(\rho_0) - h(\rho_0)) \ge (g_0)_*(\rho_0). \tag{13.54}$$

REMARK 13.22. Let $\zeta \in C^2(R)$ be nondecreasing and satisfy $\zeta(u) = u$, $u \le 1$, $\zeta'' \le 0$, and $\zeta(u) = 2$, $u \ge 3$. Define $\zeta_n(x) = n\zeta(n^{-1}|x|^2)$. If $\{\rho_n\}$ satisfies $\sup_n \int_{R^d} \zeta_n(x)\rho_n(dx) < \infty$, then $\{\rho_n\}$ is relatively compact in $(\mathcal{P}(R^d), \widehat{r})$. Since by Fatou's lemma, any limit point will be in $E = \mathcal{P}_2(R^d)$, $\{\rho_n\}$ is relatively compact in $(E, \widehat{r})$.

PROOF. Let $(f_n, g_n) \in H_\dagger$ satisfy (a) through (c), and let $\epsilon_n > 0$ and $\epsilon_n \to 0$. There exist $\rho_n \in E$ such that

$$\overline{f}(\rho_n) - f_n(\rho_n) \ge \sup_\rho(\overline{f}(\rho) - f_n(\rho)) - \epsilon_n$$

and $\alpha^{-1}(\overline{f}(\rho_n) - h(\rho_n)) - g_n^*(\rho_n) \le \epsilon_n$. Note that

$$f_n(\rho_n) \le 2\|\overline{f}\| - \inf_\rho f_0(\rho) + \epsilon_n$$

and

$$g_n^*(\rho_n) \ge -\alpha^{-1}(\|\overline{f}\| + \|h\|) - \epsilon_n.$$

It follows that $\{\rho_n\}$ is relatively compact in $(E, \widehat{r})$, and since $\overline{f} - f_n$ is a decreasing sequence of upper semicontinuous functions, by Lemma A.4, if $\rho_0 \in E$ is a $\widehat{r}$-limit point of $\{\rho_n\}$, then

$$\lim_{n\to\infty} \overline{f}(\rho_n) - f_n(\rho_n) = \overline{f}(\rho_0) - f_0(\rho_0) = \sup_\rho(\overline{f}(\rho) - f_0(\rho)). \tag{13.55}$$

For simplicity, assume that $\rho_n \to \rho_0$ in $(E, \widehat{r})$. (Otherwise, consider a subsequence.) Since f_n is lower semicontinuous in $(E, \widehat{r})$, f_0 is also, and the upper semicontinuity of $\overline{f}$ and the lower semicontinuity of f_0 imply that $\lim_{n\to\infty}\overline{f}(\rho_n) = \overline{f}(\rho_0)$ and $\lim_{n\to\infty} f_n(\rho_n) = f_0(\rho_0)$. In addition,

$$f_0(\rho_0) = \lim_{\delta\to 0+}\lim_{n\to\infty}\inf_{\widehat{r}(\rho,\rho_0)\le\delta} f_n(\rho) = \sup_{\delta>0}\sup_n \inf_{\widehat{r}(\rho,\rho_0)\le\delta} f_n(\rho). \tag{13.56}$$

Then (13.55), (13.56), and (13.49) imply

$$\begin{aligned}
&\overline{f}(\rho_0) - f_0(\rho_0) - h(\rho_0)\\
&\quad = \limsup_{n\to\infty}(\overline{f}(\rho_n) - f_n(\rho_n) - h(\rho_n)) \le \limsup(\alpha g_n^*(\rho_n) + \alpha\epsilon_n - f_n(\rho_n))\\
&\quad \le \lim_{\delta\to 0+}\limsup_{n\to\infty}(\alpha(g_n^*(\rho_n) - \inf_{\widehat{r}(\rho,\rho_0)\le\delta} f_n(\rho))\\
&\quad \le \alpha g_0^*(\rho_0) - f_0(\rho_0).
\end{aligned}$$

Consequently, (13.50) and (13.51) hold for ρ_0 and (7.2) and (7.3) hold for $y_n \equiv \rho_0$.

The proof for supersolutions is similar. □

LEMMA 13.23. *Suppose $\varphi \in C^1(R^\infty)$ (R^∞ with the usual product topology), $\inf_y \varphi(y) > -\infty$, $\partial_i\varphi \ge 0$, $\partial_1\varphi > 0$, $\lim_{r_1\to\infty}\varphi(r_1, r_2, \ldots) = \infty$ for each $r_2, r_3, \ldots \in R$, and*

$$K \equiv \sup_{y\in R^\infty}\sum_{i=1}^\infty \partial_i\varphi(y) < \infty. \tag{13.57}$$

For $k = 1, 2, \ldots$, let $p_k \in C^2(R^d)$. Suppose that there exist $a_0, b_0, a_1, b_1 > 0$ such that for each k, $a_1|x|^2 - b_1 \le p_k(x) \le a_0 + b_0|x|^2$ and

$$\sup_\rho B(\rho)p_k(x) + \frac{K}{2}|\nabla p_k(x)|^2 \le a_0 + b_0|x|^2. \tag{13.58}$$

Define

$$f_0(\rho) = \varphi(\langle\mathbf{p}, \rho\rangle) = \varphi(\langle p_1, \rho\rangle, \langle p_2, \rho\rangle, \ldots)$$

and

$$g_0(\rho) = \langle\rho, \sum_{i=1}^\infty \partial_i\varphi(\langle\mathbf{p},\rho\rangle)B(\rho)p_i\rangle + \frac{1}{2}\int_{R^d}|\sum_{i=1}^\infty \partial_i\varphi(\langle\mathbf{p},\rho\rangle)\nabla p_i|^2 d\rho.$$

Let H be defined by (13.45) and $H_\dagger \supset H$. Let $\overline{f}$ be a subsolution of $f - \alpha H_\dagger f = h$ such that $\overline{f}$ is upper semicontinuous on $(E, \widehat{r})$. Then, setting $\widehat{H}_\dagger = H_\dagger \cup \{(f_0, g_0)\}$, $\overline{f}$ is a subsolution of $f - \alpha\widehat{H}_\dagger f = h$ and there exists $\rho_0 \in E$ such that

$$\overline{f}(\rho_0) - f_0(\rho_0) = \sup_{\rho\in E}(\overline{f}(\rho) - f_0(\rho)), \qquad \alpha^{-1}(\overline{f}(\rho_0) - h(\rho_0)) \le g_0^*(\rho_0).$$

The analogous result for supersolutions is obtained by replacing φ by $-\varphi$, but keeping the same conditions on the p_i.

PROOF. For $k > n$, let $p_k^n(x) = r_k = \inf_x p_k(x)$, and for $k \le n$, let $p_k^n = n\zeta(n^{-1}p_k)$. Define

$$f_n(\rho) = \varphi(\langle\mathbf{p}^n, \rho\rangle) = \varphi(\langle p_1^n, \rho\rangle, \langle p_2^n, \rho\rangle, \ldots)$$

and

$$g_n(\rho) = \langle\rho, \sum_{i=1}^n \partial_i\varphi(\langle\mathbf{p}^n,\rho\rangle)B(\rho)p_i^n\rangle + \frac{1}{2}\int_{R^d}|\sum_{i=1}^n \partial_i\varphi(\langle\mathbf{p}^n,\rho\rangle)\nabla p_i^n|^2 d\rho.$$

Then $(f_n, g_n) \in H \subset H_\dagger$ and Conditions (a) and (b) of Lemma 13.21 are satisfied. To see that (c) is satisfied, note that

$$\sup_n f_n(\rho_n) \geq \sup_n \varphi(\langle p_1^n, \rho_n\rangle, r_2, r_3, \ldots).$$

Consequently, $\sup_n f_n(\rho_n) < \infty$ implies $\sup_n \langle p_1^n, \rho_n\rangle < \infty$, so by Remark 13.22, $\{\rho_n\}$ is relatively compact in $(E, \widehat{r})$. For simplicity, assume that $\rho_n \to \rho_0$ in $(E, \widehat{r})$. (Otherwise, consider a subsequence.) By Fatou's lemma, $\liminf_{n\to\infty} \langle p_i^n, \rho_n\rangle \geq \langle p_i, \rho_0\rangle$, and hence, if $f_n(\rho_n) \to f_0(\rho_0)$, by the monotonicity and continuity of φ,

$$\begin{aligned}
(13.59)\quad f_0(\rho_0) &= \lim_{n\to\infty} f_n(\rho_n) \\
&\geq \lim_{n\to\infty} \varphi(\inf_{m\geq n} \langle p_1^m, \rho_m\rangle, \langle p_2, \rho_0\rangle \wedge \inf_{m\geq n} \langle p_2^n, \rho_n\rangle, \langle p_3, \rho_0\rangle \wedge \inf_{m\geq n} \langle p_3^n, \rho_n\rangle, \ldots) \\
&\geq \liminf_{n\to\infty} \varphi(\langle p_1^n, \rho_n\rangle, \langle p_2, \rho_0\rangle, \langle p_3, \rho_0\rangle, \ldots) \\
&\geq f_0(\rho_0).
\end{aligned}$$

It follows that $\langle p_1^n, \rho_n\rangle \to \langle p_1, \rho_0\rangle$, which in turn implies

$$\int_{R^d} g(x)(1 + \zeta_n(x))\rho_n(dx) \to \int_{R^d} g(x)(1 + |x|^2)\rho_0(dx), \quad g \in C_b(R^d).$$

Consequently, observing that for $i \leq n$,

$$B(\rho)p_i^n = \zeta'(n^{-1}p_i)B(\rho)p_i + \frac{1}{2n}\zeta''(n^{-1}p_i)|\nabla p_i|^2 \leq \zeta'(n^{-1}p_i)B(\rho)p_i$$

and

$$g_n(\rho_n) = \int_{R^d} \left(\sum_{i=1}^n \partial_i\varphi(\langle \mathbf{p}^n, \rho_n\rangle)B(\rho_n)p_i^n + \frac{1}{2}|\sum_{i=1}^n \partial_i\varphi(\langle \mathbf{p}^n, \rho_n\rangle)\nabla p_i^n|^2 \right) d\rho_n,$$

by (13.57) and (13.58), the integrand is bounded by

$$\begin{aligned}
&\sum_{i=1}^n \partial_i\varphi(\langle \mathbf{p}^n, \rho_n\rangle)B(\rho_n)p_i^n(x) + \frac{1}{2}|\sum_{i=1}^n \partial_i\varphi(\langle \mathbf{p}^n, \rho_n\rangle)\nabla p_i^n(x)|^2 \\
&\qquad \leq \sum_{i=1}^n \partial_i\varphi(\langle \mathbf{p}^n, \rho_n\rangle)(B(\rho_n)p_i^n(x) + \frac{K}{2}|\nabla p_i^n(x)|^2) \\
&\qquad \leq K\zeta'(n^{-1}(a_1|x|^2 - b_1))(a_0 + b_0|x|^2) \\
&\qquad \leq Ka_0 + Kb_0(|x|^2 \wedge \frac{3n + b_1}{a_1}) \\
&\qquad \leq K_0(1 + \zeta_n(x)),
\end{aligned}$$

and Fatou's lemma gives $\limsup_{n\to\infty} g_n(\rho_n) \leq g_0(\rho_0)$.

To obtain the result for supersolutions, one needs a lower bound on the integrand in order to apply Fatou's lemma. But that actually requires weaker conditions on the p_i.

$$\begin{aligned}
-\sum_{i=1}^n \partial_i\varphi(\langle \mathbf{p}^n,\rho_n\rangle)B(\rho_n)p_i^n(x) + \frac{1}{2}|\sum_{i=1}^n \partial_i\varphi(\langle \mathbf{p}^n,\rho_n\rangle)\nabla p_i^n(x)|^2 \\
\geq -\sum_{i=1}^n \partial_i\varphi(\langle \mathbf{p}^n,\rho_n\rangle)\zeta'(n^{-1}p_i)B(\rho_n)p_i(x) \\
\geq -K\zeta'(n^{-1}(a_1|x|^2-b_1))(a_0+b_0|x|^2) \\
\geq -Ka_0 + Kb_0(|x|^2\wedge\frac{3n+b_1}{a_1}) \\
\geq -K_0(1+\zeta_n(x))
\end{aligned}$$

□

The following lemma gives the existence of useful $\{p_i\}$ satisfying the conditions of Lemma 13.23.

LEMMA 13.24. *Suppose Condition 13.14 holds, and let $\psi_0(x) = 1+|x|^2$. Then there exists $C_1 > 0$ such that*

$$\sup_\rho B(\rho)\psi_0(x) + \frac{1}{2}|\nabla\psi_0(x)|^2 = d - 2\nabla(\Psi+\Phi*\rho)(x)\cdot x + 2|x|^2 \leq C_1(1+|x|^2).$$

PROOF. The semiconvexity condition implies

$$-\nabla\Psi(x)\cdot x \leq -\nabla\Psi(0)\cdot x - \lambda_\Psi|x|^2,$$

and that observation and the boundedness of $\nabla\Phi$ give the lemma. □

Let $a>0$, $\alpha_k\geq 0$, $\sum_{k=1}^\infty \alpha_k = 1$, and for $n=1,2,\ldots$, define

$$\varphi_n(y) = \frac{a}{2^n}\log\sum_{k=1}^\infty \alpha_k\exp\{2^n y_k\}. \tag{13.60}$$

Then

$$\partial_i\varphi_n(y) = a\frac{\alpha_i\exp\{2^n y_i\}}{\sum_{k=1}^\infty \alpha_k\exp\{2^n y_k\}},$$

so $\sum_i|\partial_i\varphi(y)| = a$ and (13.57) is satisfied with $K=a$. By Jensen's inequality,

$$(\sum_{k=1}^\infty \alpha_k\exp\{2^n y_k\})^2 \leq \sum_{k=1}^\infty \alpha_k\exp\{2^{n+1}y_k\},$$

and it follows that $\varphi_1\leq\varphi_2\leq\cdots$. Finally, note that

$$\lim_{n\to\infty}\varphi_n(y) = a\sup_k y_k.$$

LEMMA 13.25. *Let $\theta>0$, $\psi_0(x) = 1+|x|^2$, and $p_1,\ldots,p_m\in C^2(R^d)$, and suppose that the conditions of Lemma 13.23 are satisfied with p_i replaced by $p_i+\theta\psi_0$. Define*

$$f_0(\rho) = a\theta\langle\psi_0,\rho\rangle + a\sup_{1\leq k\leq m}\langle p_k,\rho\rangle,$$

and

$$g_0(\rho) = a\theta\langle\rho, B(\rho)\psi_0\rangle + \max\Big\{a\langle\rho, B(\rho)p_i\rangle + \frac{a^2}{2}\int_{R^d}|\theta\nabla\psi_0 + \nabla p_i|^2 d\rho : \\ \langle\rho, p_i\rangle = \max_{1\le k\le m}\langle\rho, p_k\rangle\Big\}.$$

Let H be defined by (13.45), and $H_\dagger \supset H$. Let $\overline{f}$ be a subsolution of $f - \alpha H_\dagger f = h$ such that $\overline{f}$ is upper semicontinuous on $(E, \widehat{r})$. Then, setting $\widehat{H}_\dagger = H_\dagger \cup \{(f_0, g_0)\}$, $\overline{f}$ is a subsolution of $f - \alpha\widehat{H}_\dagger f = h$ and there exists $\rho_0 \in E$ such that

$$\overline{f}(\rho_0) - f_0(\rho_0) = \sup_{\rho\in E}(\overline{f}(\rho) - f_0(\rho)), \qquad \alpha^{-1}(\overline{f}(\rho_0) - h(\rho_0)) \le g_0^*(\rho_0).$$

The analogous result holds for supersolutions with a replaced by $-a$ and g_0 given by

$$\begin{aligned} g_0(\rho) &= -a\theta\langle\rho, B(\rho)\psi_0\rangle \\ &\quad + \min\Big\{-a\langle\rho, B(\rho)\sum\beta_i p_i\rangle + \frac{a^2}{2}\int_{R^d}|\theta\nabla\psi_0 + \nabla\sum\beta_i p_i|^2 d\rho : \\ &\qquad \beta_i \ge 0, \sum\beta_i = 1, \sum\beta_i\langle\rho, p_i\rangle = \max_{1\le k\le m}\langle\rho, p_k\rangle\Big\}. \end{aligned}$$

PROOF. First consider the case for subsolutions.

Let φ_n be given by (13.60) with $\alpha_i = 0$, $i > m$. Define

$$f_n(\rho) = \varphi_n(\langle\theta\psi_0 + p_1, \rho\rangle, \ldots, \langle\theta\psi_0 + p_m, \rho\rangle)$$

and

$$\begin{aligned} g_n(\rho) &= a\theta\langle\rho, B(\rho)\psi_0\rangle + \sum_{i=1}^m \partial_i\varphi_n(\langle\mathbf{p}, \rho\rangle)\langle\rho, B(\rho)p_i\rangle \\ &\quad + \frac{1}{2}\int_{R^d}|\sum_{i=1}^m \partial_i\varphi_n(\langle\mathbf{p}, \rho\rangle)(\theta\nabla\psi_0 + \nabla p_i)|^2 d\rho, \end{aligned}$$

and note that

$$g_n(\rho) \le a\theta\langle\rho, B(\rho)\psi_0\rangle + \sum_{i=1}^m \partial_i\varphi_n(\langle\mathbf{p}, \rho\rangle)(\langle\rho, B(\rho)p_i\rangle + \frac{a}{2}\int_{R^d}|\theta\nabla\psi_0 + \nabla p_i|^2 d\rho)$$

By Lemma 13.23, we can assume that $(f_n, g_n) \in H_\dagger$. Noting that $\lim_{n\to\infty} f_n(\rho) = f_0(\rho)$, if $\rho_n \to \rho$ in $(E, \widehat{r})$ and $f_n(\rho_n) \to f_0(\rho)$, then $\langle\psi_0, \rho_n\rangle \to \langle\psi_0, \rho\rangle$. (See (13.59).) It follows that $\rho_n \to \rho$ in (E, r) and consequently that $\langle p_i, \rho_n\rangle \to \langle p_i, \rho\rangle$. If $\langle p_i, \rho\rangle < \max\langle p_k, \rho\rangle$, $\partial_i\varphi_n(\langle\mathbf{p}, \rho_n\rangle) \to 0$ and $\limsup_{n\to\infty} g_n(\rho_n) \le g_0(\rho)$. Consequently, the lemma follows by Lemma 13.21.

The proof for supersolution is similar. □

LEMMA 13.26. *Let $\{p_i\} \subset C^2(R^d)$, and assume that for each i, there exist $a_i, b_i > 0$ such that*

$$-a_i \le p_i(x) \le a_i + b_i|x|^2$$

and

$$\sup_\rho B(\rho)p_i(x) + \frac{a}{2}|\nabla p_i(x)|^2 \le a_i + b_i|x|^2. \tag{13.61}$$

Let $a, \theta > 0$, and define

$$f_0(\rho) = a\theta\langle\psi_0, \rho\rangle + a\sup_k(\langle p_k, \rho\rangle),$$

where $\psi_0(x) = 1 + |x|^2$. Setting

$$h(\rho, p) = a\langle\rho, B(\rho)p\rangle + \frac{a^2}{2}\int_{R^d}|\theta\nabla\psi_0 + \nabla p|^2 d\rho, \tag{13.62}$$

define

$$\begin{aligned} g_0(\rho) &= a\theta\langle\rho, B(\rho)\psi_0\rangle \\ &\quad + \lim_{\eta\to 0+}\limsup_{m\to\infty}\sup_{r(\rho',\rho)\le\eta}\max\Big\{h(\rho', p_i) : \langle\rho', p_i\rangle = \max_{1\le k\le m}\langle\rho', p_k\rangle\Big\}. \end{aligned}$$

Let H be defined by (13.45), and $H_\dagger \supset H$. Let $\overline{f}$ be a subsolution of $f - \alpha H_\dagger f = h$ such that $\overline{f}$ is upper semicontinuous on $(E, \widehat{r})$. Then, setting $\widehat{H}_\dagger = H_\dagger \cup \{(f_0, g_0)\}$, $\overline{f}$ is a subsolution of $f - \alpha\widehat{H}_\dagger f = h$ and there exists $\rho_0 \in E$ such that

$$\overline{f}(\rho_0) - f_0(\rho_0) = \sup_{\rho\in E}(\overline{f}(\rho) - f_0(\rho)), \qquad \alpha^{-1}(\overline{f}(\rho_0) - h(\rho_0)) \le g_0^*(\rho_0).$$

The analogous result holds for supersolutions with a replaced by $-a$ and g_0 given by

$$\begin{aligned} g_0(\rho) = -a\theta\langle\rho, B(\rho)\psi_0\rangle + \lim_{\eta\to 0+}\liminf_{m\to\infty}\inf_{r(\rho',\rho)\le\eta}\min\Big\{h(\rho', \sum_{i=1}^m \beta_i p_i) : \\ \beta_i \ge 0, \sum_{i=1}^m \beta_i = 1, \sum_{i=1}^m \beta_i\langle\rho', p_i\rangle = \max_{1\le k\le m}\langle\rho', p_k\rangle\Big\}. \end{aligned}$$

PROOF. For the subsolution case, define

$$f_n(\rho) = a\theta\langle\psi_0, \rho\rangle + a\sup_{1\le k\le n}(\langle p_k, \rho\rangle)$$

and

$$\begin{aligned} g_n(\rho) = a\theta\langle\rho, B(\rho)\psi_0\rangle + \max\Big\{a\langle\rho, B(\rho)p_i\rangle + \frac{a^2}{2}\int_{R^d}|\theta\nabla\psi_0 + \nabla p_i|^2 d\rho : \\ i \le n, \langle\rho, p_i\rangle = \max_{1\le k\le n}\langle\rho, p_k\rangle\Big\}. \end{aligned}$$

By Lemma 13.25, we can assume that $(f_n, g_n) \in H_\dagger$. Noting that $\lim_{n\to\infty} f_n(\rho) = f_0(\rho)$, if $\rho_n \to \rho$ in $(E, \widehat{r})$ and $f_n(\rho_n) \to f_0(\rho)$, then $\langle\psi_0, \rho_n\rangle \to \langle\psi_0, \rho\rangle$. (See (13.59).) It follows that $\rho_n \to \rho$ in (E, r) and, by the definition of g_0,

$$\limsup_{n\to\infty} g_n(\rho_n) \le g_0(\rho).$$

Consequently, the lemma follows by Lemma 13.21.

The proof for supersolutions is similar. □

COROLLARY 13.27. *For $\psi_0(x) = 1 + |x|^2$, let $C_{\psi_0} = \{p \in C(R^d) : p/\psi_0 \in C_b(R^d)\}$ be the Banach space of continuous functions with norm*

$$\|p\|_{\psi_0} = \sup_x |p(x)/\psi_0(x)|.$$

Let $\Gamma \subset C_{\psi_0} \cap C^2(R^d)$ be a compact, convex subset of C_{ψ_0} such that for each $p \in \Gamma$, $\inf_x p(x) > -\infty$ and there exist $a_p, b_p > 0$, such that

$$\sup_\rho B(\rho)p(x) + \frac{a}{2}|\nabla p(x)|^2 \leq a_p + b_p|x|^2.$$

Let $a, \theta > 0$. Define

$$f_\Gamma(\rho) = a\theta\langle\psi_0, \rho\rangle + a \sup_{p\in\Gamma}\langle p, \rho\rangle + b \tag{13.63}$$

and

$$\begin{aligned} g_\Gamma(\rho) &= a\theta\langle\rho, B(\rho)\psi_0\rangle \\ &\quad + \lim_{\eta\to 0+} \sup\Big\{h(\rho', p) : r(\rho', \rho) \leq \eta, p \in \Gamma, \langle\rho', p\rangle \geq \sup_{p'\in\Gamma}\langle\rho', p'\rangle - \eta\Big\}. \end{aligned}$$

Let H be defined by (13.45), and $H_\dagger \supset H$. Let $\overline{f}$ be a subsolution of $f - \alpha H_\dagger f = h$ such that $\overline{f}$ is upper semicontinuous on $(E, \widehat{r})$. Then, setting $\widehat{H}_\dagger = H_\dagger \cup \{(f_\Gamma, g_\Gamma)\}$, $\overline{f}$ is a subsolution of $f - \alpha\widehat{H}_\dagger f = h$ and there exists $\rho_\Gamma \in E$ such that

$$\overline{f}(\rho_\Gamma) - f_\Gamma(\rho_\Gamma) = \sup_{\rho\in E}(\overline{f}(\rho) - f_\Gamma(\rho)), \qquad \alpha^{-1}(\overline{f}(\rho_\Gamma) - h(\rho_\Gamma)) \leq g_\Gamma^*(\rho_\Gamma).$$

The analogous result holds for supersolutions with a replaced by $-a$ and g_Γ given by

$$\begin{aligned} g_\Gamma(\rho) &= -a\theta\langle\rho, B(\rho)\psi_0\rangle \\ &\quad + \lim_{\eta\to 0+} \inf\Big\{h(\rho', p) : r(\rho', \rho) \leq \eta, p \in \Gamma, \langle\rho', p\rangle \geq \sup_{p'\in\Gamma}\langle\rho', p'\rangle - \eta\Big\}. \end{aligned}$$

PROOF. Let $\{p_i\}$ be dense in Γ, and note that $\{p_i\}$ satisfies the conditions of Lemma 13.26. Defining (f_0, g_0) as in Lemma 13.26, the density of $\{p_i\}$ ensures that $f_\Gamma = f_0$ and that for each ρ and $\eta > 0$, for m sufficiently large, $\max_{1\leq k\leq m}\langle\rho', p_k\rangle > \sup_{p'\in\Gamma}\langle\rho', p'\rangle - \eta$ for all ρ' satisfying $r(\rho', \rho) \leq \eta$. That observation implies $g_\Gamma(\rho) \geq g_0(\rho)$ ($g_\Gamma(\rho) \leq g_0(\rho)$ in the supersolution case) and the corollary follows. □

Let $c \in C_b(R^d \times R^d)$ satisfy $c(x, y) \geq 0$, and assume that c is uniformly continuous. Define $F_c(\rho, \gamma)$ as in (D.22),

$$F_c(\rho, \gamma) = \inf\{\int c d\pi : \pi \in \mathbf{\Pi}(\rho, \gamma)\},$$

and let J_ϵ be the mollifier defined in (D.4). In particular, $J_\epsilon(x) = \epsilon^{-d}J(\epsilon^{-1}x)$, where $J \in C_c^\infty(R^d)$ is nonnegative and $\int_{R^d} J(x)dx = 1$, and

$$J_\epsilon * \rho(dx) = \int_{R^d} J_\epsilon(x - y)\rho(dy)dx.$$

Note that if Z is an R^d-valued random variable with density function J, then ϵZ has density J_ϵ.

We will also write

$$J_\epsilon * p(x) = \int_{R^d} J_\epsilon(x - y)p(y)dy$$

and note that $\langle p, J_\epsilon * \rho\rangle = \langle J_\epsilon * p, \rho\rangle$.

Let $\widehat{\Gamma}_c^\gamma$ be defined in (D.24), and note that by Corollary D.21

$$F_c(\rho, \gamma) = \sup_{p\in\widehat{\Gamma}_c^\gamma}\langle p, \rho\rangle.$$

By Remark D.19, the functions in $\widehat{\Gamma}_c^\gamma$ satisfy $\|p\|_\infty \le \|c\|_\infty$ and have a common modulus of continuity. Consequently $\widehat{\Gamma}_c^\gamma$ is compact in $C_b(R^d)$ in the topology of uniform convergence on compact sets and is compact in C_{ψ_0} in the norm topology. In addition, $\Gamma_{c,\epsilon}^\gamma = \{J_\epsilon * p : p \in \widehat{\Gamma}_c^\gamma\}$ will have the same compactness properties.

COROLLARY 13.28. *For $a_0, k_0 > 0$, let $c \in C_b(R^d \times R^d)$ satisfy $c(x,y) = a_0$ for $|x| + |y| \ge k_0$, and let $\Gamma_{c,\epsilon}^\gamma$ be defined as above. Define*

$$f_{c,\epsilon}^\gamma(\rho) = \theta\langle \psi_0, \rho\rangle + \frac{1}{2\delta}\sup_{p\in\Gamma_{c,\epsilon}^\gamma}\langle p, \rho\rangle = \theta\langle\psi_0,\rho\rangle + \frac{1}{2\delta}F_c(J_\epsilon * \rho, \gamma). \tag{13.64}$$

Setting

$$h_{\delta,\theta}(\rho, p) = \frac{1}{2\delta}\langle \rho, B(\rho)p\rangle + \frac{1}{8\delta^2}\int_{R^d}|2\delta\theta\nabla\psi_0 + \nabla p|^2 d\rho, \tag{13.65}$$

there exists $p_{\epsilon,\rho}^\gamma \in \widehat{\Gamma}_c^\gamma$ such that

$$f_{c,\epsilon}^\gamma(\rho) = \theta\langle\psi_0,\rho\rangle + \frac{1}{2\delta}\langle p_{\epsilon,\rho}^\gamma, J_\epsilon * \rho\rangle = \theta\langle\psi_0,\rho\rangle + \frac{1}{2\delta}F_c(J_\epsilon * \rho, \gamma)$$

and

$$\begin{aligned} g_{c,\epsilon}^\gamma(\rho) &= \theta\langle\rho, B(\rho)\psi_0\rangle \\ &\quad + \lim_{\eta\to 0+}\sup\left\{h_{\delta,\theta}(\rho', p) : r(\rho',\rho)\le\eta, \langle\rho', p\rangle \ge \sup_{p'\in\Gamma_{c,\epsilon}^\gamma}\langle\rho', p'\rangle - \eta\right\} \\ &= \theta\langle\rho, B(\rho)\psi_0\rangle + h_{\delta,\theta}(\rho, J_\epsilon * p_{\epsilon,\rho}^\gamma) \end{aligned}$$

Let H be defined by (13.45), and $H_\dagger \supset H$. Let $\overline{f}$ be a subsolution of $f - \alpha H_\dagger f = h$ such that $\overline{f}$ is upper semicontinuous on $(E, \widehat{r})$. Then, setting $\widehat{H}_\dagger = H_\dagger \cup \{(f_{c,\epsilon}^\gamma, g_{c,\epsilon}^\gamma)\}$, $\overline{f}$ is a subsolution of $f - \alpha\widehat{H}_\dagger f = h$ and there exists $\rho_0 \in E$ such that

$$\overline{f}(\rho_0) - f_{c,\epsilon}^\gamma(\rho_0) = \sup_{\rho\in E}(\overline{f}(\rho) - f_{c,\epsilon}^\gamma(\rho)), \qquad \alpha^{-1}(\overline{f}(\rho_0) - h(\rho_0)) \le (g_{c,\epsilon}^\gamma)^*(\rho_0).$$

The analogous result holds for supersolutions with

$$f_{c,\epsilon}^\gamma(\rho) = -\theta\langle\psi_0,\rho\rangle - \frac{1}{2\delta}\langle p_{\epsilon,\rho}^\gamma, J_\epsilon * \rho\rangle = -\theta\langle\psi_0,\rho\rangle - \frac{1}{2\delta}F_c(J_\epsilon * \rho, \gamma)$$

and

$$\begin{aligned} g_{c,\epsilon}^\gamma(\rho) &= -\theta\langle\rho, B(\rho)\psi_0\rangle \\ &\quad + \lim_{\eta\to 0+}\inf\left\{h_{\delta,\theta}(\rho', p) : r(\rho',\rho)\le\eta, \langle\rho', p\rangle \ge \sup_{p'\in\Gamma_{c,\epsilon}^\gamma}\langle\rho', p'\rangle - \eta\right\} \\ &= -\theta\langle\rho, B(\rho)\psi_0\rangle + h_{-\delta,\theta}(\rho, J_\epsilon * p_{\epsilon,\rho}^\gamma) \end{aligned}$$

where

$$h_{-\delta,\theta}(\rho, p) = -\frac{1}{2\delta}\langle\rho, B(\rho)p\rangle + \frac{1}{8\delta^2}\int_{R^d}|2\delta\theta\nabla\psi_0 + \nabla p|^2 d\rho$$

PROOF. Taking $a = 1/(2\delta)$ and replacing θ by $2\delta\theta$ in the definition of f_Γ in (13.63), the Corollary follows by Corollary 13.27, the compactness of $\widehat{\Gamma}_c^\gamma$, and the fact that $J_\epsilon * p(x)$ is constant for $|x| \ge k_0 + \epsilon$ (see (D.21)). □

For $\theta > 0$, let

$$
\begin{aligned}
(13.66)\qquad f_0(\rho) &= \theta\langle\psi_0,\rho\rangle + \sup_{p\in\Gamma^\gamma_{c_0,\epsilon}}\langle p,\rho\rangle \\
&= \theta\langle\psi_0,\rho\rangle + \theta\langle\psi_0,\gamma\rangle + \frac{1}{2\delta}d^2(J_\epsilon * \rho,\gamma).
\end{aligned}
$$

We again apply Lemma 13.21 to extend the operator to functions of this form. For ζ defined as in Remark 13.22, and $n = 1,2,\ldots$, let

$$c_n(x,y) = n\zeta(n^{-1}|x-y|^2 + n^{-1}2\delta\theta\psi_0(y))$$

and

$$c_0(x,y) = |x-y|^2 + 2\delta\theta\psi_0(y).$$

Then c_n satisfies the conditions of Corollary 13.28, and setting $g_n = g^\gamma_{c_n,\epsilon}$, $f_n = f^\gamma_{c_n,\epsilon}$, $\lim_{n\to\infty} f_n = f_0$. Since $c_1 \le c_2 \le \ldots$, $f_1 \le f_2 \le \cdots$, Conditions (a) and (b) of the first part of Lemma 13.21 are satisfied. It remains to identify g_0 satisfying Condition (c).

Suppose $\widehat{r}(\rho_n,\rho) \to 0$, $f_n(\rho_n) \to f_0(\rho)$, and

$$(13.67)\qquad \inf_n g_n^*(\rho_n) > -\infty.$$

By Fatou's lemma

$$\liminf_{n\to\infty} F_{c_n}(J_\epsilon * \rho_n,\gamma) \ge F_{c_0}(J_\epsilon * \rho,\gamma), \quad \liminf_{n\to\infty}\langle\psi_0,\rho_n\rangle \ge \langle\psi_0,\rho\rangle,$$

so convergence of $f_n(\rho_n)$ to $f_0(\rho)$ implies convergence of both sequences. In particular, $\langle\psi_0,\rho_n\rangle \to \langle\psi_0,\rho\rangle$ which in turn implies $r(\rho_n,\rho) \to 0$.

LEMMA 13.29. *Assume that γ has compact support K_γ. Let ρ_n be as above, and let $p_n = p^\gamma_{\epsilon,\rho_n}$. Then for $x,\widehat{x} \in R^d$,*

$$(13.68)\qquad |p_n(x) - p_n(\widehat{x})| \le \sup_{y\in K_\gamma}(|x| + |\widehat{x}| + 2|y|)|x-\widehat{x}|,$$

and $\{p_n\}$ is bounded in C_{ψ_0} and relatively compact in the topology of uniform convergence on compact sets. Any limit point $p_{\epsilon,\rho}$ of p_n satisfies

$$
\begin{aligned}
(13.69)\qquad f_0(\rho) &= \theta\langle\psi_0,\rho\rangle + \frac{1}{2\delta}\langle p_{\epsilon,\rho}, J_\epsilon * \rho\rangle \\
&= \theta\langle\psi_0,\rho\rangle + \theta\langle\psi_0,\gamma\rangle + \frac{1}{2\delta}d^2(J_\epsilon * \rho,\gamma)
\end{aligned}
$$

and

$$(13.70)\qquad |\nabla p_{\epsilon,\rho}(x)| \le 2|x| + \sup_{y\in K_\gamma} 2|y|, \quad a.e.$$

PROOF. The inequality (13.68) follows as in (D.21). Since the right side of (13.68) is independent of n and $p_n(x) \le \int_{R^d} c_0(x,y)\gamma(dy)$, for each compact set $K \subset R^d$,

$$\sup_n \sup_{x\in K} p_n(x) < \int_{R^d} c_0(\widehat{x},y)\gamma(dy) + \sup_{x\in K}\sup_{y\in K_\gamma}(|x| + |\widehat{x}| + 2|y|)|x-\widehat{x}| < \infty$$

and either

$$\liminf_{n\to\infty}\sup_{x\in K} p_n(x) = -\infty$$

or

(13.71)
$$\liminf_{n\to\infty}\inf_{x\in K} p_n(x) > \liminf_{n\to\infty}\sup_{x\in K} p_n(x) - \sup_{x,\widehat{x}\in K}\sup_{y\in K_\gamma}(|x|+|\widehat{x}|+2|y|)|x-\widehat{x}| > -\infty.$$

But

$$\langle p_n, J_\epsilon * \rho_n\rangle \to 2\delta\theta\langle\psi_0,\gamma\rangle + d^2(J_\epsilon*\rho,\gamma) > -\infty,$$

so (13.71) must hold. Consequently, $\sup_n |p_n(0)| < \infty$ and by (13.68),

$$|p_n(x)| \le \sup_n |p_n(0)| + \sup_{y\in K_\gamma}(|x|+2|y|)|x|,$$

and the second assertion of the lemma follows. (13.69) follows by the dominated convergence theorem, and (13.70) follows from (13.68). □

Since $\{p_n\}$ bounded in C_{ψ_0} implies $\{\Delta J_\epsilon * p_n\}$ is bounded in C_{ψ_0} and the estimate on $|\nabla p_n|$ implies $\{|\nabla J_\epsilon * p_n|^2\}$ is bounded in C_{ψ_0}, p_n converging to $p_{\epsilon,\rho}$ implies that all the terms in

$$g_n(\rho_n) = \theta\langle\rho_n, B(\rho_n)\psi_0\rangle + \frac{1}{2\delta}\langle\rho_n, B(\rho_n)J_\epsilon * p_n\rangle + \frac{1}{8\delta^2}\int_{R^d}|2\delta\theta\nabla\psi_0 + \nabla J_\epsilon * p_n|^2 d\rho_n$$

converge to the corresponding terms in

$$\theta\langle\rho, B(\rho)\psi_0\rangle + \frac{1}{2\delta}\langle\rho, B(\rho)J_\epsilon * p_{\epsilon,\rho}\rangle + \frac{1}{8\delta^2}\int_{R^d}|2\delta\theta\nabla\psi_0 + \nabla J_\epsilon * p_{\epsilon,\rho}|^2 d\rho$$

with the possible exceptions of $-\frac{1}{2\delta}\langle\rho_n, \nabla\Psi\cdot\nabla J_\epsilon * p_n\rangle$ and $-\theta\langle\rho_n,\nabla\Psi\cdot\nabla\psi_0\rangle$. The limit in (13.37) and Fatou's lemma imply that

$$\limsup_{n\to\infty}(-\theta\langle\rho_n,\nabla\Psi\cdot\nabla\psi_0\rangle) \le -\theta\langle\rho,\nabla\Psi\cdot\nabla\psi_0\rangle.$$

Let X_n have distribution ρ_n and let Z be independent of X_n and have distribution with density function J. Then $X_n^\epsilon = X_n + \epsilon Z$ has distribution $J_\epsilon * \rho_n$. By Theorem D.20 and Corollary D.21, there exists a distribution $\pi_n^\epsilon \in \mathbf{\Pi}(J_\epsilon * \rho_n, \gamma)$ and q_n such that $p_n(x) + q_n(y) \le c_n(x,y)$ with equality holding almost surely π_n^ϵ. It follows that (as in the proof of Lemma D.23)

$$\nabla p_n(x) = \nabla_x c_n(x,y) = \zeta'(n^{-1}|x-y|^2 + n^{-1}\delta\theta\psi_0(y))2(x-y) \quad a.s.\ \pi_n^\epsilon.$$

There exists a probability space with random variables (X_n, Y_n, Z_n) such that X_n and Z_n are independent, X_n has distribution ρ_n, Z_n has density J, and the joint distribution of $(X_n + \epsilon Z_n, Y_n)$ is π_n^ϵ. We then have

$$\begin{aligned}
&-\langle\rho_n, \nabla\Psi\cdot\nabla J_\epsilon * p_n\rangle\\
&\quad = -E[\nabla\Psi(X_n)\cdot\nabla p_n(X_n+\epsilon Z_n)]\\
&\quad = E[2\zeta'(n^{-1}|X_n+\epsilon Z_n - Y_n|^2 + n^{-1}\delta\theta\psi_0(Y_n))\nabla\Psi(X_n)\cdot(Y_n - X_n - \epsilon Z_n)]
\end{aligned}$$

As in the proof of Lemma 13.16, the semiconvexity of Ψ implies

$$\nabla\Psi(x)\cdot(y-x) \le \Psi(y) - \Psi(x) - \frac{\lambda_\Psi}{2}|x-y|^2.$$

Consequently, setting $V_n = \zeta'(n^{-1}|X_n + \epsilon Z - Y_n|^2 + n^{-1}\delta\theta\psi_0(Y_n))$

(13.72)
$$\begin{aligned}
&-\langle\rho_n, \nabla\Psi\cdot\nabla J_\epsilon * p_n\rangle\\
&\quad \le 2E[V_n(\Psi(Y_n - \epsilon Z_n) - \Psi(X_n) - \frac{\lambda_\Psi}{2}|X_n + \epsilon Z_n - Y_n|^2)].
\end{aligned}$$

By (13.67) and the earlier proved fact that $\{\Delta J_\epsilon * p_n\}$ is bounded in C_{ψ_0}, it follows that $\sup_n E[\Psi(X_n)] = \sup_n \langle \rho_n, \Psi\rangle < \infty$. By (13.36) and the compactness of the support of Y_n and Z_n, there exists a constant $a > 0$ such that the random variable on the right is bounded above by $a\psi_0(X_n)$. Let (X, Y, Z) be the limit in distribution of a subsequence of $\{(X_n, Y_n, Z_n)\}$. Then Y has distribution γ, Z has density J, $X + \epsilon Z$ has distribution $J_\epsilon * \rho$, and

$$E[|X_n + \epsilon Z_n - Y_n|^2] = F_{c_n}(J_\epsilon * \rho_n, \gamma) \to d^2(J_\epsilon * \rho, \gamma) = E[|X + \epsilon Z - Y|^2].$$

Fatou's lemma then implies

$$\limsup_{n\to\infty}(-\langle \rho_n, \nabla\Psi \cdot \nabla J_\epsilon * p_n\rangle) \le 2E[\Psi(Y - \epsilon Z)] - 2\langle \rho, \Psi\rangle - \lambda_\Psi d^2(J_\epsilon * \rho, \gamma).$$

Letting π^ϵ denote the joint distribution of $(X + \epsilon Z, Y)$, $p_{\epsilon,\rho}$ is locally Lipschitz and as in Lemma D.23

$$\nabla p_{\epsilon,\rho}(x) = 2(x - y) \quad a.s.\ \pi^\epsilon.$$

Since X_n and Z_n are independent, X and Z are independent, and we have

$$\begin{aligned}
\int_{R^d} |2\delta\theta\nabla\psi_0 + \nabla J_\epsilon * p_{\epsilon,\rho}|^2 d\rho &= E[|4\delta\theta X + E[\nabla p_{\epsilon,\rho}(X + \epsilon Z)|X]|^2] \\
&= E[|4\delta\theta X + 2E[X + \epsilon Z - Y|X]|^2] \\
&= E[|4\delta\theta X + 2X - 2E[Y|X]|^2] \\
&\le 4\Big(\sqrt{E[|X + \epsilon Z - Y|^2]} \\
&\qquad + \sqrt{E[|2\delta\theta X - \epsilon Z + (Y - E[Y|X])|^2]}\Big)^2 \\
&= 4(d(J_\epsilon * \rho, \gamma) + \sqrt{E[R(X, Y, Z, \epsilon, \delta, \theta)^2]})^2,
\end{aligned}$$

where we define

$$R(X, Y, Z, \epsilon, \delta, \theta) = |2\delta\theta X - \epsilon Z + (Y - E[Y|X])|.$$

Similarly,

$$\int_{R^d} |2\delta\theta\nabla\psi_0 + \nabla J_\epsilon * p_{\epsilon,\rho}|^2 d\rho \ge 4(d(J_\epsilon * \rho, \gamma) - \sqrt{E[R(X, Y, Z, \epsilon, \delta, \theta)^2]})^2.$$

Summarizing these estimates and applying Lemma 13.21, we have the following.

LEMMA 13.30. *For $\theta, \delta > 0$, let*

$$f_\epsilon(\rho) = \theta\langle \psi_0, \rho\rangle + \theta\langle \psi_0, \gamma\rangle + \frac{1}{2\delta} d^2(J_\epsilon * \rho, \gamma)$$

and

$$\begin{aligned}
g_\epsilon(\rho) &= \theta\langle \rho, B(\rho)\psi_0\rangle \\
&\quad + \frac{1}{2\delta}\Big(\frac{1}{2}\langle \rho, \Delta J_\epsilon * p_{\epsilon,\rho}\rangle - \langle \rho, \nabla\Phi * \rho \cdot J_\epsilon * \nabla p_{\epsilon,\rho}\rangle\Big) \\
&\quad + \frac{1}{\delta}\Big(E[\Psi(Y_\rho - \epsilon Z_\rho)] - \langle \rho, \Psi\rangle - \frac{\lambda_\Psi}{2} d^2(J_\epsilon * \rho, \gamma)\Big) \\
&\quad + \frac{1}{2\delta^2}(d(J_\epsilon * \rho, \gamma) + \sqrt{E[R(X_\rho, Y_\rho, Z_\rho, \epsilon, \delta, \theta)^2]})^2,
\end{aligned}$$

where (X_ρ, Y_ρ, Z_ρ) is the limit in distribution of (X_n, Y_n, Z_n) (possibly along a subsequence).

Let H be defined by (13.45), and $H_\dagger \supset H$. Let $\overline{f}$ be a subsolution of $f-\alpha H_\dagger f = h$ such that $\overline{f}$ is upper semicontinuous on $(E,\widehat{r})$. Then, setting $\widehat{H}_\dagger = H_\dagger \cup \{(f_\epsilon, g_\epsilon)\}$, $\overline{f}$ is a subsolution of $f-\alpha \widehat{H}_\dagger f = h$ and there exists $\rho_\epsilon \in E$ such that

$$\overline{f}(\rho_\epsilon) - f_\epsilon(\rho_\epsilon) = \sup_{\rho\in E}(\overline{f}(\rho) - f_\epsilon(\rho)), \qquad \alpha^{-1}(\overline{f}(\rho_\epsilon) - h(\rho_\epsilon)) \le g_\epsilon^*(\rho_\epsilon).$$

The analogous result holds for supersolutions with

$$f_\epsilon(\rho) = -\theta\langle\psi_0,\rho\rangle - \theta\langle\psi_0,\gamma\rangle - \frac{1}{2\delta}d^2(J_\epsilon * \rho,\gamma)$$

and

$$\begin{aligned} g_\epsilon(\rho) &= -\theta\langle\rho, B(\rho)\psi_0\rangle \\ &\quad -\frac{1}{2\delta}\Big(\frac{1}{2}\langle\rho,\Delta J_\epsilon * p_{\epsilon,\rho}\rangle - \langle\rho,\nabla\Phi*\rho\cdot J_\epsilon*\nabla p_{\epsilon,\rho}\rangle\Big) \\ &\quad -\frac{1}{\delta}\Big(E[\Psi(Y_\rho-\epsilon Z_\rho)] - \langle\rho,\Psi\rangle - \frac{\lambda_\Psi}{2}d^2(J_\epsilon*\rho,\gamma)\Big) \\ &\quad +\frac{1}{2\delta^2}(d(J_\epsilon*\rho,\gamma) - \sqrt{E[R(X_\rho,Y_\rho,Z_\rho,\epsilon,\delta,\theta)^2]})^2. \end{aligned}$$

Set $S(\rho) = \int_{R^d}\rho(x)\log\rho(x)dx$, if ρ has a Lebesgue density and the integral exists. Otherwise, set $S(\rho)=+\infty$. The Donsker-Varadhan variational principle (e.g., Lemma 1.4.3 of [**35**]) implies

$$S(\rho) = \sup\{\int_{R^d}(\log f(x))\rho(dx) : f\in C_b(R^d), f\ge 0, \int_{R^d} f(x)dx = 1\},$$

so $S(\rho)$ is lower semicontinuous, and taking $f(x) = ce^{-|x|^2}$,

$$\int_{R^d}\rho(x)\log\rho(x)dx \ge -\log\int_{R^d}e^{-|x|^2}dx - \int_{R^d}|x|^2\rho(x)dx. \tag{13.73}$$

If γ has a Lebesgue density and $S(\gamma)<\infty$, then by Lemma D.53,

$$\frac{1}{2}\langle\rho,\Delta J_\epsilon * p_{\epsilon,\rho}\rangle \le S(\gamma) - S(J_\epsilon*\rho).$$

As in (13.72)

$$\begin{aligned} -\langle\rho,\nabla\Phi*\rho\cdot J_\epsilon*\nabla p_{\epsilon,\rho}\rangle &= -E[\nabla\Phi(X_\rho - x)\rho(dx)\cdot\nabla p_{\epsilon,\rho}(X_\rho+\epsilon Z_\rho)] \\ &= 2E[\int_{R^d}\nabla\Phi(X_\rho-x)\rho(dx)\cdot(Y_\rho - X_\rho-\epsilon Z_\rho)] \\ &\le 2E[\int_{R^d}\Phi(Y_\rho-\epsilon Z_\rho - x)\rho(dx)] \\ &\quad -2\int_{R^d\times R^d}\Phi(\widehat{x}-x)\rho(dx)\rho(d\widehat{x}) - \lambda_\Phi d^2(J_\epsilon*\rho,\gamma) \end{aligned}$$

Recalling the definition of $\mathcal{E}$ and (9.96)

$$\mathcal{E}(\rho) = \frac{1}{2}S(\rho) + \langle\Psi,\rho\rangle + \frac{1}{2}\langle\Phi*\rho,\rho\rangle + e_0,$$

where $e_0 = \frac{1}{2}\int_{R^d} e^{-2\Psi(x)}dx$, we have

$$\begin{aligned} g_\epsilon(\rho) \;\le\;& \theta\langle\rho, B(\rho)\psi_0\rangle + \frac{1}{2\delta}(S(\gamma) - S(J_\epsilon * \rho)) \\ &+\frac{1}{\delta}\Big(E[\int_{R^d}\Phi(Y_\rho - \epsilon Z_\rho - x)\rho(dx)] \\ &-\int_{R^d\times R^d}\Phi(\widehat{x}-x)\rho(dx)\rho(d\widehat{x}) - \frac{1}{2}\lambda_\Phi d^2(J_\epsilon * \rho,\gamma)\Big) \\ &+\frac{1}{\delta}\Big(E[\Psi(Y_\rho - \epsilon Z_\rho)] - \langle\rho,\Psi\rangle - \frac{\lambda_\Psi}{2}d^2(J_\epsilon*\rho,\gamma)\Big) \\ &+\frac{1}{2\delta^2}(d(J_\epsilon*\rho,\gamma) + \sqrt{E[R(X_\rho,Y_\rho,Z_\rho,\epsilon,\delta,\theta)^2]})^2, \end{aligned} \tag{13.74}$$

LEMMA 13.31. *Suppose $S(\gamma) < \infty$. Let*

$$f_0(\rho) = \theta\langle\psi_0,\rho\rangle + \theta\langle\psi_0,\gamma\rangle + \frac{1}{2\delta}d^2(\rho,\gamma)$$

and

$$\begin{aligned} g_0(\rho) \;=\;& \theta\langle\rho, B(\rho)\psi_0\rangle \\ &+\frac{1}{2\delta}\Big(S(\gamma) - S(\rho) + 2(\langle\Phi*\rho,\gamma\rangle - \langle\Phi*\rho,\rho\rangle)\Big) \\ &+\frac{1}{\delta}\Big(\langle\gamma,\Psi\rangle - \langle\rho,\Psi\rangle - \frac{\lambda_\Psi+\lambda_\Phi}{2}d^2(\rho,\gamma)\Big) \\ &+\frac{1}{2\delta^2}(d(\rho,\gamma) + 2\delta\theta\sqrt{\langle\psi_0 - 1,\rho\rangle})^2 \\ =\;& \theta\langle\rho, B(\rho)\psi_0\rangle \\ &+\frac{1}{\delta}\Big(\mathcal{E}(\gamma) - \mathcal{E}(\rho) + (\langle\Phi*\rho,\gamma\rangle - \frac{1}{2}\langle\Phi*\gamma,\gamma\rangle - \frac{1}{2}\langle\Phi*\rho,\rho\rangle)\Big) \\ &-\frac{\lambda_\Psi+\lambda_\Phi}{2\delta}d^2(\rho,\gamma) + \frac{1}{2\delta^2}(d(\rho,\gamma) + 2\delta\theta\sqrt{\langle\psi_0-1,\rho\rangle})^2, \end{aligned}$$

Let H be defined by (13.45), and $H_\dagger \supset H$. Let $\overline{f}$ be a subsolution of $f - \alpha H_\dagger f = h$ such that $\overline{f}$ is upper semicontinuous on $(E,\widehat{r})$. Then, setting $\widehat{H}_\dagger = H_\dagger \cup \{(f_0, g_0)\}$, $\overline{f}$ is a subsolution of $f - \widehat{H}_\dagger f = h$ and there exists $\rho_0 \in E$ such that

$$\overline{f}(\rho_0) - f_0(\rho_0) = \sup_{\rho\in E}(\overline{f}(\rho) - f_0(\rho)), \qquad \alpha^{-1}(\overline{f}(\rho_0) - h(\rho_0)) \le g_0^*(\rho_0).$$

The conclusion also holds with $\theta = 0$.

The analogous result holds for supersolutions with

$$f_0(\rho) = -\theta\langle\psi_0,\rho\rangle - \theta\langle\psi_0,\gamma\rangle - \frac{1}{2\delta}d^2(\rho,\gamma)$$

and

$$\begin{aligned} g_0(\rho) =\;& \theta\langle\rho, B(\rho)\psi_0\rangle \\ &-\frac{1}{\delta}\Big(\mathcal{E}(\gamma) - \mathcal{E}(\rho) + (\langle\Phi*\rho,\gamma\rangle - \frac{1}{2}\langle\Phi*\gamma,\gamma\rangle - \frac{1}{2}\langle\Phi*\rho,\rho\rangle)\Big) \\ &+\frac{\lambda_\Psi+\lambda_\Phi}{2\delta}d^2(\rho,\gamma) + \frac{1}{2\delta^2}(d(\rho,\gamma) - 2\delta\theta\sqrt{\langle\psi_0-1,\rho\rangle})^2. \end{aligned}$$

PROOF. Denote the right side of (13.74) by $\widehat{g}_\epsilon(\rho)$. Then the conclusion of Lemma 13.30 holds with g_ϵ replaced by $\widehat{g}_\epsilon$. Let ρ_ϵ satisfy

$$\overline{f}(\rho_\epsilon) - f_\epsilon(\rho_\epsilon) = \sup_{\rho\in E}(\overline{f}(\rho) - f_\epsilon(\rho)), \qquad \alpha^{-1}(\overline{f}(\rho_\epsilon) - h(\rho_\epsilon)) \le \widehat{g}^*_\epsilon(\rho_\epsilon).$$

Then $\sup_\epsilon f_\epsilon(\rho_\epsilon) < \infty$ and $\inf_\epsilon \widehat{g}^*(\rho_\epsilon) \ge -\alpha^{-1}(\|\overline{f}\|+\|h\|)$. It follows that $\sup_\epsilon\langle\Psi,\rho_\epsilon\rangle < \infty$ and $\sup_\epsilon S(J_\epsilon * \rho_\epsilon) < \infty$. Consequently, $\{\rho_\epsilon\}$ is relatively compact in (E,r). Assume that $r(\rho_\epsilon,\rho_0) \to 0$ (otherwise, select a convergent subsequence). Then $S(\rho_0) < \infty$ and ρ_0 has a Lebesgue density. Selecting a convergent subsequence, if necessary, $(X_{\rho_\epsilon}, Y_{\rho_\epsilon}) \Rightarrow (X_0, Y_0)$ in distribution, and $E[|X_0 - Y_0|^2] = d^2(\rho,\gamma)$. The convergence $(X_{\rho_\epsilon}, Y_{\rho_\epsilon}) \Rightarrow (X_0, Y_0)$ implies

$$\begin{aligned}
\limsup_{\epsilon\to 0} E[(Y_{\rho_\epsilon} - E[Y_{\rho_\epsilon}|X_{\rho_\epsilon}])^2] &\le \inf_{g\in C_b(R^d)} \lim_{\epsilon\to 0} E[(Y_{\rho_\epsilon} - g(X_{\rho_\epsilon}))^2] \\
&= \inf_{g\in C_b(R^d)} E[(Y_0 - g(X_0))^2] \\
&= E[(Y_0 - E[Y_0|X_0])^2],
\end{aligned}$$

and Theorem D.25 implies the right side is zero.

The upper semicontinuity of $\overline{f}$ implies $\overline{f}(\rho_0) - f_0(\rho_0) = \sup_{\rho\in E}(\overline{f}(\rho) - f_0(\rho))$ and $\overline{f}(\rho_\epsilon) \to \overline{f}(\rho_0)$. The convergence of ρ_ϵ implies $r(J_\epsilon * \rho_\epsilon, \rho_0) \to 0$, and the continuity and semicontinuity properties of the terms in $\widehat{g}_\epsilon$ imply $\limsup_{\epsilon\to 0}\widehat{g}_\epsilon(\rho_\epsilon) \le g_0(\rho_0)$. It follows that

$$\alpha^{-1}(\overline{f}(\rho_0) - h(\rho_0)) \le \widehat{g}^*_0(\rho_0)$$

giving the lemma.

For $\theta > 0$, rename (f_0, g_0, ρ_0), $(f_\theta, g_\theta, \rho_\theta)$. The above argument now applies as $\theta \to 0$.

The proof for supersolutions is similar. □

13.3.4. The comparison principle.

THEOREM 13.32. *Let H be defined by (13.45), and let $h \in D_0$ and $\alpha > 0$. Suppose that $\overline{f}$ is a subsolution and $\underline{f}$ is a supersolution to*

$$(I - \alpha H)f = h. \tag{13.75}$$

Then $\overline{f} \le \underline{f}$.

REMARK 13.33. By Remark 13.20, we can assume, without loss of generality, that $\overline{f}$ is upper semicontinuous on $(E,\widehat{r})$ and $\underline{f}$ is lower semicontinuous on $(E,\widehat{r})$.

By Theorem 9.41, Theorem 13.32 is a consequence of the following lemma. Let ω_h denote the modulus of continuity of h with respect to d, that is,

$$h(\rho) - h(\gamma) \le \omega_h(d(\rho,\gamma)), \quad \rho,\gamma\in E.$$

LEMMA 13.34. *Let $\widehat{H}_\dagger$ and $\widehat{H}_\ddagger$ be defined as in (9.107) and (9.108). For $\delta > 0$, define*

$$\begin{aligned}
\overline{f}_\delta(\rho) &= \lambda_{0,\delta}\sup_{\gamma\in E}\{\overline{f}(\gamma) - \frac{1}{2\delta}d^2(\rho,\gamma)\}, \\
\underline{f}_\delta(\gamma) &= \lambda_{1,\delta}\inf_{\rho\in E}\{\underline{f}(\rho) + \frac{1}{2\delta}d^2(\rho,\gamma)\},
\end{aligned}$$

where

$$\lambda_{0,\delta} = (1 + 2\delta(|\lambda_\Psi| + 2|\lambda_\Phi|)), \quad \lambda_{1,\delta} = (1 - 2\delta(|\lambda_\Psi| + 2|\lambda_\Phi|))^{-1},$$

and

$$h_{0,\delta} = \lambda_{0,\delta}\Big(h + \omega_h(2\sqrt{\delta \sup_{\rho\in E} |\overline{f}(\rho)|})\Big)$$

and

$$h_{1,\delta} = \lambda_{1,\delta}\Big(h - \omega_h(2\sqrt{\delta \sup_{\rho\in E} |\underline{f}(\rho)|})\Big).$$

Then $\overline{f}_\delta$ is a continuous subsolution of

$$(I - \alpha\widehat{H}_\dagger)f = h_{0,\delta} \tag{13.76}$$

and $\underline{f}_\delta$ is a continuous supersolutions of

$$(I - \alpha\widehat{H}_\ddagger)f = h_{1,\delta}. \tag{13.77}$$

PROOF. We prove the subsolution case. The proof of the supersolution case is similar.

Since

$$\begin{aligned}&\overline{f}_\delta(\rho) - \overline{f}_\delta(\rho') \\ &\quad\le \frac{\lambda_{0,\delta}}{2\delta}\sup\{d^2(\rho',\gamma) - d^2(\rho,\gamma) : \gamma\in E, d^2(\rho,\gamma) \le 2\delta(\sup_{\gamma'}\overline{f}(\gamma') - \overline{f}(\rho))\},\end{aligned}$$

$\overline{f}_\delta$ is continuous.

Let f_0 be of the form (9.111):

$$f_0(\rho) = ad^2(\rho,\tau_0) + b\mathcal{E}(\rho) + cG(\rho,\sigma_0), \tag{13.78}$$

where $a, c > 0$, $0 < b < 1$, $\tau_0, \sigma_0 \in E$, $\mathcal{E}$ is defined in (9.94), and G in (9.110).

Since $\mathcal{E}$ has compact level set in (E, r) (Lemma 9.39) and since $\overline{f}_\delta - f_0$ is upper semicontinuous, there exists $\rho_0 \in E$ such that

$$(\overline{f}_\delta - f_0)(\rho_0) = \sup_{\rho,\gamma\in E}\Big(\lambda_{0,\delta}(\overline{f}(\gamma) - \frac{1}{2\delta}d^2(\rho,\gamma)) - f_0(\rho)\Big). \tag{13.79}$$

Note that ρ_0 must satisfy $f_0(\rho_0) < \infty$ and $\mathcal{E}(\rho_0) < \infty$ and have a Lebesgue density. Since $\overline{f}$ is a viscosity subsolution for (13.75), by Lemma 13.31, there exists $\gamma_0 \in E$ satisfying

(13.80)

$$\lambda_{0,\delta}(\overline{f}(\gamma_0) - \frac{1}{2\delta}d^2(\rho_0,\gamma_0)) - f_0(\rho_0) = \sup_{\rho,\gamma\in E}\Big(\lambda_{0,\delta}(\overline{f}(\gamma) - \frac{1}{2\delta}d^2(\rho,\gamma)) - f_0(\rho)\Big)$$

and

$$\begin{aligned}\frac{\overline{f} - h}{\alpha}(\gamma_0) &\le H_0(\frac{1}{2\delta}d^2(\rho_0,\cdot) + \frac{1}{\lambda_{0,\delta}}f_0(\rho_0))(\gamma_0) \\ &= \frac{1}{\delta}\Big(\mathcal{E}(\rho_0) - \mathcal{E}(\gamma_0) + (\langle\Phi*\gamma_0,\rho_0\rangle - \frac{1}{2}\langle\Phi*\gamma_0,\gamma_0\rangle - \frac{1}{2}\langle\Phi*\rho_0,\rho_0\rangle)\Big) \qquad (13.81)\\ &\qquad -\frac{\lambda_\Psi + 2\lambda_\Phi}{2\delta}d^2(\rho_0,\gamma_0) + \frac{1}{2\delta^2}d(\rho_0,\gamma_0). \\ &\le \frac{1}{\delta}(\mathcal{E}(\rho_0) - \mathcal{E}(\gamma_0)) + \frac{1}{2\delta^2}(1 - \delta(\lambda_\Psi + 2\lambda_\Phi))d(\rho_0,\gamma_0).\end{aligned}$$

To complete the proof, we must show

$$\begin{aligned}\overline{f}_\delta(\rho_0) &= \lambda_{0,\delta}\sup_{\gamma\in E}\{\overline{f}(\gamma)-\frac{1}{2\delta}d^2(\rho_0,\gamma)\}\\ &= \lambda_{0,\delta}(\overline{f}(\gamma_0)-\frac{1}{2\delta}d^2(\rho_0,\gamma_0))\\ &\le \lambda_{0,\delta}\Big(h(\rho_0)+\omega_h(2\sqrt{\delta\sup_{\rho\in E}|\overline{f}(\rho)|})\Big)+\alpha\widehat{H}_\dagger f_0(\rho_0),\end{aligned}$$

which is the consequence of the next two lemmas. □

LEMMA 13.35. *Let ρ_0 satisfy (13.79). Then $I^{\#}(\rho_0)<\infty$.*

PROOF. Noting that $\mathcal{E}(\rho_0)<\infty$, from (D.31), (D.32), and the definition of f_0 in (13.78),

$$\begin{aligned}\text{(13.82)}\quad \operatorname{grad} f_0(\rho_0) &= b\operatorname{grad}\mathcal{E}(\rho_0)+a\operatorname{grad}_{\rho_0}d^2(\rho_0,\tau_0)+c\operatorname{grad}_{\rho_0}G(\rho_0,\sigma_0)\\ &= -b(\frac{1}{2}\Delta\rho_0+\nabla\cdot(\rho_0\nabla(\Psi+\rho_0*\Phi)))\\ &\quad -2c\nabla\cdot(\rho_0\nabla(\sum_{k=1}^{\infty}\frac{\langle e_k,\rho_0-\sigma_0\rangle}{2^k(1+m_k^2)}e_k))\\ &\quad -2a\nabla\cdot(\rho_0\nabla p_{\rho_0,\tau_0}),\end{aligned}$$

where p_{ρ_0,τ_0} is as in Theorem D.25.

With reference to Appendix D.5 for the definition and properties of $H_\rho^{-1}(R^d)$,

$$\|\operatorname{grad}_{\rho_0}\frac{1}{2}d^2(\rho_0,\tau_0)\|_{-1,\rho_0}^2=d^2(\rho_0,\tau_0)<\infty,$$

by (D.46), and $\|\operatorname{grad}_{\rho_0}G(\rho_0,\sigma_0)\|_{-1\rho_0}<\infty$, so

$$\begin{aligned}\|\operatorname{grad} f_0(\rho_0)\|_{-1,\rho_0}^2 &= a^2\|\operatorname{grad}_{\rho_0}d^2(\rho_0,\tau_0)\|_{-1,\rho_0}^2+b^2\|\operatorname{grad}\mathcal{E}(\rho_0)\|_{-1,\rho_0}^2\\ &\quad +c^2\|\operatorname{grad}_{\rho_0}G(\rho_0,\sigma_0)\|_{-1,\rho_0}^2\\ \text{(13.83)}\qquad &\quad +2ab\langle\operatorname{grad}_{\rho_0}d^2(\rho_0,\tau_0),\operatorname{grad}\mathcal{E}(\rho_0)\rangle_{-1,\rho_0}\\ &\quad +2ac\langle\operatorname{grad}_{\rho_0}d^2(\rho_0,\tau_0),\operatorname{grad}_{\rho_0}G(\rho_0,\sigma_0)\rangle_{-1,\rho_0}\\ &\quad +2bc\langle\operatorname{grad}_{\rho_0}d^2(\rho_0,\tau_0),\operatorname{grad}_{\rho_0}G(\rho_0,\sigma_0)\rangle_{-1,\rho_0}\end{aligned}$$

is finite if and only if $\|\operatorname{grad}\mathcal{E}(\rho_0)\|_{-1,\rho_0}<\infty$. Since (13.80) implies that

$$f_0(\rho)-f_0(\rho_0)\ge-\lambda_{0,\delta}\frac{1}{2\delta}d^2(\rho,\gamma_0)+\lambda_{0,\delta}\frac{1}{2\delta}d^2(\rho_0,\gamma_0),\tag{13.84}$$

by Lemma D.55 and (D.31),

$$\operatorname{grad} f_0(\rho_0)=\operatorname{grad}_{\rho=\rho_0}\frac{-\lambda_{0,\delta}}{2\delta}d^2(\rho,\gamma_0)=\lambda_{0,\delta}\frac{\nabla\cdot(\rho_0\nabla p_{\rho_0,\gamma_0})}{\delta}.\tag{13.85}$$

By (D.46),

$$\|\operatorname{grad} f_0(\rho_0)\|_{-1,\rho_0}^2=\|\operatorname{grad}_{\rho=\rho_0}\frac{\lambda_{0,\delta}}{2\delta}d^2(\rho,\gamma_0)\|_{-1,\rho_0}^2=\frac{\lambda_{0,\delta}^2}{\delta^2}d^2(\rho_0,\gamma_0)<\infty.\tag{13.86}$$

Consequently, by (13.83),

$$I^{\#}(\rho_0)=4\|\operatorname{grad}\mathcal{E}(\rho_0)\|_{-1,\rho_0}^2<\infty.\tag{13.87}$$

□

Next, using $I^{\#}(\rho_0) < \infty$ and the mass transport inequality (D.75), we connect the right side of (13.81) with $\widehat{H}_\dagger f_0(\rho_0)$.

LEMMA 13.36.

$$\begin{aligned} \lambda_{0,\delta}\alpha^{-1}(\overline{f} - h)(\gamma_0) &\leq -\langle \operatorname{grad}\mathcal{E}(\rho_0), \operatorname{grad} f_0(\rho_0)\rangle_{-1,\rho_0} + \frac{1}{2}\|\operatorname{grad} f_0(\rho_0)\|^2_{-1,\rho_0} \\ &= \widehat{H}_\dagger f_0(\rho_0). \end{aligned} \tag{13.88}$$

PROOF. By (D.75),

$$\begin{aligned} \alpha^{-1}(\overline{f} - h)(\gamma_0) &\leq \frac{1}{\delta}\Big(\int_{R^d} \nabla p_{\rho_0,\gamma_0}\cdot(\frac{1}{2}\frac{\nabla\rho_0}{\rho_0} + \nabla\Psi + \nabla(\Phi * \rho_0))d\rho_0\Big) \\ &\quad + \frac{\lambda_{0,\delta}}{2\delta^2}d^2(\rho_0,\gamma_0), \end{aligned}$$

where $\lambda_{0,\delta} = 1 + 2\delta(|\lambda_\Psi| + 2|\lambda_\Phi|)$. Note that by (13.85) and Lemma D.46,

$$\begin{aligned} &\frac{\lambda_{0,\delta}}{\delta}\Big(\int_{R^d} \nabla p_0(x)\cdot(\frac{1}{2}\frac{\nabla\rho_0(x)}{\rho_0(x)} + \nabla\Psi + \nabla(\Phi * \rho_0))\rho_0(x)dx \\ &\quad = \langle -\operatorname{grad}\mathcal{E}(\rho_0), \operatorname{grad} f_0(\rho_0)\rangle_{-1,\rho_0}. \end{aligned}$$

Therefore, applying (13.86), the lemma follows. □

PROOF. (Lemma 13.34) By (13.80),

$$\begin{aligned} (\overline{f}_\delta - f_0)(\rho_0) &= \lambda_{0,\delta}(\overline{f}(\gamma_0) - \frac{1}{2\delta}d^2(\rho_0,\gamma_0)) - f_0(\rho_0) \\ &= \sup_{\rho,\gamma\in E}(\lambda_{0,\delta}(\overline{f}(\gamma) - \frac{1}{2\delta}d^2(\rho,\gamma)) - f_0(\rho)) \\ &= \sup_{\rho\in E}(\overline{f}_\delta(\rho) - f_0(\rho)), \end{aligned}$$

which implies that

$$\begin{aligned} \lambda_{0,\delta}(\overline{f} - h)(\gamma_0) &\geq \lambda_{0,\delta}\overline{f}(\gamma_0) - \frac{\lambda_{0,\delta}}{2\delta}d^2(\rho_0,\gamma_0) - \lambda_{0,\delta}h(\gamma_0) \\ &= \overline{f}_\delta(\rho_0) - \lambda_{0,\delta}h(\gamma_0). \end{aligned}$$

In addition, (13.80) implies

$$d^2(\rho_0,\gamma_0) \leq 2\delta(\overline{f}(\gamma_0) - \overline{f}(\rho_0)) \leq 4\delta\sup_{\rho\in E}|\overline{f}(\rho)|,$$

which further implies that

$$h(\rho_0) - h(\gamma_0) \leq \omega_h(2\sqrt{\delta\sup_{\rho\in E}|\overline{f}(\rho)|}).$$

Therefore, from (13.88), we have

$$\begin{aligned} \widehat{H}_\dagger f_0(\rho_0) &\geq \alpha^{-1}(\overline{f}_\delta(\rho_0) - \lambda_{0,\delta}h(\gamma_0)) \\ &= \alpha^{-1}\Big(\overline{f}_\delta(\rho_0) - \lambda_{0,\delta}(h(\rho_0) - h(\gamma_0) + h(\rho_0))\Big) \\ &\geq \alpha^{-1}(\overline{f}_\delta - h_{0,\delta})(\rho_0), \end{aligned}$$

and we conclude that $\overline{f}_\delta$ is a subsolution to (13.76). □

PROOF. (Theorem 13.32) By Lemma 13.34 and Theorem 9.41,

$$\sup_{\rho\in E}\left(\overline{f}_\delta(\rho)-\underline{f}_\delta(\rho)\right)\le\sup_{\rho\in E}\left(h_{0,\delta}(\rho)-h_{1,\delta}(\rho)\right).$$

Letting $\delta\to 0+$, we conclude $\overline{f}\le\underline{f}$. □

13.3.5. Variational representation. Let $\rho\in\mathcal{P}(R^d)$ and $p\in C_c^\infty(R^d)$. Then

$$\begin{aligned}\frac{1}{2}\int_{R^d}|\nabla p(x)|^2\rho(dx) &= \sup_{u\in L^2_\rho(R^d)}\{\int_{R^d}u(x)\cdot\nabla p(x)\rho(dx)-\frac{1}{2}\int_{R^d}|u(x)|^2\rho(dx)\}) \\ &= \sup_{u\in L^2_{\rho,\nabla}(R^d)}\{\int_{R^d}u(x)\cdot\nabla p(x)\rho(dx)-\frac{1}{2}\int_{R^d}|u(x)|^2\rho(dx)\},\end{aligned}\tag{13.89}$$

where $L^2_{\rho,\nabla}(R^d)$ is the closure of $\{\nabla p : p\in C_c^\infty(R^d)\}$ in $L^2_\rho(R^d)$. The second equality follows from the fact that $\int_{R^d}u(x)\cdot\nabla p(x)\rho(dx)=\int_{R^d}P_{\rho,\nabla}u(x)\cdot\nabla p(x)\rho(dx)$ and $\int_{R^d}|u(x)|^2\rho(dx)\ge\int_{R^d}|P_{\rho,\nabla}u(x)|^2\rho(dx)$, where $P_{\rho,\nabla}: L^2_\rho(R^d)\mapsto L^2_{\rho,\nabla}(R^d)$ is the projection onto $L^2_{\rho,\nabla}(R^d)$. This identity suggests a variational form for H in which the controls are R^d-valued functions; however, the compactness conditions of Chapter 8 are difficult to obtain using that approach. Instead, let the control space U be the subspace of $\mu\in\mathcal{P}(R^d\times R^d)$ such that $\int(1+|x|^2+|z|)\mu(dx\times dz)<\infty$, with the topology in which $\mu_n\to\mu$ if and only if

$$\int f(x,z)(1+|x|^2+|z|)\mu_n(dx\times dz)\to\int f(x,z)(1+|x|^2+|z|)\mu(dx\times dz)$$

for all $f\in C_b(R^d\times R^d)$. Let the set Γ that determines the admissible controls be the collection of $(\rho,\mu)\in E\times U$ such that $\rho(dx)=\mu(dx\times R^d)$. For $(\rho,\mu)\in\Gamma$, we write $\mu(dx\times dz)=\mu(dz|x)\rho(dx)$ and define

$$L(\rho,\mu)\equiv\int_{R^d\times R^d}|z|^2\mu(dx\times dz)\ge\int_{R^d}|\int_{R^d}z\mu(dz|x)|^2\rho(dx),\tag{13.90}$$

where the inequality follows by Jensen's inequality. Consequently, by (13.89) and the inequality in (13.90), we have the variational representation for the H in (13.45):

$$\begin{aligned}Hf(\rho) &= \sup_{\mu\in\Gamma_\rho}\{\int_{R^d\times R^d}(B(\rho)+z\cdot\nabla)\frac{\delta f}{\delta\rho}(x)\mu(dx\times dz)-L(\rho,\mu)\} \\ &= \sup_{u\in L^2_\rho(R^d)}\{\int_{R^d}(B(\rho)+u(x)\cdot\nabla)\frac{\delta f}{\delta\rho}(x)\rho(dx)-\int_{R^d}|u(x)|^2\rho(dx)\} \\ &= \sup_{u\in L^2_{\rho,\nabla}(R^d)}\{\int_{R^d}(B(\rho)+u(x)\cdot\nabla)\frac{\delta f}{\delta\rho}(x)\rho(dx)-\int_{R^d}|u(x)|^2\rho(dx)\},\end{aligned}$$

where $f\in\mathcal{D}(H)$, $\delta f/\delta\rho$ is defined by (13.44), and the second expression is obtained from the first by setting $\mu(dx\times dz)=\delta_{u(x)}(dz)\rho(dx)$. Then, for $f\in\mathcal{D}(H)$ and $(\rho,\mu)\in\Gamma$, define

$$Af(\rho,\mu)=\int_{R^d\times R^d}(B(\rho)+z\cdot\nabla)\frac{\delta f}{\delta\rho}(x)\mu(dx\times dz),\tag{13.91}$$

so

$$Hf(\rho)=\sup_{\mu\in\Gamma_\rho}(Af(\rho,\mu)-L(\rho,\mu)).\tag{13.92}$$

The control equation becomes

$$f(\rho(t)) = f(\rho(0)) + \int_0^t (B(\rho(s)) + u_\mu(s,x)\cdot\nabla)\frac{\delta f}{\delta\rho}(x)\rho(s,dx)ds,$$

where $u_\mu(s,x) = \int z\mu(dz|x,s)$ and we require

$$\int_0^t |(B(\rho(s)) + u_\mu(s,x)\cdot\nabla)\frac{\delta f}{\delta\rho}(x)\rho(s,dx)|ds < \infty, \quad f\in\mathcal{D}(H).$$

The control equation is equivalent to requiring

$$(13.93)\quad \langle q,\rho(t)\rangle = \langle q,\rho(0)\rangle + \int_0^t \langle B(\rho(s))q + u_\mu(s,\cdot)\cdot\nabla q,\rho(s)\rangle ds, \quad q\in C_c^\infty(R^d).$$

The continuity condition, Condition 8.9.(1), follows immediately from the definition of the topology on U. Setting $\mu(dx\times dz) = \rho(dx)\delta_0(dz)$, the control problem reduces to (1.44) and Condition 8.9.(2) follows, as does Condition 8.10.

Let $K\subset E$ be compact and set

$$\Gamma_{K,c} = \{(\rho,\mu)\in\Gamma : L(\rho,\mu)\le c\}\cap(K\times U).$$

Then $\{\mu(\cdot\times R^d) : (\rho,\mu)\in\Gamma_{K,c}\} = K$ is compact in E by assumption and $\{\mu(R^d\times\cdot) : (\rho,\mu)\in K_{K,c}\}$ is compact in the topology determined by the requirement that $\int(1+|z|)f(z)\nu_n(dz)\to\int(1+|z|)f(z)\nu(dz)$, $f\in C_b(R^d)$. Compactness of $\Gamma_{K,c}$ then follows by a minor modification of Prohorov's theorem.

By the Schwartz inequality, Condition 8.9.(5) follows with $\psi_{f,K}(r)$ of the form $C_{f,K}(1+\sqrt{|r|})$, for some constant $C_{f,K}$.

For $f(\rho) = \varphi(\langle\mathbf{p},\rho\rangle)\in\mathcal{D}(H)$, the supremum in (13.92) is achieved by taking $\mu_{f,\rho}(dx\times dz) = \rho(dx)\delta_{u_{f,\rho}(x)}(dz)$ with $u_{f,\rho}(x) = \sum_{i=1}^l \partial_i\varphi(\langle\mathbf{p},\rho\rangle)\nabla p_i(x)$. Taking $\mathbf{H} = \mathbf{H}_\dagger = \mathbf{H}_\ddagger = H$, Condition 8.11 follows from the existence of a solution of

$$(13.94)\quad \langle q,\rho(t)\rangle = \langle q,\rho(0)\rangle + \int_0^t \langle B(\rho(s))q + u_{f,\rho}\cdot\nabla q,\rho(s)\rangle ds, \quad q\in C_c^\infty(R^d).$$

A solution can be constructed as the limit of a system

$$dX_{n,i}(t) = (-\nabla\Psi(X_{n,i}(t)) - \int_{y\in R^d}\nabla\Phi(X_{n,i}(t)-y)\rho_n(t,dy)$$
$$+\sum_{i=1}^k \partial_i\varphi(\langle\mathbf{p},\rho_n\rangle)\nabla p_i(X_{n,i}(t)))dt + dW_i(t),$$

where

$$\rho_n(t) = \frac{1}{n}\sum_{i=1}^n \delta_{X_{n,i}(t)}$$

and $\{X_{n,i}(0)\}$ are independent with distribution $\rho(0)$. This system is another example of a McKean-Vlasov system. Typically, these systems are considered under Lipschitz assumptions (see, for example, Méléard [**85**]) that are not assumed here; however, the semi-convexity assumption on Ψ gives growth estimates on the $X_{n,i}$ which in turn ensure the relative compactness of the system. Any limit point will give a solution of

$$dX_i(t) = (-\nabla\Psi(X_i(t)) - \int_{y\in R^d} \nabla\Phi(X_i(t) - y)\rho(t, dy)$$
$$+ \sum_{i=1}^{k} \partial_i\varphi(\langle \mathbf{p}, \rho\rangle)\nabla p_i(X_i(t)))dt + dW_i(t),$$

where $\rho(t) = \lim_{k\to\infty} \frac{1}{k}\sum_{i=1}^{k} \delta_{X_i(t)}$ satisfies (13.94). Systems of this form have been considered by Kurtz and Protter [**76**] and Kurtz and Xiong [**78**].

Finally, to verify Condition 8.9.(4), note that $K \subset E$ is compact if and only if $\{|Z_\rho|^2 : \rho \in K\}$ is uniformly integrable, where Z_ρ is a random variable with distribution ρ.

Let (ρ, μ) be a solution of (13.93). Then by Corollary 1.12 of Kurtz and Stockbridge [**77**], there exists a stochastic process $\{(X(t), \Lambda(t))\}$ with values in $R^d \times \mathcal{P}(R^d)$ such that $E[\int_{R^d} h(X(t), z)\Lambda(s, dz)] = \int h(x, z)\mu(s, dx \times dz)$, $h \in C_b(R^d \times R^d)$, and setting $U(t) = \int z\Lambda(s, dz)$,

$$q(X(t)) - q(X(0)) - \int_0^t (B(\rho(s))q(X(s)) + U(s) \cdot \nabla q(X(s)))ds$$

is a $\{\mathcal{F}_t^{X,\Lambda}\}$-martingale for each $q \in C_c^\infty(R^d)$. It follows that there exists a Brownian motion so that

$$X(t) = X(0) + \int_0^t (-\nabla\Psi(X(s)) - \int_{y\in R^d} \nabla\Phi(X(s) - y)\rho(s, dy) + U(s))ds + W(t).$$

Let $K \subset E$ be compact. Then the corresponding uniform integrability implies that there exists a convex, C^2 function $\gamma : [0, \infty) \to [0, \infty)$ such that $\lim_{r\to\infty} r^{-1}\gamma(r) = \infty$ and $\sup_{\rho\in K} \int \gamma(|x|^2)\rho(dx) < \infty$. Without loss of generality, we can assume that $\gamma'' \geq 0$ is nonincreasing, $\gamma(0) = \gamma'(0) = 0$, and, recalling Lemma 13.16, that γ' grows slowly enough so that

$$\lim_{|x|\to\infty} \frac{\gamma'(|x|^2)|x|^2}{\nabla\Psi(x) \cdot x} = \lim_{|x|\to\infty} \frac{\gamma'(|x|^2)}{\nabla\Psi(x) \cdot x/|x|^2} = 0. \tag{13.95}$$

Setting $\tau_c = \inf\{t : |X(t)| \geq c\}$ and applying Itô's formula, we have

$$\begin{aligned}
&E[\gamma(|X(t \wedge \tau_c|^2)]\\
&= E[\gamma(|X(0)|^2)] + E\Big[\int_0^{t\wedge\tau_c} \Big(2\gamma'(|X(s)|^2)X(s)\\
&\qquad \cdot(-\nabla\Psi(X(s)) - \rho * \nabla\Phi(X(s)) + U(s))\\
&\qquad +2\gamma''(|X(s)|^2)|X(s)|^2 + \gamma'(|X(s)|^2)\Big)ds\Big]\\
&\leq E[\gamma(|X(0)|^2)] + E\Big[\int_0^{t\wedge\tau_c} \Big(|U(s)|^2 + C_1 + C_2(1 + \gamma(|X(s)|^2))\Big)ds\Big],
\end{aligned}$$

where

$$C_1 = \sup_x(\gamma'(|x|^2)^2|x|^2 - 2\gamma'(|x|^2)x \cdot \nabla\Psi(x))$$

and

$$C_2 = \sup_{x,\rho} \frac{2\gamma''(|x|^2)|x|^2 + \gamma'(|x|^2) - 2\gamma'(|x|^2)x \cdot \rho * \nabla\Phi(x)}{1 + \gamma(|x|^2)}.$$

Note that $C_1 < \infty$ by (13.95) and $C_2 < \infty$ by the fact that $\gamma''(r)r \leq \gamma'(r)$ and $\gamma'(r)r \leq 2\gamma(r)$. Gronwall's inequality gives

$$E[\gamma(|X(t \wedge \tau_c|^2)] \leq (E[\gamma(|X(0)|^2)] + C_1 + C_2 + E[\int_0^t |U(s)|^2 ds])e^{C_2 t}.$$

Letting $c \to \infty$, we have

$$\int \gamma(|x|^2)\rho(t, dx) \leq (\int \gamma(|x|^2)\rho(0, dx) + C_1 + C_2 + \int_0^t L(\rho(s), \mu(s))ds)e^{C_2 t},$$

and Condition 8.9.(4) follows with

$$\widehat{K}(K, T, M) = \{\rho : \int \gamma(|x|^2)\rho(dx) \leq (\sup_{\rho' \in K} \int \gamma(|x|^2)\rho'(dx) + C_1 + C_2 + M)e^{C_2 T}\}.$$

The control equation (13.93) is the weak (in the Schwartz distributional sense) formulation of the controlled partial differential equation

$$\partial_t \rho = \frac{1}{2}\Delta\rho + \nabla \cdot (\rho\nabla(\Psi + \rho * \Phi)) - \nabla \cdot (\rho u_\mu). \tag{13.96}$$

By (8.18) and (13.90)

$$\begin{aligned} I(\rho) &= I_0(\rho(0)) + \inf\{\int_0^\infty \int |u_\mu(s, x)|^2 \rho(s, dx))ds : (\rho, u_\mu) \text{ satisfies } (13.93)\} \\ &= I_0(\rho(0)) + \inf\{\frac{1}{2}\int_0^\infty \|u_\mu(s)\|^2_{0,\rho(s)} ds : (\rho, u_\mu) \text{ satisfies } (13.93)\} \\ &= I_0(\rho(0)) + \inf\{\frac{1}{2}\int_0^\infty \|P_{\rho(s),\nabla} u_\mu(s)\|^2_{0,\rho(s)} ds : (\rho, u_\mu) \text{ satisfies } (13.93)\}. \end{aligned}$$

Let $q \in C_c^\infty((0, \infty) \times R^d)$, and write $q(s) = q(s, \cdot) \in C_c^\infty(R^d)$. Then for any solution of (13.93),

$$\int_0^\infty \langle B(\rho(t))q(t) + \partial_t q(t), \rho(t)\rangle dt = -\int_0^\infty \langle u_\mu(t, \cdot)\nabla q(t), \rho(t)\rangle dt.$$

Taking the supremum over all $q \in C_c^\infty((0, \infty) \times R^d)$,

$$\begin{aligned} &\frac{1}{2}\int_0^\infty \|P_{\rho(s),\nabla} u_\mu(s)\|^2_{0,\rho(s)} ds \\ &= \sup_q\{-\int_0^\infty \langle u_\mu(t, \cdot)\nabla q(t), \rho(t)\rangle dt - \frac{1}{2}\int_0^\infty \int_{R^d} |\nabla q(t, x)|^2 \rho(t, dx)\} \\ &= \sup_q\{\int_0^\infty \langle B(\rho(t))q(t) + \partial_t q(t), \rho(t)\rangle dt - \frac{1}{2}\int_0^\infty \int_{R^d} |\nabla q(t, x)|^2 \rho(t, dx)\} \\ &= \frac{1}{2}\int_0^\infty \|\partial_t \rho - \frac{1}{2}\Delta\rho - \nabla \cdot (\rho\nabla(\Psi + \rho * \Phi))\|^2_{-1,\rho(t)} dt. \end{aligned}$$

13.3.6. The large deviation theorem.

THEOREM 13.37. *Let Ψ, Φ satisfy Conditions 13.14, and let $(E, r) = (\mathcal{P}_2(R^d), d_2)$, where d_2 is the 2-Wasserstein metric on the space of probability measures with finite second moments $\mathcal{P}_2(R^d)$. See Definition D.13.*

Let $\rho_n, n = 1, 2, \ldots$ be the empirical measures (1.43) of the $X_{n,k}$ satisfying (1.42). Assume that $\{\rho_n(0)\}$, as a sequence of E-valued random variables, satisfies *the large deviation principle with good rate function* I_0. *Then* $\{\rho_n(\cdot) : n = 1, 2, \ldots\}$ *satisfies a large deviation principle in* $C_E[0, \infty)$ *with good rate function*

$$I(\rho) = I_0(\rho(0)) + \frac{1}{2}\int_0^\infty \|\frac{\partial}{\partial t}\rho - \frac{1}{2}\Delta\rho - \nabla \cdot (\rho\nabla(\Psi + \rho * \Phi))\|_{-1,\rho(t)}^2 dt.$$

Appendix

APPENDIX A

Operators and convergence in function spaces

Let E be a metric space.

A.1. Semicontinuity

DEFINITION A.1. $f : E \mapsto [-\infty, \infty]$ is *lower semicontinuous* if for each $x \in E$, $f(x) \leq \liminf_{y \to x} f(y)$; $f : E \mapsto [-\infty, \infty]$ is *upper semicontinuous* if for each $x \in E$, $f(x) \geq \limsup_{y \to x} f(y)$.

The proofs of the following standard lemmas are left as exercises.

LEMMA A.2. *For $f : E \mapsto [-\infty, \infty]$, define $f^*(x) = \lim_{\epsilon \to \infty} \sup_{y \in B_\epsilon(x)} f(y)$. Then f^* is an upper semicontinuous function called the* upper semicontinuous regularization *of f. Similarly, $f_*(x) = \lim_{\epsilon \to \infty} \inf_{y \in B_\epsilon(x)} f(y)$ is a lower semicontinuous function called the* lower semicontinuous regularization *of f. If f is upper semicontinuous, then $f^* = f$, and if f is lower semicontinuous, $f_* = f$.*

LEMMA A.3. *Suppose $f_1 \leq f_2 \leq \cdots$ are lower semicontinuous. Then the pointwise limit f of $\{f_n\}$ is lower semicontinuous. Suppose $f_1 \geq f_2 \geq \cdots$ are upper semicontinuous. Then the pointwise limit f of $\{f_n\}$ is upper semicontinuous.*

LEMMA A.4. *Let $\{f_n\}$ be a decreasing sequence of upper semicontinuous functions with limit f. Suppose that $x_n \in E$ and $\epsilon_n > 0$ satisfy $\sup_{x \in E} f_n(x) \leq f_n(x_n) + \epsilon_n$, $\epsilon_n \to 0$, and $x_n \to x_0$. Then*

$$\lim_{n \to \infty} f_n(x_n) = f(x_0) = \sup_{x \in E} f(x).$$

The analogous result holds for increasing sequences of lower semicontinuous functions.

REMARK A.5. See Proposition 2.42 of Attouch [**5**] for much more general results of this type.

PROOF. Let $a = \lim_{\epsilon \to 0} \lim_{n \to \infty} \sup_{y \in B_\epsilon(x_0)} f_n(y)$. Then

$$a = \lim_{n \to \infty} f_n(x_n) = \lim_{n \to \infty} \sup_{y \in E} f_n(y),$$

and since

$$f_n(x_n) + \epsilon_n \geq f_n(x_0) \geq \limsup_{m \to \infty} f_n(x_m) \geq \limsup_{m \to \infty} f_m(x_m) = a,$$

it follows that

$$\lim_{n \to \infty} f_n(x_n) = f(x_0) = a = \sup_{x \in E} f(x).$$

□

A.2. General notions of convergence

DEFINITION A.6. (*buc*-Convergence)

a) A sequence $\{f_n\}$ of functions on E converges *boundedly and uniformly on compacts* to f if and only if $\sup_n \|f_n\| < \infty$ and for each compact $K \subset E$,

$$\lim_{n\to\infty} \sup_{x\in K} |f_n(x) - f(x)| = 0.$$

We will denote this convergence by

$$f = buc\text{-}\lim_{n\to\infty} f_n.$$

b) Let E be a metric space, and for $n = 1, 2, \ldots$, let E_n be a metric space and $\eta_n : E_n \mapsto E$ be Borel measurable. Define $\eta_n : B(E) \mapsto B(E_n)$ by $\eta_n f = f \circ \eta_n$. A sequence $\{f_n\}$ with $f_n \in E_n$ *buc*-converges to $f \in E$ if and only if $\sup_n \|f_n\| < \infty$ and for each compact $K \subset E$

$$\lim_{n\to\infty} \sup_{y\in\eta_n^{-1}(K)} |f_n(y) - \eta_n f(y)| = 0.$$

We write $f = buc\text{-}\lim_{n\to\infty} f_n$.

c) A function $f \in C_b(E)$ is *buc-approximable* by a set $B \subset C_b(E)$ if there exists a constant $C_{f,B}$ such that for every $\epsilon > 0$ and every compact $K \subset E$, there exists $g_{\epsilon,K} \in B$ such that $\|g_{\epsilon,K}\| \le C_{f,B}$ and $\sup_{x\in K} |f(x) - g_{\epsilon,K}(x)| \le \epsilon$.

d) A set $B \subset C_b(E)$ is *buc-closed* if B contains every $f \in C_b(E)$ that is *buc*-approximable by B.

e) C is the *buc-closure* of B if C is the smallest *buc*-closed set containing B. B is *buc-dense* in C, if C is the *buc*-closure of B.

Note that *buc*-closure is a legitimate notion of closure (that is, if $\{B_\alpha\}$ is a collection of *buc*-closed sets then $\cap_\alpha B_\alpha$ is *buc*-closed and a finite union of *buc*-closed sets is *buc*-closed). We will refer to the corresponding topology as the *buc topology*. The buc topology is closely related to the strict topology introduced by Buck [**12**] for E locally compact and extended to more general E by a variety of authors [**14, 55, 59**]. (In fact, we have not ruled out the possibility that the two topologies are equivalent. Wheeler [**130**] gives a survey of a collection of related topologies.) For compact $K \subset E$, let p_K be the seminorm given by $p_K(f) = \sup_{x\in K} |f(x)|$. The collection of seminorms $\{p_K : K \subset E \text{ compact}\}$ generates the compact uniform topology on $C(E)$. For a sequence of compact sets $\{K_n\}$ and a sequence $\{a_n\} \subset (0, \infty)$ satisfying $\lim_{n\to\infty} a_n = 0$, let $p_{\{(K_n,a_n)\}}(f) = \sup_n a_n p_{K_n}(f)$. The strict topology is generated by the collection of seminorms of this form. Clearly, any strictly closed set is *buc*-closed (that is, the topology corresponding to *buc*-closure is finer than the strict topology). Furthermore, a sequence is *buc*-convergent if and only if it is strictly convergent. (See Cooper [**14**].) Furthermore, restricted to bounded sets, both the topology corresponding to *buc*-closure and the strict topology are just the compact uniform topology.

For $r > 0$, let $B_r = \{f \in C_b(E) : \|f\| \le r\}$.

LEMMA A.7. *Let $D \subset C_b(E)$. Suppose that for each $r > 0$, $D \cap B_r$ is buc-closed. Then D is buc-closed.*

PROOF. Suppose f is *buc*-approximable by D. Then there exists a constant $r > 0$ such that for every $\epsilon > 0$ and every compact $K \subset E$, there exists $g_{\epsilon,K} \in D$ such that $\|g_{\epsilon,K}\| \leq r$ and $\sup_{x\in K} |f(x) - g_{\epsilon,K}(x)| \leq \epsilon$. But then, f is *buc*-approximable by $D \cap B_r$ and hence $f \in D \cap B_r$, so $f \in D$ and D is *buc*-closed. □

Let $D \subset C_b(E)$, D is an *algebra* if and only if for all $f_1, f_2 \in D$, $a, b \in R$, we have $af_1 + bf_2 \in D$ and $f_1 f_2 \in D$.

THEOREM A.8. *Let $D \subset C_b(E)$ be an algebra that separates points and vanishes nowhere (that is, for each $x \in E$, there exists $f \in D$ such that $f(x) \neq 0$). Then $C_b(E)$ is the buc-closure of D, and for each $r > 0$, B_r is the buc-closure of $D \cap B_r$.*

PROOF. If *buc*-closure is the same as strict closure, then this theorem is a special case of Theorem 3.1 of Giles [**55**]. If not, then the current theorem is a consequence of the earlier result. To see that, first note that the norm closure of D contains $\{ge^{-\epsilon g^2} : g \in D\}$ for each $\epsilon > 0$. Consequently, letting D^{st} denote the strict closure of D, the *buc*-closure of D contains all functions of the form $\{ge^{-\epsilon g^2} : g \in D^{st}\}$. But by Giles's theorem, $D^{st} = C_b(E)$. Letting $\epsilon \to 0$, $ge^{-\epsilon g^2}$ converges uniformly to g, so the *buc*-closure of D contains $C_b(E)$.

Let $\varphi_r(u) = (-r) \vee (r \wedge u)$. Since for $g \in D$ and $a_0, \ldots, a_m \in R$, $\sum_{k=0}^m a_k g^k \in D$, and any continuous $\varphi : R \mapsto R$ can be approximated uniformly on bounded intervals by polynomials, it follows that the *buc*-closure $\overline{D}_r$ of $D_r = D \cap B_r$ contains $\varphi_r(g)$ for every $g \in D$. Since $g \mapsto \varphi_r(g)$ is continuous, it in turn follows that $\overline{D}_r$ contains $\varphi_r(g)$ for every $g \in C_b(E)$. Consequently, $\overline{D}_r = B_r$. □

COROLLARY A.9. *If E is separable, then there exists a countable set $D_0 \subset C_b(E)$ such that $C_b(E)$ is the buc-closure of $\{g_i\}$.*

REMARK A.10. Note that for $a < b \in R$, $\{f \in C_b(E) : a \leq f \leq b\}$ is the *buc*-closure of $\{a \vee (f \wedge b) : f \in D_0$.

PROOF. Let $\{x_j\}$ be dense in E and let D_0 be the collection of functions of the form

$$f(y) = \sum_{k=1}^m a_k \prod_{j \in J_k} e^{-d(x_j, y)},$$

where a_k is rational and $J_k \subset \{1, 2, \ldots\}$ is finite. Then D_0 is countable, linear, and closed under multiplication. The norm closure D of D_0 is an algebra that separates points and vanishes nowhere. Consequently, $C_b(E)$ is the *buc*-closure of D and hence of D_0. □

LEMMA A.11. *Let $Q : D \subset C_b(E) \mapsto C_b(E)$. Suppose that for each $\delta > 0$, $r > 0$, and compact $K \subset E$, there exist $C_0(r) > 0$, $C_1(\delta, r) > 0$ and compact $K'(\delta, r, K) \subset E$ that are nondecreasing in δ^{-1}, r, and K, such that*

$$\sup_{x\in K} |Qh_1(x) - Qh_2(x)| \leq \delta C_0(r) + C_1(\delta, r) \sup_{x \in K'(\delta,r,K)} |h_1(x) - h_2(x)| \tag{A.1}$$

for all $h_1, h_2 \in D$ satisfying $\|h_1\| \vee \|h_2\| \leq r$. Let $B_r = \{h \in C_b(E) : \|h\| \leq r\}$ and $D_r = D \cap B_r$. Then for each $r > 0$, there exists a unique extension $\widehat{Q}_r$ of $Q_r \equiv Q|_{D_r}$ to the buc-closure *$\overline{D}_r$ of D_r such that the mapping $\widehat{Q}_r : \overline{D}_r \mapsto C_b(E)$ is continuous in the buc topology and (A.1) is satisfied. If $r_1 < r_2$, then $\widehat{Q}_{r_2}$ is an extension of $\widehat{Q}_{r_1}$.*

PROOF. Let $L_1 \subset B_r \times C_b(E)$ be the collection of (f, g) such that

$$\sup_{x\in K} |Qh(x) - g(x)| \leq \delta C_0(r) + C_1(\delta, r) \sup_{x\in K'(\delta,r,K)} |h(x) - f(x)|, \tag{A.2}$$

for all $h \in D_r$. Then L_1 is closed in the product *buc*-topology on $B_r \times C_b(E)$ and $L_1 \supset Q_r$. Let $L_2 \subset B_r \times C_b(E)$ be the collection of $(\widehat{f}, \widehat{g})$ such that

$$\sup_{x\in K} |\widehat{g}(x) - g(x)| \leq \delta C_0(r) + C_1(\delta, r) \sup_{x\in K'(\delta,r,K)} |\widehat{f}(x) - f(x)|, \tag{A.3}$$

for all $(f, g) \in L_1$. Then L_2 is also closed, and $L_2 \subset L_1$, since $Q_r \subset L_1$ and $Q_r \subset L_2$ by the definition of L_1. If $(f, g_1), (f, g_2) \in L_2$, then $g_1 = g_2$, since for each compact K, $\sup_{x\in K} |g_1(x) - g_2(x)| \leq \delta C_0(r)$ for all $\delta > 0$. Consequently, L_2 is a (single-valued) closed mapping that extends Q_r. Setting $\widehat{Q}_r = \{(f, g) \in L_2 : f \in \overline{D}_r\}$, we claim the $\mathcal{D}(\widehat{Q}_r)$ is a *buc*-closed subset of B_r. Note that if $h \in D_r$ and $f \in \mathcal{D}(\widehat{Q}_r)$, then (A.2) implies

$$\|\widehat{Q}_r f\| \leq \|Q_r h\| + \delta C_0(r) + 2rC_1(\delta, r), \tag{A.4}$$

that is, the range of $\widehat{Q}_r$ is uniformly bounded.

Suppose that f is *buc*-approximable by $\mathcal{D}(\widehat{Q}_r)$. Let $\delta_n > 0$ decrease to zero, and let $f_n^K \in \mathcal{D}(\widehat{Q}_r)$ satisfy

$$\sup_{x\in K'(\delta_n,r,K)} |f(x) - f_n^K(x)| \leq \frac{\delta_n}{C_1(\delta_n, r)}.$$

By assumption, for $m > n$, $C_1(\delta_n, r) \leq C_1(\delta_m, r)$ and $K'(\delta_n, r, K) \subset K'(\delta_m, r, K)$. Consequently, for $m > n$,

$$\begin{aligned}
&\sup_{x\in K} |\widehat{Q}_r f_n^K(x) - \widehat{Q}_r^K f_m(x)| \\
&\leq \delta_n C_0(r) + C_1(\delta_n, r) \sup_{x\in K'(\delta_n,r,K)} |f_n^K(x) - f_m^K(x)| \\
&\leq (C_0(r) + 2)\delta_n.
\end{aligned}$$

It follows that $\{\widehat{Q}_r f_n|_K\}$ is Cauchy in $C(K)$, and hence, there exist $g_K \in C(K)$ such that $\lim_{n\to\infty} \sup_{x\in K} |\widehat{Q}_r f_n(x) - g_K(x)| = 0$. Furthermore, if $K_1 \subset K_2$, then $g_{K_1}(x) = g_{K_2}(x)$ for $x \in K_1$, so there is a unique g defined on E such that $g(x) = g_K(x)$ for $x \in K$. Since g_K is continuous on K and bounded by the right side of (A.4), $g \in C_b(E)$. It follows that (f, g) is in the *buc*-closure of $\widehat{Q}_r$ and hence in $\widehat{Q}_r$, so $\mathcal{D}(\widehat{Q}_r) = \overline{D}_r$. □

In several of the examples discussed in the Introduction, the generators H_n are defined for functions on a different state space than the limiting operator H. The following definition makes precise the type of convergence we will need in these settings.

DEFINITION A.12. Let E be a metric space, and for $n = 1, 2, \ldots$, let E_n be a metric space and $\eta_n : E_n \mapsto E$ be Borel measurable. Define $\eta_n : B(E) \mapsto B(E_n)$ by $\eta_n f = f \circ \eta_n$. Let $H_n \subset B(E_n) \times B(E_n)$. Then the *extended limit* of $\{H_n\}$ is the collection of $(f, g) \in B(E) \times B(E)$ such that there exist $(f_n, g_n) \in H_n$ satisfying

$$\lim_{n\to\infty} (\|f_n - \eta_n f\| + \|g_n - \eta_n g\|) = 0.$$

We will denote the convergence by

$$ex\text{-}\lim_{n\to\infty} H_n.$$

$(f,g) \in B(E) \times B(E)$ is in the *extended buc-limit* of $\{H_n\}$, if there exist $(f_n, g_n) \in H_n$ satisfying $\sup_n(\|f_n\| + \|g_n\|) < \infty$ and for each compact $K \subset E$,

$$\lim_{n\to\infty} \sup_{y\in\eta_n^{-1}(K)} (|f_n(y) - \eta_n f(y)| + |g_n(y) - \eta_n g(y)|) = 0.$$

We will denote this convergence by

$$ex\text{-}buc\text{-}\lim_{n\to\infty} H_n.$$

It is, in fact, useful to consider even more general notions of convergence of $f_n \in J_n \subset B(E_n)$ to a function $f \in J \subset B(E)$. We denote such a notion by LIM, and we assume that LIM satisfies the following conditions (cf. [**36**], Section 1.6):

CONDITION A.13.

(1) $\text{LIM} f_n = f$ *and* $\text{LIM} g_n = g$ *imply*

$$\text{LIM}(af_n + bg_n) = a\text{LIM}f_n + b\text{LIM}g_n, \quad a, b \in R.$$

(2) $\text{LIM} f_n^{(k)} = f^{(k)}$, $k = 1, 2, \ldots$, *and* $\lim_{k\to\infty} \sup_n \|f_n^{(k)} - f_n\| \vee \|f^{(k)} - f\| = 0$ *imply* $\text{LIM} f_n = f$.

(3) *There exists* $K > 0$ *such that for each* $f \in J$, *there exist* $f_n \in J_n$ *satisfying* $\sup_n \|f_n\| \leq K\|f\|$ *and* $\text{LIM} f_n = f$.

The following examples satisfy Condition A.13.

EXAMPLE A.14. a) (*buc*-convergence) $\text{LIM} f_n = f$ if and only if $\sup_n \|f_n\| < \infty$ and for each compact $K \subset E$

$$\lim_{n\to\infty} \sup_{y\in\eta_n^{-1}(K)} |f_n(y) - \eta_n f(y)| = 0.$$

b) ($\mathcal{Q}$-limit) Let $\mathcal{Q}$ be an index set. For each $q \in \mathcal{Q}$ and $n = 1, 2, \ldots$, let $K_n^q \subset E_n$. Assume that for $q_1, q_2 \in \mathcal{Q}$, there exists $q_3 \in \mathcal{Q}$ such that $K_n^{q_1} \cup K_n^{q_2} \subset K_n^{q_3}$ and $\cup_{q\in\mathcal{Q}} \cap_m \text{cl} \cup_{n\geq m} \eta_n(K_n^q) = E$. $\text{LIM} f_n = f$ if and only if $\sup_n \|f_n\| < \infty$ and for each $q \in \mathcal{Q}$

$$\lim_{n\to\infty} \sup_{y\in K_n^q} |f_n(y) - \eta_n f(y)| = 0.$$

Let $\mathcal{L}$ denote the Banach space $(\prod_n J_n) \times J$ with $\|(\{f_n\}, f)\| = \sup_n \|f_n\| \vee \|f\|$ and define

$$\mathcal{L}_0 = \{(\{f_n\}, f) \in \mathcal{L} : \text{LIM} f_n = f\}.$$

Note that LIM determines a bounded linear operator from $\mathcal{L}_0$ to J.

If $H_n \subset J_n \times J_n$, we define

$$ex\text{-}\text{LIM}H_n = \{(f,g) : \exists (f_n, g_n) \in H_n \ni \text{LIM} f_n = f, \text{LIM} g_n = g\}. \tag{A.5}$$

A.3. Dissipativity of operators

Let B be a general Banach space. An operator $H \subset B \times B$ is *dissipative* if for $(f_i, g_i) \in H$, $i = 1, 2$,

$$\|(f_1 - \alpha g_1) - (f_2 - \alpha g_2)\| \geq \|f_1 - f_2\|,$$

for all $\alpha > 0$. In Hilbert space, dissipativity is equivalent to the requirement that

$$\langle f_1 - f_2, g_1 - g_2 \rangle \leq 0, \tag{A.6}$$

where $\langle \cdot, \cdot \rangle$ is the usual inner product. The following brackets give a type of generalization of the Hilbert inner product to Banach spaces:

$$[f, g]_+ \equiv \inf_{\epsilon > 0} \frac{\|f + \epsilon g\| - \|f\|}{\epsilon},$$

and

$$[f, g]_- \equiv \sup_{\epsilon < 0} \frac{\|f + \epsilon g\| - \|f\|}{\epsilon},$$

for $f, g \in B$.

The following lemma extends (A.6) to the Banach space setting. (See Corollary 2.7 of Chapter 2 in Miyadera [**86**].)

LEMMA A.15. *An operator $H \subset B \times B$ is dissipative if and only if for $(f_i, g_i) \in H$, $i = 1, 2$,*

$$[f_1 - f_2, g_1 - g_2]_- \leq 0.$$

The analogous inequality for $[\cdot, \cdot]_+$ gives a slightly stronger condition.

DEFINITION A.16. [Strongly dissipative operators] Let B be a Banach space, and $H \subset B \times B$. H is *strongly dissipative* if $(f_i, g_i) \in \mathcal{H}$, $i = 1, 2$, implies

$$[f_1 - f_2, g_1 - g_2]_+ \leq 0.$$

One can show that the strong infinitesimal generator of a possibly nonlinear contraction semigroup is strongly dissipative (Theorem 3.1 of Sinestrari [**110**]). For function spaces with the supremum norm, strong dissipativity corresponds to a version of the positive maximum principle.

Let E be an arbitrary set and $\widetilde{B}(E)$ be the space of bounded scalar functions on E with the supremum norm. Then for $f, g \in \widetilde{B}(E)$ with $f \not\equiv 0$, $[f, g]_+$ and $[f, g]_-$ have the following representation (see Example 6.3 of Sato [**105**]).

$$[f, g]_+ = \inf_{\epsilon > 0} \frac{\|f + \epsilon g\| - \|f\|}{\epsilon} = \lim_{\epsilon \to 0+} \sup_{x \in E(f, \epsilon)} (\text{sgn} f(x)) g(x), \tag{A.7}$$

$$[f, g]_- = \sup_{\epsilon < 0} \frac{\|f + \epsilon g\| - \|f\|}{\epsilon} = \lim_{\epsilon \to 0+} \inf_{x \in E(f, \epsilon)} (\text{sgn} f(x)) g(x), \tag{A.8}$$

where $E(f, \epsilon) = \{x : |f(x)| > \|f\| - \epsilon\}$.

LEMMA A.17. *An operator $H \subset \widetilde{B}(E) \times \widetilde{B}(E)$ is dissipative if and only if for $(f_i, g_i) \in H$, $i = 1, 2$, there exist x_n such that $\lim_{n \to \infty} |f_1(x_n) - f_2(x_n)| = \|f_1 - f_2\| > 0$ and $\limsup_{n \to \infty} (g_1(x_n) - g_2(x_n) \text{sgn}(f_1(x_n) - f_2(x_n)) \leq 0$.*

LEMMA A.18. *An operator $H \subset \widetilde{B}(E) \times \widetilde{B}(E)$ is strongly dissipative if and only if for $(f_i, g_i) \in H$, $i = 1, 2$, $\lim_{n \to \infty} f_1(x_n) - f_2(x_n) = \|f_1 - f_2\| > 0$ implies $\limsup_{n \to \infty} (g_1(x_n) - g_2(x_n) \leq 0$.*

LEMMA A.19. *Let $H \subset \widetilde{B}(E) \times \widetilde{B}(E)$ be dissipative, and assume that for $f \in \mathcal{D}(H)$ and $K \in R$, $f + K \in \mathcal{D}(H)$ and $H(f+K) = Hf$. Then for $(f_i, g_i) \in H$, $i = 1, 2$, there exist $\{x_n\} \subset E$ such that $\lim_{n\to\infty}(f_1(x_n) - f_2(x_n)) = \sup_x(f_1(x) - f_2(x))$ and $\limsup_{n\to\infty}(g_1(x_n) - g_2(x_n)) \leq 0$, and there exist z_n such that $\lim_{n\to\infty}(f_1(z_n) - f_2(z_n)) = \inf_x(f_1(x) - f_2(x))$ and $\liminf_{n\to\infty}(g_1(z_n) - g_2(z_n)) \geq 0$.*

COROLLARY A.20. *Let $A \subset \widetilde{B}(E) \times \widetilde{B}(E)$ be linear and strongly dissipative and $(1, 0) \in A$. Let $(f, g) \in A$ and $\{x_n\} \subset E$. If $\lim_{n\to\infty} f(x_n) = \sup_x f(x)$, then $\limsup_{n\to\infty} g(x_n) \leq 0$, and if $\lim_{n\to\infty} f(x_n) = \inf_x f(x)$, then $\liminf_{n\to\infty} g(x_n) \geq 0$.*

APPENDIX B

Variational constants, rate of growth and spectral theory for the semigroup of positive linear operators

Let (E, r) be a complete, separable metric space, $\mathcal{M}_f(E)$ be the space of finite (positive) Borel measures on E, $\mathcal{P}(E) \subset \mathcal{M}_f(E)$, the space of probability measures on E, $K(E) \subset C_b(E)$, the collection of nonnegative, bounded, continuous functions, $K_0(E) \subset K(E)$, the collection of strictly positive, bounded, continuous functions, and $K_1(E) \subset K_0(E)$, the collection of bounded continuous functions satisfying $\inf_{x\in E} f(x) > 0$. For a general operator $A \subset M(E) \times M(E)$, we denote $\mathcal{D}(A) = \{f \in M(E) : \exists g \in M(E) \ni (f,g) \in A\}$, $\mathcal{D}^+(A) = \mathcal{D}(A) \cap K_0(E)$, and $\mathcal{D}^{++}(A) = \mathcal{D}(A) \cap K_1(E)$. We are interested in finding explicit conditions under which inequalities such as (11.16) and (11.56) hold. These and the related inequalities of Chapter 11 are implied by the following conditions.

(1) Let $P(y, dz)$ be a transition function for a discrete time Markov chain Y on E, and define $Pg(y) = \int_E g(z)P(y, dz)$ for $g \in C_b(E)$. Let ψ be upper semicontinuous on E and bounded above, and for $c \in R$, assume with $\{y : \psi(y) \geq c\}$ compact.

For $V \in C_b(E)$,

$$\inf_{g\in K_1(E)} \sup_{y\in E}(V(y) + \log(\frac{Pg(y)}{g(y)})) \leq \sup_{g\in K_1(E)} \inf_{y\in E}(V(y) + \log(\frac{Pg(y)}{g(y)})). \tag{B.1}$$

or

$$\begin{aligned}\inf_{0<\kappa\leq 1} \inf_{g\in K_1(E)} \sup_{y\in E}(V(y) + (1-\kappa)\log\frac{Pg(y)}{g(y)} + \kappa\psi(y)) \\ \leq \sup_{k>0} \sup_{g\in K_1(E)} \inf_{y\in E}(V(y) + (1+\kappa)\log\frac{Pg(y)}{g(y)} - \kappa\psi(y)).\end{aligned} \tag{B.2}$$

(2) Let B be the infinitesimal generator of a continuous time Markov process Y in the sense of (11.31), and let ψ be as above. For $V \in C_b(E)$,

$$\inf_{g\in\mathcal{D}^{++}(B)} \sup_{y\in E}(V(y) + \frac{Bg(y)}{g(y)}) \leq \sup_{g\in\mathcal{D}^{++}(B)} \inf_{y\in E}(V(y) + \frac{Bg(y)}{g(y)}) \tag{B.3}$$

or

$$\begin{aligned}\inf_{0<\kappa\leq 1} \inf_{g\in\mathcal{D}^{++}(B)} \sup_{y\in E}(V(y) + (1-\kappa)\frac{Bg(y)}{g(y)} + \kappa\psi(y)) \\ \leq \sup_{\kappa>0} \sup_{g\in\mathcal{D}^{++}(B)} \inf_{y\in E}(V(y) + (1+\kappa)\frac{Bg(y)}{g(y)} - \kappa\psi(y))\end{aligned} \tag{B.4}$$

There are a number of results in the literature giving these inequalities. A variant of Condition F in Freidlin-Wentzell [**52**] (Conditions B.7 for continuous time and B.15 for discrete time) implies (B.2) and (B.4) (Theorems B.12 and B.20). This condition can be verified using a condition of Deuschel and Stroock [**30**] on the transition probability of the process Y (Conditions B.8, B.16 and Lemmas B.9 and B.17).

B.1. Relationship to the spectral theory of positive operators

Note that (B.1) holds if there exists $g \in K_1(E)$, such that $V(y) + \log \frac{Pg(y)}{g(y)}$ is a constant. Exponentiating, suppose that there exist $\lambda_V \in R$ and $g_0 \in K_1(E)$ such that

$$e^V P g_0 = e^{\lambda_V} g_0. \tag{B.5}$$

Then

$$\inf_{g \in K_0(E)} \sup_y (V(y) + \log(\frac{Pg(y)}{g(y)})) \leq \lambda_V \leq \sup_{g \in K_0(E)} \inf_y (V(y) + \log(\frac{Pg(y)}{g(y)})).$$

Of course, (B.5) says that e^{λ_V} is a positive eigenvalue for the operator $e^V P$ with a positive eigenfunction g_0. In particular, if E is finite and P is the transition operator for an irreducible Markov chain, then the existence of g_0 and e^{λ_V} follows by the Perron-Frobenius theorem.

Similarly, in the continuous time case (B.3), the analogous condition is that there exist $\lambda_V \in R$ and $g_0 \in \mathcal{D}^{++}(B)$ such that

$$(B + V) g_0 = \lambda_V g_0. \tag{B.6}$$

Then

$$\inf_{g \in \mathcal{D}^{++}(B)} \sup_{y \in E} (V(y) + \frac{Bg(y)}{g(y)}) \leq \lambda_V \leq \sup_{g \in \mathcal{D}^{++}(B)} \inf_{y \in E} (V(y) + \frac{Bg(y)}{g(y)})).$$

Again, if E is finite, the Perron-Frobenius theorem gives the desired result.

Extensions of Perron-Frobenius theory for eigenvalue problems of the forms (B.5) and (B.6) exist in the literature. Such extensions are usually proved under some type of irreducibility and compactness hypotheses on P or the resolvent of B.

B.1.1. Continuous time case.

Throughout this section, we assume that Y is an E-valued, cadlag Markov process with semigroup $S(t)f(y) = E[f(Y(t))|Y(0) = y]$ satisfying $S(t) : C_b(E) \mapsto C_b(E)$ and weak infinitesimal generator B. Denote $P(t, x, dy) = P(Y(t) \in dy | Y(0) = x)$ and for $V \in C_b(E)$ define

$$T(t)f(y) = E[f(Y(t)) e^{\int_0^t V(Y(s)) ds} | Y(0) = y], \quad f \in C_b(E).$$

The weak infinitesimal generator for $\{T(t)\}$ is $Af = Bf + Vf$, $f \in \mathcal{D}(B)$.

CONDITION B.1.

(1) *There exists a measurable function $\rho(t, x, y) : [0, \infty) \times E \times E \mapsto [0, \infty)$ such that for each compact $K \subset E$,*

$$\lim_{\epsilon \to 0+} \sup_{x \in K} \sup_{r(x,y) \leq \epsilon} \rho(t, x, y) = 0,$$

and

$$\sup_{A \in \mathcal{B}(E)} |P(t, x, A) - P(t, y, A)| \leq \rho(t, x, y),$$

for each $t > 0$ and $x, y \in E$.

(2) *P is irreducible in the sense that for each $y \in E$ and open $A \subset E$, there exists $t_0 > 0$ such that $P(Y(t_0) \in A | Y(0) = y) > 0$.*

LEMMA B.2. *For each $t > 0$, there exists $\omega_t(x, y)$ such that*

$$|T(t)f(x) - T(t)f(y)| \leq \|f\|\omega_t(x, y)$$

and

$$\lim_{\epsilon \to 0} \sup_{x \in K} \sup_{r(x,y) \leq \epsilon} \omega_t(x, y) = 0.$$

PROOF. Let E_x denote the expectation under the assumption that $Y(0) = x$, and assume that $\underline{v} \leq V \leq \overline{v}$. Then for $0 < \delta \leq t$ and $f \geq 0$,

$$e^{\underline{v}\delta} E_x[T(t-\delta)f(Y(\delta)] \leq T(t)f(x) = E_x[e^{\int_0^t V(Y(s))ds} f(Y(t))] \leq e^{\overline{v}\delta} E_x[T(t-\delta)f(Y(\delta)]$$

and hence

$$\begin{aligned} & T(t)f(x) - T(t)f(y) \\ & \quad \leq (e^{\overline{v}\delta} - e^{\underline{v}\delta})e^{\overline{v}(t-\delta)}\|f\| + e^{\underline{v}\delta}(E_x[T(t-\delta)f(Y(\delta)] - E_y[T(t-\delta)f(Y(\delta)]) \\ & \quad \leq (e^{\overline{v}\delta} - e^{\underline{v}\delta})e^{\overline{v}(t-\delta)}\|f\| + e^{\underline{v}\delta}e^{\overline{v}(t-\delta)}\|f\|\rho(\delta, x, y). \end{aligned}$$

It follows that we can take

$$\omega_t(x, y) = \inf_{0 \leq \delta \leq t} [(e^{\overline{v}\delta} - e^{\underline{v}\delta})e^{\overline{v}(t-\delta)} + e^{\underline{v}\delta}e^{\overline{v}(t-\delta)}\rho(\delta, x, y)].$$

□

THEOREM B.3. *Suppose that Condition B.1 is satisfied, that E is compact, and that $V \in C(E)$. Let*

$$\lambda_V = \lim_{t \to \infty} \frac{1}{t} \sup_{y \in E} \log T(t)1(y).$$

Then there exists $g_0 \in \mathcal{D}^{++}(B)$ such that (B.6) holds. Furthermore, the pair (λ_V, g_0) satisfying (B.6) is unique up to multiplication of g_0 by a positive constant.

PROOF. First assume

$$\sup_y V(y) = \overline{v} = -\beta_1 < 0. \tag{B.7}$$

Then

$$Sg \equiv -\int_0^\infty T(t)g dt \tag{B.8}$$

is a well-defined operator on $C(E)$. Since

$$\begin{aligned} |Sg(y)| &= \left| \int_0^\infty E[e^{\int_0^t V(Y(s))ds} g(Y(t)) | Y(0) = y] dt \right| \\ &\leq \|g\| \int_0^\infty E[e^{\int_0^t V(Y(s))ds} | Y(0) = y] dt \\ &\leq \|g\|\beta_1^{-1}, \end{aligned}$$

$\|S\| \leq \beta_1^{-1}$. For each $g \in \mathcal{D}(B)$,

$$g = T(t)g - \int_0^t T(s)(B + V)g ds$$

and $\lim_{t\to\infty} T(t)g = 0$. It follows that

$$S(B+V)g = g, \quad g \in \mathcal{D}(B),$$

and therefore, $(B+V)^{-1} = S$.

We can assume that ρ in Condition B.1.1 is bounded by 1 (otherwise, replace ρ by $\rho \wedge 1$), and hence $\omega_t(x,y) \leq e^{\overline{v}t} = e^{-\beta_1 t}$. Then

$$|Sf(x) - Sf(y)| \leq \|f\| \int_0^\infty \omega_t(x,y)dt,$$

and it follows that $S = (B+V)^{-1}$ is compact.

By Condition B.1.2 and the representation

$$-(B+V)^{-1}g(y) = \int_0^\infty E[e^{\int_0^t V(Y(s))ds} g(Y(t))|Y(0) = y]dt,$$

$-(B+V)^{-1}$ is a strictly positive operator in the sense that $g \in C(E)$, $g \geq 0$ and g not identically zero implies $-(B+V)^{-1}g(y) > 0$ for all $y \in E$.

Setting $\beta_0 = -\inf_y V(y)$, the spectral radius satisfies

$$\begin{aligned} r((B+V)^{-1}) &\equiv \lim_{n\to\infty} \|(B+V)^{-n}\|^{1/n} \\ &\geq \lim_{n\to\infty} (\sup_y |(B+V)^{-n}1(y)|)^{1/n} \\ &\geq (\frac{1}{\beta_0^n})^{1/n} = \frac{1}{\beta_0}, \end{aligned}$$

and by Corollary 1 of Karlin [**64**], $r((B+V)^{-1})$ is an eigenvalue for $-(B+V)^{-1}$ with nonnegative eigenfunction g_0. The strict positivity of $-(B+V)^{-1}$ implies that the eigenfunction is strictly positive, and (B.6) follows with $\lambda_V = -1/r((B+V)^{-1})$.

In general, without assuming $\sup_y V(y) < 0$, let $c > \|V\|$. Then $-(B+V-c)^{-1}$ has a unique (up to constant multiples), strictly positive eigenfunction g_0, and setting $\lambda_{V-c} = -1/r((B+V-c)^{-1})$, we have

$$(B+V-c)g_0 = \lambda_{V-c}g_0,$$

so with $\lambda_V = \lambda_{V-c} + c$,

$$(B+V)g_0 = \lambda_V g_0.$$

By (B.6),

$$T(t)g_0 = e^{-t\lambda_V} g_0.$$

Since there exist constants $0 < \inf_y g_0(y) \leq \sup_y g_0(y) < \infty$,

$$(\inf_y g_0(y))T(t)1 \leq T(t)g_0 \leq (\sup_y g_0(y))T(t)1,$$

and

$$\lambda_V = \lim_{t\to\infty} \sup_{y\in E} \frac{1}{t} \log T(t)1(y).$$

□

B.1.2. Discrete time case. Let $\{Y(0), Y(1), \ldots\}$ be an E-valued, discrete time, Feller, Markov chain with transition probability $P(x, dy)$.

CONDITION B.4.

(1) *There exists a function* $\rho : E \times E \mapsto [0, \infty)$ *such that for each compact* $K \subset E$,
$$\lim_{\epsilon \to 0} \sup_{x \in K} \sup_{r(x,y) \le \epsilon} \rho(x, y) = 0$$
and
$$\sup_{A \in \mathcal{B}(E)} |P(x, A) - P(y, A)| \le \rho(x, y).$$

(2) P *is irreducible in the sense that for each* $y \in E$ *and open* $A \subset E$, *there exists* $n_0 > 0$ *such that* $P(Y(n_0) \in A | Y(0) = y) > 0$.

Let $Tf(y) = e^{V(y)} Pf(y)$.

THEOREM B.5. *Suppose that Condition B.4 is satisfied, that* E *is compact, and that* $V \in C(E)$. *Let*

$$\lambda_V = \lim_{n \to \infty} \frac{1}{n} \log \sup_{y \in E} T^n 1(y). \tag{B.9}$$

Then there exists $g_0 \in K_1(E)$ *such that (B.5) holds, and* (λ_V, g_0) *is unique up to multiplication of* g_0 *by a positive constant.*

PROOF. By uniform continuity of e^V on E and Condition B.4, T is a compact operator on $C(E)$. Let $\rho > 0$ satisfy $\rho \|T\| < 1$, and define $S_\rho = \sum_{k=0}^{\infty} \rho^k T^k = (I - \rho T)^{-1}$. Then S_ρ is a strictly positive, compact operator. As in the proof of Theorem B.3, S_ρ has a positive eigenvalue λ_ρ and corresponding strictly positive eigenfunction g_0. Consequently,
$$(I - \rho T)^{-1} g_0 = \lambda_\rho g_0$$
and
$$T g_0 = \frac{\lambda_\rho - 1}{\rho \lambda_\rho} g_0 \equiv e^{\lambda_V} g_0.$$
As before $0 < C_1 \le g_0 \le C_2$, and
$$C_1 \|T^n\| \le T^n g_0 = e^{n \lambda_V} g_0 \le C_2 \|T^n\|.$$
Therefore
$$\lambda_V = \lim_{n \to \infty} \frac{1}{n} \log \|T^n\|.$$
□

B.2. Relationship to some variational constants

As we have already noted, the validity of (B.4) and (B.2) is closely related to the eigenvalues of the operators involved and to variational representations of these eigenvalues. In this section, we develop this relationship further. A key tool is Sion's theorem.

THEOREM B.6. *Let* M *and* N *be convex subsets of possibly different topological vector spaces. Suppose that* $F : M \times N \mapsto R$ *satisfies the following conditions:*

a) *For each* $c \in R$ *and* $\nu \in N$, $\{\mu \in M : f(\mu, \nu) \ge c\}$ *is closed and convex.*
b) *For each* $c \in R$ *and* $\mu \in M$, $\{\nu \in N : f(\mu, \nu) \le c\}$ *is closed and convex.*

If either N or M is compact, then

$$\inf_{\nu\in N}\sup_{\mu\in M} f(\mu,\nu) = \sup_{\mu\in M}\inf_{\nu\in N} f(\mu,\nu).$$

PROOF. This result is Corollary 3.3 of Sion [**111**]. □

B.2.1. Continuous time case. Assume that $S(t) : C_b(E) \mapsto C_b(E)$ and $V \in C_b(E)$. Define

$$T(t)f(y) = E[f(Y(t))e^{\int_0^t V(Y(s))ds}|Y(0) = y]. \tag{B.10}$$

Then $\{T(t)\}$ is a semigroup with weak generator $A = B + V$, where B is the generator in the sense of (11.31) and V denotes multiplication by V. Note that $(1, V) \in A$. Positivity implies $\|T(t)\| = \|T(t)1\| = \sup_y T(t)1(y)$ and

$$\|T(t)\| = \sup_y E[e^{\int_0^t V(Y(s))ds}|Y(0) = y]. \tag{B.11}$$

For $\nu \in \mathcal{P}(E)$ and g satisfying $\int_E g^+ d\nu < \infty$ or $\int_E g^- d\nu < \infty$ (so $\int_E g d\nu$ is unambiguously defined, although possible $\pm\infty$), define

$$\langle g, \nu\rangle = \int_E g d\nu,$$

and let ψ be as in (B.4). We will develop results based upon the following hypotheses.

CONDITION B.7. *There exists a constant c_V such that*

$$\limsup_{t\to\infty} \frac{1}{t}\log E_\nu[e^{\int_0^t V(Y(s))ds}] = c_V \tag{B.12}$$

for each $\nu \in \mathcal{P}(E)$ satisfying $\langle\psi, \nu\rangle > -\infty$.

Condition B.7 is a variant of Condition F of Freidlin and Wentzell (Chapter 7, (4.1), page 234 of [**52**]). It can be verified (Lemma B.9) under the following condition on the transition probability of Y. This criteria is an adaptation of Condition $\widetilde{U}$ in Deuschel and Stroock [**30**].

CONDITION B.8. *For $\nu_1, \nu_2 \in \mathcal{P}(E)$ satisfying $\langle\nu_i, \psi\rangle > -\infty$, there exists $t_0 > 0$, $\rho_1, \rho_2 \in \mathcal{P}((0, t_0])$ and $1 \le M < \infty$ such that*

$$\int_{(0,t_0]}\int_E P(t,x,A)\nu_1(dx)\rho_1(dt) \le M\int_{(0,t_0]}\int_E P(t,y,A)\nu_2(dy)\rho_2(dt), \tag{B.13}$$

for each $A \in \mathcal{B}(E)$.

LEMMA B.9. *Condition B.8 implies Condition B.7.*

PROOF. Let $\nu_1, \nu_2 \in \mathcal{P}(E)$. Then

$$\begin{aligned}
&E_{\nu_1}[e^{\int_0^t V(Y(r))dr}] \\
&\leq e^{2t_0\|V\|} \int_{(0,t_0]} E_{\nu_1}[e^{\int_s^{s+t} V(Y(r))dr}]\rho_1(ds) \\
&= e^{2t_0\|V\|} \int_{(0,t_0]} \int_{x\in E} \int_{z\in E} E[e^{\int_0^t V(Y(u))du}|Y(0)=z]P(s,x,dz)\nu_1(dx)\rho_1(ds) \\
&\leq e^{2t_0\|V\|} M \int_{(0,t_0]} \int_{y\in E} \int_{z\in E} E[e^{\int_0^t V(Y(u))du}|Y(0)=z]P(s,y,dz)\nu_2(dy)\rho_2(ds) \\
&= e^{2t_0\|V\|} M \int_{(0,t_0]} \int_{y\in E} E[e^{\int_s^{t+s} V(Y(r))dr}|Y(0)=y]\nu_2(dy)\rho_2(ds) \\
&\leq e^{2t_0\|V\|} M \int_{(0,t_0]} E_{\nu_2}[e^{\int_0^t V(Y(r))dr}]e^{2t_0\|V\|}\rho_2(ds) \\
&= Me^{4t_0\|V\|}E_{\nu_2}[e^{\int_0^t V(Y(r))dr}].
\end{aligned}$$

Therefore

$$\frac{1}{t}\log E_{\nu_1}[e^{\int_0^t V(Y(r))dr}] \leq \frac{1}{t}\log E_{\nu_2}[e^{\int_0^t V(Y(r))dr}] + \frac{\log M + 4t_0\|V\|}{t}, \tag{B.14}$$

implying that $\limsup_{t\to\infty}\frac{1}{t}\log E_\nu[e^{\int_0^t V(Y(r))dr}]$ is independent of ν. $\square$

We now identify the relation between c_V and (B.4). Define

$$\begin{aligned}
c_V^* &= \inf_{\nu\in\mathcal{P}(E),\langle\psi,\nu\rangle>-\infty} \limsup_{t\to\infty}\frac{1}{t}\log E_\nu[e^{\int_0^t V(Y(s))ds}] \\
c_V^{**} &= \sup_{\nu\in\mathcal{P}(E),\langle\psi,\nu\rangle>-\infty} \limsup_{t\to\infty}\frac{1}{t}\log E_\nu[e^{\int_0^t V(Y(s))ds}]
\end{aligned}$$

Condition B.7 is just the statement that $c_V^* = c_V^{**}$.

LEMMA B.10. *Assume that $S(t): C_b(E)\mapsto C_b(E)$, $V\in C_b(E)$, and $K_c = \{y : \psi(y)\geq -c\}$ is compact for each $c\in R$. Then*

$$\sup_{\kappa>0}\sup_{g\in\mathcal{D}^{++}(B)}\inf_{y\in E}(V(y) + (1+\kappa)\frac{Bg(y)}{g(y)} - \kappa\psi(y)) \geq c_V^*. \tag{B.15}$$

PROOF. Denote the left side of the inequality by H_2^V. For $\lambda < c_V^*$, let

$$R_\lambda^t g = \int_0^t e^{-\lambda s}T(s)gds,$$

and let Γ be the collection of functions of the form

$$g_\gamma = \int_0^\infty R_\lambda^t 1\gamma(dt), \quad \gamma\in\mathcal{P}([0,\infty)),$$

such that $g_\gamma\in C_b(E)$. The dominated convergence theorem implies that if g_γ is continuous, then so is $\int_0^\infty R_\lambda^t V\gamma(dt)$, and it follows that

$$(V+B)g_\gamma = \int_0^\infty R_\lambda^t V\gamma(dt).$$

Note that $\|(V+B)g_\gamma\| \le \|V\|\|g_\gamma\|$ and

$$\left|\frac{(V+B)g_\gamma}{g_\gamma}\right| = \left|\frac{\int_0^\infty R_\lambda^t V\gamma(dt)}{\int_0^\infty R_\lambda^t 1\gamma(dt)}\right| \le \|V\|.$$

We topologize Γ so that $g_{\gamma_n} \mapsto g_\gamma$ if and only if $\|g_{\gamma_n} - g_\gamma\| + \|Bg_{\gamma_n} - Bg_\gamma\| \to 0$ and $\gamma_n \Rightarrow \gamma$. Then setting $\mathcal{K}_c = \{\nu \in \mathcal{P}(E) : \nu(K_c) = 1\}$,

$$\begin{aligned}
H_2^V &\ge \lim_{\kappa\to 0+}\lim_{c\to\infty}\sup_{g_\gamma\in\Gamma}\inf_{,y\in K_c}(V(y) + (1+\kappa)\frac{Bg_\gamma(y)}{g_\gamma(y)} - \kappa\psi(y)) \\
&= \lim_{\kappa\to 0+}\lim_{c\to\infty}\sup_{g_\gamma\in\Gamma}\inf_{\nu\in\mathcal{K}_c}\frac{\langle -\kappa(V+\psi)g_\gamma + (1+\kappa)(V+B)g_\gamma,\nu\rangle}{\langle g_\gamma,\nu\rangle} \\
&= \lim_{\kappa\to 0+}\inf_{\nu\in\cup_c\mathcal{K}_c}\sup_{g_\gamma\in\Gamma}\frac{\langle -\kappa(V+\psi)g_\gamma + (1+\kappa)(V+B)g_\gamma,\nu\rangle}{\langle g_\gamma,\nu\rangle} \\
&\ge \inf_{\nu\in\mathcal{P}(E),\langle\psi,\nu\rangle>-\infty}\sup_{g_\gamma\in\Gamma}\frac{\langle (V+B)g_\gamma,\nu\rangle}{\langle g_\gamma,\nu\rangle}.
\end{aligned}$$

The first inequality follows from the fact that $\Gamma \subset \mathcal{D}^{++}(B)$ and the inequality

$$V(y) + (1+\kappa)\frac{Bg_\gamma(y)}{g_\gamma(y)} - \kappa\psi(y) \ge -(1+2\kappa)\|V\| - \kappa\psi(y).$$

Since $\mathcal{K}_c$ is compact and $\mathcal{K}_c$ and Γ are convex, the interchange of the inf and sup is justified by Sion's theorem, Theorem B.6. Take $N = \mathcal{K}_c$ and $M = \Gamma$.

Note that

$$(\lambda - (B+V))R_l^t 1 = 1 - e^{-\lambda t}T(t)1 \tag{B.16}$$

so that

$$\frac{\langle (V+B)R_\lambda^t 1,\nu\rangle}{\langle R_\lambda^t 1,\nu\rangle} = \lambda - \frac{1 - \langle e^{-\lambda t}T(t)1,\nu\rangle}{\langle R_\lambda^t 1,\nu\rangle}, \tag{B.17}$$

and that for ν satisfying $\langle\psi,\nu\rangle > -\infty$, $\limsup_{t\to\infty} e^{-\lambda t}\langle T(t)1,\nu\rangle = \infty$. For $\epsilon > 0$, there exists ν_ϵ with $\langle\psi,\nu_\epsilon\rangle > -\infty$ such that

$$\begin{aligned}
H_2^V &\ge \sup_{g_\gamma\in\Gamma}\frac{\langle (V+B)g_\gamma,\nu_\epsilon\rangle}{\langle g_\gamma,\nu_\epsilon\rangle} - \epsilon \\
&\ge \lambda + \limsup_{t\to\infty}\frac{\langle e^{-\lambda t}T(t)1,\nu_\epsilon\rangle - 1}{\langle R_\lambda^t 1,\nu_\epsilon\rangle} - \epsilon \\
&\ge \lambda - \epsilon.
\end{aligned}$$

Since $\epsilon > 0$ and $\lambda < c_V^*$ are arbitrary, (B.15) follows. □

LEMMA B.11. *For $V \in C_b(E)$,*

$$\sup_{\nu\in\mathcal{P}(E),\langle\psi,\nu\rangle>-\infty}\ \inf_{g\in\mathcal{D}^{++}(B)}\int (V + \frac{Bg}{g})d\nu \le c_V^{**} \tag{B.18}$$

PROOF. By (B.16),

$$\inf_{g\in\mathcal{D}^{++}(B)}\int_E \frac{(B+V)g}{g}d\nu \le \lambda - \int_E \frac{1 - e^{-\lambda t}T(t)1}{R_\lambda^t 1}d\nu.$$

Let $\lambda > c_V^{**}$ and $t > 1$. Then $R_\lambda^t 1 \geq \int_0^1 e^{-(\lambda - \inf_y V(y))s} ds \equiv \delta_0 > 0$, and

$$\inf_{g \in \mathcal{D}^{++}(B)} \int_E \frac{(B+V)g}{g} d\nu \leq \lambda + \delta_0^{-1} e^{-\lambda t} E_\nu[e^{\int_0^t V(Y(r))dr}].$$

Assuming $\langle \psi, \nu \rangle > -\infty$ and letting $t \to \infty$, the right side converges to λ giving (B.18). □

Putting these lemmas together, we have the following.

THEOREM B.12. *Assume that $S(t) : C_b(E) \to C_b(E)$, $V \in C_b(E)$, and $K_c = \{y : \psi(y) \geq -c\}$ is compact for each $c \in R$. If B.7 holds, then (B.4) holds.*

PROOF. As in Lemma 11.35, the left side of (B.4) equals the left side of (B.18). Since Condition B.7 is just the assertion that $c_V^* = c_V^{**}$, (B.4) follows by B.15 and B.18. □

Lemma B.10 also gives the following.

THEOREM B.13. *Assume that $S(t) : C_b(E) \mapsto C_b(E)$, $V \in C_b(E)$, and $K_c = \{y : \psi(y) \geq -c\}$ is compact for each $c \in R$. Define I_B as in Lemma 11.35, that is,*

$$I_B(\nu) = - \inf_{g \in \mathcal{D}^{++}(B)} \int_E \frac{Bg}{g} d\nu \wedge \int_E \psi d\nu.$$

If for each $y \in E$ such that $\psi(y) > -\infty$,

$$\liminf_{t\to\infty} \frac{1}{t} \log E_y[e^{\int_0^t V(Y(s))ds}] \geq \sup_{\mu \in \mathcal{P}(E)} (\langle V, \mu \rangle - I_B(\mu)), \tag{B.19}$$

then (B.4) holds.

PROOF. As in Lemma 11.35, the right side of (B.19) equals the left side of (B.4). Noting that

$$\frac{1}{t} \log E_\nu[e^{\int_0^t V(Y(s))ds}] \geq \int_E \frac{1}{t} \log E_y[e^{\int_0^t V(Y(s))ds}] \nu(dy),$$

Fatou's lemma implies

$$c_V^* \geq \liminf_{t\to\infty} \frac{1}{t} \log E_\nu[e^{\int_0^t V(Y(s))ds}] \geq \sup_{\mu \in \mathcal{P}(E)} (\langle V, \mu \rangle - I_B(\mu)),$$

Consequently, (B.4) follows by (B.15). □

EXAMPLE B.14. Consider the Ornstein-Uhlenbeck process in R^d satisfying

$$Y(t) = Y(0) + W(t) - \alpha \int_0^t Y(s) ds, \quad \alpha > 0. \tag{B.20}$$

Then

$$Bf(y) = \frac{1}{2} \Delta f(y) - \alpha y \nabla f(y), \quad f \in D_d.$$

Define

$$\mu_\infty(dy) = \left(\frac{\alpha}{\pi}\right)^{d/2} e^{-\alpha |y|^2} dy.$$

Then μ_∞ is the stationary distribution for Y and Y is reversible. We can therefore define a symmetric Dirichlet form

$$\mathcal{E}(f,g) = -\int f B g d\mu_\infty = \int \nabla f \cdot \nabla g d\mu_\infty, \quad \nabla f, \nabla g \in L^2(\mu_\infty), f, g \in L^2(\mu_\infty),$$

where $\nabla f, \nabla g$ are given in the Schwartz distribution sense. By Theorem 7.44 on page 152 of Stroock [**115**],

$$I_B(\mu) = \mathcal{E}(\sqrt{\frac{d\mu}{d\mu_\infty}}, \sqrt{\frac{d\mu}{d\mu_\infty}}).$$

Indeed, this representation can also be directly verified by a computation similar to that of Lemma D.44. Let $\Gamma \subset \mathcal{P}(R^d)$ be the collection of measures satisfying

$$\frac{d\mu}{d\mu_\infty} = \frac{e^{2g}}{\int e^{2g} d\mu_\infty}, \tag{B.21}$$

for some $g \in C_c^2(R^d)$. As in Lemma D.44, by the lower semicontinuity of I_B and by convexity, each $\mu \in \mathcal{P}(R^d)$ satisfying $I_B(\mu) < \infty$ can be approximated by a sequence of $\{\mu_n\} \subset \Gamma$ in the sense that $\lim_{n\to\infty} I(\mu_n) = I(\mu)$ and $\lim_{n\to\infty} \mu_n = \mu$ in the weak topology. Therefore, (B.19) is implied by

$$\liminf_{t\to\infty} \frac{1}{t} \log E_y[e^{\int_0^t V(Y(s))ds}] \geq \int V d\mu - \mathcal{E}(\sqrt{\frac{d\mu}{d\mu_\infty}}, \sqrt{\frac{d\mu}{d\mu_\infty}}) \tag{B.22}$$

for all μ satisfying (B.21).

We now introduce a class of perturbations of B given by

$$B^g = \frac{1}{2}\Delta - \alpha y \cdot \nabla + \nabla g \cdot \nabla, \qquad g \in C_c^2(R^d).$$

Using stochastic calculus, we know that

$$dY(t) = (\nabla g(Y(t)) - \alpha Y(t))dt + dW(t)$$

defines a process Y which solves the martingale problem of B^g. Let $P^g, P \in \mathcal{P}(C_{R^d}[0,\infty))$ denote the probability measures solving the martingale problems for B^g and B, and let $\{\mathcal{F}_t\}$ be the natural filtration on $C_{R^d}[0,\infty)$. Then, through the Girsanov transformation,

$$\frac{dP^g}{dP}|_{\mathcal{F}_t} = e^{g(Y(t)) - g(Y(0)) - \int_0^t e^{-g} B e^g (Y(s)) ds}.$$

Finally, Y has a unique stationary distribution under P^g as well, which is given by

$$d\mu^g = \frac{e^{2g}}{\int e^{2g} d\mu_\infty} d\mu_\infty.$$

For every $y \in E$ and $g \in D$,

$$\begin{aligned}
&\liminf_{t\to\infty} \frac{1}{t} \log E_y^P[e^{\int_0^t V(Y(s))ds}] \\
&\geq \lim_{t\to\infty} \frac{1}{t} E_y^{P^g}[\int_0^t V(Y(s))ds - (g(Y(t)) - g(Y(0)) - \int_0^t e^{-g} B e^g (Y(s)) ds)] \\
&= \int V d\mu^g + \int e^{-g} B e^g d\mu^g \\
&= \int V d\mu^g - \mathcal{E}(\sqrt{\frac{d\mu^g}{d\mu_\infty}}, \sqrt{\frac{d\mu^g}{d\mu_\infty}}) \\
&= \int V d\mu^g - I(\mu^g),
\end{aligned}$$

where the first inequality follows by Jensen's inequality and the concavity of the log and the first equality follows from the ergodicity of the process Y under P^g. By the arbitrariness of g, (B.22) follows. Consequently by Theorem B.13, (B.4) holds for the Ornstein-Uhlenbeck process. (Of course, (B.4) also follows by Lemma 11.32 and the calculations in Example 11.33.

B.2.2. Discrete time case. Let $\{Y(0), Y(1), \ldots\}$ be an E-valued Markov chain with transition probability $P(x, dy)$. Define $Pf(x) = \int f(y)P(x, dy)$, and assume

$$P : C_b(E) \mapsto C_b(E). \tag{B.23}$$

Let $P^{(l)}(x, A)$ denote the l step transition probability of Y, and let $V \in C_b(E)$. Define

$$Tf(y) = e^{V(y)} \int f(z)P(y, dz).$$

Then

$$T^n f(y) = E[f(Y(n))e^{\sum_{k=0}^n V(Y(k))}|Y(0) = y].$$

We verify (B.2) under discrete time versions of Conditions B.7 and B.8.

CONDITION B.15. *There exists a constant c_V such that*

$$\limsup_{n\to\infty} \frac{1}{n} \log E_\nu[e^{\sum_{k=0}^n V(Y(k))}] = c_V, \tag{B.24}$$

for each $\nu \in \mathcal{P}(E)$ satisfying $\langle \psi, \nu \rangle > -\infty$.

The following is a variant of Condition (U) on page 100 in Deuschel and Stroock [**30**]

CONDITION B.16. *For $\nu_1, \nu_2 \in \mathcal{P}(E)$ satisfying $\langle \nu_i, \psi \rangle > -\infty$, there exist positive integers l, N, M with $1 \le l \le N$ and $M \ge 1$ such that*

$$\int P^{(l)}(x, A)\nu_1(dx) \le \frac{M}{N} \sum_{m=1}^{N} \int P^{(m)}(y, A)\nu_2(dy), \quad A \in \mathcal{B}(E). \tag{B.25}$$

As in Lemma B.9, we have

LEMMA B.17. *Condition B.16 implies Condition B.15.*

We now identify the relation between c_V and (B.2). Define

$$\begin{aligned} c_V^* &= \inf_{\nu\in\mathcal{P}(E),\langle\psi,\nu\rangle>-\infty} \limsup_{n\to\infty} \frac{1}{n} \log E_\nu[e^{\sum_{k=0}^n V(Y(k))}] \\ c_V^{**} &= \sup_{\nu\in\mathcal{P}(E),\langle\psi,\nu\rangle>-\infty} \limsup_{t\to\infty} \frac{1}{n} \log E_\nu[e^{\sum_{k=1}^n V(Y(k))}] \end{aligned}$$

Condition B.15 is just the statement that $c_V^* = c_V^{**}$.

LEMMA B.18. *Assume that $P : C_b(E) \mapsto C_b(E)$, $V \in C_b(E)$, and $K_c = \{y : \psi(y) \ge -c\}$ is compact for each $c \in R$. Then*

$$\sup_{k>0} \sup_{g\in K_1(E)} \inf_{y\in E} (V(y) + (1+\kappa)\log\frac{Pg(y)}{g(y)} - \kappa\psi(y)) \ge c_V^* \tag{B.26}$$

PROOF. Denote the left side of the inequality by H_2^V. For $\lambda < c_V^*$, let

$$R_\lambda^m g = \sum_{k=0}^{m} e^{-\lambda k} T^k g,$$

and let Γ be the collection of functions of the form

$$g_\gamma = \sum_{m=0}^{\infty} R_\lambda^m 1 \gamma_m, \quad \gamma_m \geq 0, \sum_{m=0}^{\infty} \gamma_m = 1,$$

such that $g_\gamma \in C_b(E)$. Note that

$$TR_\lambda^m g = R_\lambda^m Tg = e^\lambda (R_\lambda^m g + e^{-\lambda(m+1)} T^{m+1} g - g) \tag{B.27}$$

and

$$\left| \log \frac{Tg_\gamma}{g_\gamma} \right| = \left| \log \frac{\sum_{m=0}^\infty R_\lambda^m e^V \gamma_m}{\sum_{m=0}^\infty R_\lambda^m 1 \gamma_m} \right| \leq \|V\|.$$

The dominated convergence theorem implies that if g_γ is continuous, then so is

$$Tg_\gamma = \sum_{m=0}^{\infty} R_\lambda^m e^V \gamma_m.$$

We topologize Γ so that $g_{\gamma^{(n)}} \to g_\gamma$ if and only if $\|g_{\gamma^{(n)}} - g_\gamma\| \to 0$ and $\gamma_m^{(n)} \Rightarrow \gamma_m$, $m = 0, 1, 2, \ldots$. Then, as before, setting $\mathcal{K}_c = \{\nu \in \mathcal{P}(E) : \nu(K_c) = 1\}$.

$$\begin{aligned}
H_2^V &\geq \lim_{\kappa \to 0+} \lim_{c \to \infty} \sup_{g_\gamma \in \Gamma} \inf_{y \in K_c} (1+\kappa) \log \frac{e^{-\kappa(1+\kappa)^{-1}(V(y)+\psi(y))} Tg_\gamma(y)}{g_\gamma(y)} \\
&= \lim_{\kappa \to 0+} \lim_{c \to \infty} \sup_{g_\gamma \in \Gamma} \inf_{\nu \in \mathcal{K}_c} (1+\kappa) \log \frac{\langle e^{-\kappa(1+\kappa)^{-\kappa}(V+\psi)} Tg_\gamma, \nu \rangle}{\langle g_\gamma, \nu \rangle} \\
&= \lim_{\kappa \to \infty} \inf_{\nu \in \cup_c \mathcal{K}_c} \sup_{g_\gamma \in \Gamma} (1+\kappa) \log \frac{\langle e^{-\kappa(1+\kappa)^{-1}(V+\psi)} Tg_\gamma, \nu \rangle}{\langle g_\gamma, \nu \rangle} \\
&\geq \inf_{\nu \in \mathcal{P}(E), \langle \psi, \nu \rangle > -\infty} \sup_{g_\gamma \in \Gamma} \log \frac{\langle Tg_\gamma, \nu \rangle}{\langle g_\gamma, \nu \rangle}.
\end{aligned}$$

The first inequality follows from the fact that $\Gamma \subset K_1(E)$ and the inequality

$$(1+\kappa) \log \frac{e^{-\kappa(1+\kappa)^{-1}(V(y)+\psi(y))} Tg_\gamma(y)}{g_\gamma(y)} \geq -(1+2\kappa)\|V\| - \kappa\psi(y).$$

Since $\mathcal{K}_c$ is compact and $\mathcal{K}_c$ and Γ are convex, the interchange of the inf and sup is justified by Sion's theorem, Theorem B.6. Take $N = \mathcal{K}_c$ and $M = \Gamma$.

Note that

$$\frac{\langle TR_\lambda^m 1, \nu \rangle}{\langle R_\lambda^m 1, \nu \rangle} = e^\lambda \left(1 + \frac{\langle e^{-\lambda(m+1)} T^{m+1} 1, \nu \rangle - 1}{\langle R_\lambda^m 1, \nu \rangle}\right), \tag{B.28}$$

and that for ν satisfying $\langle\psi,\nu\rangle > -\infty$, $\limsup_{m\to\infty} e^{-\lambda m}\langle T^m 1,\nu\rangle = \infty$. For $\epsilon > 0$, there exists ν_ϵ with $\langle\psi,\nu_\epsilon\rangle > -\infty$ such that

$$\begin{aligned} H_2^V &\geq \sup_{g_\gamma\in\Gamma} \log\frac{\langle Tg_\gamma,\nu_\epsilon\rangle}{\langle g_\gamma,\nu_\epsilon\rangle} - \epsilon \\ &\geq \lambda + \limsup_{m\to\infty}\log\Big(1+\frac{\langle e^{-\lambda(m+1)}T^{m+1}1,\nu_\epsilon\rangle - 1}{\langle R_\lambda^m 1,\nu_\epsilon\rangle}\Big) - \epsilon \\ &\geq \lambda - \epsilon. \end{aligned}$$

Since $\epsilon > 0$ and $\lambda < c_V^*$ are arbitrary, (B.26) follows. □

LEMMA B.19. *For $V \in C_b(E)$,*

$$\sup_{\nu\in\mathcal{P}(E),\langle\psi,\nu\rangle>-\infty}\ \inf_{g\in K_1(E)} \langle V + \log\frac{Pg}{g},\nu\rangle \leq c_V^{**}. \tag{B.29}$$

PROOF. By (B.27) and the concavity of log,

$$\inf_{g\in K_1(E)}\langle \log\frac{Tg}{g},\nu\rangle \leq \lambda + \log\Big(1+\langle\frac{e^{-\lambda(m+1)}T^{m+1}1-1}{R_\lambda^m 1},\nu\rangle\Big).$$

Let $\lambda > c_V^{**}$. Then $R_\lambda^m 1 \geq 1$ and $\lim_{m\to\infty}\langle e^{-\lambda m}T^m 1,\nu\rangle = 0$, so

$$\inf_{g\in K_1(E)}\langle\log\frac{Tg}{g},\nu\rangle \leq \lambda,$$

giving (B.29). □

Putting these lemmas together, we have the following.

THEOREM B.20. *Assume that $P : C_b(E) \mapsto C_b(E)$, $V \in C_b(E)$, and $K_c = \{y : \psi(y) \geq -c\}$ is compact for each $c \in R$. If Condition B.15 is satisfied, then (B.2) holds.*

PROOF. As in Lemma 11.12, the left side of (B.2) equals the left side of (B.29). Since Condition B.15 is just the assertion that $c_V^* = c_V^{**}$, (B.2) follows by (B.26) and (B.29). □

Lemma B.18 also gives the following.

THEOREM B.21. *Let*

$$I_P(\nu) = -\inf_{g\in K_1(E)}\int_E \log\frac{Pg}{g}d\nu \wedge \int_E \psi d\nu.$$

Suppose that for each $y \in E$ satisfying $\psi(y) > -\infty$,

$$\liminf_{n\to\infty}\frac{1}{n}\log E_y[e^{\sum_{k=0}^{n-1}V(Y(k))}] \geq \sup_{\nu\in\mathcal{P}(E)}(\langle V,\nu\rangle - I_P(\nu)). \tag{B.30}$$

Then (B.2) holds.

PROOF. The proof is essentially the same as the proof of Theorem B.13. □

Dupuis and Ellis [**35**] (Part (a) of Proposition 8.6.1) show that (B.30) holds under the following condition.

CONDITION B.22.

(1) *There exist positive integers* m_0, n_0 *such that for all* $x, y \in E$,

$$\sum_{i=m_0}^{\infty} \frac{1}{2^i} P^{(i)}(x, dz) << \sum_{j=n_0}^{\infty} \frac{1}{2^j} P^{(j)}(y, dz).$$

(2) $P(x, dz)$ *has an invariant measure.*

APPENDIX C

Spectral properties for discrete and continuous Laplacians

Let $\mathcal{O} = [0,1)^d$ with periodic boundary, and let $\Lambda_m = \{(i_1/m, \ldots, i_d/m) : 0 \le k_j < m, j = 1, \ldots, d\}$. We identify $L^2(\Lambda_m)$ with $R^{|\Lambda_m|}$, endowed with the norm and inner product

$$\|\rho\|^2_{L^2(\Lambda_m)} = \|\rho\|_m^2 = \sum_{x\in\Lambda_m} \rho^2(x) m^{-d}, \qquad \langle \rho, \gamma\rangle_m = \sum_{x\in\Lambda_m} \rho(x)\gamma(x) m^{-d}.$$

If ρ and γ are vector-valued with the same dimension, we will write

$$\langle \rho, \gamma\rangle_m = \sum_{x\in\Lambda_m} \rho(x)\cdot\gamma(x) m^{-d}.$$

We are interested in the way $L^2(\Lambda_m)$ approximates $L^2(\mathcal{O})$, so we define $\pi_m : L^2(\mathcal{O}) \mapsto L^2(\Lambda_m)$ and $\widehat{\eta}_m : L^2(\Lambda_m) \mapsto L^2(\mathcal{O})$ by

$$\text{(C.1)} \qquad \pi_m\rho(\frac{i_1}{m}, \ldots, \frac{i_m}{m}) \equiv m^d \int_{y_1=\frac{i_1}{m}}^{\frac{i_1+1}{m}} \cdots \int_{y_d=\frac{i_d}{m}}^{\frac{i_d+1}{m}} \rho(y_1, \ldots, y_d) dy_d \ldots dy_1,$$

and

$$\text{(C.2)} \qquad \widehat{\eta}_m\rho(x) \equiv \sum_{i_1,\ldots,i_d=0}^{m-1} \rho(\frac{i_1}{m}, \ldots, \frac{i_d}{m}) \mathbf{1}_{(\frac{i_1}{m}\le x_1 < \frac{i_1+1}{m}, \ldots, \frac{i_d}{m} \le x_d < \frac{i_d+1}{m})}.$$

Note that for $\rho \in L^2(\Lambda_m)$, $\pi_m\widehat{\eta}_m\rho = \rho$, and for $\rho \in L^2(\mathcal{O})$, $\widehat{\eta}_m\pi_m\rho$ is the orthogonal projection of ρ onto $\widehat{\eta}_m(L^2(\Lambda_m)) \subset L^2(\mathcal{O})$. In particular, for $\rho, \gamma \in L^2(\Lambda_m)$,

$$\langle \rho, \gamma\rangle_m = \langle \widehat{\eta}_m\rho, \widehat{\eta}_m\gamma\rangle.$$

Let

$$\nabla_m^i \rho(x) = \frac{m}{2}(\rho(x + m^{-1}e_i) - \rho(x - m^{-1}e_i)), \quad \nabla_m\rho(x) = (\nabla_m^1\rho(x), \ldots, \nabla_m^d\rho(x))$$

and

$$\Delta_m\rho(x) = \nabla_m\cdot\nabla_m\rho(x) = (\frac{m}{2})^2 \sum_{k=1}^d \Big(\rho(x + 2m^{-1}e_k) - 2\rho(x) + \rho(x - 2m^{-1}e_k)\Big),$$

and note that for scalar functions ρ, γ,

$$\langle \nabla_m\rho, \nabla_m\gamma\rangle_m = -\langle \Delta_m\rho, \gamma\rangle_m.$$

Let Δ be the Laplacian on $L^2(\mathcal{O})$ with periodic boundary conditions. We collect a number of useful facts about the spectral properties of Δ_m and Δ.

C.1. The case of $d = 1$

By direct verification

$$\{\alpha_0(x) = 1, \alpha_{2k-1}(x) = \sqrt{2}\cos(2\pi kx), \alpha_{2k}(x) = \sqrt{2}\sin(2\pi kx) : k = 1, 2, \cdots\}$$

form a complete orthonormal system for $L^2(\mathcal{O})$, and the α_j are eigenfunctions for $-\Delta$ with corresponding eigenvalues

$$\lambda_0 = 0, \quad \lambda_{2k-1} = \lambda_{2k} = 4\pi^2 k^2,$$

or equivalently,

$$\lambda_j = 4\pi^2[\frac{j+1}{2}]^2, \quad j = 0, 1, \ldots$$

Let $\alpha_{j,m}$ denote α_j restricted to Λ_m. Note that $\alpha_{j,m} \neq \pi_m \alpha_j$, but that

$$\lim_{m\to\infty} \|\alpha_{j,m} - \pi_m \alpha_j\|_m = \lim_{m\to\infty} \|\widehat{\eta}_m \alpha_{j,m} - \alpha_j\| = 0.$$

Using trigonometric identities, one can verify that $\{\alpha_{j,m}, 0 \leq j \leq m-1\}$ gives a complete orthonormal system of eigenfunctions for $-\Delta_m$. The corresponding eigenvalues depend on m and are given by

$$\lambda_{j,m} = m^2 \sin^2(\frac{2\pi}{m}[\frac{j+1}{2}]) = \frac{m^2}{2}\left(1 - \cos(\frac{4\pi}{m}[\frac{j+1}{2}])\right), \quad 0 \leq j \leq m-1,$$

and satisfy $\lim_{m\to\infty} \lambda_{j,m} = \lambda_j$.

Since $\lambda_{j,m} \leq m^2$,

$$\sum_{i=1}^{m-1} \lambda_{i,m} \leq m^{1+2} = m^3. \tag{C.3}$$

In addition, using the elementary inequality

$$\sin\theta \geq \frac{2}{\pi}\theta, \quad 0 \leq \theta \leq \frac{\pi}{2},$$

we have

$$\sup_m \sum_{i=1}^{m-2} (\lambda_{i,m})^{-1} \leq 4 \sup_m \sum_{i=1}^{m/4} (\lambda_{i,m})^{-1} \leq 4 \sum_{i=1}^{\infty} (16 i^2)^{-1} < \infty. \tag{C.4}$$

Note that $\lambda_{0,m} = 0$ and that, when m is even, $\lambda_{m-1,m} = 0$.

C.2. The case of $d > 1$

For $d > 1$,

$$\{e_{k_1,\cdots,k_d}(x_1, \cdots, x_d) = \alpha_{k_1}(x_1) \cdots \alpha_{k_d}(x_d) : k_1, \cdots, k_d = 0, 1, 2, \cdots\}$$

forms a complete, orthonormal system for $L^2(\mathcal{O})$, and $e_{k_1,\ldots,k_d}$ is an eigenfunction for $-\Delta$ with eigenvalue

$$\lambda_{k_1,\cdots,k_d} = \lambda_{k_1} + \cdots + \lambda_{k_d}.$$

A similar relation holds for Δ_m with

$$\{e_{k_1,\cdots,k_d;m}(x_1, \cdots, x_d) = \alpha_{k_1,m}(x_1) \cdots \alpha_{k_d,m}(x_d)\}$$

and

$$\{\lambda_{k_1,\ldots,k_d;m} = \lambda_{k_1,m} + \cdots + \lambda_{k_d,m} : k_j = 0, 1, 2, \ldots, m-1; j = 1, \ldots, d\}.$$

The following lemma gives a sense in which the orthonormal systems for $L^2(\Lambda_m)$ converge to the orthonormal system for $L^2(\mathcal{O})$.

LEMMA C.1. *Let $\rho_m \in L^2(\Lambda_m)$ satisfy*

$$\lim_{m\to\infty} \|\rho_m - \pi_m \rho\|_m = 0.$$

Then

$$\lim_{m\to\infty} \langle \rho_m, e_{k_1,\ldots,k_d;m}\rangle_m = \langle \rho, e_{k_1,\ldots,k_d}\rangle,$$

and hence

$$\begin{aligned}
\liminf_{m\to\infty} \langle \rho_m, -\Delta_m \rho_m\rangle_m &= \liminf_{m\to\infty} \langle \nabla_m \rho_m, \nabla_m \rho_m\rangle_m \\
&= \liminf_{m\to\infty} \sum_{k_1=0,\ldots,k_d=0}^{m-1} \lambda_{k_1,\ldots,k_d;m} \langle \rho_m, e_{k_1,\ldots,k_d;m}\rangle_m^2 \\
&\geq \sum_{k_1=0,\ldots,k_d=0}^{\infty} \lambda_{k_1,\ldots,k_d} \langle \rho, e_{k_1,\ldots,k_d}\rangle^2 \\
&\equiv \langle \nabla \rho, \nabla \rho\rangle.
\end{aligned}$$

Let $B_m = (I - \Delta_m)^{-1}$ on $L^2(\Lambda_m)$, and $B = (I - \Delta)^{-1}$ on $L^2(\mathcal{O})$.

LEMMA C.2. *Let $\rho_m \in L^2(\Lambda_m)$ and $\rho \in L^2(\mathcal{O})$ satisfy*

$$\lim_{m\to\infty} \|\widehat{\eta}_m \rho_m - \rho\| = 0,$$

or equivalently, $\lim_{m\to\infty} \|\rho_m - \pi_m \rho\|_m = 0$. Then for $s = 1, 2, \ldots$,

$$\lim_{m\to\infty} \|\widehat{\eta}_m B_m^s \rho_m - B^s \rho\| = \lim_{m\to\infty} \|B_m^s \rho_m - \pi_m B^s \rho\| = 0.$$

PROOF. By induction, it is enough to check convergence for $s = 1$, and since B and B_m are bounded operators with norm 1, that

$$\lim_{m\to\infty} \|B_m \pi_m \rho - \pi_m B \rho\|_m = 0$$

holds for a dense set of $\rho \in L^2(\mathcal{O})$. Consequently, it is enough to check convergence for $\rho = \alpha_j$, but this convergence is immediate from

$$\|B_m \pi_m \alpha_j - \pi_m B \alpha_j\|_m \leq \|B_m(\pi_m \alpha_j - \alpha_{j,m})\| + \|(1+\lambda_{j,m})^{-1} \alpha_{j,m} - (1+\lambda_j)^{-1} \pi_m \alpha_j\|.$$

□

C.3. $E = L^2(\mathcal{O}) \cap \{\rho : \int \rho dx = 0\}$

In Example 1.13, because of the restriction to functions satisfying $\int \rho dx = 0$, and in the discrete case $\sum_{x\in\Lambda_m} \rho(x) = 0$, E and E_m in (1.38) become $E_m = \text{span}\{e_{k_1,\ldots,k_d;m} : (k_1, \ldots, k_d) \neq (0, \ldots, 0)\}$ and $E = \text{span}\{e_{k_1,\ldots,k_d} : (k_1, \ldots, k_d) \neq (0, \ldots, 0)\}$. In this context, if we let $B = (-\Delta)^{-1}$ and, for m odd, $B_m = (-\Delta_m)^{-1}$, then analogs of the results of the previous section hold. In particular, the conclusion of Lemma C.2 still holds with $L^2(\mathcal{O})$ replaced by E and the limit taken through m odd.

C.4. Other useful approximations

Let $\rho \in L^2(\mathcal{O})$. It follows from Jensen's inequality that

$$\|\pi_m \rho\|_m \leq \|\rho\|_{L^2(\mathcal{O})}.$$

Moreover, since $\nabla_m \pi_m \rho = \pi_m \nabla_m \rho$,

$$\|\nabla_m \pi_m \rho\|_m = \|\pi_m \nabla_m \rho\|_m \leq \|\nabla_m \rho\|_{L^2(\mathcal{O})} \leq \|\nabla \rho\|_{L^2(\mathcal{O})},$$

where the last inequality follows from by Lemma 7.23 of Gilbarg and Trudinger [**54**].

Let $F \in C^2(R)$ and $\sup_r F''(r) < \infty$ so that there exist $c_0, c_1 > 0$ such that $F(r) \leq c_0 + c_1 r^2$. As in (2.14), define a discrete free energy function $\mathcal{E}_m$ by

$$\mathcal{E}_m(\rho) \equiv \frac{1}{2} \sum_{x \in \Lambda_m} |\nabla_m \rho(x)|^2 m^{-d} + \sum_{x \in \Lambda_m} F(\rho(x)) m^{-d}, \qquad \rho \in E_m \equiv R^{|\Lambda_m|}. \tag{C.5}$$

Then

$$\mathcal{E}_m(\pi_m \rho) \leq c_0 + c_1 \|\rho\|^2_{L^2(\mathcal{O})} + \|\nabla \rho\|^2_{L^2(\mathcal{O})}, \quad \rho \in L^2(\mathcal{O}). \tag{C.6}$$

APPENDIX D

Results from mass transport theory

In Example 9.35 of Section 9.4 and in Section 13.3, we proved the comparison principle (Theorems 9.41 and 13.32) for a Hamilton-Jacobi equation with probability-measure-valued state space. A prominent feature of this equation is that the control problem associated with it is a mass-conserving flow determined by $\dot{\rho} + \nabla \cdot (\rho \mathbf{v}) = 0$, where $\mathbf{v}$ is a vector-valued function modeling the velocity field and $\nabla\cdot$ denotes the divergence. In other words, letting $z_{\mathbf{v}}(t,x)$ denote the solution of $\dot{z}_{\mathbf{v}}(t,x) = \mathbf{v}(z_{\mathbf{v}}(t,x),t)$ with $z_{\mathbf{v}}(0,x) = x$, $\rho(t)$ is the probability measure determined by the identity

$$\int_{R^d} f(x)\rho(t,dx) = \int_{R^d} f(z_{\mathbf{v}}(t,x))\rho_0(dx),$$

where $\rho(0) = \rho_0$ is the initial mass distribution. This observation naturally leads us to the consideration of mass transport techniques which we summarize in this appendix. Much of this material is taken from Villani [**127**].

Throughout this appendix, we use the convention that $0/0 = 0$. We assume that $\Psi, \Phi \in C^2(R^d)$, that Φ is even, and that there exist $\lambda_\Psi, \lambda_\Phi \in R$ (not necessarily nonnegative) such that

$$(\nabla\Psi(x) - \nabla\Psi(y)) \cdot (x-y) \geq \lambda_\Psi |x-y|^2 \tag{D.1}$$

and similarly for Φ. We use ρ or γ or σ to denote typical elements in $\mathcal{P}(R^d)$. The *support* $\mathrm{supp}(\rho)$ of $\rho \in \mathcal{P}(R^d)$ is the closed set defined by

$$\mathrm{supp}(\rho) = \{x \in R^d : \rho(U) > 0 \text{ for each neighborhood } U \text{ of } x\}. \tag{D.2}$$

As before, if $\rho \in \mathcal{P}(R^d)$ has a Lebesgue density, we use $\rho(x)$ to denote its density and $\rho(dx)$ to denote the measure. That is, $d\rho = \rho(dx) = \rho(x)dx$.

DEFINITION D.1 (Push forward of a probability measure). Let $T : R^d \to R^d$ be a measurable map and $\rho_0 \in \mathcal{P}(R^d)$. The image measure of ρ_0 by T, denoted as $(T\#\rho_0)(A) \equiv \rho_0(T^{-1}(A))$ for all Borel measurable $A \subset R^d$, is called the *push-forward* measure of ρ_0 by T.

D.1. Distributional derivatives

We will use the notion of weak derivatives in the (Schwartz) space of distributions. Let $\mathcal{O} \subset R^d$ be open and let $C_c^\infty(\mathcal{O})$ be the space of infinitely differentiable functions on $\mathcal{O}$ whose support is contained in a compact subset of $\mathcal{O}$. For $\alpha = (\alpha_1, \ldots, \alpha_d)$, α_i a nonnegative integer, let $D^\alpha f = \partial_{x_1}^{\alpha_1} \cdots \partial_{x_d}^{\alpha_d} f$. For each compact $K \subset \mathcal{O}$, let $C_c^\infty(K) = \{\varphi \in C_c^\infty(\mathcal{O}) : \mathrm{supp}(\varphi) \subset K\}$ topologized by the Frechet topology determined by the seminorms $\|D^\alpha\varphi\|_\infty = \sup_x |D^\alpha\varphi(x)|$ for all nonnegative, integer-valued vectors α. Let $\mathcal{D}(\mathcal{O})$ be the set $C_c^\infty(\mathcal{O})$ with the topology $\mathcal{U}$ such that $U \in \mathcal{U}$ if and only if $U \cap C_c^\infty(K)$ is open in $C_c^\infty(K)$ for each compact

$K \subset \mathcal{O}$. Then $\mathcal{D}'(\mathcal{O})$ is defined to be the space of continuous linear functionals on $\mathcal{D}(\mathcal{O})$. (See, for example, Reed and Simon [**102**], Section V.4.) $L^p_{loc}(\mathcal{O})$ will denote the collection of measurable functions f on $\mathcal{O}$ such that

$$\int_K |f(x)|^p dx < \infty, \quad \forall \text{ compact } K \subset \mathcal{O},$$

topologized by the seminorms $\left(\int_K |f(x)|^p dx\right)^{1/p}$. Note that if $f \in L^1_{loc}(\mathcal{O})$, then $\varphi \to \langle f, \varphi \rangle \equiv \int_{\mathcal{O}} \varphi f dx$ is in $\mathcal{D}'(\mathcal{O})$.

Let ν be a signed measure on $\mathcal{O}$. Then for each multi-index α, $D^\alpha \nu$ is the $l \in \mathcal{D}'(\mathcal{O})$ such that

$$l(\varphi) = (-1)^{|\alpha|} \int_{\mathcal{O}} D^\alpha \varphi d\nu, \quad \forall \varphi \in C_c^\infty(\mathcal{O}).$$

We will write

$$\langle D^\alpha \nu, \varphi \rangle = (-1)^{|\alpha|} \int_{\mathcal{O}} D^\alpha \varphi d\nu, \quad \forall \varphi \in C_c^\infty(\mathcal{O}).$$

$D^\alpha \nu$ is (can be identified as) a signed measure, if there exists a signed measure μ such that

$$(-1)^{|\alpha|} \int_{\mathcal{O}} D^\alpha \varphi d\nu = \int_{\mathcal{O}} \varphi d\mu, \quad \forall \varphi \in C_c^\infty(\mathcal{O}),$$

and we write

$$D^\alpha \nu = \mu \text{ in } \mathcal{D}'(\mathcal{O}).$$

We say $D^\alpha \nu \in L^p_{loc}(\mathcal{O})$ if there exists a $\xi \in L^p_{loc}(\mathcal{O})$, such that

$$(-1)^{|\alpha|} \int_{\mathcal{O}} D^\alpha \varphi d\nu = \int_{\mathcal{O}} \varphi(x) \xi(x) dx, \quad \forall \varphi \in C_c^\infty(\mathcal{O})$$

in which case we write

$$D^\alpha \nu = \xi \text{ in } \mathcal{D}'(\mathcal{O}).$$

If $\nu(dx) = f(x)dx$, $f \in L^1_{loc}(\mathcal{O})$, we write $D^\alpha f = D^\alpha \nu$.

Let

$$J(x) = C \exp\{(|x|^2 - 1)^{-1}\}, \text{ if } |x| < 1; \qquad = 0, \text{ if } |x| \geq 1, \tag{D.3}$$

where the constant $C > 0$ is selected so that $\int_{R^d} J(x)dx = 1$, and define $J_\epsilon(z) = \epsilon^{-d} J(\epsilon^{-1} z)$. For $f \in L^1_{loc}(R^d)$, define

$$J_\epsilon * f(x) = \int_{R^d} J_\epsilon(x-y) f(y) dy$$

and if ν is a finite signed measure,

$$J_\epsilon * \nu(dx) \equiv \int_{R^d} J_\epsilon(x-y) \nu(dy) dx. \tag{D.4}$$

Note that $J_\epsilon * \nu \in C^\infty(R^d)$. Let $J_\epsilon^{\mathcal{O}} * \nu$ denote the restriction of $J_\epsilon * \nu$ to $\mathcal{O}$.

LEMMA D.2. *Let ν be a finite signed measure on $\mathcal{O}$ and α a multi-index. Then $\lim_{\epsilon \to \infty} D^\alpha J_\epsilon^{\mathcal{O}} * \nu = D^\alpha \nu$ in the sense that*

$$\lim_{\epsilon \to 0} \langle D^\alpha J_\epsilon^{\mathcal{O}} * \nu, \varphi \rangle = \langle D^\alpha \nu, \varphi \rangle, \quad \varphi \in \mathcal{D}(\mathcal{O}).$$

If $D^\alpha \nu = \xi \in L^1_{loc}(R^d)$, then

$$D^\alpha J_\epsilon * \nu(x) = \int_{R^d} J_\epsilon(x-y) \xi(y) dy. \tag{D.5}$$

For $\epsilon > 0$, let $\mathcal{O}_\epsilon = \{x \in \mathcal{O} : \sup_{y \in \mathcal{O}^c} |x - y| > \epsilon$. If $D^\alpha \nu = \xi \in L^1_{loc}(\mathcal{O})$, then

$$D^\alpha J_\epsilon^{\mathcal{O}_\epsilon} * \nu(x) = \int_{\mathcal{O}} J_\epsilon(x-y)\xi(y)dy \quad \text{in } \mathcal{D}'(\mathcal{O}_\epsilon), \tag{D.6}$$

PROOF. For $\varphi \in \mathcal{D}(\mathcal{O})$, $\langle D^\alpha J_\epsilon^{\mathcal{O}} * \nu, \varphi\rangle = \langle D^\alpha J_\epsilon * \nu, \varphi\rangle$, and for $\epsilon > 0$ sufficiently small, $J_\epsilon * \varphi \in \mathcal{D}(\mathcal{O})$. Integrating by parts,

$$\begin{aligned}\langle D^\alpha J_\epsilon * \nu, \varphi\rangle &= (-1)^{|\alpha|}\langle J_\epsilon * \nu, D^\alpha \varphi\rangle \\ &= (-1)^{|\alpha|}\langle \nu, D^\alpha J_\epsilon * \varphi\rangle \\ &= \langle D^\alpha \nu, J_\epsilon * \varphi\rangle \to \langle D^\alpha \nu, \varphi\rangle,\end{aligned}$$

where the convergence follows from the fact that $J_\epsilon * \varphi$ converges to φ in $\mathcal{D}(\mathcal{O})$.

The identity (D.5) follows from the fact that for $\varphi \in C_c^\infty(R^d)$, $J_\epsilon * \varphi \in C_c^\infty(R^d)$ and integration by parts. See, for example, Evans [**38**], page 250.

Similarly, if $\varphi \in \mathcal{D}(\mathcal{O}_\epsilon)$, then $J_\epsilon * \varphi \in \mathcal{D}(\mathcal{O})$, so

$$\begin{aligned}\langle D^\alpha J_\epsilon^{\mathcal{O}_\epsilon} * \nu, \varphi\rangle &= (-1)^{|\alpha|}\langle \nu, D^\alpha J_\epsilon * \varphi\rangle \\ &= \langle \xi, J_\epsilon * \varphi\rangle \\ &= \langle J_\epsilon * \xi, \varphi\rangle\end{aligned}$$

□

LEMMA D.3. *If $f \in L^1_{loc}(\mathcal{O})$, then for each compact $K \subset \mathcal{O}$,*

$$\lim_{\epsilon \to 0} \int_K |J_\epsilon * f - f| dx = 0.$$

PROOF. If f is continuous, the result follows by uniform continuity. The general statement follows from the fact that the continuous functions are dense in $L^1(K)$. □

LEMMA D.4. *If $f \in L^1_{loc}(\mathcal{O})$, $\partial_{x_i} f = \xi \in L^1_{loc}(\mathcal{O})$, and $G : R \mapsto R$ has a bounded continuous derivative, then*

$$\partial_{x_i} G(f) = G'(f)\xi \quad \text{in } \mathcal{D}'(\mathcal{O}). \tag{D.7}$$

If $g \in C^1(R^d)$, then

$$\partial_{x_i}(fg) = g\xi + f\partial_{x_i} g. \tag{D.8}$$

PROOF. The ordinary chain rule and (D.6) gives

$$\partial_{x_i} G(J_\epsilon * f) = G'(J_\epsilon * f) J_\epsilon * \xi$$

in $\mathcal{D}'(\mathcal{O}_\epsilon)$, and (D.7) follows by Lemma D.3. The proof of (D.8) is similar. □

DEFINITION D.5. For $k \geq 0$, $H^k(\mathcal{O})$ is the collection of $f \in L^2(\mathcal{O})$ such that $D^\alpha f \in L^2(\mathcal{O})$ for all α, $|\alpha| \leq k$, and

$$\|f\|_{H^k(\mathcal{O})} = \left(\sum_{|\alpha| \leq k} \int_{\mathcal{O}} |D^\alpha f|^2 dx\right)^{1/2}.$$

In particular, $H^0(\mathcal{O}) = L^2(\mathcal{O})$.

Recall that for each k, $H^k(\mathcal{O})$ is a Hilbert space. We will need the following result regarding the relationship between $H^1(\mathcal{O})$ and $H^0(\mathcal{O})$.

THEOREM D.6. *Suppose that $\mathcal{O}$ is bounded and the boundary $\partial\mathcal{O}$ is C^1. Then bounded subsets of $H^1(\mathcal{O})$ are relatively compact in $H^0(\mathcal{O})$.*

PROOF. The result is a special case of the Rellich-Kondrachov Compactness Theorem. See Adams [**1**], Theorem 6.2. □

LEMMA D.7. *Let $f \geq 0$, $f \in L^1(\mathcal{O})$, and $\nabla f \in L^1_{loc}(\mathcal{O})$. If*

$$\int_{\mathcal{O}} \frac{|\nabla f|^2}{f} dx < \infty, \tag{D.9}$$

then $\sqrt{f} \in H^1(\mathcal{O})$, and

$$\nabla\sqrt{f} = \frac{\nabla f}{2\sqrt{f}}. \tag{D.10}$$

Conversely, if $f \geq 0$ and $\sqrt{f} \in H^1(\mathcal{O})$, then $f \in L^1(\mathcal{O})$, $\nabla f \in L^1(\mathcal{O})$, and (D.10) holds.

REMARK D.8. Note that (D.9) implies $\nabla f \in L^1(\mathcal{O})$, since

$$\int_{\mathcal{O}} |\nabla f| dx \leq \sqrt{\int_{\mathcal{O}} f dx \int_{\mathcal{O}} \frac{|\nabla f|^2}{f} dx} < \infty.$$

PROOF. Let J be defined as in (D.3), $\zeta_0(x) = \mathbf{1}_{[-2,2]}(|x|)$, and

$$\zeta = J * \zeta_0 \tag{D.11}$$

Then $\zeta \in C_c^\infty(R^d)$, $\zeta : R^d \to [0,1]$, $\zeta(x) = 1$, $|x| \leq 1$, $\zeta(x) = 0$, $|x| \geq 3$, and $\sup_x |\nabla\zeta(x)|^2/\zeta(x) < \infty$. Define $\zeta_n(x) = \zeta(n^{-1}x)$. Setting $f_n = \zeta_n f$,

$$\nabla f_n = \zeta_n \nabla f + f n^{-1} \nabla\zeta(n^{-1}x),$$

and

$$\int_{\mathcal{O}} \frac{|\nabla f_n|^2}{f_n} dx \leq 2\int_{\mathcal{O}} \frac{|\nabla f|^2}{f} dx + \frac{2}{n^2}\int_{\mathcal{O}} f(x)\frac{|\nabla\zeta(n^{-1}x)|^2}{\zeta(n^{-1}x)} dz < \infty.$$

Since $G_\delta(r) = \sqrt{r+\delta} - \sqrt{\delta}$ has a bounded, continuous derivative for $r \geq 0$,

$$\nabla G_\delta(f_n) = \frac{\nabla f_n}{2\sqrt{f_n + \delta}},$$

and by the monotone convergence theorem,

$$\lim_{\delta\to 0} \int_{\mathcal{O}} |\nabla G_\delta(f_n)|^2 dx = \int_{\mathcal{O}} \frac{|\nabla f_n|^2}{f_n} dx,$$

and in turn, by the dominated convergence theorem,

$$\lim_{\delta\to 0} \|\nabla G_\delta(f_n) - \frac{\nabla f_n}{2\sqrt{f_n}}\|_{L^2(\mathcal{O})} = 0.$$

Since $\sqrt{r+\delta} - \sqrt{\delta} \leq \sqrt{r}$, $G_\delta(f_n) \to \sqrt{f_n}$ in $L^2(\mathcal{O})$ by the dominated convergence theorem. Consequently, $G_\delta(f_n) \to \sqrt{f_n}$ in $H^1(\mathcal{O})$. Finally, $\sqrt{f_n} \to \sqrt{f}$ in $L^2(\mathcal{O})$ by the dominated convergence theorem, and writing

$$\frac{\nabla f_n}{2\sqrt{f_n}} - \frac{\nabla f}{2\sqrt{f}} = (\sqrt{\zeta_n} - 1)\frac{\nabla f}{2\sqrt{f}} + \frac{\sqrt{f}\nabla\zeta(n^{-1}\cdot)}{n\sqrt{\zeta_n}},$$

the first term on the right goes to zero in $L^2(\mathcal{O})$ by the definition of ζ_n and the dominated convergence theorem, and the $L^2(\mathcal{O})$ norm of the second term on the right is bounded by $n^{-1} \sup_x |\nabla\zeta(x)|/\sqrt{\zeta(x)}$. It follows that $\sqrt{f_n} \to \sqrt{f}$ in $H^1(\mathcal{O})$ and (D.10) holds.

Conversely, suppose $f \geq 0$ and $\sqrt{f} \in H^1(\mathcal{O})$. For $\epsilon > 0$, let $f_\epsilon = fe^{-\epsilon f}$. Setting $G_\epsilon(r) = r^2 e^{-\epsilon r^2}$, $f_\epsilon = G_\epsilon(\sqrt{f})$ and

$$\nabla f_\epsilon = 2\sqrt{f}e^{-\epsilon f}\nabla\sqrt{f} - 2\epsilon f^{3/2}e^{-\epsilon f}\nabla\sqrt{f}.$$

The second term is dominated by $4\sqrt{f}|\nabla\sqrt{f}|$, and hence converges to $2\sqrt{f}\nabla\sqrt{f}$ in $L^1(\mathcal{O})$ as $\epsilon \to 0$. It follows that $\nabla f = 2\sqrt{f}\nabla\sqrt{f}$ giving (D.10). □

D.2. Convex functions

We recall a few facts about convex functions that will be used below without further comment. The standard reference on convex functions is [**104**].

Let φ denote a convex function on R^d, and define $\mathrm{Dom}(\varphi) = \{x \in R^d : \varphi(x) < \infty\}$. If φ is proper (i.e. not identically $+\infty$), it is continuous and locally Lipschitz on the interior of $\mathrm{Dom}(\varphi)$, $\mathrm{int}(\mathrm{Dom}(\varphi))$.

THEOREM D.9. *Let φ be a convex function, and assume that Dom(φ) has nonempty interior $\mathcal{O} = \mathrm{int}(\mathrm{Dom}(\varphi))$. Then the classical gradient $\widetilde{\nabla}\varphi$ is well-defined almost everywhere in $\mathcal{O}$ and agrees with the distributional gradient*

$$\nabla\varphi = \widetilde{\nabla}\varphi \in L^\infty_{loc}(\mathcal{O}).$$

More precisely, φ is twice differentiable almost everywhere in $\mathcal{O}$ in the sense that for almost every $x \in \mathcal{O}$, there is a nonnegative definite $d \times d$ matrix D^2_A such that

$$\varphi(x + y) = \varphi(x) + \widetilde{\nabla}\varphi(x) \cdot y + y^T \cdot D^2_A\varphi(x) \cdot y + o(|y|^2), \quad y \in R^d.$$

PROOF. Since φ is locally Lipschitz in $\mathcal{O}$, Rademacher's theorem gives existence of the gradient. See Evans and Gariepy [**39**], Section 3.1.2. The second order properties are given by Aleksandrov's theorem. See Theorem 5 in Section 5.8.3 of Evans [**38**]. Furthermore, the distributional Hessian $D^2\varphi$ is a matrix-valued measure, and $D^2_A\varphi$ is just the absolutely continuous part of $D^2\varphi$. (See the comments on page 58 of [**127**]). □

For an arbitrary function ψ, let ψ^* denote the *Legendre-Fenchel transform*

$$\psi^*(x) \equiv \sup_y\{x \cdot y - \psi(y)\}.$$

Then ψ^* is a lower semicontinuous, convex function. (Do not confuse this notation with the notation for the upper semicontinuous regularization of a function.)

LEMMA D.10. *A proper, convex function φ is lower semicontinuous if and only if $\varphi = \varphi^{**}$.*

PROOF. See Theorem 12.2 of [**104**]. □

LEMMA D.11. *If φ is a lower semicontinuous, convex function, then*

$$\varphi(x) = x \cdot \nabla\varphi(x) - \varphi^*(\nabla\varphi(x)), \text{ for almost every } x \in \mathrm{int}(\mathrm{Dom}(\varphi)).$$

LEMMA D.12. *Let φ be a lower semicontinuous, convex function, and let $\psi = \varphi^*$. For $r > 0$, define*

$$\varphi_r(x) = \sup_{|y| \leq r}(x \cdot y - \psi(y)).$$

Then φ_r is finite on R^d, $\varphi_r \leq \varphi$, $|\nabla\varphi_r| \leq r$, and $\varphi_r(x) = \varphi(x)$ and $\nabla\varphi_r(x) = \nabla\varphi(x)$ for almost every x such that $|\nabla\varphi(x)| \leq r$.

PROOF. The lemma follows from Theorems 23.5 and 25.1 or [**104**]. □

D.3. The p-Wasserstein metric space

For $p \geq 1$, let

$$\mathcal{P}_p(R^d) = \{\rho \in \mathcal{P}(R^d) : \int_{R^d} |x|^p \rho(dx) < \infty\}. \tag{D.12}$$

For $\rho, \gamma \in \mathcal{P}(R^d)$, define

$$\mathbf{\Pi}(\rho, \gamma) \equiv \{\pi \in \mathcal{P}(R^d \times R^d) : \pi(\cdot \times R^d) = \rho, \pi(R^d \times \cdot) = \gamma\}. \tag{D.13}$$

DEFINITION D.13. The p-*Wasserstein metric* on the space $\mathcal{P}_p(R^d)$ is

$$d_p(\rho, \gamma) = \Big(\inf\{\int_{R^d \times R^d} |x - y|^p \pi(dx, dy) : \pi \in \mathbf{\Pi}(\rho, \gamma)\}\Big)^{1/p}. \tag{D.14}$$

for $\rho, \gamma \in \mathcal{P}_p(R^d)$.

LEMMA D.14. *For $\rho, \gamma \in \mathcal{P}(R^d)$, $\mathbf{\Pi}(\rho, \gamma)$ is compact in the weak topology on $\mathcal{P}(R^d \times R^d)$ and the infimum in the definition of d_p is achieved. If $d_p(\gamma_n, \gamma_0) \to 0$, $d_p(\rho_n, \rho_0) \to 0$ and $\pi_n \in \mathbf{\Pi}(\rho_n, \gamma_n)$ satisfies*

$$d_p(\rho_n, \gamma_n) = \Big(\int_{R^d \times R^d} |x - y|^p \pi_n(dx, dy)\Big)^{1/p},$$

then $\{\pi_n\}$ is relatively compact in the weak topology and any limit point π_0 of the sequence $\{\pi_n\}$ satisfies

$$\Big(\int_{R^d \times R^d} |x - y|^p \pi_0(dx, dy)\Big)^{1/p} = d_p(\rho_0, \gamma_0). \tag{D.15}$$

In particular, if there is a unique π_0 satisfying (D.15), then π_n converges weakly to π_0.

PROOF. If $\pi \in \mathbf{\Pi}(\rho, \gamma)$, then

$$\pi((K \times K)^c) \leq \rho(K^c) + \gamma(K^c)$$

and compactness of $\mathbf{\Pi}(\rho, \gamma)$ in the weak topology follows. The lower semicontinuity of the integral in the definition (D.14) ensures that the infimum is achieved. Similarly, tightness for $\{\pi_n\}$ follows from the tightness of $\{\rho_n\}$ and $\{\gamma_n\}$ and

$$\begin{aligned}\Big(\int_{R^d \times R^d} |x - y|^p \pi_0(dx, dy)\Big)^{1/p} &\leq \liminf_{n \to \infty} \Big(\int_{R^d \times R^d} |x - y|^p \pi_n(dx, dy)\Big)^{1/p} \\ &= d_p(\rho_0, \gamma_0),\end{aligned}$$

so (D.15) follows. □

$(\mathcal{P}_p(R^d), d_p)$ is a complete, separable metric space for $p \geq 1$ (Theorem 7.3 of [**127**]). The case $p = 2$ will be of special interest to us, and we will simply use $d = d_2$ to denote the 2-Wasserstein metric.

LEMMA D.15. *For $\rho, \gamma \in \mathcal{P}_2(R^d)$, there exists $\pi \in \mathbf{\Pi}(\rho, \gamma)$ such that*

$$d^2(\rho, \gamma) = \int_{R^d \times R^d} |x - y|^2 \pi(dx, dy). \tag{D.16}$$

If ρ or γ is absolutely continuous with respect to Lebesgue measure, then π satisfying (D.16) is unique.

REMARK D.16. We denote the collection of $\pi \in \mathbf{\Pi}(\rho, \gamma)$ satisfying (D.16) by $\mathbf{\Pi}^{opt}(\rho, \gamma)$.

PROOF. As noted above, $\mathbf{\Pi}(\rho, \gamma)$ is compact in the weak topology and the mapping $\pi \mapsto \int |x - y|^2 \pi(dx, dy)$ is lower semicontinuous. Consequently, the infimum is achieved. The uniqueness is given in Theorem D.25. □

We have the following characterization of convergence and relative compactness in $(\mathcal{P}_p(R^d), d_p)$.

LEMMA D.17. *For $p \geq 1$, $(\mathcal{P}_p(R^d), d_p)$ is a complete, separable metric space. For $\rho_n, \rho \in \mathcal{P}_p(R^d)$, the following are equivalent:*

(1) $\lim_{n\to\infty} d_p(\rho_n, \rho) = 0$.

(2) *For all $\varphi \in C(R^d)$ satisfying $|\varphi(x)| \leq C(1 + |x|^p)$ for some $C > 0$,*

$$\lim_{n\to\infty} \int_{R^d} \varphi d\rho_n = \int_{R^d} \varphi d\rho.$$

(3) *$\rho_n \Rightarrow \rho$ in the sense of weak convergence of probability measures and*

$$\lim_{n\to\infty} \int_{R^d} |x|^p \rho_n(dx) = \int_{R^d} |x|^p \rho(dx).$$

(4) *$\rho_n \Rightarrow \rho$ in the sense of weak convergence of probability measures and*

$$\lim_{N\to\infty} \sup_n \int_{|x|>N} |x|^p \rho_n(dx) = 0.$$

Let $K \subset \mathcal{P}_p(R^d)$. The following are equivalent:

(1) *K is relatively compact.*

(2) *K is p-uniformly integrable in the sense that*

$$\lim_{N\to\infty} \sup_{\rho\in K} \int_{|x|>N} |x|^p \rho(dx) = 0.$$

(3) *There exists $\varphi : [0, \infty) \mapsto [0, \infty)$ satisfying $\varphi(0) = 0$, φ convex and increasing, and φ superlinear in the sense that*

$$\lim_{|x|\to\infty} \frac{\varphi(x)}{|x|} = \infty,$$

such that

$$\sup_{\rho\in K} \int_{R^d} \varphi(|x|^p) \rho(dx) < \infty.$$

PROOF. The results follow easily from standard results on weak convergence and uniform integrability. The convergence results can be found in Villani [**127**], Theorem 7.12. The completeness and separability of the space, as well as the relative compactness results can be found in Proposition 7.4.2. of Ambrosio, Gigli and Savaré [**3**]. The use of the superlinear function φ follows from Theorem 22 and its proof on page 24 of Dellacherie [**28**].

Note that for probability measures on R^d, p-uniform integrability ($p \geq 1$) implies tightness. □

D.4. The Monge-Kantorovich problem

Many of the results on the minimization problem defining the p-Wasserstein metric (D.14) can be extended to bivariate functions c more general than $|x-y|^p$. We collect a number of results that will be useful in the main text.

Let $c : R^d \times R^d \to [0, \infty]$ be lower semicontinuous. Following Section 3.3 of Rachev and Rüschendorf [**101**] and Section 2.4 of Villani [**127**], the usual concept of concavity can be generalized as follows.

DEFINITION D.18 (c-transform, c-concavity).

a) For $p : R^d \mapsto \overline{R}$, the *$c$-transform* $p^c : R^d \mapsto \overline{R}$ is defined by

$$p^c(y) = \inf_{x \in R^d} (c(x,y) - p(x))$$

with the convention that the difference is $+\infty$ when $c(x,y) = +\infty$ and $p(x) = +\infty$. Similarly,

$$q^c(x) = \inf_{y \in R^d} (c(x,y) - q(y)).$$

b) A function q is *c-concave* if $q = p^c$ for some p; equivalently, q is c-concave if there exists some $\{(x_i, t_i)\}_{i \in I} \subset R^d \times \overline{R}$ such that

$$q(y) = \inf_{i \in I} (c(x_i, y) + t_i), \quad \forall y \in R^d.$$

It follows that $p^{cc} \equiv (p^c)^c \geq p$ and equality holds if and only if p is c-concave (Exercise 2.35 of [**127**]). If c is continuous, then any c-concave function is upper semicontinuous.

Taking $c(x,y) = |x-y|^2/2$, let

$$\varphi(x) = \frac{|x|^2}{2} - p(x), \quad \psi(y) = \frac{|y|^2}{2} - q(x).$$

Then $p(x) + q(y) \leq c(x,y)$ if and only if $\varphi(x) + \psi(y) \geq x \cdot y$. In particular, $p = q^c$ if and only if $\varphi = \psi^*$, where ψ^* is the Legendre-Fenchel transform

$$\psi^*(x) \equiv \sup_y \{x \cdot y - \psi(y)\}.$$

A function $p : R^d \mapsto \overline{R}$ is *proper* if there exists $x \in R^d$ such that $p(x) \in R$. Let $\rho, \gamma \in \mathcal{P}(R^d)$, and define

$$\text{(D.17)} \qquad \Phi_c = \{(p,q) : p, q \text{ are proper and } p = q^c, q = p^c\},$$

$$\text{(D.18)} \qquad \Phi_{c,0} = \{(p,q) \in \Phi_c : \inf_y q(y) \leq 0 \leq \sup_y q(y)\},$$

$$\text{(D.19)} \qquad \Phi_{c,0}(\gamma) = \{(p,q) \in \Phi_c : q \in L^1(d\gamma), \int_{R^d} q(y)\gamma(dy) = 0\},$$

$$\text{(D.20)} \qquad \Phi_c(\rho,\gamma) = \Phi_c \cap (L^1(d\rho) \times L^1(d\gamma)),$$

REMARK D.19. Let c be uniformly continuous with modulus of continuity ω_c, that is,

$$|c(\widehat{x},\widehat{y}) - c(x,y)| \le \omega_c(|\widehat{x}-x| + |\widehat{y}-y|).$$

If $(p,q) \in \Phi_c$, then $|p(x)-p(\widehat{x})| \le \omega_c(|\widehat{x}-x|)$ and $|q(y)-q(\widehat{y})| \le \omega_c(|y-\widehat{y}|)$. For example,

$$p(x)-p(\widehat{x}) = \inf_y(c(x,y)-q(y)) - \inf_y(c(\widehat{x},y)-q(y)) \le \sup_y(c(x,y)-c(\widehat{x},y)). \tag{D.21}$$

If c is bounded, then

$$-\sup_y q(y) \le p(x) \le \|c\|_\infty - \sup_y q(y),$$

and hence

$$q(y) = \inf_x(c(x,y)-p(x)) \ge \sup_{y'} q(y') - \|c\|_\infty$$

and similarly for p. Consequently, if $(p,q) \in \Phi_{c,0}$, then $\|p\|_\infty, \|q\|_\infty \le \|c\|_\infty$.

If c is bounded and uniformly continuous, then by Ascoli's Theorem (e.g., Theorem 17 in Chapter 7 of Kelley [**65**]), $\{p : (p,q) \in \Phi_{c,0}(\gamma)\}$ and $\{p : (p,q) \in \Phi_{c,0}\}$ are compact subsets of $C_b(R^d)$ under the topology of uniform convergence over compact subsets of R^d.

If in addition there exist k_0 and c_0 such that $c(x,y) = c_0$ for $|x| + |y| \ge k_0$, then $(p,q) \in \Phi_c$ implies $p(x) = c_0 - \sup_y q(y)$ for $|x| \ge k_0$, and $\Phi_{c,0}$ is compact in $C_b(R^d)$ under the uniform topology. For example, take

$$c(x,y) = (|x-y|^2 + \theta|y|^2) \wedge M, \quad \theta, M > 0.$$

For $\rho, \gamma \in \mathcal{P}(R^d)$, define

$$F_c(\rho,\gamma) = \inf\{\int_{R^d\times R^d} c(x,y)\pi(dx,dy) : \pi \in \mathbf{\Pi}(\rho,\gamma)\} \tag{D.22}$$

THEOREM D.20. *[Kantorovich duality] Let $c : R^d \times R^d \mapsto [0,\infty]$ be lower semicontinuous, and let $\rho,\gamma \in \mathcal{P}(R^d)$ satisfy $\int_{R^d\times R^d} c(x,y)\rho(dx)\gamma(dy) < \infty$. Then*

a) *There exist $\pi_0 \in \mathbf{\Pi}(\rho,\gamma)$ and $(p_0,q_0) \in \Phi_{c,0}(\gamma)$, such that*

$$\begin{aligned} F_c(\rho,\gamma) &= \int_{R^d\times R^d} c(x,y)\pi_0(dx,dy) \\ &= \int_{R^d} p_0(x)\rho(dx) \\ &= \sup\{\int_{R^d} p d\rho : (p,q) \in \Phi_{c,0}(\gamma)\} \\ &= \sup\{\int_{R^d} p d\rho + \int_{R^d} q d\gamma : (p,q) \in \Phi_c(\rho,\gamma)\}, \end{aligned} \tag{D.23}$$

b) *$\pi_0 \in \mathbf{\Pi}(\rho,\gamma)$ satisfies (D.23) if and only if*

$$p_0(x) + q_0(y) = c(x,y), \quad (x,y) - \pi_0 - a.e.$$

PROOF. Existence of π_0 follows from the compactness of $\mathbf{\Pi}(\rho,\gamma)$ and the lower semicontinuity of the mapping $\pi \mapsto \int_{R^d\times R^d} c(x,y)\pi(dx,dy)$, and Part (a) follows from Theorem 7.2.4 of [**3**].

Part (b) follows from the fact that $p_0(x) + q_0(y) \le c(x,y)$. $\square$

COROLLARY D.21. *Let $K_\gamma \subset R^d$ denote the support of γ, and define*

$$\widehat{\Gamma}_c^\gamma = \{\widehat{p}(\cdot) = \inf_{y \in K_\gamma} c(\cdot, y) - q(y) : (p,q) \in \Phi_{c,0}(\gamma)\}. \tag{D.24}$$

Then

$$F_c(\rho,\gamma) = \sup\{\int_{R^d} p d\rho : p \in \widehat{\Gamma}_c^\gamma\}, \tag{D.25}$$

and for π_0 and (p_0, q_0) satisfying (D.23), the supremum in (D.25) is achieved at $\widehat{p}_0(\cdot) = \inf_{y \in K_\gamma} c(\cdot, y) - q_0(y)$ and

$$\widehat{p}_0(x) + q_0(y) = c(x,y) \quad a.s.\ \pi_0.$$

REMARK D.22. If c is continuous and K_γ is compact, then, as in (D.21), $\widehat{\Gamma}_c^\gamma$ is an equicontinuous collection of functions, and the functions in $\widehat{\Gamma}_c^\gamma$ are bounded above by $\int_{R^d} c(x,y)\gamma(dy)$. It follows that for each $\alpha \in R$, $\{p \in \widehat{\Gamma}_c^\gamma : \int_{R^d} p d\rho \geq \alpha\}$ is relatively compact in the topology of uniform convergence on compact sets.

PROOF. For $(p,q) \in \Phi_{c,0}(\gamma)$, $p(x) + q(y) \leq \widehat{p}(x) + q(y) \leq c(x,y)$ a.s. π_0. Consequently,

$$\begin{aligned} F_c(\rho,\gamma) &= \sup\{\int_{R^d} p d\rho : (p,q) \in \Phi_{c,0}(\gamma)\} \\ &\leq \sup\{\int_{R^d} \widehat{p} d\rho : \widehat{p} \in \widehat{\Gamma}_c(\gamma)\} \\ &\leq \int_{R^d \times R^d} c(x,y)\pi_0(dx \times dy). \end{aligned}$$

□

We define

$$\Phi_c^{opt}(\rho,\gamma) = \{(p_0,q_0) \in \Phi_c(\rho,\gamma) : (p_0,q_0) \text{ satisfies } (D.23)\} \tag{D.26}$$

and

$$\mathbf{\Pi}_c^{opt}(\rho,\gamma) = \{\pi_0 \in \mathbf{\Pi}(\rho,\gamma) : \pi_0 \text{ satisfies } (D.23)\}. \tag{D.27}$$

LEMMA D.23. *Let c be continuously differentiable with bounded gradient, and let ρ have a Lebesgue density. Suppose that $\pi_0 \in \mathbf{\Pi}_c^{opt}(\rho,\gamma)$ and $(p_0,q_0) \in \Phi_c^{opt}(\rho,\gamma)$. Then ∇p_0 exists Lebesgue almost everywhere (hence almost everywhere with respect to ρ) and*

$$\nabla p_0(x) = \nabla_x c(x,y), \quad \pi_0 - a.e. \tag{D.28}$$

Consequently, there exist R^d-valued random variables X, Y such that (X,Y) has joint distribution π_0, and

$$\nabla p_0(X) = \nabla_x c(X,Y), \quad \text{almost surely.}$$

REMARK D.24. This lemma is essentially Part b) of Remark 3.3.14 of Rachev and Rüschendorf [**101**].

PROOF. By part (b) of Theorem D.20,

$$p_0(x) + q_0(y) = c(x,y) \quad a.s.\ \pi_0.$$

By Remark D.19, p_0 is Lipschitz and hence ∇p_0 exists Lebesgue almost everywhere. Since $c(x,y) - p_0(x) - q_0(y) \geq 0$, for each (x,y) such that $p_0(x) + q_0(y) = c(x,y)$ and $\nabla p_0(x)$ exists, we have $\nabla_x(c(x,y) - p_0(x) - q_0(y)) = 0$ and (D.28) follows. □

Results from convex analysis can be used to extend Lemma D.23 to $c(x,y) = \frac{1}{2}|x-y|^2$.

THEOREM D.25. *[Brenier's optimal transport map] Let $\rho, \gamma \in \mathcal{P}_2(R^d)$, $c(x,y) = \frac{1}{2}|x-y|^2$, and $\pi_{\rho,\gamma} \equiv \pi_0$ and $(p_{\rho,\gamma}, q_{\rho,\gamma}) \equiv (p_0, q_0) \in \Phi_{c,0}(\gamma)$ satisfy (D.23).*

Suppose $\rho(dx) = \rho(x)dx$ has a Lebesgue density, and let $\mathcal{O}$ be the interior of the convex hull of the support of ρ. Then $\nabla p_{\rho,\gamma} \in \mathcal{D}'(\mathcal{O})$ exists and agrees with the classical gradient almost everywhere in $\mathcal{O}$. The optimal $\pi_{\rho,\gamma} \equiv \pi_0$ in (D.23) is unique and is given by

$$\pi_{\rho,\gamma}(dx,dy) = \rho(dx)\delta_{T_{\rho,\gamma}(x)}(dy), \tag{D.29}$$

where

$$T_{\rho,\gamma}(x) = x - \nabla p_{\rho,\gamma}(x).$$

In particular, γ is the push forward of ρ by $T_{\rho,\gamma}$ and

$$\int_{R^d} |\nabla p_{\rho,\gamma}(x)|^2 \rho(dx) = \int_{R^d \times R^d} |x-y|^2 \pi_{\rho,\gamma}(dx,dy) = d^2(\rho,\gamma). \tag{D.30}$$

REMARK D.26. Note that while $\nabla p_{\rho,\gamma}$ is only defined in $\mathcal{O}$, $\rho(\mathcal{O}) = 1$, so the left side of (D.30) is well-defined. More generally, taking $\rho\nabla p_{\rho,\gamma}$ to be zero off the support of ρ, the finiteness of (D.30) implies $\rho\nabla p_{\rho,\gamma} \in L^1_{loc}(R^d)$ and $\nabla \cdot (\rho\nabla p_{\rho,\gamma}) \in \mathcal{D}'(R^d)$.

PROOF. For uniqueness of $\pi_{\rho,\gamma}$, see Brenier's theorem, Theorem 2.12 of [**127**]. Let $\varphi_{\rho,\gamma}(x) = \sup_y(x \cdot y - \frac{1}{2}|y|^2 + q_{\rho,\gamma}(y))$. Then

$$p_{\rho,\gamma}(x) = \frac{1}{2}|x|^2 - \varphi_{\rho,\gamma}(x),$$

and hence $\nabla p_{\rho,\gamma}$ exists Lebesgue almost everywhere in $\text{int}(\text{Dom}(\varphi_{\rho,\gamma}))$ (see Theorem D.9) and agrees with the classical gradient. Since $\int_{R^d} p_{\rho,\gamma} d\rho = d(\rho,\gamma)$, $\text{int}(\text{Dom}(\varphi))$ must contain $\mathcal{O}$. Consequently, (D.28) holds giving $\nabla p_{\rho,\gamma}(x) = (x-y)$ a. s. $\pi_{\rho,\gamma}$. □

REMARK D.27. Following [**127**], we call $T_{\rho,\gamma}$ in the above theorem *Brenier's optimal transport map* .

In Definition 9.36, we defined a gradient for a function on $\mathcal{P}_2(R^d)$. We can now identify this quantity for two useful functions.

THEOREM D.28. *Suppose that $\rho, \gamma \in \mathcal{P}_2(R^d)$ both have Lebesgue densities. Let $\mathcal{E}$ be defined according to (9.94), and let d be the 2-Wasserstein metric. Then in the sense of Definition 9.36,*

$$\text{grad}_\rho d^2(\rho,\gamma) = -2\nabla \cdot (\rho\nabla p_{\rho,\gamma}), \tag{D.31}$$

where $p_{\rho,\gamma}$ is as in Theorem D.25 and $\rho\nabla p_{\rho,\gamma}$ is defined as in Remark D.26. If in addition, $\mathcal{E}(\rho) < \infty$, then

$$\text{grad}\, \mathcal{E}(\rho) = -\Big(\frac{1}{2}\Delta\rho + \nabla \cdot (\rho\nabla(\Psi + \rho * \Phi))\Big) \text{ in } \mathcal{D}'(R^d). \tag{D.32}$$

PROOF. The conclusion follows from Theorem 8.13 of [**127**] and some computations following [**63**].

Let $\rho, \gamma \in \mathcal{P}_2(R^d)$ have Lebesgue densities, and let $p \in C_c^\infty(R^d)$. Define $\rho^p(t)$ as in (9.83), that is, ρ^p is a weak solution of

$$\dot{\rho}^p + \nabla \cdot (\rho^p \nabla p) = 0, \quad \rho^p(0) = \rho.$$

By Theorem 8.13 of [**127**],

$$\frac{d}{dt} d^2(\rho^p(t), \gamma)\Big|_{t=0} = \int_{R^d} 2\nabla p_{\rho,\gamma}(x) \cdot \nabla p(x) \rho(dx),$$

and (D.31) follows by Definition 9.36.

Similarly, (D.32) follows from (37) and (38) of [**63**] and the following computation.

Let $\rho \in \mathcal{P}_2(R^d), p \in C_c^\infty(R^d)$ and $\mathcal{E}(\rho) < \infty$. Then, with reference to (9.83),

$$\int_{R^d \times R^d} \Phi(x-y)\rho^p(t,dx)\rho^p(t,dy) = \int_{R^d \times R^d} \Phi(z_p(t,x) - z_p(t,y))\rho_0(dx)\rho_0(dy),$$

and

$$\begin{aligned} &\frac{d}{dt} \int_{R^d \times R^d} \Phi(x-y)\rho(t,dx)\rho(t,dy)\Big|_{t=0} \\ &= \int_{R^d \times R^d} \nabla\Phi(x-y) \frac{d}{dt}(z_p(t,x) - z_p(t,y))\Big|_{t=0} \rho_0(dx)\rho_0(dy) \\ &= 2\int \nabla(\rho_0 * \Phi)\nabla p d\rho_0. \end{aligned}$$

□

D.5. Weighted Sobolev spaces $H^1_\mu(R^d)$ and $H^{-1}_\mu(R^d)$

For $\mu \in \mathcal{P}(R^d)$, define

$$\|p\|^2_{1,\mu} \equiv \int_{R^d} |\nabla p|^2 d\mu, \quad \forall p \in C_c^\infty(R^d),$$

and let

$$H^1_\mu(R^d) \equiv \text{the completion of } C_c^\infty(R^d) \text{ under } \|\cdot\|_{1,\mu}.$$

For $u \in \mathcal{D}'(R^d)$, define

$$\|u\|^2_{-1,\mu} \equiv \sup_{p \in C_c^\infty(R^d)} (2\langle u,p\rangle - \|p\|^2_{1,\mu}) = \sup_{p \in C_c^\infty(R^d)} \frac{\langle u,p\rangle^2}{\|p\|^2_{1,\mu}}, \tag{D.33}$$

and let

$$H^{-1}_\mu(R^d) \equiv \{\text{equivalence classes of } u \in \mathcal{D}'(R^d) : \|u\|^2_{-1,\mu} < \infty\}.$$

Noting that

$$|\langle u,p\rangle| \le \|u\|_{-1,\mu}\|p\|_{1,\mu},$$

we see that $H^{-1}_\mu(R^d)$ gives a representation of the dual for $H^1_\mu(R^d)$.

Let $L^2_\mu(R^d)$ be the class of R^d-valued functions which are componentwise in $L^2(\mu)$, and denote

$$\|\xi\|^2_{0,\mu} \equiv \int_{R^d} |\xi|^2 d\mu, \quad \xi \in L^2_\mu(R^d).$$

Direct verification shows that $(H^1_\mu(R^d), \|\cdot\|_{1,\mu})$, $(L^2_\mu(R^d), \|\cdot\|_{0,\mu})$ and $(H^{-1}_\mu(R^d), \|\cdot\|_{-1,\mu})$ are Hilbert spaces, and their inner products can be represented using the polarization identity:

$$\langle u,v\rangle_{k,\mu} = \frac{1}{4}(\|u+v\|^2_{k,\mu} - \|u-v\|^2_{k,\mu}). \tag{D.34}$$

Geometrically, these spaces can be viewed as equivalent model spaces for the tangent bundle to $\mathcal{P}_2(R^d)$. Below, we examine the relationships among the three spaces.

Let $p, q \in H_\mu^1(R^d)$. For each q, $l_q(p) \equiv \langle p, q\rangle_{1,\mu}$ is a linear functional in p satisfying

$$|\langle p, q\rangle_{1,\mu}| \le \|q\|_{1,\mu}\|p\|_{1,\mu} = \|q\|_{1,\mu}\|\nabla p\|_{L_\mu^2(R^d)}, \quad \forall p \in C_c^\infty(R^d). \tag{D.35}$$

We can use this functional to extend the notion of ∇q from $q \in C_c^\infty(R^d)$ to all $q \in H_\mu^1(R^d)$. We denote the extension by $\widehat{\nabla} q$. In general, the extension may depend strongly on μ, and we need to clarify its relationship with the distributional derivative ∇q, if ∇q exists.

Let

$$L = \{\nabla p : p \in C_c^\infty(R^d)\} \subset (L_\mu^2(R^d), \|\cdot\|_{0,\mu}),$$

and let $L^2_{\mu,\nabla}(R^d)$ denote the closure of L in $L_\mu^2(R^d)$. Since $p \leftrightarrow \nabla p$ is a one-to-one correspondence between $C_c^\infty(R^d)$ and L, and $\|p\|_{1,\mu} = \|\nabla p\|_{0,\mu}$, we see that $H_\mu^1(R^d)$ is isomorphic to $L^2_{\mu,\nabla}(R^d)$. By the Riesz representation theorem, there exists a unique $\xi_q \in L^2_{\mu,\nabla}(R^d)$ such that

$$l_q(p) \equiv \langle p, q\rangle_{1,\mu} = \int_{R^d} \xi_q \cdot \nabla p d\mu, \quad \forall p \in C_c^\infty(R^d). \tag{D.36}$$

We define $\widehat{\nabla} q = \xi_q$. Note that $\|q\|_{1,\mu}^2 = \|\widehat{\nabla} q\|_{0,\mu}^2$ and more generally that

$$\langle p, q\rangle_{1,\mu} = \int_{R^d} \widehat{\nabla} p \cdot \widehat{\nabla} q d\mu \equiv \langle \widehat{\nabla} p, \widehat{\nabla} q\rangle_{0,\mu}.$$

LEMMA D.29. *Suppose $\xi \in L_\mu^2(R^d)$. Then $\xi \in L^2_{\mu,\nabla}(R^d)$ if and only if*

$$\|\xi + \eta\|_{L_\mu^2(R^d)} \ge \|\xi\|_{L_\mu^2(R^d)}, \quad \forall \eta \in L_\mu^2(R^d) \ni \nabla \cdot (\mu\eta) = 0,$$

or equivalently

$$\langle \xi, \eta\rangle_{0,\mu} = \int_{R^d} \xi \cdot \eta d\mu = 0, \quad \forall \eta \in L_\mu^2(R^d) \ni \nabla \cdot (\mu\eta) = 0. \tag{D.37}$$

PROOF. If $\xi \in L^2_{\mu,\nabla}(R^d)$, then by definition, there exist $p_n \in C_c^\infty(R^d)$ such that $\|\xi - \nabla p_n\|_{L_\mu^2(R^d)} \to 0$. Suppose $\nabla \cdot (\mu\eta) = 0$ Then $\langle \nabla p_n, \eta\rangle_{0,\mu} = 0$, and hence, $\langle \xi, \eta\rangle_{0,\mu} = 0$.

Conversely, for $\eta \in L_\mu^2(R^d)$, $\nabla \cdot (\mu\eta) = 0$ implies

$$\eta \in \left(\overline{\{\nabla p : p \in C_c^\infty(R^d)\}}_{L_\mu^2(R^d)}\right)^\perp = \left(L^2_{\mu,\nabla}(R^d)\right)^\perp.$$

Therefore (D.37) implies $\xi \in \left((L^2_{\mu,\nabla}(R^d))^\perp\right)^\perp = L^2_{\mu,\nabla}(R^d)$. □

Let $\xi \in L_\mu^2(R^d)$, and write

$$\xi = \xi_1 + \xi_2, \tag{D.38}$$

where ξ_1 is the projection of ξ on $L^2_{\mu,\nabla}(R^d)$ and ξ_2 is the projection of ξ on $\left(L^2_{\mu,\nabla}(R^d)\right)^\perp$. Note that $-\nabla \cdot (\mu\xi)$ exists as a Schwartz distribution, and

$$\langle -\nabla \cdot (\mu\xi), p\rangle = \langle \xi_1, \nabla p\rangle_{0,\mu}$$

for each $p \in C_c^\infty(R^d)$. Since

$$\begin{aligned} \|-\nabla\cdot(\mu\xi)\|_{-1,\mu}^2 &= \sup_{p\in C_c^\infty(R^d)} (2\int_{R^d}\nabla p\cdot\xi_1 d\mu - \int_{R^d}|\nabla p|^2 d\mu) \\ &= \|\xi_1\|_{0,\mu}^2 \le \|\xi\|_{0,\mu}^2, \end{aligned} \tag{D.39}$$

$-\nabla\cdot(\mu\xi) \in H_\mu^{-1}(R^d)$.

Let $\mu \in \mathcal{P}(R^d)$. If $q \in C_c^\infty(R^d) \subset H_\mu^1(R^d)$, then the classical gradient $\nabla q \in L^2_{\mu,\nabla}(R^d)$. By the defining relation (D.36), $\widehat{\nabla} q = \nabla q$. More generally, we have the following.

LEMMA D.30. *Suppose $q \in C^1(R^d)$ and $\int_{R^d}|\nabla q|^2 d\mu < \infty$. Then $q \in H_\mu^1(R^d)$ and $\widehat{\nabla} q = \nabla q$.*

PROOF. Let $\zeta \in C_c^\infty(R^d)$ satisfy $0 \le \zeta \le 1$ and $\zeta(x) = 1$ for $|x| \le 1$. Define $\zeta_n(x) = \zeta(n^{-1}x)$. Let $\beta \in C^\infty(R)$ be nondecreasing and satisfy $\beta(r) = r$, $|r| \le 1$, and $|\beta(r)| = 2$, $|r| \ge 3$. Define

$$\beta_m(r) = m\beta(m^{-1}r). \tag{D.40}$$

Let J_ϵ be the mollifier defined in (D.3). Define

$$q_{n,m,\epsilon} = \zeta_n J_\epsilon * (\beta_m \circ q) \in C_c^\infty(R^d). \tag{D.41}$$

Then

$$\nabla q_{n,m,\epsilon} = n^{-1}J_\epsilon * (\beta_m\circ q)\nabla\zeta(n^{-1}\cdot) + \zeta_n J_\epsilon * (\beta_m'\circ q\nabla q)$$

and

$$\begin{aligned} \int_{R^d}|\nabla q_{n,m,\epsilon} - \nabla q|^2 d\mu &\le 3n^{-2}\int_{R^d}|J_\epsilon * (\beta_m\circ q)\nabla\zeta(n^{-1}\cdot)|^2 d\mu \\ &\quad + 3\int_{R^d}|\zeta_n - 1|^2|\nabla q|^2 d\mu \\ &\quad + 3\int_{R^d}|\zeta_n|^2|\nabla q - J_\epsilon * (\beta_m'\circ q\nabla q)|^2 d\mu. \end{aligned}$$

First let $\epsilon \to 0$ eliminating the mollification, then $n \to \infty$ taking the first two terms to zero and the last term to

$$3\int_{R^d}|1-\beta_m'\circ q|^2|\nabla q|^2 d\mu.$$

Finally, this term goes to zero as $m \to \infty$, and the lemma follows. □

LEMMA D.31. *Let $\mu \in \mathcal{P}(R^d)$ have a Lebesgue density. Let $q \in L^1_{loc}(R^d)$, and assume that the gradient of q in the distributional sense satisfies $\nabla q \in L^1_{loc}(R^d)$ and $\int_{R^d}|\nabla q|^2 d\mu < \infty$. Then $q \in H_\mu^1(R^d)$ and $\widehat{\nabla} q = \nabla q$. In particular,*

$$\|q\|_{1,\mu}^2 = \int_{R^d}|\nabla q|^2 d\mu.$$

PROOF. Note that by (D.5), the distributional gradient ∇q satisfies $\nabla J_\epsilon * q = J_\epsilon * \nabla q$ and the claim follows by the same approximation argument as in Lemma D.30. □

LEMMA D.32. *Let β_m be as defined in (D.40), and let $q : R^d \mapsto R$ and $v : R^d \mapsto R^d$ be measurable. Suppose $\nabla\beta_m(q) = \beta_m'(q)v$ for each $m = 1, 2, \ldots$. If $\int_{R^d}|v|^2 d\mu < \infty$, then $q \in H_\mu^1(R^d)$ with $\|q\|_{1,\mu}^2 = \int_{R^d}|v|^2 d\mu$.*

PROOF. Let $q_{n,m,\epsilon}$ be as in (D.41). Then

$$\nabla q_{n,m,\epsilon} = n^{-1}J_\epsilon * (\beta_m \circ q)\nabla\zeta(n^{-1}\cdot) + \zeta_n J_\epsilon * (\beta_m' \circ qv),$$

and the lemma follows as in the proof of Lemma D.30 □

LEMMA D.33. *Let $\mu \in \mathcal{P}(R^d)$ have a Lebesgue density. Let φ be a lower semi-continuous, convex function with $\mu(\text{int}(\text{Dom}(\varphi))) = 1$. Assume that $\int_{R^d} |\nabla\varphi|^2 d\mu < \infty$. Then $\varphi \in H^1_\mu(R^d)$ and $\widehat{\nabla}\varphi = \nabla\varphi$.*

PROOF. Let φ_r be defined as in Lemma D.12. Then $\varphi_r \in L^\infty_{loc}(R^d)$, and since $|\nabla\varphi_r(x)| \leq |\nabla\varphi(x)|$, φ_r satisfies the conditions of Lemma D.31. Consequently, $\varphi_r \in H^1_\mu(R^d)$, and since $|\nabla\varphi(x)| < r$ implies $\nabla\varphi(x) = \nabla\varphi_r(x)$, the dominated convergence theorem gives

$$\lim_{r\to\infty} \int_{R^d} |\nabla\varphi - \nabla\varphi_r|^2 d\mu = 0,$$

giving the desired result. □

Assuming $\rho, \gamma \in \mathcal{P}_2(R^d)$ and $\rho(dx) = \rho(x)dx$, let $p_{\rho,\gamma}$ be as in Theorem D.25. Since $p_{\rho,\gamma}$ is the difference of two convex functions, by Lemma D.33, $p_{\rho,\gamma} \in H^1_\rho(R^d)$, $\widehat{\nabla} p_{\rho,\gamma} = \nabla p_{\rho,\gamma} \in L^2_{\rho,\nabla}(R^d)$, and

$$\|p_{\rho,\gamma}\|^2_{1,\rho} = \|\nabla p_{\rho,\gamma}\|^2_{0,\rho} = \int_{R^d} |\nabla p_{\rho,\gamma}|^2 d\rho = d^2(\rho,\gamma). \tag{D.42}$$

Since $H^{-1}_\mu(R^d)$ gives the dual for $H^1_\mu(R^d)$ and $H^1_\mu(R^d)$ is a Hilbert space, there must be a natural isomorphism between the spaces.

LEMMA D.34. (1) *For each $u \in H^{-1}_\mu(R^d)$, there exists a unique $p_u \in H^1_\mu(R^d)$ such that*

$$u = -\nabla \cdot (\mu\widehat{\nabla} p_u)$$

in the sense that

$$\int_{R^d} \nabla p \cdot \widehat{\nabla} p_u d\mu = \langle u, p\rangle, \quad \forall p \in C^\infty_c(R^d). \tag{D.43}$$

(2) *Let $p, q \in H^1_\mu(R^d)$, and define*

$$u = -\nabla \cdot (\mu\widehat{\nabla} p), \quad v = -\nabla \cdot (\mu\widehat{\nabla} q). \tag{D.44}$$

Then

$$\langle u, v\rangle_{-1,\mu} = \langle p, q\rangle_{1,\mu} = \int_{R^d} \widehat{\nabla} p \cdot \widehat{\nabla} q d\mu = \langle u, q\rangle = \langle v, p\rangle. \tag{D.45}$$

PROOF. For $u \in H^{-1}_\mu(R^d)$, $\langle u, p\rangle$, $p \in C^\infty_c(R^d)$, extends to a unique continuous linear functional on $H^1_\mu(R^d)$. Consequently, there must exist $p_u \in H^1_\mu(R^d)$ such that

$$\langle u, p\rangle = \langle p_u, p\rangle_{1,\mu} = \int_{R^d} \nabla p \cdot \widehat{\nabla} p_u d\mu, \quad p \in C^\infty_c(R^d).$$

It follows that

$$\|u\|^2_{-1,\mu} = \|p_u\|^2_{1,\mu} = \int_{R^d} |\widehat{\nabla} p_u|^2 d\mu,$$

and (D.34) then implies (D.45). □

By Theorem D.28, if $\rho, \gamma \in \mathcal{P}_2(R^d)$ have Lebesgue densities, the gradient (in the sense of Definition 9.36) satisfies

$$\operatorname{grad}_\rho d^2(\rho, \gamma) = -2\nabla \cdot (\rho \nabla p_{\rho,\gamma}).$$

By Lemma D.33 and (D.45),

$$\|\operatorname{grad}_\rho d^2(\rho, \gamma)\|_{-1,\rho}^2 = 4 \int_{R^d} |\nabla p_{\rho,\gamma}|^2 d\rho = 4d^2(\rho, \gamma). \tag{D.46}$$

This identity is the mass transport version of the Euclidean space identity

$$|\nabla_x |x - y|^2|^2 = 4|x - y|^2, \quad x, y \in R^d.$$

We use it in a critical way to prove the comparison principle for the nonlinear equation in Example 9.35.

D.6. Fisher information and its properties

Let μ_∞ be defined as in (9.92):

$$\mu_\infty(dx) = Z^{-1} e^{-2\Psi(x)} dx.$$

The *Fisher information* for $\rho \in \mathcal{P}(R^d)$ is

$$I(\rho) \equiv \begin{cases} \int_{R^d} \frac{|\nabla \frac{d\rho}{d\mu_\infty}|^2}{\frac{d\rho}{d\mu_\infty}} d\mu_\infty & \text{when } \rho(dx) = \rho(x)dx, \nabla \frac{d\rho}{d\mu_\infty} \in L^1_{loc}(R^d), \\ \infty & \text{otherwise.} \end{cases}$$

In the above definition, we use the convention $0/0 = 0$ and $a/0 = \infty$ for $a \in (0, \infty]$. The requirement that $\nabla \frac{d\rho}{d\mu_\infty} \in L^1_{loc}(R^d)$ only ensures that the distributional gradient is given by a function. The integral defining $I(\rho)$ may still be infinite. Note that if $\rho(dx) = \rho(x)dx$, $d\rho/d\mu_\infty(x) = Z\rho(x)e^{2\Psi(x)}$. Consequently, by (D.8),

$$\nabla(\rho e^{2\Psi}) = e^{2\Psi}\nabla\rho + 2e^{2\Psi}\rho\nabla\Psi \in \mathcal{D}'(R^d), \tag{D.47}$$

and the requirement that $\nabla(\rho e^{2\Psi}) \in L^1_{loc}(R^d)$ is equivalent to $\nabla\rho \in L^1_{loc}(R^d)$. Therefore, we also have

$$I(\rho) \equiv \begin{cases} \int_{R^d} \frac{|\nabla\rho(x) + 2\rho(x)\nabla\Psi(x)|^2}{\rho(x)} dx & \text{when } \rho(dx) = \rho(x)dx, \nabla\rho \in L^1_{loc}(R^d) \\ \infty & \text{otherwise.} \end{cases}$$

Let

$$I_1(\rho) \equiv \begin{cases} 4\int_{R^d} |\nabla\sqrt{\frac{d\rho}{d\mu_\infty}}|^2 d\mu_\infty & \text{when } \rho(dx) = \rho(x)dx \text{ and } \nabla\sqrt{\frac{d\rho}{d\mu_\infty}} \in L^1_{loc}(R^d) \\ \infty & \text{otherwise.} \end{cases}$$

Note that by Lemma D.7, $I_1(\rho) = I(\rho)$. As in (D.47),

$$\nabla(\sqrt{\rho(x)}e^{\Psi(x)}) = e^{\Psi(x)}(\nabla\sqrt{\rho(x)} + \sqrt{\rho(x)}\nabla\Psi(x)),$$

so $\nabla(\sqrt{\rho}e^{\Psi}) \in L^1_{loc}(R^d)$ is equivalent to $\nabla\sqrt{\rho} \in L^1_{loc}(R^d)$, and

$$I_1(\rho) \equiv \begin{cases} 4\int_{R^d} |\nabla\sqrt{\rho(x)} + \sqrt{\rho(x)}\nabla\Psi(x)|^2 dx & \text{when } \rho(dx) = \rho(x)dx, \nabla\sqrt{\rho} \in L^1_{loc}(R^d) \\ \infty & \text{otherwise.} \end{cases}$$

Fisher information as defined here is related to the functional I_B defined in Section 11.2 and Appendix B. Let Y satisfy

$$dY(t) = -2\nabla\Psi(Y(t))dt + \sqrt{2}dW(t),$$

where W is standard Brownian motion in R^d. Then μ_∞ is the stationary distribution for Y. Let B be the weak infinitesimal generator (in the sense of (11.31)) of the process Y, and let B_0 be the restriction of B to D_d, that is,

$$B_0\varphi(x) = \Delta\varphi(x) - 2\nabla\Psi(x)\nabla\varphi(x), \quad \varphi \in D_d.$$

With reference to Appendix B, define

$$B^{++} = \{(\varphi, B\varphi) : \varphi \in \mathcal{D}(B), \inf_y \varphi(y) > 0\}$$

and

$$I_B(\rho) = -\inf_{\varphi\in\mathcal{D}^{++}(B)} \int \frac{B\varphi}{\varphi} d\rho = -\inf_{(\varphi,\psi)\in B^{++}} \int \frac{\psi}{\varphi} d\rho, \tag{D.48}$$

and similarly,

$$B_0^{++} = \{(\varphi, B_0\varphi) : \varphi \in D_d, \inf_y \varphi(y) > 0\}$$

and

$$I_{B_0}(\rho) = -\inf_{\varphi\in D_d, \inf_y \varphi(y)>0} \int \frac{B\varphi}{\varphi} d\rho = -\inf_{(\varphi,\psi)\in B_0^{++}} \int \frac{\psi}{\varphi} d\rho. \tag{D.49}$$

It follows that I_B and I_{B_0} are convex and lower semicontinuous in the topology of weak convergence, and $I_{B_0} \le I_B$. In fact, we claim

$$I_B = I_{B_0}. \tag{D.50}$$

For $C \subset B(E) \times B(E)$ such that $(f, g) \in C$ implies $\inf_{y\in E} f(y) > 0$, define the *strictly positive closure* of C, spcl(C), to be the smallest subset $C' \subset B(E) \times B(E)$ such that $\{(f_n, g_n)\} \subset C$, $\inf_{n,y} f_n(y) > 0$, $\sup_{n,y} g_n(y) < \infty$, $f(x) = \lim f_n(x)$, $g(x) = \lim g_n(x)$, $x \in E$, implies $(f, g) \in C'$. A simple application of Fatou's lemma implies that if $(f, g) \in \text{spcl}(B_0^{++})$, then

$$I_{B_0}(\rho) \ge \int \frac{g}{f} d\rho.$$

Consequently, to prove (D.50), it is enough to prove that $B^{++} \subset \text{spcl}(B_0^{++})$.

Let $\{T(t) : t \ge 0\}$ be the semigroup on $C_b(R^d)$ generated by B, and let

$$Lf(x) = \Delta f - 2\nabla\Psi\nabla f, \quad f \in C^2(R^d).$$

B_0 is the restriction of L to D_d. We will also need the adjoint operator

$$L^*\varphi = \Delta\varphi + 2\nabla \cdot (\varphi\nabla\Psi), \quad \varphi \in C^2(R^d).$$

LEMMA D.35. *Let $h \in C_b(R^d)$, and define $u(t) = T(t)h$, $t \ge 0$. Then u is a weak (Schwartz distributional) solution of $\partial_t u = Lu$ in the sense that for each $0 \le s < t$,*

$$\begin{aligned} \int_{R^d} u(t,x)\varphi(t,x)dx &= \int_{R^d} u(s,x)\varphi(s,x)dx \\ &\quad + \int_{(s,t]\times R^d} u(r,x)(\partial_r\varphi(r,x) + L^*\varphi(r,x))dxdr, \end{aligned} \tag{D.51}$$

for all $\varphi \in C_c^\infty([s,t] \times R^d)$, and for $\lambda > 0$, $f = \int_0^\infty e^{-\lambda t}T(t)hdt$ is a weak solution of $\lambda f - Lf = h$ in the sense that

$$\int_{R^d} f(\lambda\varphi - L^*\varphi)dx = \int_{R^d} h\varphi dx, \quad \varphi \in C_c^\infty(R^d). \tag{D.52}$$

PROOF. Let $\Psi_k = J_{k^{-1}} * (\Psi \wedge k)$. Then Ψ_k is bounded and C^∞ and all its derivatives have compact support. Let

$$Y_k(t,x) = x + \sqrt{2}W(t) - \int_0^t 2\nabla\Psi_k(Y_k(s,x))ds, \tag{D.53}$$

and define $u_k(t,x) = E[f(Y_k(t,x))]$. The differentiability of Y_k as a function of x (see, for example, Theorem 40, page 254 of Protter [**96**]) and simple moment estimates for the derivatives imply that u_k is infinitely differentiable in x and satisfies

$$u_k(t,x) = f(x) + \int_0^t L_k u_k(r,x)dr,$$

where $L_k f(x) = \Delta f - 2\nabla\Psi_k\nabla f$. Integration by parts then gives

$$\begin{aligned}\int_{R^d} u_k(t,x)\varphi(t,x)dx &= \int_{R^d} u_k(s,x)\varphi(s,x)dx \\ &\quad + \int_{(s,t]\times R^d} u_k(r,x)(\partial_r\varphi(r,x) + L_k^*\varphi(r,x))dxdr.\end{aligned}$$

As $k\to\infty$, $Y_k(\cdot,x) \Rightarrow Y(\cdot,x)$, so $u_k(t,x)\to u(t,x)$ boundedly and pointwise. Since $\nabla\Psi_k$ converges to $\nabla\Psi$ uniformly on compact sets and φ has compact support, (D.51) follows.

To see that (D.52) holds, let $\langle\varphi,g\rangle = \int_{R^d}\varphi g dx$, and observe that

$$\begin{aligned}\langle\lambda\varphi - L^*\varphi, f\rangle &= \langle\lambda\varphi, f\rangle - \int_0^\infty e^{-\lambda t}\langle L^*\varphi, u(t)\rangle dt \\ &= \langle\lambda\varphi, f\rangle - \int_0^\infty \lambda e^{-\lambda t}\int_0^t \langle L^*\varphi, u(r)\rangle drdt \\ &= \langle\lambda\varphi, f\rangle - \int_0^\infty \lambda e^{-\lambda t}(\langle\varphi, u(t)\rangle - \langle\varphi, h\rangle)dt \\ &= \langle\varphi, h\rangle.\end{aligned}$$

□

Under our basic assumptions on Ψ, we have the following regularity result for $T(t)h$ and f in Lemma D.35.

LEMMA D.36. *Let $\Psi\in C^2(R^d)$, and suppose that $h\in C_b(R^d)\cap C^1(R^d)$ and $\nabla h\in C_b(R^d)$. Then $\nabla T(t)h\in C_b(R^d)$, for $t>0$, and for $\lambda>|\lambda_\Psi|$,*

$$\nabla\int_0^\infty e^{-\lambda t}T(t)hdt = \int_0^\infty e^{-\lambda t}\nabla T(t)hdt \in C_b(R^d). \tag{D.54}$$

PROOF. Let $Y(t,x)$

$$Y(t,x) = x + \sqrt{2}W(t) - \int_0^t 2\nabla\Psi(Y(s,x))ds. \tag{D.55}$$

Then, as with Y_k, Y is continuously differentiable in x with

$$D_k^i(t,x) \equiv \frac{\partial}{\partial x_k}Y_i(t,x)$$

satisfying

$$D_k^i(t,x) = \delta_k^i - \sum_{j=1}^d \int_0^t \partial_{ij}^2\Psi(Y(s,x))D_k^j(s,x)ds.$$

Let $D_k = (D_k^1, \ldots, D_k^d)$. Then by (D.1),

$$|D_k|^2(t) \le 1 - 2\lambda_\Psi \int_0^t |D_k|^2(s)ds.$$

Therefore for each $T > 0$,

$$\sup_{0\le t\le T} |D_k(t)|^2 \le e^{2|\lambda_\Psi|T}.$$

Hence

$$\frac{\partial}{\partial x_k} T(t)h(x) = E[\nabla h(Y(t,x)) \cdot D_k(t,x)] \in C_b(R^d).$$

Noting that $\|\nabla T(t)h\|_\infty \le \|\nabla h\|_\infty e^{|\lambda_\Psi|}$, we see that (D.54) holds for $\lambda > |\lambda_\Psi|$. □

There are a variety of notions of weak solution employed in the literature on partial differential equations. For elliptic equations like $\lambda f - Lf = h$, Gilbarg and Trudinger [**54**], Chapter 8, require

$$\text{(D.56)} \qquad \int_{R^d} (\nabla\varphi \cdot \nabla f + 2\varphi\nabla\Psi \cdot \nabla f + \lambda\varphi f)dx = \int_{R^d} \varphi h dx, \quad \varphi \in C_c^1(R^d).$$

Under the conditions of Lemma D.36, (D.52) implies (D.56) by first integrating by parts and then extending to $C_c^1(R^d)$ by approximating $\varphi \in C_c^1(R^d)$ by $J_\epsilon * \varphi$.

LEMMA D.37. *Suppose $\Psi \in C^\infty(R^d)$, $\lambda > |\lambda_\Psi|$, $h \in C_b(R^d) \cap C^\infty(R^d)$, and $\nabla h \in C_b(R^d)$. Then $f = \int_0^\infty e^{-\lambda t}T(t)hdt \in C^\infty(R^d)$.*

PROOF. For any bounded open subset $\mathcal{O} \subset R^d$, the restriction of f to $\mathcal{O}$ is in $W^{1,2}(\mathcal{O}) \equiv H^1(\mathcal{O})$ and is a weak solution of $\lambda f - Lf = h$ in $\mathcal{O}$. Consequently, the lemma follows by Corollary 8.11 of Gilbarg and Trudinger [**54**]. □

LEMMA D.38. *Let $\lambda > 0$ and $f_n \in \mathcal{D}(B)$, $n = 1, 2, \ldots$, and define*

$$h_n = \lambda f_n - Bf_n,$$

Suppose that $h_0 = buc\text{-}\lim_{n\to\infty} h_n$. Then

$$buc\text{-}\lim_{n\to\infty} f_n = f_0 \equiv \int_0^\infty e^{-\lambda t}T(t)h_0.$$

PROOF. Since

$$f_n(x) = \int_0^\infty e^{-\lambda t}E[h_n(Y(t,x))]dt,$$

$\sup_n \|f_n\|_\infty \le \sup_n \lambda^{-1}\|h_n\|_\infty$. If $x_n \to x$, then the continuity of $x \mapsto Y(t,x)$ and the bounded convergence theorem implies

$$f_n(x_n) \to \int_0^\infty e^{-\lambda t}E[h_0(Y(t,x))]dt,$$

which implies uniform convergence on compact sets. □

LEMMA D.39. *Suppose $\Psi \in C^\infty(R^d)$. For each $f \in \mathcal{D}(B)$ with $\inf_y f(y) > 0$, there exist $f_n \in C^\infty(R^d) \cap \mathcal{D}(B)$ with $\nabla f_n \in C_b(R^d)$ such that*

$$f = buc\text{-}\lim_{n\to\infty} f_n, \quad Bf = buc\text{-}\lim_{n\to\infty} Lf_n,$$

and $\inf_n \inf_y f_n(y) > 0$.

PROOF. For $f \in \mathcal{D}(B)$ with $a \equiv \inf_y f(y) > 0$, let $\lambda > |\lambda_\Psi| \vee (\|Bf\|_\infty / a)$ and define

$$h = \lambda f - Bf \in C_b(R^d).$$

Let $\zeta \in C_c^\infty(R^d)$ be as in (D.11), and define $\zeta_n(x) = \zeta(nx)$. Let $h_n = \zeta_n J_{n^{-1}} * h + (1 - \zeta_n) \inf_y h(y)$. Then $\inf_y h_n(y) \geq \inf_y h(y) > 0$ and

$$h = buc\text{-} \lim_{n\to\infty} h_n. \tag{D.57}$$

Then by Lemma D.37, $f_n = \int_0^\infty e^{-\lambda t} T(t) h_n dt \in C^\infty(R^d) \cap \mathcal{D}(B)$. Applying Lemma D.38,

$$f = buc\text{-} \lim_{n\to\infty} f_n, \quad Bf = \lambda f - h = buc\text{-} \lim_{n\to\infty} (\lambda f_n - h_n) = buc\text{-} \lim_{n\to\infty} Lf_n.$$

Finally, note that

$$f_n \geq \frac{\inf_y h_n(y)}{\lambda} \geq \frac{\inf_y h(y)}{\lambda} \geq \frac{\lambda a - \|Bf\|_\infty}{\lambda} > 0.$$

□

The following lemma is an immediate consequence of Lemma D.39.

LEMMA D.40. *Suppose $\Psi \in C^\infty(R^d)$. Then*

$$I_B(\rho) = -\inf\{\int_{R^d} \frac{Lf}{f} d\rho : f \in C^\infty(R^d) \cap \mathcal{D}(B), \nabla f \in C_b(R^d), \inf_y f(y) > 0\}. \tag{D.58}$$

LEMMA D.41. *Suppose $\Psi \in C^\infty(R^d)$. Then*

$$I_B = I_{B_0}.$$

PROOF. Let $f \in C^\infty(R^d) \cap \mathcal{D}(B)$, $\nabla f \in C_b(R^d)$, and $a \equiv \inf_y f(y) > 0$. Define

$$f_n(y) = \zeta_n(y) f(y) + (1 - \zeta_n(y)) \sup_z f(z) \in D_d.$$

It follows that $\inf_y f_n(y) \geq \inf_y f(y) > 0$. Since

$$\nabla \zeta_n(x) = n^{-1} \zeta'(n^{-1}|x|) \frac{x}{|x|}, \quad \Delta \zeta_n(x) = n^{-2} \zeta''(n^{-1}|x|) + n^{-1} \zeta'(n^{-1}|x|) \frac{d-1}{|x|},$$

$$\begin{aligned} Lf_n(x) &= \zeta_n(x) Lf(x) + f(x) L\zeta_n(x) + \nabla f(x) \cdot \nabla \zeta_n(x) - \sup_z f(z) L\zeta_n(x) \\ &= \zeta_n(x) Bf(x) + (f(x) - \sup_z f(z)) \Big(n^{-2} \zeta''(n^{-1}|x|) + n^{-1} \zeta'(n^{-1}|x|) \frac{d-1}{|x|} \\ &\quad -2n^{-1} \zeta'(n^{-1}|x|) \frac{x}{|x|} \cdot \nabla \Psi(x) \Big) + n^{-1} \zeta'(n^{-1}|x|) \frac{\nabla f(x) \cdot x}{|x|}. \end{aligned}$$

Since

$$\lim_{|x|\to\infty} \frac{x \cdot \nabla \Psi(x)}{|x|^2} = \infty$$

and $\zeta' \leq 0$, for n sufficiently large,

$$(f - \sup_z f(z)) (-2n^{-1} \zeta'(n^{-1}|x|) \frac{x}{|x|} \cdot \nabla \Psi) \leq 0.$$

Therefore,

$$\limsup_{n\to\infty} \sup_{y \in R^d} Lf_n(y) \leq \|Bf\|_\infty, \tag{D.59}$$

and for each compact K, $\lim_{n\to\infty}\sup_{y\in K}|Lf_n(y)-Bf(y)|=0$. Consequently, $B^{++}\subset \mathrm{spcl}(B_0^{++})$ and the lemma follows. □

We now extend the result to $\Psi\in C^2$.

LEMMA D.42. *Let $\Psi\in C^1(R^d)$. Then for each $\epsilon>0$, there exists $\Psi_\epsilon\in C^\infty(R^d)$, such that*

$$\sup_{x\in R^d}|\nabla\Psi(x)-\nabla\Psi_\epsilon(x)|<\epsilon.$$

PROOF. Let $\{\psi_n\in C^\infty(R^d):n=0,1,\ldots\}$ be a smooth partition of unity, $1=\sum_{n=0}^\infty\psi_n(x)$, satisfying $0\le\psi_n\le 1$ and $\mathrm{supp}(\psi_n)\subset\{x\in R^d:n-1<|x|<n+1\}$. Then $\Psi(x)=\sum_{n=0}^\infty\psi_n(x)\Psi(x)$. Define

$$\Psi_\epsilon(x)=\sum_{n=0}^\infty J_{\epsilon_n}*(\psi_n\Psi)(x)\in C^\infty(R^d),$$

where $0<\epsilon_n<1$ is chosen so that

$$\int_{R^d}J_{\epsilon_n}(y)|\nabla(\psi_n\Psi)(x-y)-\nabla(\psi_n\Psi)(x)|dy<\frac{\epsilon}{4}. \tag{D.60}$$

Note that $\mathrm{supp}(J_{\epsilon_n}*(\psi_n\Psi))\subset(n-2,n+2)$, so for each $x\in R^d$, at most four of the terms in the sum are non zero. Consequently,

$$|\nabla\Psi_\epsilon(x)-\nabla\Psi(x)|\le\sum_n\Big|\int_{R^d}J_{\epsilon_n}(y)\Big(\nabla(\psi_n\Psi)(x-y)-\nabla(\psi_n\Psi)(x)\Big)dy\Big|<\epsilon.$$

□

LEMMA D.43. *For Ψ satisfying Condition 13.14, $I_B=I_{B_0}$.*

PROOF. We write $I_{B,\Psi}$ and $I_{B_0,\Psi}$ to emphasize the dependence on Ψ. Let Ψ_ϵ be as in Lemma D.42. Then

$$\begin{aligned}
&I_{B_0,\Psi}(\rho)\\
=&\sup_{\varphi\in C_c^\infty(R^d)}\Big(2\langle-\Delta\varphi+2\nabla\Psi\nabla\varphi,\rho\rangle-\int_{R^d}|\nabla\varphi|^2d\rho\Big)\\
\ge&\lim_{\epsilon\to0+}\sup_{\varphi\in C_c^\infty(R^d)}\Big(2\langle-\Delta\varphi+2\nabla\Psi_\epsilon\nabla\varphi,\rho\rangle-\int_{R^d}|\nabla\varphi|^2d\rho-4\epsilon\sqrt{\int_{R^d}|\nabla\varphi|^2d\rho}\Big)\\
\ge&\lim_{\epsilon\to0+}\sup_{\varphi\in C_c^\infty(R^d)}\Big(2\langle-\Delta\varphi+2\nabla\Psi_\epsilon\nabla\varphi,\rho\rangle-(1+\epsilon)\int_{R^d}|\nabla\varphi|^2d\rho-4\epsilon\Big)\\
=&\lim_{\epsilon\to0+}(1+\epsilon)^{-1}\sup_{\varphi_\epsilon=(1+\epsilon)^{-1}\varphi\in C_c^\infty(R^d)}\Big(2\langle-\Delta\varphi_\epsilon+2\nabla\Psi_n\nabla\varphi_\epsilon,\rho\rangle-\int_{R^d}|\nabla\varphi_\epsilon|^2d\rho\Big)\\
&\qquad-4\epsilon\\
=&\lim_{\epsilon\to0+}I_{B_0,\Psi_\epsilon}(\rho).
\end{aligned}$$

The proof of the opposite inequality is similar, and $\lim_{\epsilon\to0+}I_{B_0,\Psi_\epsilon}=I_{B_0,\Psi}$.

Since $I_{B_0,\Psi_\epsilon}=I_{B,\Psi_\epsilon}$, by Lemma D.41, we have $I_{1,\Psi_\epsilon}=4I_{B,\Psi_\epsilon}=4I_{B_0,\Psi_\epsilon}$. But by the definition of I_1,

$$(I_{1,\Psi_\epsilon}(\rho)-2\epsilon)^2\le I_{1,\Psi}(\rho)\le(I_{1,\Psi_\epsilon}(\rho)+2\epsilon)^2,$$

so $\lim_{\epsilon\to 0+} I_{1,\Psi_\epsilon}(\rho) = I_{1,\Psi}(\rho)$. Therefore

$$I_{B,\Psi}(\rho) = \lim_{\epsilon\to 0+} I_{B,\Psi_\epsilon}(\rho) = \lim_{\epsilon\to 0+} I_{B_0,\Psi_\epsilon}(\rho) = I_{B_0,\Psi}(\rho).$$

□

If we take $\varphi = e^{\psi}$ in the definition of I_{B_0}, we have

$$\begin{aligned} I_2(\rho) &\equiv 4I_{B_0}(\rho) \\ &= -4 \inf_{\psi\in C_c^\infty(R^d)} \langle B\psi + |\nabla\psi|^2, \rho\rangle \\ &= 4 \sup_{\psi\in C_c^\infty(R^d)} \langle -\Delta\psi + 2\nabla\Psi\cdot\nabla\psi - |\nabla\psi|^2, \rho\rangle \\ &= \sup_{p\in C_c^\infty(R^d)} \left(2\langle -\Delta p + 2\nabla\Psi\cdot\nabla p, \rho\rangle - \langle |\nabla p|^2, \rho\rangle\right) \\ &= \|\Delta\rho + 2\nabla(\rho\nabla\Psi)\|_{-1,\rho}^2, \end{aligned}$$

where $p = 2\psi$. The last identity follows from the definition of $\|\cdot\|_{-1,\rho}$ in (D.33). By definition, $I_2 \ge 0$ and I_2 is convex and lower semicontinuous in the topology of weak convergence. Since μ_∞ is the stationary distribution for B, $I_2(\mu_\infty) = 0$.

We also define

$$I_3(\rho) = 2 \sup_{\xi\in C_c^\infty(R^d)} \langle -\nabla\cdot\xi + 2\xi\cdot\nabla\Psi - \frac{1}{2}|\xi|^2, \rho\rangle,$$

where ξ is an R^d-valued function. Taking $\xi = \nabla p$, we see that $I_3 \ge I_2$. In fact, we will show equality for all four expressions.

LEMMA D.44. *For $\rho \in \mathcal{P}(R^d)$, $I(\rho) = I_1(\rho) = I_2(\rho) = I_3(\rho)$.*

PROOF. As already noted, $I(\rho) = I_1(\rho)$ by Lemma D.7. By Theorem 7.44 of Stroock [**115**],

$$I_1(\rho) = 4I_B(\rho) = 4I_{B_0}(\rho) = I_2(\rho), \tag{D.61}$$

and by Theorem 7.25 of [**115**], $I_1(\rho) < \infty$ implies $\rho << \mu_\infty$. Hence, ρ has a Lebesgue density.

If $\rho(dx) = \rho(x)dx$ and $\nabla\rho \in L^1_{loc}(R^d)$, then $I(\rho) = \|\frac{\nabla\rho + 2\rho\nabla\Psi}{\rho}\|_{0,\rho}^2$, and we have

$$\begin{aligned} I(\rho) &= \sup_{\xi\in C_c^\infty(R^d)} \left(2\int_{R^d} \xi(x)\cdot(\frac{\nabla\rho(x)}{\rho(x)} + 2\nabla\Psi(x))\rho(x)dx - \int_{R^d} |\xi(x)|^2\rho(x)dx\right) \\ &= \sup_{\xi\in C_c^\infty(R^d)} 2\langle -\nabla\cdot\xi + 2\xi\cdot\nabla\Psi - \frac{1}{2}|\xi|^2, \rho\rangle \\ &= I_3(\rho). \end{aligned}$$

Define $v = \frac{\nabla\rho + 2\rho\nabla\Psi}{\rho}$. If $I(\rho) = I_3(\rho) < \infty$, then $I_2(\rho)$ is the square of the norm of the projection of v onto $L^2_{\rho,\nabla}(R^d)$. Note that for β_m defined in (D.40), $\nabla\beta_m(\log(\rho e^{2\Psi})) = \beta_m'(\log(\rho e^{2\Psi}))v$. By Lemma D.32, $I_2(\rho) = \int_{R^d} |v|^2 d\rho = I(\rho)$.

To see that $I_3(\rho) < \infty$ implies $\nabla\rho \in L^1_{loc}(R^d)$, define the linear functional

$$l(\xi) \equiv \langle -\nabla\cdot\xi + 2\xi\cdot\nabla\Psi, \rho\rangle, \quad \xi \in C_c^\infty(R^d). \tag{D.62}$$

Then

$$2tl(\xi) - t^2\int_{R^d} |\xi|^2 d\rho \le I_3(\rho), \quad \xi \in C_c^\infty(R^d),\ t \ge 0,$$

which implies

$$|l(\xi)| \leq \sqrt{I_3(\rho)}\sqrt{\int_{R^d} |\xi|^2 d\rho}.$$

By the Riesz representation theorem, $l(\xi)$ can be extended to $L^2_\rho(R^d)$ and there exists an R^d-valued function $\sigma \in L^2_\rho(R^d)$, such that

$$l(\xi) = \int_{R^d} \xi \cdot \sigma d\rho. \tag{D.63}$$

Combining (D.62) with (D.63),

$$\nabla\rho + 2\rho\nabla\Psi = \rho\sigma \quad \text{in } \mathcal{D}'(R^d),$$

and noting that $|\rho\sigma| \leq \frac{1}{2}(\rho + \rho\sigma^2)$, $\nabla\rho \in L^1_{loc}(R^d)$. Consequently, $I_3(\rho) = I(\rho)$ for all $\rho \in \mathcal{P}(R^d)$ □

Similarly, for $\rho \in \mathcal{P}(R^d)$, define

$$I^\#(\rho) \equiv \begin{cases} \int_{R^d} \frac{|\nabla\rho(x)+\rho(x)\nabla(2\Psi+2\rho*\Phi)|^2}{\rho(x)} dx & \text{when } \rho(dx) = \rho(x)dx, \nabla\rho \in L^1_{loc}(R^d) \\ \infty & \text{otherwise.} \end{cases}$$

Let $\mathcal{E}$ be the free energy function defined by (9.94), and let the gradient for functions on $\mathcal{P}(R^d)$ be defined according to Definition 9.36. (The definition is not restricted to $\mathcal{P}_2(R^d)$.)

THEOREM D.45. *$I^\#$ is lower semicontinuous and satisfies*

$$I^\#(\rho) = \|\Delta\rho + \nabla(\rho\nabla(2\Psi + 2\rho * \Phi))\|^2_{-1,\rho}. \tag{D.64}$$

If $\mathcal{E}(\rho) < \infty$, then by (D.32),

$$I^\#(\rho) = 4\|\text{grad}\mathcal{E}(\rho)\|^2_{-1,\rho}. \tag{D.65}$$

In view of Theorem D.28, (D.65), and (D.46), we expect

$$\begin{aligned} &\langle -\text{grad}\mathcal{E}(\rho), \text{grad}_\rho \frac{1}{2} d^2(\rho, \gamma)\rangle_{-1,\rho} \\ &= -\int_{R^d} \nabla p_{\rho,\gamma}(x) \cdot (\frac{1}{2}\frac{\nabla\rho(x)}{\rho(x)} + \nabla\Psi + \nabla(\Phi * \rho))\rho(x)dx \end{aligned} \tag{D.66}$$

whenever $I^\#(\rho) < \infty$ and $\gamma \in \mathcal{P}_2(R^d)$.

LEMMA D.46. *If $I^\#(\rho) < \infty$, then (D.66) holds.*

PROOF. Let $\mathcal{O}$ be as in Theorem D.25, and define $p_{\rho,\gamma}(x) = -\infty$ for $x \notin \mathcal{O}$. Define $p_{n,m,\epsilon}$ as in (D.41) with q replaced by $p_{\rho,\gamma}$. Selecting n_k, m_k and ϵ_k appropriately, there exists a sequence $p_k \in C_c^\infty(R^d)$ such that $\lim_{n\to\infty} \|p_k - p_{\rho,\gamma}\|_{1,\rho} = 0$, or equivalently (by (D.45))

$$\lim_{n\to\infty} \| - \nabla(\rho\nabla p_k) - \text{grad}\frac{1}{2}d^2(\rho, \gamma)\|_{-1,\rho} = 0.$$

Since $-\nabla(\rho\nabla p_k) \in H^{-1}_\rho(R^d)$ (Lemma D.30), replacing Ψ by $\Psi + p_k$ and applying (D.65), using the definition of $I^\#$,

$$-\int_{R^d} |\frac{1}{2}\frac{\nabla\rho(x)}{\rho(x)} + \nabla(\Psi + p_k) + \nabla(\Phi * \rho)|^2\rho(x)dx = \| - \text{grad}\mathcal{E}(\rho) + (-\nabla(\rho\nabla p_k))\|^2_{-1,\rho}.$$

Similarly, replacing Ψ by $\Psi - p_k$,

$$-\int_{R^d} |\frac{1}{2}\frac{\nabla\rho(x)}{\rho(x)} + \nabla(\Psi - p_k) + \nabla(\Phi * \rho)|^2 \rho(x)dx = \| - \mathrm{grad}\mathcal{E}(\rho) + (-\nabla(\rho\nabla p_k))\|_{-1,\rho}^2.$$

Therefore, by the polarization identity,

$$\langle -\mathrm{grad}\mathcal{E}(\rho), -\nabla(\rho\nabla p_k)\rangle_{-1,\rho} = -\int_{R^d} \nabla p_k(x) \cdot (\frac{1}{2}\frac{\nabla\rho(x)}{\rho(x)} + \nabla\Psi + \nabla(\Phi * \rho))\rho(x)dx.$$

Sending $n \to \infty$, we arrive at (D.66). □

D.7. Mass transport inequalities

Let $\Psi, \Phi \in C^2(R^d)$ satisfy Condition 13.14, and recall the definition of the relative entropy and free energy functionals in (9.93) and (9.94):

$$\mathcal{R}(\rho\|\mu_\infty) = \int_{R^d} \log \frac{d\rho}{d\mu_\infty} d\rho,$$

and

$$\begin{aligned} \mathcal{E}(\rho) &= \frac{1}{2}\mathcal{R}(\rho\|\mu_\infty) + \frac{1}{2}\int_{R^d\times R^d} \Phi(x-y)\rho(dx)\rho(dy) \\ &= \frac{1}{2}\int_{R^d} \rho\log\rho dx + \int_{R^d} \Psi\rho dx + \frac{1}{2}\int_{R^d\times R^d} \Phi(x-y)\rho(dx)\rho(dy). \end{aligned} \tag{D.67}$$

Let $\zeta \in C_c^\infty(R^d)$, $\zeta : R^d \to [0,1]$, $\zeta(x) = 1$, $|x| \le 1$, $\zeta(x) = 0$, $|x| \ge 3$, and $\sup_x |\nabla\zeta(x)|^2/\zeta(x) < \infty$ (see (D.11)), and define $\zeta_n(x) = \zeta(n^{-1}x)$. Let β_m be defined as in (D.40).

LEMMA D.47. *Suppose $\rho \in \mathcal{P}_2(R^d)$ has a Lebesgue density, and define*

$$\rho_n = c_n\zeta_n\beta_{m_n}(\rho)$$

where $c_n > 0$ is selected so that ρ_n is a probability density (at least for n sufficiently large) and $\lim_{n\to\infty} m_n = \infty$. Then ρ_n is bounded and has compact support and $\lim_{n\to\infty}\mathcal{E}(\rho_n) = \mathcal{E}(\rho)$.

If, in addition, $\nabla\rho \in L^1_{loc}(R^d)$, then for $m_n \to \infty$ sufficiently rapidly,

$$\lim_{n\to\infty} I(\rho_n) = I(\rho). \tag{D.68}$$

PROOF. Since $c_n \to 1$, the monotone convergence theorem implies

$$\lim_{n\to\infty}\frac{1}{2}\int_{R^d\times R^d}\Phi(x-y)\rho_n(dx)\rho_n(dy) = \frac{1}{2}\int_{R^d\times R^d}\Phi(x-y)\rho(dx)\rho(dy).$$

It follows from (13.73), that

$$\int_{\{\rho<1\}} |\rho\log\rho| dx < \infty,$$

and since Ψ is bounded below, the first two terms in

$$\begin{aligned} \mathcal{R}(\rho_n\|\mu_\infty) = \frac{1}{2}\int_{R^d} c_n\zeta_n\beta_{m_n}(\rho)\log\beta_{m_n}(\rho)dx + \int_{R^d}\Psi c_n\zeta_n\beta_{m_n}(\rho)dx \\ + \int_{R^d} c_n\zeta_n\beta_{m_n}(\rho)\log c_n\zeta_n dx, \end{aligned}$$

converge (possibly to $+\infty$) by the dominated convergence and monotone convergence theorems. The last term converges to zero.

To prove (D.68), we only need to consider ρ with $I(\rho) < \infty$. Then $|\nabla\rho|^2/\rho \in L^2_{loc}(R^d)$, and since ζ_n has compact support, for $m_n \to \infty$ sufficiently rapidly,

$$\lim_{n\to\infty} \int_{R^d} \left|\frac{\beta'_{m_n}(\rho(x))\nabla\rho(x)}{\beta_{m_n}(\rho(x))} - \frac{\nabla\rho(x)}{\rho(x)}\right|^2 c_n\zeta_n(x)\beta_{m_n}(\rho(x))dx = 0. \tag{D.69}$$

Consequently,

$$\begin{aligned}\lim_{n\to\infty} I(\rho_n) &= \lim_{n\to\infty} \int_{R^d} \left|\frac{\beta'_{m_n}(\rho(x))\nabla\rho(x)}{\beta_{m_n}(\rho(x))} + \frac{\nabla\zeta_n(x)}{\zeta_n(x)} + 2\nabla\Psi(x)\right|^2 c_n\zeta_n(x)\beta_{m_n}(\rho(x))dx \\ &= \lim_{n\to\infty} \int_{R^d} \left|\frac{\nabla\rho(x)}{\rho(x)} + 2\nabla\Psi(x)\right|^2 c_n\zeta_n(x)\beta_{m_n}(\rho(x))dx \\ &= I(\rho),\end{aligned}$$

where the last equality follows by the monotone convergence theorem. □

LEMMA D.48. *Suppose $\rho_n, \rho_0 \in \mathcal{P}_2(R^d)$ satisfy $\sup_n I(\rho_n) < \infty$ and $\lim_{n\to\infty} d(\rho_n, \rho_0) = 0$ and that $\gamma_n, \gamma_0 \in \mathcal{P}_2(R^d)$ have Lebesgue densities and satisfy $\lim_{n\to\infty} d(\gamma_n, \gamma_0) = 0$. Let $p_n \equiv p_{\rho_n,\gamma_n}, p_0 \equiv p_{\rho,\gamma}$ be the functions in (D.23) of Theorem D.20 such that*

$$d^2(\rho_n, \gamma_n) = \int_{R^d} p_n(x)\rho_n(x)dx = \int_{R^d} |\nabla p_n(x)|^2 \rho_n(x)dx.$$

Then

$$\lim_{n\to\infty} \|\sqrt{\rho_n}\nabla p_n - \sqrt{\rho_0}\nabla p_0\|_0 = 0 \tag{D.70}$$

and

$$\lim_{n\to\infty} \int_{R^d} \frac{\nabla\rho_n + 2\rho_n\nabla\Psi}{\sqrt{\rho_n}} \cdot \nabla p_n\sqrt{\rho_n}dx = \int_{R^d} \frac{\nabla\rho_0 + 2\rho_0\nabla\Psi}{\sqrt{\rho_0}} \cdot \nabla p_0\sqrt{\rho_0}dx. \tag{D.71}$$

REMARK D.49. By the lower semicontinuity of I, $I(\rho_0) < \infty$. The finiteness of $I(\rho_n)$ and $\int |\nabla p_n|^2 \rho_n dx = d^2(\rho_n, \gamma_n)$ implies the existence of the integrals in (D.71).

PROOF. To simplify notation, let

$$\xi_n = \frac{\nabla\rho_n + 2\rho_n\nabla\Psi}{\sqrt{\rho_n}} = 2(\nabla\sqrt{\rho_n} + \sqrt{\rho_n}\nabla\Psi) \in L^2(R^d)$$

and

$$\xi_0 = \frac{\nabla\rho_0 + 2\rho_0\nabla\Psi}{\sqrt{\rho_0}} = 2(\nabla\sqrt{\rho_0} + \sqrt{\rho_0}\nabla\Psi) \in L^2(R^d).$$

Since

$$\begin{aligned}&\left|\int_{R^d} \xi_n \cdot \nabla p_n\sqrt{\rho_n}dx - \int_{R^d} \xi_0 \cdot \nabla p_0\rho_0 dx\right| \\ &\quad\leq \left|\int_{R^d} \xi_n \cdot (\nabla p_n\sqrt{\rho_n} - \nabla p_0\sqrt{\rho_0})dx\right| + \left|\int_{R^d} (\xi_n - \xi_0) \cdot \nabla p_0\sqrt{\rho_0}dx\right|,\end{aligned}$$

$$\sup_n \int |\xi_n|^2 dx = \sup_n I(\rho_n) < \infty$$

and

$$\sup_n \int |\nabla p_n\sqrt{\rho_n}|^2 dx = \sup_n d^2(\rho_n, \gamma_n) < \infty,$$

(D.71) holds if $\{\xi_n\}$ converges weakly to ξ_0 in $L^2(R^d)$, that is,

$$\lim_{n\to\infty} \langle \xi_n, q\rangle = \langle \xi_0, q\rangle, \quad q \in L^2(R^d), \tag{D.72}$$

and (D.70) holds. Indeed, since

$$\int_{R^d} |\sqrt{\rho_n}\nabla p_n|^2 dx = d^2(\rho_n, \gamma_n) \to d^2(\rho_0, \gamma_0) = \int_{R^d} |\sqrt{\rho_0}\nabla p_0|^2 dx,$$

and

$$\|\sqrt{\rho_n}\nabla p_n - \sqrt{\rho_0}\nabla p_0\|_0^2 = \|\sqrt{\rho_n}\nabla p_n\|^2 + \|\sqrt{\rho_0}\nabla p_0\|^2 - 2\langle\sqrt{\rho_n}\nabla p_n, \sqrt{\rho_0}\nabla p_0\rangle,$$

(D.70) is implied by the seemingly weaker statement that $\sqrt{\rho_n}\nabla p_n$ converge weakly to $\sqrt{\rho_0}\nabla p_0$.

Setting $B_m(0) = \{x \in R^d : |x| < m\}$, $m = 1, 2, \ldots$,

$$\begin{aligned}\|\nabla \rho_n/\sqrt{\rho_n}\|_{L^2(B_m(0))} &\leq \|\nabla \rho_n/\sqrt{\rho_n} + \sqrt{\rho_n}\nabla\Psi\|_{L^2(B_m(0))} + \|\sqrt{\rho_n}\nabla\Psi\|_{L^2(B_m(0))} \\ &\leq \sqrt{I(\rho_n)} + \sup_{x\in B_m(0)} |\nabla\Psi(x)|,\end{aligned}$$

so

$$\sup_n \|\nabla\rho_n/\sqrt{\rho_n}\|_{L^2(B_m(0))} < \infty. \tag{D.73}$$

Consequently, Lemma D.7 implies $\sqrt{\rho_n} \in H^1(B_m(0))$, and

$$\nabla\sqrt{\rho_n} = \frac{\nabla\rho_n}{2\sqrt{\rho_n}}.$$

Selecting subsequences if necessary, by Lemma D.6, there exists $\eta \in L^2(B_m(0))$ such that

$$\lim_{n\to\infty} \|\sqrt{\rho_n} - \eta\|_{L^2(B_m(0))} = 0.$$

The boundedness in (D.73) implies the existence of a further subsequence along which $\nabla\sqrt{\rho_n}$ converges weakly in $L^2(B_m(0))$. Along the subsequence, for $\varphi \in \mathcal{D}(B_m(0))$

$$\langle\varphi, \partial_{x_i}\sqrt{\rho_n}\rangle = -\langle\partial_{x_i}\varphi, \sqrt{\rho_n}\rangle \to -\langle\partial_{x_i}\varphi, \eta\rangle = \langle\varphi, \partial_{x_i}\eta\rangle.$$

But $\mathcal{D}(B_m(0))$ is dense in $L^2(B_m(0))$, so $\nabla\sqrt{\rho_n}$ converges weakly to $\nabla\eta$. In addition

$$\int_{O_m} |\rho_n - \eta^2| dx \leq \|\sqrt{\rho_n} - \eta\|_{L^2(B_m(0))}\|\sqrt{\rho_n} + \eta\|_{L^2(B_m(0))} \to 0,$$

so since ρ_n converges to ρ_0 in the 2-Wasserstein metric, we must have $\eta^2 = \rho_0$. Hence $\nabla\sqrt{\rho_n}$ converges weakly to $\nabla\sqrt{\rho_0}$ in $L^2(B_m(0))$. Since m is arbitrary and $\cup_m L^2(B_m(0))$ is dense in $L^2(R^d)$ (extend $f \in L^2(B_m(0))$ by setting $f \equiv 0$ on $B_m(0)^c$), (D.72) follows.

Since

$$\|\sqrt{\rho_n}\nabla p_n\|_{L^2(R^d)}^2 = \int_{R^d} |\nabla p_n|^2 d\rho_n = d^2(\rho_n, \gamma_n),$$

$\{\sqrt{\rho_n}\nabla p_n\}$ is uniformly bounded in $L^2(R^d)$. Let $\xi \in L^2(R^d)$ be a weak limit point. Then along the appropriate subsequence

$$\lim_{n\to\infty} \int_{R^d} \nabla p_n \sqrt{\rho_n} q dx = \int_{R^d} \xi q dx, \quad q \in L^2(R^d).$$

In particular, for each R^d-valued $\psi \in C_c^\infty(R^d)$,

$$\lim_{n\to\infty} \int_{R^d} \psi \cdot \nabla p_n \sqrt{\rho_n}\sqrt{\rho_0} dx = \int_{R^d} \psi \cdot \xi\sqrt{\rho_0} dx.$$

Since $\|\sqrt{\rho_n}-\sqrt{\rho_0}\|_{L^2(B_m(0))}\to 0$, this implies

$$\lim_{n\to\infty}\int_{R^d}\psi\cdot\nabla p_n\sqrt{\rho_n}\sqrt{\rho_n}dx=\int_{R^d}\psi\cdot\xi\sqrt{\rho_0}dx. \tag{D.74}$$

Since, by Lemma D.14, $\pi_{\rho_n,\gamma_n}\Rightarrow\pi_{\rho,\gamma}$ and $\sup_n\int|x-y|^2\pi_{\rho_n,\gamma_n}(dx,dy)<\infty$,

$$\begin{aligned}\int_{R^d}\nabla p_0(x)\cdot\psi(x)\rho_0(x)dx &= \int_{R^d\times R^d}(x-y)\cdot\psi(x)d\pi_{\rho,\gamma}(dx,dy)\\ &= \lim_{n\to\infty}\int_{R^d\times R^d}(x-y)\cdot\psi(x)d\pi_{\rho_n,\gamma_n}(dx,dy)\\ &= \lim_{n\to\infty}\int_{R^d}\nabla p_n(x)\cdot\psi(x)\rho_n(x)dx\\ &= \int_{R^d}\xi(x)\sqrt{\rho_0(x)}\cdot\psi(x)dx, \qquad \psi\in C_c^\infty(R^d),\end{aligned}$$

where the last step follows from (D.74). Hence $\xi=\nabla p_0\sqrt{\rho_0}$, giving the weak convergence. □

Consider Δ on $L^2(\mathcal{O})$, $\mathcal{O}=[0,1)^d$, with periodic boundary conditions. Then Δ is dissipative with

$$\langle\Delta\rho-\Delta\gamma,\rho-\gamma\rangle=-\|\nabla\rho-\nabla\gamma\|^2_{L^2(\mathcal{O})}\le 0,\quad \rho,\gamma\in\mathcal{D}(\Delta)\subset L^2(\mathcal{O}).$$

Let $\rho(t),\gamma(t)$ be solutions to

$$\dot\rho=\Delta\rho,\qquad \dot\gamma=\Delta\gamma.$$

Then

$$\frac{d}{dt}\frac{1}{2}\|\rho(t)-\gamma(t)\|^2_{L^2(\mathcal{O})}\le-\|\nabla(\rho(t)-\gamma(t))\|^2_{L^2(\mathcal{O})}.$$

Estimates based on similar observations played an important role in the proof of the comparison principle for Example 9.29 (see also Section 13.1.3)). Similarly, in Example 9.32 and the related large deviation problem in Section 13.2, we used the dissipative property of $-\Delta^2$ with respect to the $\|\cdot\|_{-1}$ norm to prove another comparison principle.

The inequalities (D.76) and (D.78) in the next theorem are mass transport versions of the above inequality (see Remark D.51).

THEOREM D.50 (Entropy inequalities). *Let $\rho,\gamma\in\mathcal{P}_2(R^d)$ have Lebesgue densities, and let $p_{\rho,\gamma}$ and $q_{\rho,\gamma}$ be the functions in Theorem D.25.*

If $I^\#(\rho)<\infty$, then

$$\begin{aligned}\mathcal{E}(\gamma) &\ge \mathcal{E}(\rho)-\int_{R^d}\nabla p_{\rho,\gamma}(x)\cdot(\frac{1}{2}\frac{\nabla\rho(x)}{\rho(x)}+\nabla\Psi(x)+\nabla\Phi*\rho(x))\rho(x)dx\\ &\qquad +\frac{\lambda_\Psi+\lambda_\Phi}{2}d^2(\rho,\gamma)-\frac{\lambda_\Phi}{2}|\int x\rho(dx)-\int y\gamma(dy)|^2.\end{aligned}\tag{D.75}$$

If $I^\#(\rho)+I^\#(\gamma)<\infty$, then

$$\begin{aligned}&\int_{R^d}\nabla p_{\rho,\gamma}\cdot\frac{\nabla\rho+2\rho\nabla\Psi+2\rho\nabla\Phi*\rho}{\rho}(x)\rho(x)dx\\ &\quad+\int_{R^d}\nabla p_{\gamma,\rho}\cdot\frac{\nabla\gamma+2\gamma\nabla\Psi+2\gamma\nabla\Phi*\gamma}{\gamma}(y)\gamma(y)dy\\ &\quad\ge 2(\lambda_\Psi+\lambda_\Phi)d^2(\rho,\gamma)-2\lambda_\Phi|\int xd\rho-\int yd\gamma|^2.\end{aligned}\tag{D.76}$$

PROOF. If $\mathcal{E}(\gamma) = \infty$, (D.75) is immediate, so we may as well assume that $\mathcal{E}(\gamma) < \infty$. For ρ and γ with compact support, (D.75) is Theorem 4.1 of Cordero-Erausquin, Gangbo and Houdré [**15**]. With ζ_n and β_{m_n} as in the proof of Lemma D.47, define $\rho_n = c_n\zeta_n\beta_{m_n}(\rho)$ and $\gamma_n = c_n\zeta_n\beta_{m_n}(\gamma)$, where m_n is selected so that (D.69) holds for ρ and with ρ replaced by γ. Then (D.75) holds with ρ and γ replaced by ρ_n and γ_n. Letting $n \to \infty$, $\mathcal{E}(\rho_n)$ and $\mathcal{E}(\gamma_n)$ converge by Lemma D.47, and the last two terms on the right also converge. Setting $p_n = p_{\rho_n,\gamma_n}$, the second term on the right can be written as

$$\begin{aligned}
-\int_{R^d} &\nabla p_n(x)\cdot\Big(\frac{\beta_{m_n}(\rho(x)\nabla\rho(x))}{2\beta_{m_n}(\rho(x))\rho(x)} + \frac{\nabla\zeta_n(x)}{2\zeta_n(x)} + \nabla\Psi(x) + \nabla\Phi * \rho_n(x)\Big)\rho_n(x)dx \\
&= -\int_{R^d} \nabla p_n(x)\cdot\Big(\frac{\nabla\rho(x)}{2\rho(x)} + \nabla\Psi(x)\Big)\rho_n(x)dx - \int_{R^d}\nabla p_n(x)\cdot\frac{\nabla\zeta_n(x)}{2\zeta_n(x)}\rho_n(x)dx \\
&\qquad - \int_{R^d}\nabla p_n(x)\cdot\Big(\frac{\beta_{m_n}(\rho(x)\nabla\rho(x))}{2\beta_{m_n}(\rho(x))\rho(x)} - \frac{\nabla\rho(x)}{\rho(x)}\Big)\rho_n(x)dx \\
&\qquad - \int_{R^d}(y-x)\cdot\nabla\Phi * \rho_n(x)\pi_{\rho_n,\gamma_n}(dx,dy) \\
&\to -\int_{R^d}\nabla p_{\rho,\gamma}(x)\cdot\Big(\frac{\nabla\rho(x)}{2\rho(x)} + \nabla\Psi(x)\Big)\rho(x)dx \\
&\qquad - \int_{R^d}\nabla p_{\rho,\gamma}(x)\cdot\nabla\Phi * \rho(x)\rho(x)dx,
\end{aligned}$$

where the first term converges by Lemma D.48, the second and third goes to zero by

$$\sup_n \int_{R^d} |\nabla p_n(x)|^2\rho_n(x)dx < \infty,$$

the properties of ζ_n, and the definition of m_n, and the fourth converges by the weak convergence of $\nu_n \equiv \pi_{\rho_n,\gamma_n}\times\rho_n$ and the uniform integrability implied by

$$\sup_n \int_{R^d\times R^d\times R^d} |(x-y)\cdot\nabla\Phi(x-z)|^2\nu_n(dx.dy,dz) < \infty.$$

□

REMARK D.51. If $\mathcal{E}(\rho) < \infty$, by Lemma D.46, (D.75) implies

$$\begin{aligned}
\text{(D.77)}\qquad &\langle -\mathrm{grad}\mathcal{E}(\rho), \mathrm{grad}_\rho\frac{1}{2}d^2(\rho,\gamma)\rangle_{-1,\rho} \\
&\le \mathcal{E}(\gamma) - \mathcal{E}(\rho) - \frac{\lambda_\Psi+\lambda_\Phi}{2}d^2(\rho,\gamma) + \frac{\lambda_\Phi}{2}\Big|\int_{R^d}xd\rho - \int_{R^d}yd\gamma\Big|^2 \\
&\le \mathcal{E}(\gamma) - \mathcal{E}(\rho) + \frac{-\lambda_\Psi + 2|\lambda_\Phi|}{2}d^2(\rho,\gamma)
\end{aligned}$$

and

$$\begin{aligned}
\text{(D.78)}\quad &\langle -\mathrm{grad}\mathcal{E}(\rho), \mathrm{grad}_\rho\frac{1}{2}d^2(\rho,\gamma)\rangle_{-1,\rho} + \langle -\mathrm{grad}\mathcal{E}(\gamma), \mathrm{grad}_{\gamma_0}\frac{1}{2}d^2(\rho_0,\gamma)\rangle_{-1,\gamma} \\
&\le -(\lambda_\Psi+\lambda_\Phi)d^2(\rho,\gamma) + \lambda_\Phi\Big|\int_{R^d}xd\rho - \int_{R^d}yd\gamma\Big|^2. \\
&\le (-\lambda_\Psi + 2|\lambda_\Phi|)d^2(\rho,\gamma).
\end{aligned}$$

This inequality gives the contraction property with respect to the 2-Wasserstein metric d for solutions of

$$\dot{\rho} = -\text{grad}\mathcal{E}(\rho) = \frac{1}{2}\Delta\rho + \nabla\cdot(\rho\nabla(\Psi + \rho * \Phi)). \tag{D.79}$$

In particular, uniqueness holds for (D.79).

In Sections 9.35 and 13.3, we make critical use of the above inequalities to prove a comparison principle for some Hamilton-Jacobi equations with probability-measure-valued state space.

In the mass transport literature, various interpolation inequalities such as the HWI inequality (Theorem 9.17 of Villani [**127**]) and the log-Sobolev inequality are special cases of (D.75).

COROLLARY D.52 (Interpolation inequality). *Suppose* $\int \Psi e^{-2\Psi}dx < \infty$, *and* $\rho, \gamma \in \mathcal{P}(R^d)$ *satisfy* $\int |x|^2 d\rho + \int |y|^2 d\rho < \infty$. *Then*

$$\mathcal{E}(\rho) \leq \mathcal{E}(\gamma) + \frac{1}{2}d(\rho,\gamma)\sqrt{I^{\#}(\rho)} - \frac{\lambda_\Psi + \lambda_\Phi}{2}d^2(\rho,\gamma) + \frac{\lambda_\Phi}{2}|\int_{R^d} x d\rho - \int_{R^d} y d\gamma|^2. \tag{D.80}$$

In particular, taking $\Phi = 0$ *and* $\lambda_\Phi = 0$,

$$\mathcal{R}(\rho\|\mu_\infty) \leq \mathcal{R}(\gamma\|\mu_\infty) + d(\rho,\gamma)\sqrt{I(\rho)} - \lambda_\Psi d^2(\rho,\gamma). \tag{D.81}$$

Taking $\gamma = \mu_\infty$,

$$\mathcal{R}(\rho\|\mu_\infty) \leq d(\rho,\mu_\infty)\sqrt{I(\rho)} - \lambda_\Psi d^2(\rho,\mu_\infty). \tag{D.82}$$

If Ψ *is strictly convex (that is,* $\lambda_\Psi > 0$*), then*

$$\mathcal{R}(\rho\|\mu_\infty) \leq \frac{1}{2\lambda_\Psi}I(\rho). \tag{D.83}$$

PROOF. (D.80) follows by applying the Cauchy-Schwartz inequality to the right side of (D.75). Similarly, (D.83) follows from (D.82). □

A variant of (D.75) holds even without the assumption $I^{\#}(\rho) < \infty$. The next proof is a modification of that in Otto [**92**], where ρ is required to be smooth and have compact support. See also Theorem 5.15 of Villani [**127**].

LEMMA D.53. *Suppose that* $\rho, \gamma \in \mathcal{P}_2(R^d)$ *have Lebesgue densities and that*

$$\int \gamma(y)\log\gamma(y)dy < \infty.$$

Recalling that $p_{\rho,\gamma}$ *is the difference of two convex functions, let* $D_A^2 p_{\rho,\eta}$ *denote the Hessian in the Aleksandrov sense, and* $\Delta_A p_{\rho,\gamma} \equiv \text{tr}D_A^2 p_{\rho,\gamma}$*, the Laplacian. Then*

$$\int_{R^d}\gamma(y)\log\gamma(y)dy - \int_{R^d}\rho(x)\log\rho(x)dx \geq \int_{R^d}\Delta_A p_{\rho,\gamma}(x)\rho(x)dx. \tag{D.84}$$

Noting that $\Delta_A p_{\rho,\gamma} \leq d$*, the right side is well-defined.*

REMARK D.54. Again, $\Delta_A p_{\rho,\gamma}$ may only be defined in $\mathcal{O}$, the interior of the convex hull of the support of ρ; however, $\rho(\mathcal{O}) = 1$, so the right side of D.84 is well defined. See McCann [**84**] for regularity regarding $D_A^2 p_{\rho,\gamma}$.

PROOF. Let

$$\phi_t(x) \equiv \frac{|x|^2}{2} - tp_{\rho,\gamma}(x), \quad \pi_t(dx,dy) \equiv \rho(dx)\delta_{\nabla\phi_t(x)}(dy).$$

Then by Proposition 1.3 of [**84**], $\rho_t(dy) = \pi_t(R^d, dy), 0 \leq t \leq 1$ defines a function in $\mathcal{P}(R^d)$ with $\rho_0 = \rho, \rho_1 = \gamma$. Furthermore, $\rho_t(dy) = \rho_t(y)dy$ has a Lebesgue density.

Let $U(r) = r\log r$, for $r \geq 0$. By the change of variable formula in Theorem 4.4 of [**84**], $D_A^2\phi_t(x)$ is defined, finite, and invertible, for x almost everywhere with respect to $\rho(x)dx$. Moreover,

$$\text{(D.85)}\quad \begin{aligned} S(\rho_t) &\equiv \int_{R^d} U(\rho_t(y))dy = \int_{R^d} U(\frac{\rho(x)}{\det[D_A^2\phi_t(x)]})\det[D_A^2\phi_t(x)]dx \\ &= \int_{R^d} \rho(x)(\log\rho(x) - \log\det[D_A^2\phi_t(x)])dx, \end{aligned}$$

where we allow the possibility of $\infty = \infty$.

Let $T(x) = D_A^2 p_{\rho,\gamma}(x)$. Then $T(x) \leq I_{d\times d}$ (that is, $I_{d\times d} - T(x)$ is nonnegative definite) and $D_A^2\phi_t(x) = I_{d\times d} - tT(x)$. By Lemma 5.21 of [**127**],

$$(\det[D_A^2\phi_t(x)])^{1/d} = (\det[I_{d\times d} - tT(x)])^{1/d} \text{ is concave in } t \in [0,1].$$

That is, for x almost everywhere with respect to $\rho(x)dx$, $-\log(\det[D_A^2\phi_t(x)])$ is finite and is convex in $t \in [0,1]$. Consequently,

$$h_t(x) \equiv -\frac{1}{t}\log(\det[D_A^2\phi_t(x)]) = -\frac{1}{t}\Big(\log(\det[D_A^2\phi_t(x)]) - \log(\det[D_A^2\phi_0(x)])\Big)$$

is nondecreasing in t and

$$-\frac{1}{t}\log(\det[D_A^2\phi_t(x)]) = h_t(x) \leq h_1(x) = -\log(\det[D_A^2\phi_1(x)]).$$

Since

$$\lim_{t\to 0+} h_t(x) = -\frac{d}{dt}\log\Big(\det[I_{d\times d} - tD_A^2 p_{\rho,\gamma}(x)]\Big)\Big|_{t=0} = \mathrm{tr} D_A^2 p_{\rho,\gamma}(x) = \Delta_A p(x),$$

for $0 < t \leq 1$,

$$\Delta_A p_{\rho,\gamma}(x) \leq h_t(x) \leq h_1(x), \quad x - \text{a.e. } \rho(dx).$$

The conclusion of the lemma (D.84) trivially holds if the right side is $-\infty$. Therefore, we may assume that

$$\text{(D.86)}\quad -\infty < \int_{R^d} \Delta_A p(x)\rho(x)dx \leq \int_{R^d} h_t d\rho = -\int_{R^d} \frac{1}{t}\log\Big(\det[D_A^2\phi_t(x)]\Big)\rho(x)dx.$$

By (D.85),

$$S(\rho_t) = S(\rho) - \int_{R^d} \rho(x)\log\det[D_A^2\phi_t(x)])dx.$$

Taking $t = 1$ and noting $S(\rho_1) = S(\gamma) < \infty$ by assumption, we have $S(\rho) < \infty$, and by (D.86),

$$S(\gamma) - S(\rho) = \int_{R^d} h_1 d\rho \geq \int_{R^d} \Delta_A p(x)\rho(x)dx.$$

□

D.8. Miscellaneous

Let $f, g \in C^1(R)$. Then by elementary calculus, it is an easy to verify that

$$f(x) - f(x_0) \leq g(x) - g(x_0)$$

implies grad$f(x_0)$=grad$g(x_0)$. We give the mass transport version of this result next.

LEMMA D.55. *Let f and g be lower semicontinuous functions on $(\mathcal{P}_2(R^d), d)$. Let $\rho_0 \in \mathcal{P}_2(R^d)$. Suppose that both* grad$f(\rho_0)$ *and* grad$g(\rho_0)$ *exist as elements in $\mathcal{D}'(R^d)$, and that*

$$f(\rho) - f(\rho_0) \leq g(\rho) - g(\rho_0), \quad \forall \rho \in \mathcal{P}_2(R^d).$$

Then

$$\text{grad}f(\rho_0) = \text{grad}g(\rho_0).$$

PROOF. Let $p \in C_c^\infty(R^d)$, and define $\{\rho(t) : -\infty < t < \infty\}$ by

$$\dot{\rho} + \nabla \cdot (\rho \nabla p) = 0, \quad \rho(0) = \rho_0.$$

Then

$$\begin{aligned} \langle \text{grad}f(\rho_0), p\rangle &= \lim_{t \to 0+} \frac{1}{t}(f(\rho(t)) - f(\rho(0))) \\ &\leq \lim_{t \to 0+} \frac{1}{t}(g(\rho(t)) - g(\rho(0))) = \langle \text{grad}g(\rho_0), p\rangle. \end{aligned}$$

By the arbitrariness of p, the conclusion follows. □

Bibliography

1. Adams, Robert A. *Sobolev spaces. Pure and Applied Mathematics*, Vol. 65. Academic Press, New York-London, 1975.
2. Aldous, David. Stopping times and tightness. *Ann. Probab.* **6** (1984), 335-340.
3. Ambrosio, Luigi, Gigli, Nicola and Savaré, Giuseppe. *Gradient flows in metric spaces and in the space of probability measures.* Lectures in Mathematics ETH Zrich. Birkhuser Verlag, Basel, (2005).
4. Anderson, Robert F. and Orey, Steven. Small random perturbation of dynamical systems with reflecting boundary. *Nagoya Math. J.* **60** (1976), 189-216.
5. Attouch, Hédy. *Variational Convergence for Functions and Operators*, Pitman Advanced Publishing Program, Boston, London, 1984.
6. Baldi, Paolo. Large deviations for diffusion processes with homogenization and applications. *Ann. Probab.* **19** (1991), 509-524.
7. Barles, G, and Perthame, B. Exit time problems in optimal control and vanishing viscosity method. *SIAM J. Control Optim.* **26** (1988), 1133-1148.
8. Barles, G, and Perthame, B. Comparison results in Dirichlet type first-order Hamilton-Jacobi Equations. *Appl. Math. Optim.* **21** (1990) 21-44.
9. Barles, G, and Souganidis, Panagiotis.E. Convergence of approximation schemes for fully nonlinear second order equations. *Asymptotic Analysis.* **4** (1991) 271-283.
10. Bertini, L., Landim, C. and Olla, S. Derivation of Cahn-Hilliard equations from Ginzburg-Landau models. *J. Stat. Phys.* 88, 365-381, (1997)
11. Borovkov, Alexandr. A. Boundary-value problems for random walks and large deviations in function spaces. *Theory Probab. Appl.* **12** (1967) 575-595.
12. Buck, R. Creighton. Bounded continuous functions on a locally compact space. *Michigan Math. J.* **5** (1958), 95-104.
13. Chen, Hong and Yao, David D. *Fundamentals of Queueing Networks* Springer, New York, 2001.
14. Cooper, J. B. The strict topology and spaces with mixed topologies. *Proc. Amer. Math. Soc.* **30** (1971),583-592.
15. Cordero-Erausquin, Dario; Gangbo, Wilfrid; and Houdré, Christian. Inequalities for generalized entropy and optimal transportation. *Recent advances in the theory and applications of mass transport*, 73–94, *Contemp. Math.*, **353**, Amerocan Mathematical Society, Providence, RI, 2004
16. Cramér, H. Sur un nouveau théoreme-limite de la théorie des probabilités. *Acta. Sci. et Ind.* 736 (1938), 5-23.
17. Crandall, Michael G., Ishii, Hitoshi and Lions, Pierre-Louis. User's Guide to Viscosity Solutions of Second Order Partial Differential Equations. *Bulletin A. M. S., N. S.* **27** (1992), 1-67.
18. Crandall, Michael G. and Liggett, Thomas M. Generation of Semigroups of Nonlinear Transfomations on General Banach Spaces. *Amer. J. Math.* **93** (1971), 265-298.
19. Crandall, Michael G. and Lions, Pierre-Louis. Viscosity solutions of Hamilton-Jacobi Equations. *Trans. A. M. S.* **277** (1983), 1-42.
20. Crandall, Michael G. and Lions, Pierre-Louis. Hamilton-Jacobi Equations in Infinite Dimensions, Part VI: Nonlinear A and Tataru's Method Refined. *Evolution equations, Control theory and Biomathematics (Han Sur Lasse, 1991) 51-89* Lecture Notes in Pure and Applied Math.155. Dekker, New York (1994).
21. Crandall, Michael G. and Lions, Pierre-Louis. Hamilton-Jacobi Equations in Infinite Dimensions, PART I *J. Funct. Anal.* **62** (1985) 3. 379-396. PART II *J. Funct. Anal.* **65** (1986) 3.

368-405. PART III *J. Funct. Anal.* **68** (1986) 2. 214-247. PART IV *J. Funct. Anal.* **90** (1990) 2. 237-283. PART V *J. Funct. Anal.* **97** (1991) 2. 417-465. PART VII *J. Funct. Anal.* **125** (1994) 1. 111-148.

22. Dawson, Donald. A. and Gärtner, Jürgen. Large Deviations from the McKean-Vlasov limit for weakly interacting diffusions. *Stochastics* **20**. 247-308, 1987.
23. Dawson, Donald. A. and Gărtner, J. Large deviations, free energy functional and quasi-potential for a mean field model of interacting diffusions. *Memoirs of the American Mathematics Society* **Vol 78**, No. 398, March 1989.
24. de Acosta, Alejandro. Exponential Tightness and Projective Systems in Large Deviation Theory. *Festschrift for Lucien Le Cam. Springer, New York.* 143-156, 1997.
25. de Acosta, Alejandro. Large deviations for vector-valued Lévy processes. *Stochastic Process. Appl.* **51** (1994), 75-115.
26. de Acosta, Alejandro. A general non-convex large deviation result with applications to stochastic equations. *Probab. Theory Relat. Fields* 118 (2000), 483-521.
27. de Acosta, Alejandro. A general non-convex large deviation result II. *Ann. Probab.* (to appear)
28. Dellacherie, Claude and Meyer, Paul-André. *Probabilities and Potential*, (North-Holland mathematics studies, 29; Translation of Probabilitiés et potentiel.) North-Holland Publishing, New York, 1978.
29. Dembo, Amir and Zeitouni, Ofer. *Large Deviations Techniques and Applications*, Jones and Bartlett Publishers, Boston, 1993.
30. Deuschel, Jean-Dominique and Stroock, Daniel W. *Large Deviations.* Academic Press, Boston, 1989.
31. Diestel, J.; Uhl, J. J., Jr. The Radon-Nikodym theorem for Banach space valued measures. *Rocky Mountain J. Math.* **6** (1976), 1-46.
32. Donsker, Monroe D. and Varadhan, S. R. S. On a variational formula for the principal eigenvalue for operators with maximum principle *Proc. Nat. Acad. Sci. USA* **72** No.3 (1975), 780-783.
33. Donsker, Monroe D. and Varadhan, S. R. S. Asymptotic evaluation of Markov process expectations for large time, I,II,III *Comm. Pure Appl. Math.* textbf27 (1975), 1-47, **28** (1975), 279-301, **29** (1976), 389-461.
34. Doss, Halim and Priouret, Pierre. Petites perturbations de systemes dynamiques avec reflexion. *Seminar on probability, XVII*, 353–370, Lecture *Notes in Math.*, 986, Springer, Berlin, 1983.
35. Dupuis Paul and Ellis, Richard S. *A Weak Convergence Approach to the Theory of Large Deviations.* Wiley, New York, 1997.
36. Ethier, Stewart N. and Kurtz, Thomas G. *Markov Processes*, John Wiley and Sons, New York, 1986.
37. Evans, Lawrence C. The perturbed test function method for viscosity solutions of nonlinear PDE. *Proc. Roy. Soc. Edinburgh Sect A.* **111** (1989), 359-375.
38. Evans, Lawrence. *Partial Differential Equations.* Graduate Studies in Mathematics, Volume 19, American Mathematical Society. Providence, Rhode Island 1998.
39. Evans, Lawrence and Gariepy, Ronald. *Measure theory and fine properties of functions.* Studies in Advanced Mathematics, CRC Press, London 1992.
40. Evans, Lawrence C. and Ishii, Hitoshi. A PDE approach to some asymptotic problems concerning random differential equations with small noise intensities. *Ann. Inst. H. Poincaré Anal. Non Linéaire.* **2** (1985), 1-20.
41. Feng, Jin. Martingale problems for large deviations of Markov processes. *Stochastic Process. Appl.* **81** (1999), no. 2, 165–216.
42. Feng, Jin. Large deviation for a stochastic Cahn-Hilliard equation. *Methods of functional analysis and topology***9** (2003), no. 4, 333-356.
43. Feng, Jin and Katsoulakis, Markos A Hamilton-Jacobi theory for controlled gradient flows in infinite dimensions. *Submitted* 2003.
44. Fleming, Wendell. H. Exit probabilities and optimal stochastic control. *Applied Math. Optimiz.* **4** 329-346, 1978.
45. Fleming, Wendell H. A stochastic control approach to some large deviation problems. *Recent mathematical methods in dynamic programmming (Rome 1984). Lecture Notes in Math.* **1119**, Springer, Berlin-New York, 1985. 52-66.

46. Fleming, Wendell H. and Soner, H. Mete. Asymptotic expansions for Markov processes with Lévy generators. *Appl. Math. Optim.* **19** (1989), 203-223.
47. Fleming, Wendell H. and Soner, H.Mete. *Controlled Markov processes and viscosity solutions.* Springer-Verlag, New York, 1991.
48. Fleming, Wendell H. and Souganidis, Panagiotis E. PDE-viscosity solution approach to some problems of large deviations. *Ann. Scuola Norm. Sup. Pisa Cl. Sci.* **13** (1986), 171-192.
49. Freidlin, Mark. I. Fluctuations in dynamical systems with averaging. *Soviet Math. Dokl.* **17** (1976), 104-108.
50. Freidlin, Mark. I. The averaging principle and theorems on large deviations. *Russion Math. Surveys* **33** (1978), 117-176.
51. Freidlin, Mark I., and Sowers, Richard B. A comparison of homogenization and large deviations, with applications to wavefront propagation. *Stochastic Process. Appl.* **82** (1999), 23-52.
52. Freidlin,Mark I. and Wentzell, Alexander D. *Random perturbations of dynamical systems.* Second Edition, Springer-Verlag, New York, 1998.
53. Garcia, Jorge. An extension of the contraction principle. *Journal of Theoretical Probability* **17**., no. 2, (2004), 403-434.
54. Gilbarg, David and Trudinger, Neil S. *Elliptic Partial Differential Equations of Second Order.* Classics in Mathematics, Springer-Verlag, New York, 1998 edition.
55. Giles, Robin. A generalization of the strict topology. *Trans. Amer. Math. Soc.* **161** (1971), 467-474.
56. Guillin, A. Averaging principle of SDE with small diffusion: moderate deviations. *Ann. Probab.* **31** (2003), 413-443.
57. Gulinsky, O. V.; Veretennikov, A. Yu. *Large deviations for discrete-time processes with averaging.* VSP, Utrecht, 1993.
58. Graham, Carl. McKean-Vlasov Itô-Skorohod equations and nonlinear diffusions with discrete jump sets. *Stochastic Process. Appl.* **40** (1992), 69-82.
59. Hoffmann-Jorgensen, J., A generalization of the strict topology *Math. Scand.* **30** (1972), 313-323.
60. Ishii, Hitoshi. On uniqueness and existence of viscosity solutions of fully nonlinear second-order elliptic PDE's. *Comm. Pure Appl. Math.* **42** (1989), 14-45.
61. Ishii, Hitoshi and Lions, Pierre-Louis. Viscosity solutions of fully nonlinear second-order elliptic partial differential equations. *J. Differential Equations* **83** (1990), 26-78.
62. Jakubowski, Adam. On the Skorohod topology. *Ann. Inst. H. Poincaré B* **22** (1986), 263-285.
63. Jordan, Richard, Kinderlehrer, David and Otto, Felix. The variational formulation of the Fokker-Planck equation. *SIAM J. Math. Anal.* **Vol. 29** 1998, No.1, 1-17.
64. Karlin, Samuel. Positive operators *J. Math. Mech.* **8** (1959), 907-937.
65. Kelley, John, *General Topology.* Springer, New York, 1975 (Reprint of the 1955 ed. published by Van Nostrand).
66. Khas'minskii, R. Z. A limit theorem for the solutions of differential equations with random right-hand sides. *Theory Probab. Appl.* **11** (1966) 390-406.
67. Kontoyiannis, Ioannis. and Meyn, Sean. P. Spectral theory and limit theorems for geometrically ergodic Markov processes. *Ann. Appl. Probability* **13** (2003), no.1, 3 04-362.
68. Kontoyiannis, Ioannis. and Meyn, Sean. P. Large deviations asymptotics and the spectral theory of multiplicatively regular Markov processes. *Preprint* 2003.
69. Kurtz, Thomas G. Extensions of Trotter's operator semigroup approximation theorems. *J. Functional Analysis* **3** (1969), 354-375.
70. Kurtz, Thomas G. A general theorem on the convergence of operator semigroups. *Trans. Amer. Math. Soc.* **148** (1970), 23-32.
71. Kurtz, Thomas G. A limit theorem for perturbed operator semigroups with applications to random evolutions *J. Funct. Anal.* **12** (1973), 55-67.
72. Kurtz, Thomas G., Convergence of sequences of semigroups of nonlinear operators with an application to gas kinetics. *Trans. Amer. Math. Soc.* **186**. 259–272, 1973.
73. Kurtz, Thomas G. Semigroups of conditioned shifts and approximation of Markov processes. *Ann. Probab.* **3** (1975), 618-642.
74. Kurtz, Thomas G. A variational formula for the growth rate of a positive operator semigroup *SIAM J. Math. Anal.* **10** (1979), 112-117.

75. Kurtz, Thomas G. Martingale problems for controlled processes *Springer Lecture Notes in Control and Information Sciences* **91**. 75-90, 1987.
76. Kurtz, Thomas G. and Protter, Philip. Weak convergence of stochastic integrals and differential equations II: Infinite dimensional case. *Probabilistic Models for Nonlinear Partial Differential Equations (Montecatini Terme, 1995)* 197–285. *Lecture Notes in Math* 1627, Springer, Berlin, 1996.
77. Kurtz, Thomas G.and Stockbridge, Richard H. Stationary solutions and forward equations for controlled and singular martingale problems. *Electron. J. Probab.* **6** (2001), no. 15, 52 pp
78. Kurtz, Thomas G. and Xiong, Jie. Particle representations for a class of nonlinear SPDEs. *Stochastic Process. Appl.* **83** (1999), no. 1, 103–126.
79. Kushner, Harold J., *Approximation and Weak Convergence Methods for Random Processes* MIT Press, Cambridge, MA, 1984.
80. Lions, Pierre-Louis. Neumann type boundary conditions for Hamilton-Jacobi equations. *Duke Math. J.* **52** (1985), 793-820.
81. Lions, Pierre-Louis, and Sznitman, Alain-Sol. Stochastic differential equations with reflecting boundary conditions. *Comm. Pure Appl. Math.* **37** (1984), 511-537.
82. Lynch, James. and Sethuraman, Jayaram. Large deviations for processes with independent increments. *Ann. Probab.* **15** (1987), 610-627.
83. McCann, J. Robert. Existence and uniqueness of monotone measure-preserving maps. *Duke Math. J.* **Vol. 80** 1995, No.2, 309-323.
84. McCann, J. Robert. A convexity principle for interacting gases. *Advances in Mathematics* **128**, 1997, 153-179.
85. Méléard, Sylvie. Asymptotic behaviour of some interacting particle systems; McKean-Vlasov and Boltzmann models. *Probabilistic models for nonlinear partial differential equations (Montecatini Terme, 1995)*, 42–95, *Lecture Notes in Math.*, **1627**, Springer, Berlin, 1996.
86. Miyadera, Isao. *Nonlinear Semigroups*, Translations of Mathematical Monographs, **109**. AMS, 1991.
87. Mogulskii, A. A. Large deviations for trajectories of multi-dimensional random walks. *Theory Probab. Appl.* **21** (1976) 300-315.
88. Mogulskii, A. A. Large deviations for processes with independent increments. *Ann. Probab.* **21** (1993) 202-215.
89. O'Brien, George L. Sequences of capcities with connections to large-deviation theory. *J. Theoret. Probab.* **9** (1996), 19-35.
90. O'Brien, George L. and Vervaat, Wim. Capacities, large deviations and loglog laws. *Stable processes and related topics* (Ithaca, NY, 1990), 43–83, *Progr. Probab.*, **25**, Birkhuser Boston, Boston, MA, 1991.
91. O'Brien, George L. and Vervaat, Wim. Compactness in the theory of large deviations. *Stochastic Process. Appl.* **57** (1995), 1-10.
92. Otto, Felix. The geometry of dissipative evolution equations: the porous medium equation. *Comm. P.D.Es* **26** (2001), no. 1-2, 101-174.
93. Papanicolaou, George C. and Varadhan, S. R. S. A limit theorem with strong mixing in Banach space and two applications to stochastic differential equations. *Comm. Pure Appl. Math.* **26** (1973), 497-524.
94. Pinsky, Mark A. *Lectures on random evolution.* World Scientific Publishing Co., Inc., River Edge, NJ, 1991.
95. Pinsky, Ross G. *Positive Harmonic Functions and Diffusions.* Cambridge Studies in Advanced Mathematics, Vol 45. Cambridge University Press, 1995.
96. Protter, Philip *Stochastic integration and differential equations. A new approach.* Applications of Mathematics, **21**. Springer-Verlag, Berlin, 1990.
97. Puhalskii, Anatolii. On functional principle of large deviations. *New Trends in Probability and Statistics. Vol. 1 (Bakuriani, 1990)* 198-219, VSP, Utrecht, 1991.
98. Puhalskii, Anatolii. The method of stochastic exponentials for large deviations. *Stochastic Process. Appl.* **54** (1994), 45-70.
99. Puhalskii, Anatolii. Large deviations of semimartingales: a maxingale problem approach. I. Limits as solutions to a maxingale problem. *Stochastic and Stochastic Reports* **61** (1997) 141-243. II. Uniqueness for the maxingale problem. *Stochastic and Stochastic Reports* **68** (1999) 65-143.

100. Puhalskii, Anatolii. *Large deviations and Idempotent Probability.* Chapman and Hall/CRC, 2001, New York.

101. Rachev, Svetlozar T. and Rüschendorf, Ludger. *Mass transportation Problems, Vol I: Theory.* Springer-Verlag, New York, 1998.

102. Reed, Michael; Simon, Barry. *Methods of modern mathematical physics. I. Functional analysis.* Second edition. Academic Press, Inc. [Harcourt Brace Jovanovich, Publishers], New York, 1980.

103. Russell, Raymond. *The large deviations of random time-changes.* PhD Thesis, University of Dublin, 1997.

104. Rockafellar, R. Tyrrell. *Convex analysis.* Princeton University Press, Princeton, NJ, 1970.

105. Sato, K. On the generators of non-negative contraction semi-groups in Banach lattices. *J. Math. Soc. Japan* **20** No.3, 1968, 423-436.

106. Schied, Alexander. Criteria for exponential tightness in path spaces. *Unpublished preprint.* 1995.

107. Schilder, Michael. Some asymptotic formulae for Wiener integrals *Trans. Amer. Math. Soc.* **125** (1966), 63-85.

108. Sheu, Shuenn Jyi. Stochastic control and exit probabilities of jump processes. *SIAM J. Control Optim.* **23** (1985), 306-328.

109. Sheu, Shuenn-Jyi. Stochastic control and exit probabilities of jump processes. *Springer Lecture Notes in Control and Information Sciences* **91**, 75-90, 1987.

110. Sinestrari, Eugenio. Accretive Differential Operators. *Bollettino U.M.I.*, **13**-B, (5), 1976, 19-31.

111. Sion, Maurice. On general minimax theorems. *Pacific J. Math.* **8** (1959), 171-176.

112. Sowers, Richard. Large deviations for a reaction-diffusion equation with non-Gaussian perturbations *Annals of Probability* 20 (1992) 504-537.

113. Spohn, H. *Large Scale Dynamics of Interacting Particles.* Texts and Monographs in Physics, Springer, (1991)

114. Stroock, Daniel W. Diffusion processes associated with Levy generators. *Z. Wahrsch. verw. Gebiete* **32** (1975), 209-244.

115. Stroock, Daniel W. *An introduction to the theory of large deviations.* Springer-Verlag, Berlin, 1984.

116. Stroock, Daniel W. and Varadhan, S. R. S. *Multidimensional diffusion processes.* Grundlehren der Mathematischen Wissenschaften, 233. Springer-Verlag, Berlin-New York, 1979.

117. Tataru, Daniel. Viscosity solutions of Hamilton-Jacobi equations with unbounded nonlinear terms. *Journal of Mathematical Anlaysis and Applications* **163** (1992), 345-392.

118. Trotter, H. F. Approximation of semi-groups of operators. *Pacific J. Math.* **8** (1958), 887-919.

119. Varadhan, S. R. S. Asymptotic probabilities and differential equations. *Comm. Pure Appl. Math.* **19** (1966), 261-286.

120. Veretennikov, A. Yu. On large deviations in averaging principle for stochastic differential equations with periodic coefficients. I. *Probability theory and mathematical statistics*, Vol. II (Vilnius, 1989), 542-551, "Mokslas", Vilnius, 1990.

121. Veretennikov, A. Yu. *On large deviations in the averaging principle for stochastic differential equations with periodic coefficients. II. Math. USSR-Izv.* **39** (1992), no. 1, 677–701

122. Veretennikov, A. Yu. On large deviations in the averaging principle for stochastic difference equations on a torus. *Proc. Steklov Inst. Math.* **202** (1994), 27-33.

123. Veretennikov, A. Yu. On large deviations in averaging principle for systems of stochastic differential equation with unbounded coefficients. *Probability theory and mathematical statistics (Vilnius, 1993)*, 735-742, TEV, Vilnius, 1994.

124. Veretennikov, A. Yu. On large deviations in the averaging principle for Markov processes (compact case, discrete time). *Probability theory and mathematical statistics (St. Petersburg, 1993)*, 235-240, Gordon and Breach, Amsterdam, 1996.

125. Veretennikov, A. Yu. On large deviations for stochastic differential equations with small diffusion and averaging. *Theory Probab. Appl.* **43** (1998), 335–337.

126. Veretennikov, A. Yu. On large deviations for SDEs with small diffusion and averaging. *Stochastic Process. Appl.* **89** (2000), 69-79.

127. Villani, Cedric. *Topics in optimal transportation.* American Mathematical Society, Providence, Rhode Island. Graduate Studies in Mathematics, Vol 58. 2003.

128. Wentzell, Alexander. D. Rough limit theorems on large deviations for Markov stochastic processes. I, II, III. *Theory Probab. Appl.* **21** (1976), 227-242, **24** (1979), 675-672, **27** (1982), 215-234.
129. Wentzell, Alexander. D. *Limit Theorems on Large Deviations for Markov Stochastic Processes.* Kluwer, Dordrecht. (1990)
130. Wheeler, Robert F., A survey of Baire measures and strict topologies. *Exposition. Math.* **1** (1983), 97-190.

Index

Titles in This Series

131 **Jin Feng and Thomas G. Kurtz,** Large deviations for stochastic processes, 2006

130 **Qing Han and Jia-Xing Hong,** Isometric embedding of Riemannian manifolds in Euclidean spaces, 2006

129 **William M. Singer,** Steenrod squares in spectral sequences, 2006

128 **Athanassios S. Fokas, Alexander R. Its, Andrei A. Kapaev, and Victor Yu. Novokshenov,** Painlevé transcendents, 2006

127 **Nikolai Chernov and Roberto Markarian,** Chaotic billiards, 2006

126 **Sen-Zhong Huang,** Gradient inequalities, 2006

125 **Joseph A. Cima, Alec L. Matheson, and William T. Ross,** The Cauchy Transform, 2006

124 **Ido Efrat, Editor,** Valuations, orderings, and Milnor K-Theory, 2006

123 **Barbara Fantechi, Lothar Göttsche, Luc Illusie, Steven L. Kleiman, Nitin Nitsure, and Angelo Vistoli,** Fundamental algebraic geometry: Grothendieck's FGA explained, 2005

122 **Antonio Giambruno and Mikhail Zaicev, Editors,** Polynomial identities and asymptotic methods, 2005

121 **Anton Zettl,** Sturm-Liouville theory, 2005

120 **Barry Simon,** Trace ideals and their applications, 2005

119 **Tian Ma and Shouhong Wang,** Geometric theory of incompressible flows with applications to fluid dynamics, 2005

118 **Alexandru Buium,** Arithmetic differential equations, 2005

117 **Volodymyr Nekrashevych,** Self-similar groups, 2005

116 **Alexander Koldobsky,** Fourier analysis in convex geometry, 2005

115 **Carlos Julio Moreno,** Advanced analytic number theory: L-functions, 2005

114 **Gregory F. Lawler,** Conformally invariant processes in the plane, 2005

113 **William G. Dwyer, Philip S. Hirschhorn, Daniel M. Kan, and Jeffrey H. Smith,** Homotopy limit functors on model categories and homotopical categories, 2004

112 **Michael Aschbacher and Stephen D. Smith,** The classification of quasithin groups II. Main theorems: The classification of simple QTKE-groups, 2004

111 **Michael Aschbacher and Stephen D. Smith,** The classification of quasithin groups I. Structure of strongly quasithin K-groups, 2004

110 **Bennett Chow and Dan Knopf,** The Ricci flow: An introduction, 2004

109 **Goro Shimura,** Arithmetic and analytic theories of quadratic forms and Clifford groups, 2004

108 **Michael Farber,** Topology of closed one-forms, 2004

107 **Jens Carsten Jantzen,** Representations of algebraic groups, 2003

106 **Hiroyuki Yoshida,** Absolute CM-periods, 2003

105 **Charalambos D. Aliprantis and Owen Burkinshaw,** Locally solid Riesz spaces with applications to economics, second edition, 2003

104 **Graham Everest, Alf van der Poorten, Igor Shparlinski, and Thomas Ward,** Recurrence sequences, 2003

103 **Octav Cornea, Gregory Lupton, John Oprea, and Daniel Tanré,** Lusternik-Schnirelmann category, 2003

102 **Linda Rass and John Radcliffe,** Spatial deterministic epidemics, 2003

101 **Eli Glasner,** Ergodic theory via joinings, 2003

100 **Peter Duren and Alexander Schuster,** Bergman spaces, 2004

99 **Philip S. Hirschhorn,** Model categories and their localizations, 2003

TITLES IN THIS SERIES

98 **Victor Guillemin, Viktor Ginzburg, and Yael Karshon,** Moment maps, cobordisms, and Hamiltonian group actions, 2002

97 **V. A. Vassiliev,** Applied Picard-Lefschetz theory, 2002

96 **Martin Markl, Steve Shnider, and Jim Stasheff,** Operads in algebra, topology and physics, 2002

95 **Seiichi Kamada,** Braid and knot theory in dimension four, 2002

94 **Mara D. Neusel and Larry Smith,** Invariant theory of finite groups, 2002

93 **Nikolai K. Nikolski,** Operators, functions, and systems: An easy reading. Volume 2: Model operators and systems, 2002

92 **Nikolai K. Nikolski,** Operators, functions, and systems: An easy reading. Volume 1: Hardy, Hankel, and Toeplitz, 2002

91 **Richard Montgomery,** A tour of subriemannian geometries, their geodesics and applications, 2002

90 **Christian Gérard and Izabella Łaba,** Multiparticle quantum scattering in constant magnetic fields, 2002

89 **Michel Ledoux,** The concentration of measure phenomenon, 2001

88 **Edward Frenkel and David Ben-Zvi,** Vertex algebras and algebraic curves, second edition, 2004

87 **Bruno Poizat,** Stable groups, 2001

86 **Stanley N. Burris,** Number theoretic density and logical limit laws, 2001

85 **V. A. Kozlov, V. G. Maz′ya, and J. Rossmann,** Spectral problems associated with corner singularities of solutions to elliptic equations, 2001

84 **László Fuchs and Luigi Salce,** Modules over non-Noetherian domains, 2001

83 **Sigurdur Helgason,** Groups and geometric analysis: Integral geometry, invariant differential operators, and spherical functions, 2000

82 **Goro Shimura,** Arithmeticity in the theory of automorphic forms, 2000

81 **Michael E. Taylor,** Tools for PDE: Pseudodifferential operators, paradifferential operators, and layer potentials, 2000

80 **Lindsay N. Childs,** Taming wild extensions: Hopf algebras and local Galois module theory, 2000

79 **Joseph A. Cima and William T. Ross,** The backward shift on the Hardy space, 2000

78 **Boris A. Kupershmidt,** KP or mKP: Noncommutative mathematics of Lagrangian, Hamiltonian, and integrable systems, 2000

77 **Fumio Hiai and Dénes Petz,** The semicircle law, free random variables and entropy, 2000

76 **Frederick P. Gardiner and Nikola Lakic,** Quasiconformal Teichmüller theory, 2000

75 **Greg Hjorth,** Classification and orbit equivalence relations, 2000

74 **Daniel W. Stroock,** An introduction to the analysis of paths on a Riemannian manifold, 2000

73 **John Locker,** Spectral theory of non-self-adjoint two-point differential operators, 2000

72 **Gerald Teschl,** Jacobi operators and completely integrable nonlinear lattices, 1999

71 **Lajos Pukánszky,** Characters of connected Lie groups, 1999

70 **Carmen Chicone and Yuri Latushkin,** Evolution semigroups in dynamical systems and differential equations, 1999

For a complete list of titles in this series, visit the AMS Bookstore at **www.ams.org/bookstore/**.